Pit Pichappan · Simon Fong · Qun Song ·
Tran Trung Duy
Editors



# 2025 Real-Time Intelligent Systems

## Proceedings of the Seventh International Conference on Real Time Intelligent Systems RTIS 2025

Springer

*Editors*
Pit Pichappan
Intelligent Information Processing Department
Digital Information Research Labs
Chennai, Tamil Nadu, India

Qun Song
Chongqing Business Technology University
Chongqing, China

Simon Fong
University of Macau
Taipa, Macao

Tran Trung Duy
Posts and Telecommunications Institute of Technology
Ho Chi Minh City, Vietnam

ISSN 2367-3370 ISSN 2367-3389 (electronic)
Lecture Notes in Networks and Systems
ISBN 978-3-032-23395-0 ISBN 978-3-032-23396-7 (eBook)
https://doi.org/10.1007/978-3-032-23396-7



This Springer imprint is published by the registered company Springer Nature Switzerland AG
The registered company address is: Gewerbestrasse 11, 6330 Cham, Switzerland

# Contents

# Enhancing Medical Image Classification with Noise-Injected Multi-Head Attention

Nguyen Nang Hung Van[1], Phuc Hao Do[2], Van Nam Hoang[3], Truong Duy Dinh[4(✉)], and Tran Duc Le[5]

[1] University of Science and Technology – The University of Danang, Da Nang, Vietnam
nguyenvan@dut.udn.vn

[2] Bonch-Bruevich Saint Petersburg State University of Telecommunications, Saint Petersburg, Russia
do.hf@sut.ru

[3] Dong A University, Da Nang, Vietnam
nam95755@donga.edu.vn

[4] Posts and Telecommunications Institute of Technology, Hanoi, Vietnam
duydt@ptit.edu.vn

[5] University of Wisconsin–Stout, Menomonie, WI, USA
let@uwstout.edu

**Abstract.** Vision Transformers (ViTs) have shown strong potential in medical image classification due to their ability to model long-range dependencies. Despite this advantage, the deterministic nature of the standard Multi-Head Attention (MHA) mechanism can lead to overfitting and reduced robustness, especially when working with noisy and heterogeneous medical datasets. To address this issue, we introduce a modified attention mechanism called Noise-Injected Multi-Head Attention (NIMHA), which integrates controlled Gaussian noise into the key and value projections of MHA. This stochastic regularization approach enhances feature learning and model generalization while maintaining computational efficiency and compatibility with existing ViT architectures. We evaluate NIMHA on two public datasets: Brain Tumor MRI and CT Kidney. Experimental results show that ViTs with NIMHA outperform baseline ViTs in classification accuracy, particularly on the more complex Brain Tumor MRI dataset. In addition, models with NIMHA exhibit more stable training behavior and faster convergence. Attention map analysis further reveals that the proposed method promotes a more distributed focus, improving the model's ability to generalize to diverse clinical data. These findings suggest that incorporating noise-based regularization into attention mechanisms is a practical strategy to enhance the robustness and reliability of ViT-based models for medical imaging tasks.

**Keywords:** Medical Image Classification · Multi-Head Attention · Noise Injection · Regularization Techniques · Vision Transformers


P. Pichappan et al. (Eds.): RTIS 2025, LNNS 1936, pp. 1–12, 2027.
https://doi.org/10.1007/978-3-032-23396-7_1

# 1 Introduction

Deep learning, especially Vision Transformers (ViTs) [7], has significantly advanced medical image analysis by modeling long-range dependencies and complex interactions, making them compelling alternatives to CNNs [16]. However, applying ViTs to clinical data faces challenges due to inherent high variability, limited sample sizes, and intrinsic noise in medical datasets [9]. This environment compromises generalization and robustness. The standard Multi-Head Attention (MHA) mechanism, being deterministic, is particularly susceptible to overfitting and feature collapse, leading to over-reliance on spurious, dataset-specific patterns [1]. Effective stochastic regularization is therefore crucial for capturing subtle diagnostic features.

To address this, we propose *Noise-Injected Multi-Head Attention* (NIMHA), a novel modification where controlled Gaussian noise is integrated into the key and value projections of MHA. Drawing inspiration from established noise injection techniques [11], NIMHA acts as an implicit regularizer, reducing the model's sensitivity to input perturbations and promoting enhanced feature diversity. This approach maintains computational efficiency and ensures compatibility with existing ViT architectures. Our study investigates NIMHA's efficacy in improving ViT performance and stability in medical imaging tasks.

The main contributions of this paper are:

- We propose NIMHA, incorporating Gaussian noise into MHA's key and value projections to enhance robustness and regularization in ViTs for medical image classification.
- We evaluate NIMHA on two challenging medical datasets (Brain Tumor MRI and CT Kidney), demonstrating improved Top-1 accuracy, stable training behavior, and faster convergence compared to standard MHA.
- We provide evidence that targeted stochastic regularization improves ViT generalization and reliability in noisy, resource-constrained medical imaging domains.

The remainder of the paper is structured as follows: Sect. 2 reviews related work. Section 3 defines the problem. Section 4 details the NIMHA methodology. Section 5 presents experimental results, followed by a discussion in Sect. 6 and conclusion in Sect. 7.

# 2 Related Work

This section situates our proposed Noise-Injected Multi-Head Attention (NIMHA) within existing research, reviewing the application of Vision Transformers (ViTs) in medical imaging, advancements in attention mechanisms, and the use of stochastic regularization.

Transformer architectures, initially designed for natural language processing, have demonstrated significant potential in medical data analysis due to their ability to model complex spatial and temporal dependencies. ViTs capture global contextual information, often outperforming traditional convolutional neural networks (CNNs) in tasks like image segmentation and diagnosis [13]. For instance, studies on MRI scans [4] and electronic health records [10] show promising results but highlight significant challenges related to small, noisy datasets and the resulting overfitting issues. These findings underscore the need for robust transformer variants that can effectively handle noise and data scarcity characteristic of clinical applications.

To enhance the performance and robustness of the core MHA mechanism, various strategies have been explored. Sparse Attention [14] reduces computational complexity for high-resolution images, while adaptive mechanisms [12] dynamically adjust attention based on input complexity. However, these enhancements generally lack specific regularization strategies to counteract the noise-induced overfitting prevalent in heterogeneous medical data.

Noise injection is a well-established technique in deep learning for enhancing generalization by introducing stochasticity [11], including methods like dropout [15] and noise addition to activations or weights. In the transformer domain, noise has been applied to input embeddings or attention weights [5] to improve robustness, and other stochastic attention mechanisms have been studied [17]. NIMHA differs from these approaches by uniquely introducing controlled Gaussian noise directly into the key ($\mathbf{K}$) and value ($\mathbf{V}$) projections of MHA. This targeted stochastic regularization acts implicitly on the feature representation level, promoting feature diversity and robust learning. Our systematic evaluation of this strategy addresses a critical gap in improving the reliability and generalization of ViT-based models for high-stakes medical image classification.

## 3 Problem Statement

ViTs excel in medical image classification by modeling global relationships, but face challenges with noisy, heterogeneous, and small medical datasets, leading to overfitting and poor generalization [9]. The standard MHA mechanism, central to ViTs, contributes to these issues due to its deterministic nature, which struggles in complex, noisy environments [1]. This section formalizes these challenges and highlights MHA's limitations addressed by our proposed NIMHA.

In ViTs (Fig. 1), images $\mathbf{x}_i \in \mathbb{R}^{H \times W \times C}$ from a dataset $\mathcal{D} = \{(\mathbf{x}_i, y_i)\}_{i=1}^{N}$ with labels $y_i \in \{1, \ldots, K\}$ are divided into patches and processed through transformer blocks. MHA computes attention over embedded patches $\mathbf{Z} \in \mathbb{R}^{N \times D}$:

$$\mathrm{MHA}(\mathbf{Z}) = \sum_{j=1}^{h} \mathrm{Softmax}\left(\frac{\mathbf{Q}_j \mathbf{K}_j^\top}{\sqrt{d_k}}\right) \mathbf{V}_j, \tag{1}$$

where $\mathbf{Q}_j = \mathbf{Z}\mathbf{W}_j^Q$, $\mathbf{K}_j = \mathbf{Z}\mathbf{W}_j^K$, $\mathbf{V}_j = \mathbf{Z}\mathbf{W}_j^V$. Despite its strengths, MHA exhibits two key limitations in medical imaging:

**Over-sensitivity to Perturbations**: MHA's Lipschitz constant, bounded by $\mathcal{L}_{\mathrm{MHA}} \geq \max_j \|\mathbf{W}_j^Q \mathbf{W}_j^{K\top}\|_2$, indicates high sensitivity to minor input variations (e.g., patient motion, scanner differences) [6]. This can destabilize learning and impair generalization, causing the model to focus on spurious correlations rather than diagnostic features.

**Feature Redundancy and Collapse**: The attention matrix

$$\mathbf{A}_j = \text{Softmax}\left(\frac{\mathbf{Q}_j\mathbf{K}_j^\top}{\sqrt{d_k}}\right) \tag{2}$$

has rank limited by

$$\min(\text{rank}(\mathbf{Q}_j), \text{rank}(\mathbf{K}_j)).$$

High feature correlation in medical images can cause low-rank attention, leading to feature collapse and missing subtle diagnostic cues [8].

To address these, we propose NIMHA, which injects controlled Gaussian noise into key and value projections to regularize attention, reduce sensitivity to perturbations, and promote diverse feature representations. This stochastic approach enhances robustness and generalization in ViTs for medical imaging while preserving computational efficiency. Subsequent sections detail NIMHA's formulation, integration, and empirical evaluation.

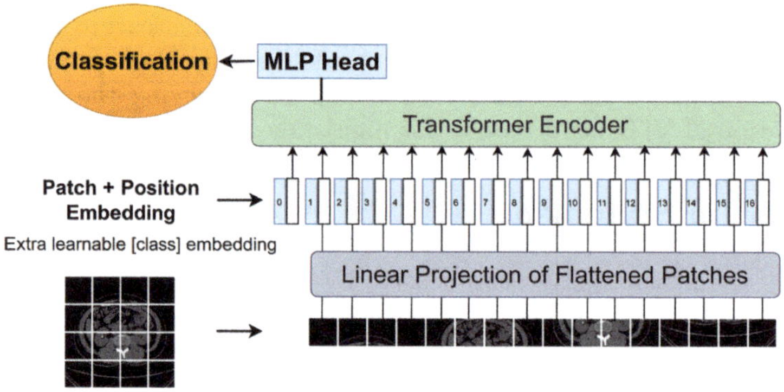


**Fig. 1.** The framework of Vision Transformer.

## 4 Methodology

This section presents the NIMHA mechanism and its integration into ViTs to enhance robustness and generalization for medical image classification under noisy, limited datasets. We outline the standard ViT architecture, introduce NIMHA, provide its theoretical basis, and summarize its implementation.

### 4.1 Vision Transformer Baseline Architecture

The standard ViT [7] processes an image $I \in \mathbb{R}^{H \times W \times C}$ by dividing it into $N = (H \cdot W)/P^2$ non-overlapping patches of size $P \times P$. Each patch $\mathbf{x}_p \in \mathbb{R}^{P^2 C}$ is linearly projected to an embedding $\mathbf{z}_p \in \mathbb{R}^D$ using a weight matrix $\mathbf{W}_E \in \mathbb{R}^{(P^2 C) \times D}$ and bias $\mathbf{b}_E \in \mathbb{R}^D$:

$$\mathbf{z}_p = \mathbf{x}_p \mathbf{W}_E + \mathbf{b}_E. \tag{3}$$

A class token $\mathbf{z}_{\text{class}} \in \mathbb{R}^D$ is prepended, and positional embeddings $\mathbf{P}_{\text{pos}} \in \mathbb{R}^{(N+1) \times D}$ are added to the sequence $\mathbf{Z}_0 = [\mathbf{z}_{\text{class}}; \mathbf{z}_1; \ldots; \mathbf{z}_N]$. This sequence is processed by $L$ Transformer encoder layers, each with MHA and a Feed-Forward Network (FFN), connected via residual connections and Layer Normalization (LN):

$$\mathbf{Z}'_\ell = \text{MHA}(\text{LN}(\mathbf{Z}_{\ell-1})) + \mathbf{Z}_{\ell-1}, \tag{4}$$

$$\mathbf{Z}_\ell = \text{FFN}(\text{LN}(\mathbf{Z}'_\ell)) + \mathbf{Z}'_\ell. \tag{5}$$

The FFN comprises two linear layers with GELU activation. The final class token output is fed to a linear classification head.

### 4.2 Noise-Injected Multi-Head Attention

**Standard MHA.** In standard MHA (Fig. 2), input $\mathbf{Z} \in \mathbb{R}^{(N+1) \times D}$ is projected into queries ($\mathbf{Q}_i$), keys ($\mathbf{K}_i$), and values ($\mathbf{V}_i$) for each head $i$:

$$\mathbf{Q}_i = \mathbf{Z}\mathbf{W}_{Q_i}, \quad \mathbf{K}_i = \mathbf{Z}\mathbf{W}_{K_i}, \quad \mathbf{V}_i = \mathbf{Z}\mathbf{W}_{V_i}, \tag{6}$$

where $\mathbf{W}_{Q_i}, \mathbf{W}_{K_i}, \mathbf{W}_{V_i} \in \mathbb{R}^{D \times d_k}$, $d_k = D/h$. The attention output is:

$$\text{AH}_i = \text{Softmax}\left(\frac{\mathbf{Q}_i \mathbf{K}_i^\top}{\sqrt{d_k}}\right) \mathbf{V}_i. \tag{7}$$

Outputs are concatenated and projected: $\text{MHA}(\mathbf{Z}) = \text{Concat}(\text{AH}_1, \ldots, \text{AH}_h)\mathbf{W}_O$.

**NIMHA Formulation.** NIMHA introduces Gaussian noise to key and value projections to regularize attention (Fig. 3):

$$\tilde{\mathbf{K}}_i = \mathbf{K}_i + \boldsymbol{\epsilon}_{K_i}, \quad \boldsymbol{\epsilon}_{K_i} \sim \mathcal{N}(\mathbf{0}, \sigma^2 \mathbf{I}), \tag{8}$$

$$\tilde{\mathbf{V}}_i = \mathbf{V}_i + \boldsymbol{\epsilon}_{V_i}, \quad \boldsymbol{\epsilon}_{V_i} \sim \mathcal{N}(\mathbf{0}, \sigma^2 \mathbf{I}), \tag{9}$$

where $\sigma$ controls noise intensity. Queries remain unchanged. The attention for head $i$ is:

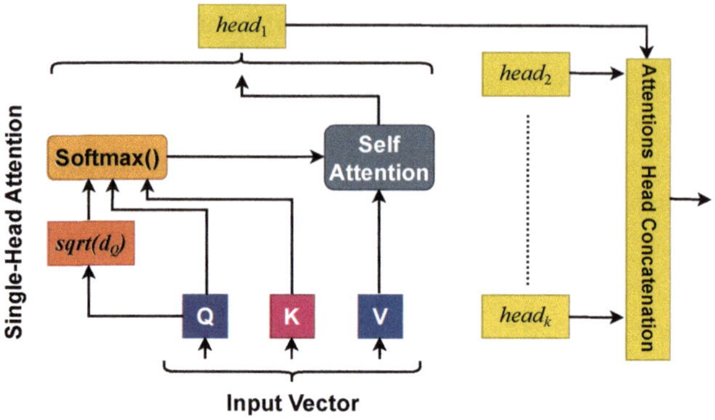


**Fig. 2.** Standard Multi-Head Attention mechanism.

$$\text{NIMHAH}_i = \text{Softmax}\left(\frac{\mathbf{Q}_i\tilde{\mathbf{K}}_i^\top}{\sqrt{d_k}}\right)\tilde{\mathbf{V}}_i. \tag{10}$$

The final output is $\text{NIMHA}(\mathbf{Z}) = \text{Concat}(\text{NIMHAH}_1, \ldots, \text{NIMHAH}_h)\mathbf{W}_O$. Gaussian noise models real-world imaging artifacts and ensures uniform regularization across feature directions.

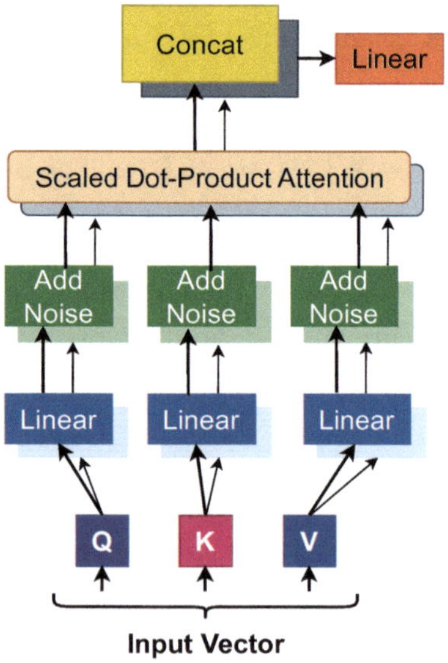


**Fig. 3.** Noise-Injected Multi-Head Attention mechanism.

# 5 Experiments and Results

This section evaluates the NIMHA mechanism within ViTs on two medical imaging datasets: Brain Tumor MRI and CT Kidney. We compare ViT-NIMHA against ViT-MHA under identical conditions to isolate NIMHA's impact on classification accuracy, robustness, and training stability, focusing on metrics like loss, Top-1 accuracy, and F1-score, alongside qualitative analysis of attention maps.

## 5.1 Dataset Description

**Brain Tumor MRI Dataset.** The Brain Tumor MRI Dataset [3] includes 7,023 MRI scans across four classes: glioma, meningioma, non-tumor, and pituitary tumor. Its heterogeneity from multiple sources (e.g., figshare, SARTAJ) introduces variability in imaging protocols. After preprocessing (intensity normalization, artifact removal, label correction), the dataset is split into 5,712 training, 571 validation, and 1,311 test images. Training uses 25 epochs with L4 GPU acceleration (Fig. 4).

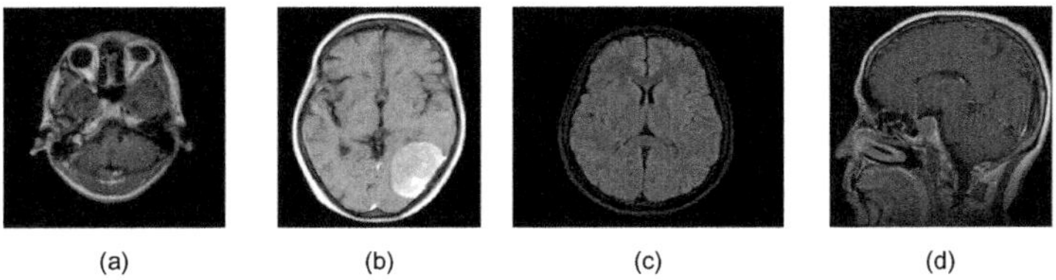


**Fig. 4.** Sample Brain Tumor MRI images: (a) Glioma, (b) Meningioma, (c) Non-tumor, (d) Pituitary tumor.

**CT Kidney Dataset.** The CT Kidney Dataset [2] comprises 12,446 CT images (normal, cyst, tumor, stone) from hospitals in Dhaka, Bangladesh, with variations in contrast and orientation. Radiologist-annotated, it is split into 9,957 training, 2,489 validation, and 1,000 test images. Training spans 15 epochs with L4 GPU acceleration (Fig. 5).

## 5.2 Experimental Methodology

We compare ViT-MHA and ViT-NIMHA using identical setups (data splits, learning rate, batch size, optimizer) to isolate NIMHA's effects. The workflow (Algorithm 1) includes preprocessing (resizing, normalization, patch extraction), training ViT-MHA and ViT-NIMHA with noise ($\sigma$) injected into key/value projections, and evaluation on test sets using accuracy, F1-score, and attention map visualization to assess robustness and feature focus.

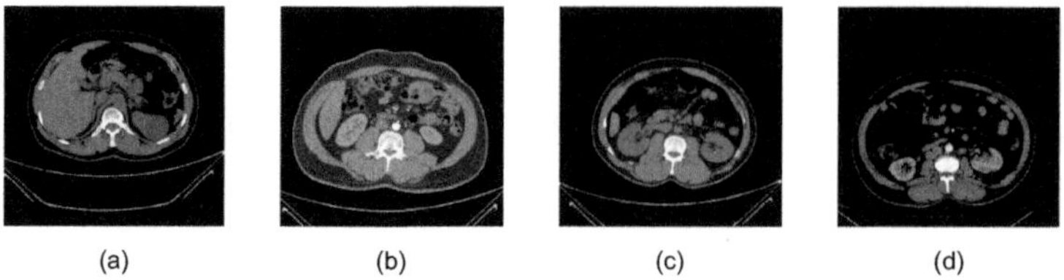


**Fig. 5.** Sample CT Kidney images: (a) Cyst, (b) Normal, (c) Stone, (d) Tumor.

---

**Algorithm 1.** Experimental Workflow for MHA vs. NIMHA

---

**Require:** Datasets $\mathcal{D}_{\text{Brain}}, \mathcal{D}_{\text{Kidney}}$, learning rate $\eta$, batch size $B$, epochs $E$, noise $\sigma$, patch size $P$, layers $L$, heads $H$
**Ensure:** Metrics (accuracy, F1-score)

Preprocess datasets: resize, normalize, split into patches
Split into $\mathcal{D}_{\text{train}}, \mathcal{D}_{\text{val}}, \mathcal{D}_{\text{test}}$
Initialize ViT with $L$ layers, $H$ heads
**for** model in {MHA, NIMHA} **do**
    **for** epoch $= 1$ to $E$ **do**
        **for** batch $\mathcal{B} \subset \mathcal{D}_{\text{train}}$ **do**
            Compute attention (NIMHA: add noise $\epsilon \sim \mathcal{N}(0, \sigma^2)$ to keys/values)
            Forward pass, compute loss $\mathcal{L}$, backpropagate
        **end for**
        Validate on $\mathcal{D}_{\text{val}}$
    **end for**
    Evaluate on $\mathcal{D}_{\text{test}}$: compute accuracy, F1-score
**end for**
Compare metrics, visualize attention maps

---

### 5.3 Performance on Brain Tumor MRI Dataset

**ViT-MHA.** Figure 6 shows ViT-MHA's training loss dropping from 1.4 to 0.22 and validation loss from 0.98 to 0.24 over 25 epochs, with fluctuations indicating overfitting. Accuracy rises from 0.41 (training) and 0.58 (validation) to 0.91 and 0.905, respectively, but validation accuracy varies, reflecting sensitivity to noise and variability.

**ViT-NIMHA.** Figure 7 shows ViT-NIMHA's training loss decreasing from 1.4 to 0.18 and validation loss from 1.1 to 0.31, with smoother curves. Accuracy improves from 0.36 (training) and 0.47 (validation) to 0.93 and 0.90, with fewer fluctuations, indicating better regularization. Test accuracy is 0.8719 (NIMHA) vs. 0.8352 (MHA).

### 5.4 Performance on CT Kidney Dataset

**ViT-MHA.** Figure 8 shows ViT-MHA's training loss falling from 1.18 to 0.02 and validation loss from 0.79 to 0.025 over 15 epochs, with oscillations. Accu-

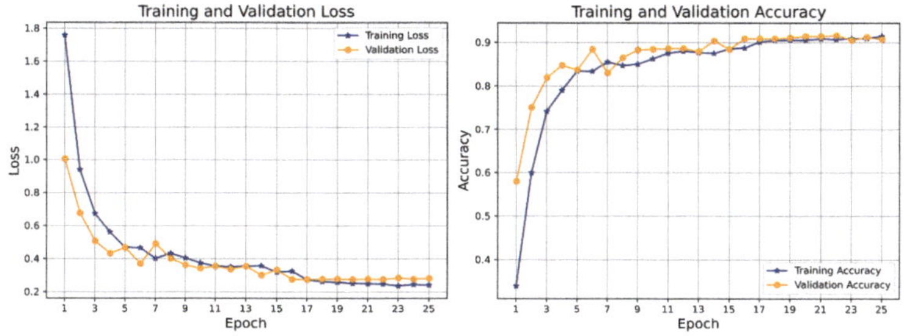


**Fig. 6.** ViT-MHA on Brain Tumor MRI: (Left) Loss, (Right) Accuracy over 25 epochs.

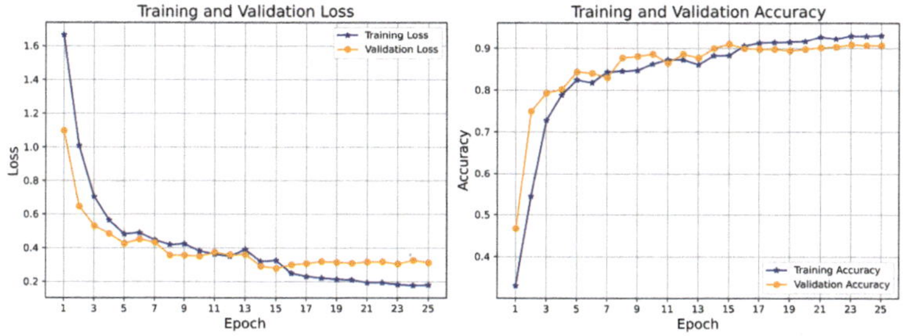


**Fig. 7.** ViT-NIMHA on Brain Tumor MRI: (Left) Loss, (Right) Accuracy over 25 epochs.

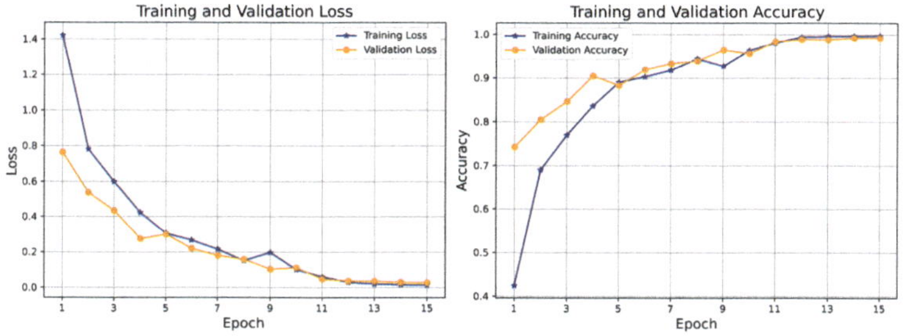


**Fig. 8.** ViT-MHA on CT Kidney: (Left) Loss, (Right) Accuracy over 15 epochs.

racy rises from 0.52 (training) and 0.74 (validation) to near 1.0, but validation fluctuations suggest overfitting.

**ViT-NIMHA.** Figure 9 shows ViT-NIMHA's training loss dropping from 1.2 to 0.02 and validation loss from 0.78 to 0.024, with smoother convergence. Accuracy rises from 0.53 (training) and 0.71 (validation) to near 1.0, with minimal oscillations. Test accuracy is 0.9936 (NIMHA) vs. 0.9919 (MHA).

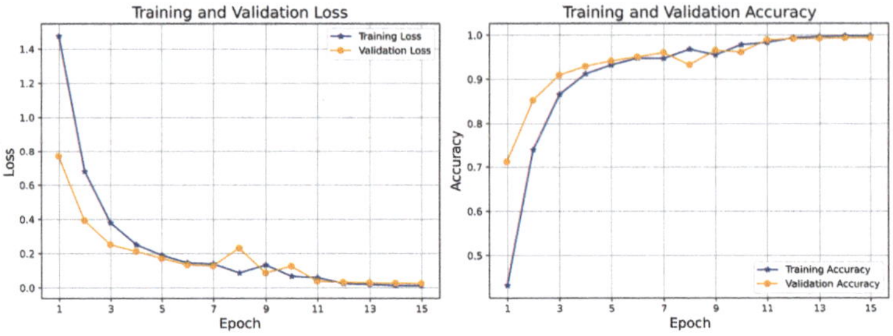


**Fig. 9.** ViT-NIMHA on CT Kidney: (Left) Loss, (Right) Accuracy over 15 epochs.

## 6 Discussion

This section discusses the clinical relevance, performance comparisons, and design insights of NIMHA in ViTs for medical imaging, emphasizing its robustness and generalization benefits.

### 6.1 Clinical Relevance and Impact

NIMHA addresses overfitting and feature collapse in Transformer-based models, critical for medical imaging where data is heterogeneous and labeled samples are limited. Validated on Brain Tumor MRI and CT Kidney datasets, NIMHA improves classification accuracy and stability, aiding early and reliable disease detection. Its efficiency supports real-time clinical use, enhancing diagnostic reliability. While focused on medical imaging, NIMHA's stochastic regularization shows potential for broader vision tasks, a direction for future exploration.

### 6.2 Comparison of Attention Mechanisms

Table 1 compares ViT and ViT+NIMHA on the Brain Tumor MRI dataset across glioma, meningioma, no tumor, and pituitary classes. ViT+NIMHA outperforms ViT, with precision improving (e.g., 0.89 vs. 0.86 for glioma, 0.80 vs. 0.74 for meningioma) and F1-scores increasing (e.g., 0.73 vs. 0.60 for meningioma). Recall for meningioma rises significantly from 0.51 to 0.67, indicating better detection of challenging cases.

Table 2 shows results on the CT Kidney dataset. ViT+NIMHA achieves perfect scores (1.00) for cyst and tumor classes, outperforming ViT (0.98 for tumor). For normal, F1-score improves from 0.99 to 1.00. Stone class shows a slight recall drop (0.97 to 0.96), but overall metrics remain high, confirming NIMHA's robustness.

Table 3 summarizes accuracy. On Brain Tumor MRI, ViT+NIMHA achieves 0.8613 (train) and 0.8718 (test) vs. ViT's 0.8390 and 0.8352, a 4.4% test accuracy gain. On CT Kidney, ViT+NIMHA reaches 0.9947 (train) and 0.99357 (test) vs. ViT's 0.9907 and 0.99196. A McNemar test ($p = 0.018$, $\chi^2 = 5.63$, $df = 1$)

**Table 1.** Comparison of Precision, Recall, and F1-score between ViT and ViT+NIMHA on Brain Tumor MRI Dataset.

| Class | Precision | | Recall | | F1-Score | |
|---|---|---|---|---|---|---|
| | ViT | ViT+NIMHA | ViT | ViT+NIMHA | ViT | ViT+NIMHA |
| Glioma | 0.86 | **0.89** | **0.85** | 0.84 | 0.86 | **0.87** |
| Meningioma | 0.74 | **0.80** | 0.51 | **0.67** | 0.60 | **0.73** |
| No Tumor | 0.81 | **0.87** | **0.99** | 0.98 | 0.89 | **0.92** |
| Pituitary | 0.92 | **0.92** | 0.94 | **0.96** | 0.93 | **0.94** |

**Table 2.** Comparison of Precision, Recall, and F1-score between ViT and ViT+NIMHA on CT Kidney Dataset.

| Class | Precision | | Recall | | F1-Score | |
|---|---|---|---|---|---|---|
| | ViT | ViT+NIMHA | ViT | ViT+NIMHA | ViT | ViT+NIMHA |
| Cyst | 1.00 | 1.00 | 1.00 | 1.00 | 1.00 | 1.00 |
| Normal | 0.99 | 0.99 | 1.00 | 1.00 | 0.99 | **1.00** |
| Stone | 0.99 | 0.99 | **0.97** | 0.96 | 0.98 | 0.98 |
| Tumor | 0.98 | **1.00** | 0.98 | **1.00** | 0.98 | **1.00** |

confirms significant improvement on CT Kidney, with a 95% confidence interval of [0.04, 0.31]

**Table 3.** Comparison of Train and Test Accuracy between ViT and ViT+NIMHA.

| Model | Brain Tumor MRI | | CT Kidney | |
|---|---|---|---|---|
| | Train Accuracy | Test Accuracy | Train Accuracy | Test Accuracy |
| ViT | 0.8390 | 0.8352 | 0.9907 | 0.99196 |
| ViT+NIMHA | **0.8613** | **0.8718** | **0.9947** | **0.99357** |

## 7 Conclusion

This study introduces Noise-Injected Multi-Head Attention, a novel regularization technique for ViTs in medical image analysis. By injecting Gaussian noise into key and value projections, NIMHA enhances robustness and generalization on heterogeneous datasets like Brain Tumor MRI and CT Kidney. Empirical results show ViT+NIMHA outperforms standard ViT-MHA, with a 4.4% Top-1 accuracy gain on Brain Tumor MRI (0.8719 vs. 0.8352) and slight improvements on CT Kidney (0.9936 vs. 0.9919). NIMHA stabilizes training and mitigates overfitting, making it promising for clinical applications.

Future work will explore adaptive noise scaling, synergy with other regularization methods, and NIMHA's application to diverse medical imaging tasks and vision tasks like object detection (e.g., DETR) and semantic segmentation (e.g., SegFormer) to further validate its generalizability.

## References

1. Abibullaev, B., Keutayeva, A., Zollanvari, A.: Deep learning in EEG-based BCIS: a comprehensive review of transformer models, advantages, challenges, and applications. IEEE Access (2023)
2. Batool, A., Byun, Y.C.: Brain tumor detection with integrating traditional and computational intelligence approaches across diverse imaging modalities–challenges and future directions. Comput. Biol. Med., 108412 (2024)
3. Chaki, J.: Brain tumor MRI dataset (2023). iEEE Dataport
4. Chen, J., Frey, E.C., He, Y., Segars, W.P., Li, Y., Du, Y.: Transmorph: transformer for unsupervised medical image registration. Med. Image Anal. **82**, 102615 (2022)
5. Dai, Y., et al.: Improving adversarial robustness of medical imaging systems via adding global attention noise. Comput. Biol. Med. **164**, 107251 (2023)
6. Gonçalves, T., Rio-Torto, I., Teixeira, L., Cardoso, J.S.: A survey on attention mechanisms for medical applications: are we moving toward better algorithms? IEEE Access **10**, 98909–98935 (2022)
7. Khan, S., Naseer, M., Hayat, M., Zamir, S.W., Khan, F.S., Shah, M.: Transformers in vision: a survey. ACM Comput. Surv. (CSUR) **54**(10s), 1–41 (2022)
8. Kurasova, O.: Semi-supervised learning with pseudo-labeling for pancreatic cancer detection on CT scans. In: 2023 18th Iberian Conference on Information Systems and Technologies (CISTI), pp. 1–6. IEEE (2023)
9. Li, J., Chen, J., Tang, Y., Wang, C., Landman, B.A., Zhou, S.K.: Transforming medical imaging with transformers? a comparative review of key properties, current progresses, and future perspectives. Med. Image Anal. **85**, 102762 (2023)
10. Li, Y., et al.: Behrt: transformer for electronic health records. Sci. Rep. **10**(1), 7155 (2020)
11. Lu, Y., Bo, Y., He, W.: Noise attention learning: enhancing noise robustness by gradient scaling. Adv. Neural Inf. Process. Syst. **35**, 23164–23177 (2022)
12. Meng, L.: Adavit: adaptive vision transformers for efficient image recognition. In: Proceedings of the IEEE/CVF Conference on Computer Vision and Pattern Recognition, pp. 12309–12318 (2022)
13. Nerella, S.: Transformers in healthcare: a survey. arXiv preprint arXiv:2307.00067 (2023)
14. Roy, A., Saffar, M., Vaswani, A., Grangier, D.: Efficient content-based sparse attention with routing transformers. Trans. Assoc. Comput. Linguistics **9**, 53–68 (2021)
15. Sanjar, K., Rehman, A., Paul, A., JeongHong, K.: Weight dropout for preventing neural networks from overfitting. In: 2020 8th International Conference on Orange Technology (ICOT), pp. 1–4. IEEE (2020)
16. Satvati, M.A., Lakestani, M., Khamnei, H.J., Allahviranloo, T.: Deblurring medical images using a new grünwald-letnikov fractional mask. Informatica **35**(4), 817–836 (2024)
17. Shi, M., Tang, Y., Zhu, X., Zhuang, Y., Lin, M., Liu, J.: Feature-attention graph convolutional networks for noise resilient learning. IEEE Trans. Cybern. **52**(8), 7719–7731 (2022)

# Implementing Piecemeal-DFS and Adversarial-DFS Algorithms on General Trees

Cuong Khong Phu, Nhat Ho Duy, Quan Dinh Viet, Tan Do Thanh, Nguyen Nguyen Khanh, and Nguyen Minh Tuan(✉)

Faculty of Information Technology, Posts and Telecommunications Institute of Technology, 11 Nguyen Dinh Chieu,
Sai Gon ward, Ho Chi Minh city 700000, Vietnam
{n23dccn076,n23dccn111,n23dccn116,n23dccn122, n23dccn109}@student.ptithcm.edu.vn, minhtuan@ptit.edu.vn

**Abstract.** In the context of optimization and search, the ability to efficiently traverse graphs plays a pivotal role in countless real-world applications. However, traditional traversal methods often face significant challenges in terms of performance or feasibility, especially in environments with strict resource constraints, such as energy, or when dealing with adversarial strategies designed to hinder the search process. To address these limitations, our research delves into the implementation and in-depth analysis of two advanced variants of the Depth-First Search (DFS) algorithm on general trees: piece-meal-DFS (PDFS) and adversarial-DFS (ADFS). The results from our implementation and experimentation demonstrate the effectiveness of the proposed algorithms. PDFS has demonstrated superior capability in achieving complete tree coverage, even under extremely stringent energy constraints. This confirms PDFS's potential as a viable solution for applications requiring comprehensive exploration with limited resources. In parallel, ADFS successfully identified specific adversarial scenarios that can significantly increase traversal costs. This deep understanding of worst-case scenarios not only provides valuable insights into the upper bounds of actual costs but also contributes to evaluating the sustainability and resilience of traversal algorithms in adversarial environments. Notably, we introduced significant improvements to PDFS through the application of an optimized energy replenishment strategy based on the potential of the next branch. This improved method is designed to minimize overall costs, allowing the algorithm to adapt more flexibly to tree structures and varying energy constraints.

**Keywords:** Piecemeal DFS · DFS · Tree Exploration · Energy Constraint

N. H. Duy, Q. D. Viet, T. D. Thanh, and N. N. Khanh—Equal contribution.


P. Pichappan et al. (Eds.): RTIS 2025, LNNS 1936, pp. 13–24, 2027.
https://doi.org/10.1007/978-3-032-23396-7_2

## 1 Introduction

The piecemeal exploration problem has previously been studied not only for trees but also for general connected graphs. However, those results are limited to visiting vertices at depth at most $B/(2(1+\beta))$, for some $\beta > 0$, with the exploration cost increasing sharply as $\beta$ approaches zero [1]. In [2] he authors also introduced implementations of both PDFS and ADFS strategies, though only at an asymptotic level of cost optimality. Furthermore, their analysis was restricted to binary trees. Therefore, in this work, aim to extend the algorithm to operate on general trees of arbitrary structure, supporting depths up to the maximum feasible limit of $B/2$, while ensuring both the minimal number of required routes and the overall cost-optimality of the exploration strategy.

There exists a large body of research on graph traversal and exploration algorithms. In this section, we focus on the most relevant results related to graph exploration by mobile agents. The problem of exploring a general graph with $n$ vertices and $m$ edges by a single agent was studied in [3], which proposed an algorithm with a total of $m + O(n)$ steps. More recent results on exploration time can be found in works such as [4,5] and [6]. Recent studies such as [7] and [8] have shifted the focus from theoretical models to practical applications, including delivery, information broadcasting, and terrain exploration, thereby emphasizing the importance of energy-efficient strategies. Meanwhile, complexity-theoretic results such as those in [9] offer new perspectives on the feasibility of simultaneously optimizing multiple objectives during exploration. In particular, adversarial models where the structure of the tree is incrementally revealed in a manner that may be unfavorable to the agent require algorithms to operate under worst-case assumptions. This setting is exemplified in [6] and [3], which show that in uncertain or adversarial environments, algorithm performance is highly sensitive to the chosen exploration strategy.

Therefore, the design and analysis of the Adversarial-DFS algorithm becomes essential for coping with adversarially revealed tree structures. However, none of the aforementioned works take energy constraints into account. The energy-constrained exploration problem was first introduced under the name Piecemeal Graph Exploration in [1], assuming that the path length satisfies $B \geq 2(1+\beta)r$, where $r$ is the maximum distance from the starting vertex to any other vertex, and $0 < \beta < 1$.That work proposed exploration algorithms for a special class of grid graphs containing rectangular obstacles. Later, [10] showed that for general graphs, there exists an energy-constrained exploration algorithm with a total cost of $O(m + n^{1+o(1)})$. To implement the Piecemeal-DFS algorithm, we begin by performing a depth-first traversal of the tree to record the full exploration path, including return steps to the root. The algorithm operates under an energy constraint, where the agent is only allowed to move along paths whose total cost does not exceed a given bound BBB [11]. To demonstrate how the PDFS algorithm functions, we apply it in the context of terrain exploration using a mobile robot. In practice, any robotic agent is subject to a limited energy budget, denoted by B, and the environment is modeled as a tree structure. As discussed in [2], a natural assumption is that every node in the tree is at most distance

$B/2$ from the root; otherwise, complete exploration would be impossible. In such settings, exploration remains feasible if the robot is allowed to recharge upon returning to its starting location. Consequently, the entire exploration process consists of a collection of routes, each of which starts and ends at the root and has length at most B.

## 2 Related Definition

To construct routes using PDFS, it is necessary to rely on a route that spans the entire tree (including the path that returns to the root). This route is defined as $R_{\mathrm{DFS}} = (v_0, v_1, \ldots, v_l)$. Note that $R_{\mathrm{D}FS}$ includes the return path to the root, so during the DFS traversal, some vertices may appear multiple times. We combine DFS and recursion to generate this route, while also storing the parent of each node. Let the tree to be traversed be $T$ with root $r$, and let $R = (v_0, v_1, \ldots, v_\ell)$ be a sequence where each $v_i$ is a vertex of $T$ such that $i \in \{0, \ldots, \ell\}$. A route is defined as a sequence of vertices that forms a path in $T$ that starts and ends at the root. The length of $R$ in Sect. 2 is defined as follows:

$$l(R) = \sum_{i=1}^{l} \omega(\{v_{i-1}, v_i\}), \tag{1}$$

where $\omega(v_{i-1}, v_i)$ is the weight of the edge that connects the vertex $v_{i-1}$ and $v_i$. Given a tree $T$ and an energy budget $B$, the routes constructed by PDFS, denoted as $S = (R_1, \ldots, R_k)$, form an exploration strategy for the tree $T$ with the energy budget $B$.
From this we derive the following theorem regarding the condition under which PDFS operates effectively:

**Theorem 1.** *Given a real number $B > 1$ and a weighted tree with maximum height at most $B/2$, an exploration strategy for the tree $T$ with energy limit $B$ consists of $k$ routes, where $k$ is the smallest integer possible.*

The routes constructed using PDFS will explore the entire tree without exceeding the energy limit B. However, PDFS is only an approximation of the optimal solution. In many cases, PDFS does not generate the most cost-effective routes.

Consider the routes constructed by the PDFS algorithm for the tree in Fig. 1a [2]

By performing a PDFS traversal, the constructed routes are the following:

- Route 1: a b c b d b a e a | Cost = 20.
- Route 2: a e f e.g. e a | Cost = 16.

The total cost for these two routes is 36. Cost-optimal routes can be constructed as follows:

- Route 1: a b c b d b a | Cost = 18.

- Route 2: a e f e.g. e a | Cost = 16.

In this case, the total cost for both routes is 34, which is lower than the initial routes.

**Reason**: During the PDFS traversal, the constraint is that the total cost of the forward and return paths must not exceed B. As a result, a worst-case scenario may occur in which, after fully exploring one branch and returning to the root with remaining energy, the algorithm continues to traverse nodes from the next branch without recharging. Therefore, we improve the PDFS algorithm to be more cost-effective.

The following theorem provides the main result for the above case.

**Theorem 2.** *Given a tree $T$, an energy limit $B$ and a root node $r$. We have $C_{OptPDFS} \leq C_{PDFS}$, where $C_{OptPDFS}$ denotes the total cost of the improved strategy, and $C_{PDFS}$ denotes the total cost of the standard strategy.*

***Proof of Theorem*** 2 In the worst-case scenario, the PDFS traversal may fail to optimize the total cost when, after exploring an entire branch and returning to the root with residual energy, the algorithm immediately continues to the next branch without recharging.

To address this issue, we enhance the PDFS algorithm by evaluating the potential of the upcoming branch upon returning to the root node $r$. This evaluation guides the decision on whether to recharge before proceeding or to continue with the remaining energy, based on the following potential function:

$$\beta(r) = B' - B(r). \tag{2}$$

Let:

- $B'$ be the remaining energy after completing the previous branch and returning to the root $r$,
- $B(r)$ be the minimum energy required to explore the next branch (including the trip to the branch and back to the root),

then applying the following rule:

- If $\beta \geq 0$: proceed to explore the next branch without recharging,
- If $\beta < 0$: recharge before exploring the next branch,

will always result in a total exploration cost that does not exceed the total cost of the original PDFS algorithm. Thus, $C_{OptPDFS} \leq C_{PDFS}$. □

# 3 Methods

## 3.1 Piecemeal-DFS Exploration on General Trees

In this section, we incorporate the theoretical foundations presented above, particularly Proof of Theorem 2, to develop an algorithm that guarantees an optimal solution. That is, we obtain a strategy $S = (R_1, \dots, R_k)$ such that the number of routes $k$ is minimized and the total energy consumption is as small as possible and the cost of $S$ is optimal. To implement the PDFS traversal algorithm with optimal cost on a general tree, we provide the following pseudocode. As a preliminary step, we first define the necessary auxiliary functions:

- **loadData()** is designed to read and parse the input data from the file.The input file contains the energy bound $B$ and the list of edges representing the tree to be explored.
- **buildDFS()** performs a full depth-first traversal of the tree, recording the parent of each node and storing the traversal path.
- **getCost()** computes and returns the total weight of the specified traversal path.
- **pathToRoot()** returns the path from a given node to the root of the tree.

The analysis of the Piecemeal-DFS algorithm yields the following complexity results:

- **Time Complexity:** The overall time cost of the algorithm is dominated by the main loop. The loop iterates $N$ times, and within each iteration, path reconstruction and cost calculation can take up to $O(N)$ time in the worst-case scenario (a degenerate tree). Therefore, the time complexity is $O(N^2)$.
- **Space Complexity:** The primary data structures occupying space are the `parent` map ($O(N)$) and the `dfsPath` vector ($O(N)$). However, the `routes` vector can potentially store multiple exploration routes. In the worst-case scenario, if the algorithm generates $O(N)$ long routes, each with a length of up to $O(N)$, the space complexity could escalate to $O(N^2)$.

### 3.2 Adversarial-DFS Exploration on General Trees

In the context of PDFS(T) with specified energy constraints, the cost of an exploration process extends beyond a mere summation of traversed steps. Instead, it is intrinsically linked to the lengths of the routes undertaken during the search. A typical route originates from the root of a subtree, traverses a sequence of nodes, and subsequently returns to its originating subtree root. The length of a route is quantified by the aggregate sum of edge weights along its path. Crucially, the cost associated with any given state or exploration sequence must reflect adherence to the prescribed energy limits. Should a route necessitate an energy expenditure exceeding these permissible thresholds (denoted as $B$ or $B'$), it is deemed invalid or assigned an infinitely high cost. This mechanism ensures that the search algorithm is inherently precluded from selecting infeasible paths.

Essentially, for the complete exploration of the entire tree, the cumulative cost of all requisite routes must satisfy the imposed energy constraints. Specifically, the first route navigating into a subtree (either from the overall tree's root or from a previously explored point) is subject to a distinct energy limit, $B'$. Conversely, all subsequent routes performed within that same subtree are governed by a common energy limit, $B$. To identify the worst-case scenario that an adversary can engineer, the ADFS(T) strategy is meticulously designed to maximize the cost incurred by the search algorithm. This implies that ADFS(T) will consistently select moves or tree traversal strategies that compel the algorithm to bear the highest possible cost to complete its exploration. In essence, the algorithm may be strategically guided toward states where subsequent routes are prone to exceeding the energy limits ($B$ or $B'$), thereby forcing the algorithm to undertake detours or explore redundant segments of the tree, which collectively escalates the total route length.

To implement the ADFS algorithm efficiently, we first define the necessary auxiliary functions before presenting the pseudocode:

- **calculateSubtreeWeight()** computes the total weight of the subtree rooted at a given node by summing the weights of all edges within that branch. It sed to determine the traversal order at the root.
- **calculatePotential()** evaluates the potential value $\varphi(v)$ based on the current energy level and the distance of the node from the root. This value is used to prioritize which child to explore next during DFS.
- **DFS()** executes the depth-first traversal under energy constraints. At each step, it checks the feasibility using the go-cost and return-cost, updates the remaining energy, and selects the next node according to subtree weight (at the root) or potential (elsewhere).

$$\omega(T') = \sum_{e \in E(T')} \omega(e),$$

$$\varphi(v) = \frac{B}{2} - d(r, v).$$

The analysis of the Piecemeal-DFS algorithm yields the following complexity results:

- **Time Complexity:** The time complexity of this algorithm is also dominated by the main loop. Since the `DFS` function is invoked repeatedly within the loop, leading to up to $O(N)$ iterations, the total time complexity is consequently $O(N^2)$.
- **Space Complexity:** Similar to the Piecemeal-DFS algorithm, the storage of routes in the `routes` vector can consume significant memory. In the worst-case scenario, the space complexity can reach $O(N^2)$, primarily due to the potential for a large number of long exploration routes.

# 4 Experimental Results

## 4.1 Results of PDFS Traversal

In order to explore all the vertices or edges of a tree under the condition that the tree height does not exceed B/2, we implemented the PDFS algorithm on a general tree and obtained the following exploration routes, shown in Table 1. The performance of the routes is depicted in Fig. 4. In advance, the result of PDFS is shown in Table 2.

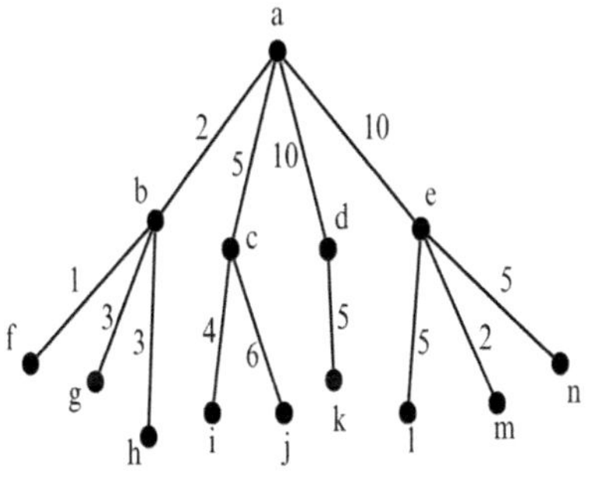


**Fig. 1.** The tree used for PDFS exploration.

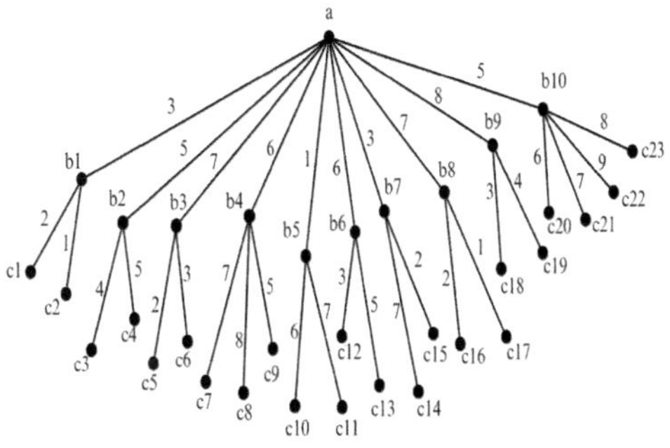


**Fig. 2.** The tree used for PDFS and ADFS exploration.

**Table 1.** Example of input and output for PDFS strategy with B = 30.

| INPUT | OUTPUT |
|---|---|
| Enter energy level B: 30 | Full DFS path: a b c d b g b a e f e j e a h k h a m n m l m p m a |
| Enter the number of tree edges: 13 | h a m n m l m p m a |
| a b 2 | PDFS routes (B = 30): |
| a c 5 | **Route 1:** a b f b g b h b a \| Cost = 18 |
| a d 10 | **Route 2:** a c i c j c a \| Cost = 30 |
| a e 10 | **Route 3:** a d k d a \| Cost = 30 |
| b f 1 | **Route 4:** a e l e a \| Cost = 30 |
| b g 3 | **Route 5:** a e m e a \| Cost = 24 |
| b h 3 | **Route 6:** a e n e a \| Cost = 30 |
| c i 4 | **Total cost of all PDFS routes: 162** |
| c j 6 | |
| d k 5 | |
| e l 5 | |
| e m 2 | |
| e n 5 | |

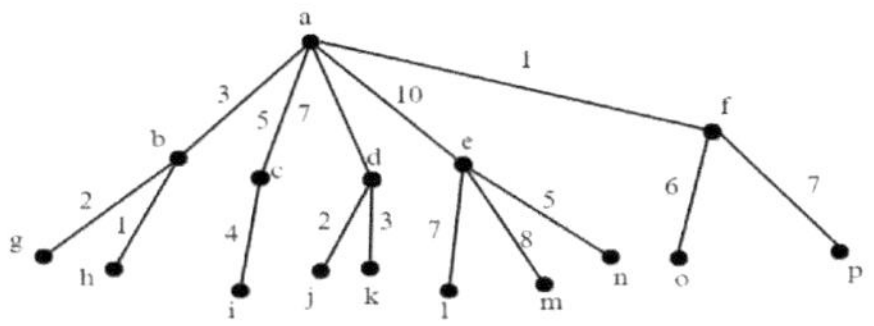


**Fig. 3.** The tree used for ADFS exploration.

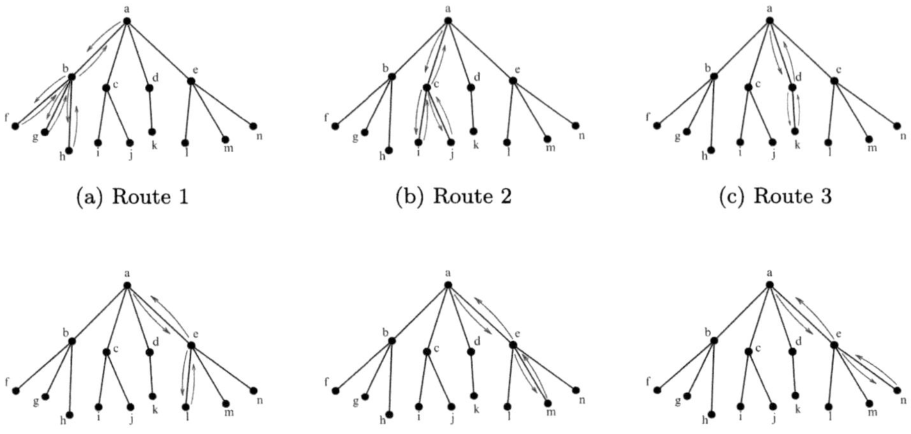

(a) Route 1 (b) Route 2 (c) Route 3

(d) Route 4 (e) Route 5 (f) Route 6

**Fig. 4.** Illustration of routes generated by the PDFS strategy.

**Table 2.** Example of input and output for PDFS strategy with B = 37 use tree in Fig. 1.

| INPUT | OUTPUT |
|---|---|
| Enter energy level B: 30 | Full DFS path: a b1 c1 b1 c2 b1 a b2 c3 b2 c4 b2 a b3 c5 b3 c6 b3 a b4 c7 b4 c8 b4 c9 b4 a b5 c10 b5 c11 b5 a b6 c12 b6 c13 b6 a b7 c14 b7 c15 b7 a b8 c16 b8 c17 b8 a b9 c18 b9 c19 b9 a b10 c20 b10 c21 b10 c22 b10 c23 b10 a |
| Enter the number of tree edges: 33 | Full DFS path: a b1 c1 b1 c2 b1 a b2 c3 b2 c4 b2 a b3 c5 b3 c6 b3 a b4 c7 b4 c8 b4 c9 b4 a b5 c10 b5 c11 b5 a b6 c12 b6 c13 b6 a b7 c14 b7 c15 b7 a b8 c16 b8 c17 b8 a b9 c18 b9 c19 b9 a b10 c20 b10 c21 b10 c22 b10 c23 b10 a |
| a b1 3 | c12 b6 c13 b6 a b7 c14 b7 c15 b7 a b8 c16 b8 c17 b8 a b9 |
| a b2 5 | c18 b9 c19 b9 a b10 c20 b10 c21 b10 c22 b10 c23 b10 a |
| a b3 7 | PDFS routes (B = 37): |
| a b4 6 | **Route 1:** a b1 c1 b1 c2 b1 a b2 c3 b2 a  \| Cost = 30 |
| a b5 1 | **Route 2:** a b2 c4 b2 a b3 a  \| Cost = 34 |
| a b6 6 | **Route 3:** a b3 c5 b3 c6 b3 a  \| Cost = 24 |
| a b7 3 | **Route 4:** a b4 c7 b4 a  \| Cost = 26 |
| a b8 7 | **Route 5:** a b4 c8 b4 a  \| Cost = 28 |
| a b9 8 | **Route 6:** a b4 c9 b4 a  \| Cost = 22 |
| a b10 5 | **Route 7:** a b5 c10 b5 c11 b5 a  \| Cost = 28 |
| b1 c1 2 | **Route 8:** a b6 c12 b6 c13 b6 a  \| Cost = 28 |
| b1 c2 1 | **Route 9:** a b7 c14 b7 c15 b7 a  \| Cost = 24 |
| b2 c3 4 | **Route 10:** a b8 c16 b8 c17 b8 a b9 a  \| Cost = 36 |
| b2 c4 5 | **Route 11:** a b9 c18 b9 c19 b9 a  \| Cost = 30 |
| b3 c5 2 | **Route 12:** a b10 c20 b10 c21 b10 a  \| Cost = 36 |
| b3 c6 3 | **Route 13:** a b10 c22 b10 a  \| Cost = 28 |
| b4 c7 7 | **Route 14:** a b10 c23 b10 a  \| Cost = 26 |
| b4 c8 8 | **Total cost of all PDFS routes: 400** |
| b4 c9 5 | |
| b5 c10 6 | |
| b5 c11 7 | |
| b6 c12 3 | |
| b6 c13 5 | |
| b7 c14 7 | |
| b7 c15 2 | |
| b8 c16 2 | |
| b8 c17 1 | |
| b9 c18 3 | |
| b9 c19 4 | |
| b10 c20 6 | |
| b10 c21 7 | |
| b10 c22 9 | |
| b10 c23 8 | |

## 4.2 Results of ADFS Traversal

In order to explore all the vertices or edges of a tree under the condition that the tree height does not exceed B/2, we implemented the ADFS algorithm on a general tree and obtained the following exploration routes. The illustration of ADFS is shown Table 3 . In advance, the performance of the algorithm is shown in Fig. 5 and Table 4.

**Table 3.** Example of input and output for ADFS strategy with B' = 20, B = 30 used for Fig. 2.

| INPUT | OUTPUT |
|---|---|
| Enter energy level B' and B : 20 30 | ADFS routes (B' = 20, B = 30): |
| Enter the number of tree edges: 15 | **Route 1:** a e n e a \| Cost = 22 |
| a b 3 | **Route 2:** a e l e a \| Cost = 26 |
| a c 5 | **Route 3:** a e m e a \| Cost = 28 |
| a d 7 | **Route 4:** a f o f p f a \| Cost = 28 |
| a e 6 | **Route 5:** a d j d k d a \| Cost = 24 |
| a f 1 | **Route 6:** a c i c a b h b g b a \| Cost = 30 |
| b g 2 | **Total cost of all ADFS routes: 158** |
| b h 1 | |
| c i 4 | |
| d j 2 | |
| d k 3 | |
| e l 7 | |
| e m 8 | |
| e n 5 | |
| f o 6 | |
| f p 7 | |

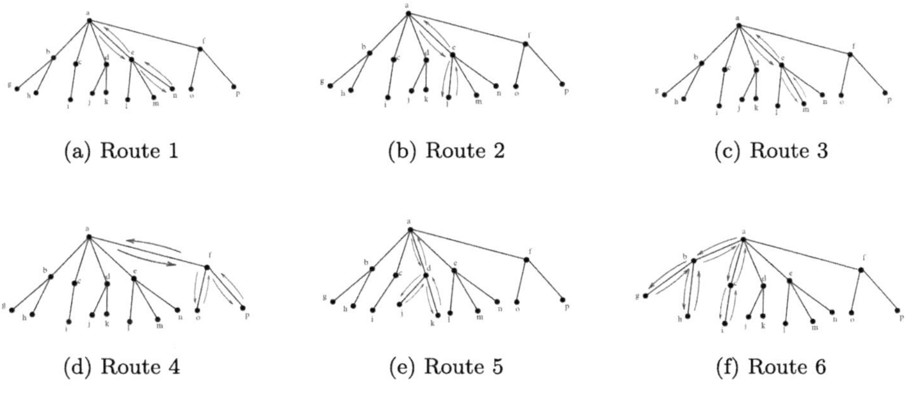

(a) Route 1 (b) Route 2 (c) Route 3

(d) Route 4 (e) Route 5 (f) Route 6

**Fig. 5.** Illustration of routes generated by the ADFS strategy.

**Table 4.** Example of input and output for ADFS strategy with B' = 30, B = 37 use tree in Fig. 3.

| INPUT | OUTPUT |
|---|---|
| Enter energy level B' and B : 30 37 | ADFS routes (B' = 30, B = 37): |
| Enter the number of tree edges: 33 | **Route 1:** a b10 c20 b10 a \| Cost = 22 |
| a b1 3 | **Route 2:** a b10 c21 b10 a \| Cost = 24 |
| a b2 5 | **Route 3:** a b10 c23 b10 a \| Cost = 26 |
| a b3 7 | **Route 4:** a b10 c22 b10 a \| Cost = 28 |
| a b4 6 | **Route 5:** a b4 c9 b4 c7 b4 a \| Cost = 36 |
| a b5 1 | **Route 6:** a b4 c8 b4 a \| Cost = 28 |
| a b6 6 | **Route 7:** a b9 c18 b9 c19 b9 a \| Cost = 30 |
| a b7 3 | **Route 8:** a b2 c3 b2 c4 b2 a b5 a \| Cost = 30 |
| a b8 7 | **Route 9:** a b5 c10 b5 c11 b5 a \| Cost = 28 |
| a b9 8 | **Route 10:** a b6 c12 b6 c13 b6 a \| Cost = 28 |
| a b10 5 | **Route 11:** a b3 c5 b3 c6 b3 a b7 c15 b7 a \| Cost = 34 |
| b1 c1 2 | **Route 12:** a b7 c14 b7 a b8 c17 b8 a \| Cost = 36 |
| b1 c2 1 | **Route 13:** a b8 c16 b8 a b1 c2 b1 c1 b1 a \| Cost = 30 |
| b2 c3 4 | **Total cost of all ADFS routes: 380** |
| b2 c4 5 | |
| b3 c5 2 | |
| b3 c6 3 | |
| b4 c7 7 | |
| b4 c8 8 | |
| b4 c9 5 | |
| b5 c10 6 | |
| b5 c11 7 | |
| b6 c12 3 | |
| b6 c13 5 | |
| b7 c14 7 | |
| b7 c15 2 | |
| b8 c16 2 | |
| b8 c17 1 | |
| b9 c18 3 | |
| b9 c19 4 | |
| b10 c20 6 | |
| b10 c21 7 | |
| b10 c22 9 | |
| b10 c23 8 | |

## 5 Conclusion

In our research, we proposed and successfully implemented an energy replenishment optimization strategy for PDFS to minimize the total traversal cost. We integrated a more intelligent decision-making mechanism based on the evaluation of the potential of the upcoming branch. This approach allows for the maximal utilization of the remaining energy, reducing the number of returns to the root and the number of new routes that need to be initiated [12, 13]. This is a critical factor in resource-constrained applications such as exploration robots, sensor networks, and IoT systems. In addition to PDFS, this study also successfully implemented the Adversarial-DFS (ADFS) algorithm. ADFS plays an essential role in modeling the worst-case scenarios that a search algorithm may encounter under energy constraints. It provides a powerful tool for identifying upper bounds on actual traversal costs, thereby enabling the assessment of the robustness and sustainability of search algorithms under adverse conditions.

## References


1. Betke, M., Rivest, R.L., Singh, M.: Piecemeal learning of an unknown environment. Mach. Learn. **18**, 231–254 (1995). https://doi.org/10.1007/BF00993411
2. Das, S., Dereniowski, D., Uzna'nski, P.: Energy constrained depth first search. Algorithmica **86**(12), 3759–3782 (2024). https://doi.org/10.1007/s00453-024-01275-8
3. Panaite, P., Pelc, A.: Exploring unknown undirected graphs. J. Algorithms **33**(2), 281–295 (1999). https://doi.org/10.1006/jagm.1999.1043
4. Dereniowski, D., Disser, Y., Kosowski, A., Pajak, D., Uzna'nski, P.: Fast collaborative graph exploration. Inf. Comput. 243, 37–49 (2015). https://doi.org/10.1016/j.ic.2014.12.005. 40th International Colloquium on Automata, Languages and Programming (ICALP 2013)
5. Disser, Y., Mousset, F., Noever, A., Skori'c, N., Steger, A.: A general lower bound for collaborative tree exploration. Special Issue Struct. Inf. Commun. Complexit **811**, 70–78 (2020). https://doi.org/10.1016/j.tcs.2018.03.006. Theoretical Computer Science
6. Megow, N., Mehlhorn, K., Schweitzer, P.: Online graph exploration: new results on old and new algorithms. Theor. Comput. Sci. **463**, 62–72 (2012). https://doi.org/10.1016/j.tcs.2012.06.034. Special Issue on Theory and Applications of Graph Searching Problems
7. Bärtschi, A., et al.: Collaborative delivery with energy-constrained mobile robots. Theoret. Comput. Sci. **810**, 2–14 (2020). https://doi.org/10.1016/j.tcs.2017.04.018
8. Czyzowicz, J., Diks, K., Moussi, J., Rytter, W.: Energy-optimal broadcast and exploration in a tree using mobile agents. Theoret. Comput. Sci. **795**, 362–374 (2019). https://doi.org/10.1016/j.tcs.2019.07.018
9. Borowiecki, P., Dereniowski, D., Osula, D.: The complexity of bicriteria treedepth. Theor. Comput. Sci. **947**, 113682 (2023). https://doi.org/10.1016/j.tcs.2022.12.032
10. Awerbuch, B., Betke, M., Rivest, R.L., Singh, M.: Piecemeal graph exploration by a mobile robot. Inf. Comput. **152**(2), 155–172 (1999). https://doi.org/10.1006/inco.1999.2795

11. M., T.N., H., S.N., V., V.C.: Piecemeal-DFS algorithm for energy-constrained depth-first search on the merging problem of two given trees (in Vietnamese). The 2nd national symposium on natural sciences and applications in the digital age. On the Proceedings of the National Scientific Conference on Natural Sciences and Applications in the Digital Age (NSA), Ha Noi, Viet Nam 46–53, ISBN 978-604-67-330-0 (2025)
12. Tuan, N.M., Meesad, P., Nguyen, H.H.C.: English–Vietnamese machine translation using deep learning for chatbot applications. SN Comput. Sci. **5**(1), 5 (2024). https://doi.org/10.1007/s42979-023-02339-2
13. Tuan, N.M., Meesad, P., Hieu, D.V., Cuong, N.H.H., Maliyaem, M.: On students' sentiment prediction based on deep learning: applied information literacy. SN Comput. Sci. **5**(7), 928 (1995). https://doi.org/10.1007/s42979-024-03281-7

# UNet-VGG19-CBAM: A Hybrid Attention-Based Multi-branch Model for Pneumonia Detection in Chest X-Ray Images

Nguyen Doan Cong Tien, Ha Tien Khoa, and Nguyen Minh Tuan(✉)

Faculty of Information Technology, Posts and Telecommunications Institute of Technology, 11 Nguyen Dinh Chieu,
Sai Gon ward, Ho Chi Minh city 700000, Vietnam
{n22dccn184,n22dccn041}@student.ptithcm.edu.vn, minhtuan@ptit.edu.vn

**Abstract.** **Purpose**: Pneumonia remains one of the leading causes of mortality worldwide. However, current deep learning methods for diagnosing pneumonia from chest X-ray (CXR) images often struggle with class imbalance, limited discrimination among pneumonia subtypes, and poor interpretability.

**Method**: Introducing UNetVGG19CBAM, a hybrid model that integrates: UNet for lung segmentation to localize critical regions, a pre-trained VGG19 encoder for extracting hierarchical features, Convolutional Block Attention Module (CBAM) inserted into skip connections to emphasize informative channels and spatial locations. The model is trained and evaluated on a chest X-ray dataset collected from public sources using metrics that include AUC and F1 score. Attention heatmaps (via CBAM) are used to visualize disease-sensitive areas.

**Results**: Our model achieves high performance, for example: AUC 0.9962, F1 score 0.9852, outperforming baseline models such as standalone VGG19 or UNet only by significant margins. The attention maps clearly highlight regions corresponding to pneumonia infiltrates, improving model transparency.

**Conclusion**: Combining segmentation, pre-trained feature extraction, and lightweight attention modules results in a highly accurate and interpretable CXR-based pneumonia screening tool. Attention heatmaps provide visual justification for predictions, supporting clinical trust and adoption.

**Keywords:** Deep learning · Pneumonia detection · Chest X-ray · UNet · VGG19 · CBAM · Attention mechanism · Medical image analysis

---

N. D. Cong, H. T. Khoa—Contributing authors.


P. Pichappan et al. (Eds.): RTIS 2025, LNNS 1936, pp. 25–35, 2027.
https://doi.org/10.1007/978-3-032-23396-7_3

# 1 Introduction

Pneumonia is a common and life-threatening respiratory disease that affects millions of people around the world each year. It remains a major cause of hospitalization and mortality. Early and accurate diagnosis is essential for effective treatment and improved patient outcomes. Chest X-ray (CXR) imaging is the most widely used and cost-effective method for detecting pneumonia. However, manual interpretation of X-ray images can be time-consuming and prone to human error, especially in regions with limited access to experienced radiologists.[1,2]

In recent years, deep learning techniques, particularly convolutional neural networks (CNNs), have shown promising results in automated pneumonia detection. These models can learn complex visual patterns and often outperform traditional machine learning approaches in classification tasks. Several studies have attempted to improve model performance by adopting advanced architectures and additional techniques. For example, Sharma et al. [3] developed a VGG16-based model that achieved high accuracy, but it lacked tools for visual explanation. Zhang et al. [4] proposed an ensemble of pre-trained CNNs, which improved sensitivity but increased computational demands. Bhattacharya et al. [5] applied vision transformers (ViT), achieving strong performance, but the model introduced complexity and reduced transparency. A comprehensive review by Ali et al. [6] highlighted that despite improvements in accuracy, many existing models still struggle with three key issues: class imbalance, difficulty distinguishing between bacterial and viral pneumonia, and lack of interpretability.

To address these limitations, we propose a hybrid deep learning model named UNet-VGG19-CBAM. The model combines three components: (1) UNet for accurate lung segmentation, allowing the model to focus on relevant regions; (2) a pre-trained VGG19 network for extracting deep and transferable features; and (3) the Convolutional Block Attention Module (CBAM), which enhances both spatial and channel-wise feature representation. Our architecture not only improves classification performance but also generates attention heatmaps that visually highlight suspicious regions, thereby improving interpretability and clinical usability.

# 2 Fundamental Definitions

**Definition 1** (Feature map). A feature map is the output of a convolutional layer in a neural network, representing the activation responses of learned filters across different spatial locations of the input. [7]

**Definition 2** (VGG19). VGG19 is a convolutional neural network architecture characterized by its simplicity, using only $3 \times 3$ convolutional layers stacked on top of each other, followed by max-pooling layers. It deepens the network by increasing the number of filters. VGG19 is commonly used as a powerful feature extractor, especially when pre-trained on large datasets like ImageNet. [8]

**Definition 3** (UNet). UNet is a convolutional network architecture primarily designed for image segmentation, notable for its U-shaped structure. It consists of a contracting path (encoder) to capture context and a symmetric expanding path (decoder) that enables precise localization. Crucially, UNet employs "skip connections" from the contracting path to the expanding path, allowing the decoder to recover fine-grained details lost during downsampling. In our model, a UNet-like encoder is utilized for feature extraction. [9]

**Definition 4** (CBAM (Convolutional Block Attention Module)). CBAM is an attention mechanism that sequentially applies channel attention and spatial attention to a feature map. It helps the model focus on informative regions and channels, improving feature representation. [10]

# 3 Materials and Methods

## 3.1 Dataset 1: Chest X-Ray Pneumonia, COVID-19 and Pneumonia Chest X-Ray

The first dataset used in this study is publicly available (https://www.kaggle.com/datasets/amanullahasraf/covid19-pneumonia-normal-chest-xray-pa-dataset). It contains a total of 5,856 chest X-ray images, including both normal and pneumonia cases, collected from the Guangzhou Women and Children's Medical Center. All images were obtained during routine clinical care, and low-quality scans were removed. The diagnostic labels were confirmed by two expert physicians [11]. The dataset is organized into `train`, `val`, and `test` splits, with class distribution maintained across subsets. The second dataset contains chest X-ray images of COVID-19, pneumonia, and normal cases, publicly available at (https://www.kaggle.com/datasets/marlesson/covid19-xray-images). This dataset comprises 6,432 images in total, organized into two main folders (`train` and `test`), each containing three subfolders corresponding to `COVID19`, `PNEUMONIA`, and `NORMAL`. Approximately 20% of the data is allocated for testing. COVID-19 cases were collected from multiple hospitals and preprocessed for consistent quality and resolution. This dataset provides a reliable benchmark for evaluating deep learning models for multi-class chest disease classification [12].

## 3.2 Deep Learning Algorithm

In this study, we propose a deep learning model that combines VGG19, a UNet-like encoder, and CBAM [13,14] to detect pneumonia from chest X-ray images. Specifically, the model utilizes VGG19 pretrained on ImageNet to extract deep, hierarchical features, leveraging its ability to learn complex representations from images. For grayscale chest X-rays, the input is replicated across three channels to match VGG19's expected input. Meanwhile, a UNet-like encoder branch extracts complementary features directly from the original single-channel image, aiming to preserve important morphological and spatial characteristics shown in Fig. 1.

The outputs from the VGG19 and the UNet-like encoder are aligned and fused using element-wise multiplication and addition, enhancing the interaction between the two branches. The fusion operation is defined as:

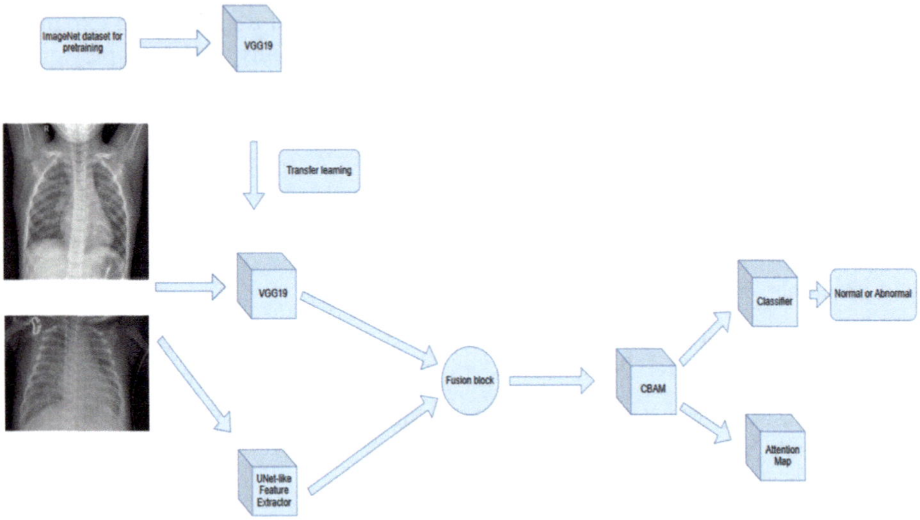


**Fig. 1.** The algorithm flow chart.

$$F_{\text{fused}} = (F_{\text{vgg}} \odot F_{\text{unet}}) + (F_{\text{vgg}} + F_{\text{unet}})$$

This fused feature map is then passed through a fusion block consisting of convolution and normalization layers, followed by CBAM, an attention mechanism composed of Channel Attention and Spatial Attention, which helps the model focus on the most relevant regions and channels in the image. Next, the output features (after CBAM) are flattened after Global Average Pooling and fed into fully connected layers to classify the image into one of two categories: pneumonia or normal. Importantly, we leverage the Spatial Attention output from CBAM as a form of visual interpretability, similar to a Class Activation Map (CAM), to highlight the areas likely to be affected by pneumonia. This architecture enables the model not only to achieve accurate classification but also to provide valuable spatial information to assist in clinical diagnosis. We summarize the characteristics of the classifiers in Table 1.

# 4 Numerical Results

## 4.1 Dataset 1: Chest X-Ray Pneumonia Dataset

To evaluate the proposed UNet-VGG19-CBAM model, we conducted comprehensive experiments on the ChestX-ray dataset. This dataset contains a total of 5,856 chest X-ray images, divided into two categories: 1,583 normal images and 4,273 pneumonia images (including both bacterial and viral cases). The images were collected from Guangzhou Women and Children's Medical Center during routine clinical care.

The dataset was randomly split into three subsets: 70% for training, 15% for validation, and 15% for testing, while maintaining class distribution across splits.

**Table 1.** UNet-VGG19 Multi-Task Classifier Architecture

| Block / Layer | Layer Type & Parameters | Output Size | Description |
|---|---|---|---|
| Input | | 1 × 128 × 128 | Grayscale input image |
| UNet Block | Conv2d(1,64,3,1,1) + ReLU | 64 × 128 × 128 | Low-level feature extraction |
| | Conv2d(64,64,3,1,1) + ReLU | 64 × 128 × 128 | Enhanced feature map |
| | AdaptiveAvgPool2d((16,16)) | 64 × 16 × 16 | Downsample features |
| | Conv2d(64,512,3,1,1) | 512 × 16 × 16 | Match dimension with VGG branch |
| VGG19 Block | Pretrained VGG19 layers (Conv + Pooling) | 512 × 16 × 16 | High-level semantic features |
| Fusion Block | Element-wise: (UNet ⊙ VGG + VGG ⊙ UNet) | 512 × 16 × 16 | Combine dual-branch features |
| | Conv2d(512,512,3,1,1) + BN + ReLU + Dropout(0.3) | 512 × 16 × 16 | Refine fused representation |
| Attention Block (CBAM) | Channel & Spatial Attention | 512 × 16 × 16 | Focus on salient regions |
| | BatchNorm2d(512) | 512 × 16 × 16 | Normalize attention output |
| Classifier Head | AdaptiveAvgPool2d((1,1)) | 512 × 1 × 1 | Global feature aggregation |
| | Flatten() | 512 | Vectorize feature map |
| | Linear(512→256) + ReLU + Dropout(0.5) | 256 | Fully connected layer |
| | Linear(256→2) | 2 | Final classification layer |
| Output | | 2 | Binary classification result |

To ensure fair evaluation, no patient overlap exists between the splits, minimizing data leakage. The model was implemented in Python using Pytorch 2.6.0. Training was performed on an NVIDIA RTX 4060 GPU with 16 GB VRAM. The network was trained with a batch size of 8 for 50 epochs using the Adam optimizer with a learning rate of $3 \times 10^{-5}$ and a weight decay of $1 \times 10^{-5}$. A loss weighting of [1.0, 1.5] was applied to balance the contribution of each loss component. Mixed-precision training was utilized to accelerate computation and reduce memory consumption. Early stopping and learning rate scheduling were applied to avoid overfitting. Data augmentation techniques, including random rotation, zoom, shift, and contrast adjustment, were employed to increase diversity and partially address class imbalance.

Our model achieved excellent performance on the test set with the following metrics: accuracy = 98.52%, precision = 98.5%, recall = 98.0%, F1 score = 0.9852, and AUC = 0.9962. Compared to standard architectures such as ResNet18, DenseNet121, EfficientNet-B0, and a Vision Transformer (ViT) variant, UNet-VGG19-CBAM consistently outperformed in all key metrics, highlighting the advantage of the multi-branch fusion and attention mechanism.

**Table 2.** Performance comparison of UNet-VGG19-CBAM with baseline models on the test set.

| Model | Accuracy | Precision | Recall | F1 |
|---|---|---|---|---|
| ResNet18 | 0.9100 | 0.9418 | 0.9852 | 0.9632 |
| DenseNet121 | 0.9613 | 0.9736 | 0.9889 | 0.9812 |
| EfficientNet-B0 | 0.9431 | 0.9599 | 0.9919 | 0.9757 |
| ViT-B16 | 0.7608 | 0.8276 | 0.8152 | 0.8214 |
| UNet-VGG19-CBAM | **0.9852** | **0.9850** | **0.9800** | **0.9852** |

The training dynamics of UNet-VGG19-CBAM over 50 epochs are illustrated in Fig. 5. The model converges smoothly, and validation accuracy closely follows training trends, indicating minimal overfitting due to the combined use of attention and fusion mechanisms. The prediction is shown in Table 2. To provide interpretability, we extracted the CBAM attention maps from test images. Figure 2 shows examples where the spatial attention correctly highlights regions affected by pneumonia, providing visual explanation consistent with radiological findings.

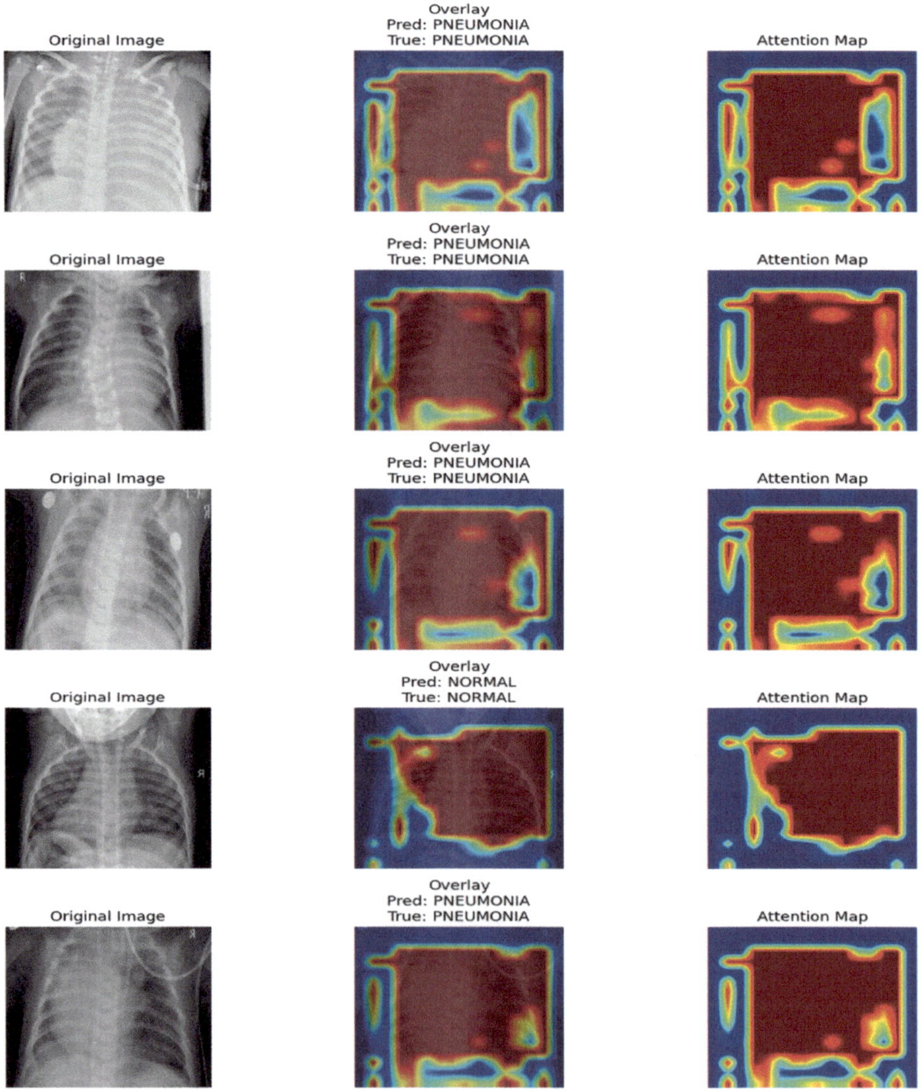


**Fig. 2.** CBAM attention maps: original image, CAM overlay, and spatial attention map.

The internal structure of the fusion block, which combines UNet and VGG19 features before applying CBAM, is summarized in Fig. 3. This design allows

| Layer (type) | Output Shape | Param # |
|---|---|---|
| Conv2d-1 | [-1, 64, 256, 256] | 640 |
| ReLU-2 | [-1, 64, 256, 256] | 0 |
| Conv2d-3 | [-1, 64, 256, 256] | 36,928 |
| ReLU-4 | [-1, 64, 256, 256] | 0 |
| Conv2d-5 | [-1, 512, 16, 16] | 295,424 |
| Conv2d-6 | [-1, 64, 256, 256] | 1,792 |
| ReLU-7 | [-1, 64, 256, 256] | 0 |
| Conv2d-8 | [-1, 64, 256, 256] | 36,928 |
| ReLU-9 | [-1, 64, 256, 256] | 0 |
| MaxPool2d-10 | [-1, 64, 128, 128] | 0 |
| Conv2d-11 | [-1, 128, 128, 128] | 73,856 |
| ReLU-12 | [-1, 128, 128, 128] | 0 |
| Conv2d-13 | [-1, 128, 128, 128] | 147,584 |
| ReLU-14 | [-1, 128, 128, 128] | 0 |
| MaxPool2d-15 | [-1, 128, 64, 64] | 0 |
| Conv2d-16 | [-1, 256, 64, 64] | 295,168 |
| ReLU-17 | [-1, 256, 64, 64] | 0 |
| Conv2d-18 | [-1, 256, 64, 64] | 590,080 |
| ReLU-19 | [-1, 256, 64, 64] | 0 |
| Conv2d-20 | [-1, 256, 64, 64] | 590,080 |
| ReLU-21 | [-1, 256, 64, 64] | 0 |
| Conv2d-22 | [-1, 256, 64, 64] | 590,080 |
| ReLU-23 | [-1, 256, 64, 64] | 0 |
| MaxPool2d-24 | [-1, 256, 32, 32] | 0 |
| Conv2d-25 | [-1, 512, 32, 32] | 1,180,160 |
| ReLU-26 | [-1, 512, 32, 32] | 0 |
| Conv2d-27 | [-1, 512, 32, 32] | 2,359,808 |
| ReLU-28 | [-1, 512, 32, 32] | 0 |
| Conv2d-29 | [-1, 512, 32, 32] | 2,359,808 |
| ReLU-30 | [-1, 512, 32, 32] | 0 |
| Conv2d-31 | [-1, 512, 32, 32] | 2,359,808 |
| ReLU-32 | [-1, 512, 32, 32] | 0 |
| MaxPool2d-33 | [-1, 512, 16, 16] | 0 |
| Conv2d-34 | [-1, 512, 16, 16] | 2,359,808 |
| ReLU-35 | [-1, 512, 16, 16] | 0 |
| Conv2d-36 | [-1, 512, 16, 16] | 2,359,808 |
| ReLU-37 | [-1, 512, 16, 16] | 0 |
| Conv2d-38 | [-1, 512, 16, 16] | 2,359,808 |
| ReLU-39 | [-1, 512, 16, 16] | 0 |
| Conv2d-40 | [-1, 512, 16, 16] | 2,359,808 |
| ReLU-41 | [-1, 512, 16, 16] | 0 |
| VGG19FeatureExtractor-42 | [-1, 512, 16, 16] | 0 |
| Conv2d-43 | [-1, 512, 16, 16] | 2,359,808 |
| BatchNorm2d-44 | [-1, 512, 16, 16] | 1,024 |
| ReLU-45 | [-1, 512, 16, 16] | 0 |
| Dropout2d-46 | [-1, 512, 16, 16] | 0 |
| Conv2d-47 | [-1, 64, 1, 1] | 32,832 |
| ReLU-48 | [-1, 64, 1, 1] | 0 |
| Conv2d-49 | [-1, 512, 1, 1] | 33,280 |
| Conv2d-50 | [-1, 64, 1, 1] | 32,832 |
| ReLU-51 | [-1, 64, 1, 1] | 0 |
| Conv2d-52 | [-1, 512, 1, 1] | 33,280 |
| Sigmoid-53 | [-1, 512, 1, 1] | 0 |
| ChannelAttention-54 | [-1, 512, 1, 1] | 0 |
| Conv2d-55 | [-1, 1, 16, 16] | 99 |
| Sigmoid-56 | [-1, 1, 16, 16] | 0 |
| SpatialAttention-57 | [-1, 1, 16, 16] | 0 |
| CBAM-58 | [-1, 512, 16, 16] | 0 |
| BatchNorm2d-59 | [-1, 512, 16, 16] | 1,024 |
| AdaptiveAvgPool2d-60 | [-1, 512, 1, 1] | 0 |
| Flatten-61 | [-1, 512] | 0 |
| Linear-62 | [-1, 256] | 131,328 |
| ReLU-63 | [-1, 256] | 0 |
| Dropout-64 | [-1, 256] | 0 |
| Linear-65 | [-1, 2] | 514 |
| Conv2d-66 | [-1, 1, 16, 16] | 99 |
| Sigmoid-67 | [-1, 1, 16, 16] | 0 |
| SpatialAttention-68 | [-1, 1, 16, 16] | 0 |

Total params: 22,983,496
Trainable params: 17,117,960
Non-trainable params: 5,865,536

**Fig. 3.** Summary of the fusion block architecture in the proposed model.

multi-scale and multi-channel information to be effectively integrated, improving classification performance.

Figure 4 presents the confusion matrix on the test set. It demonstrates that our model achieves balanced sensitivity and specificity, correctly identifying both normal and pneumonia cases with minimal misclassifications.

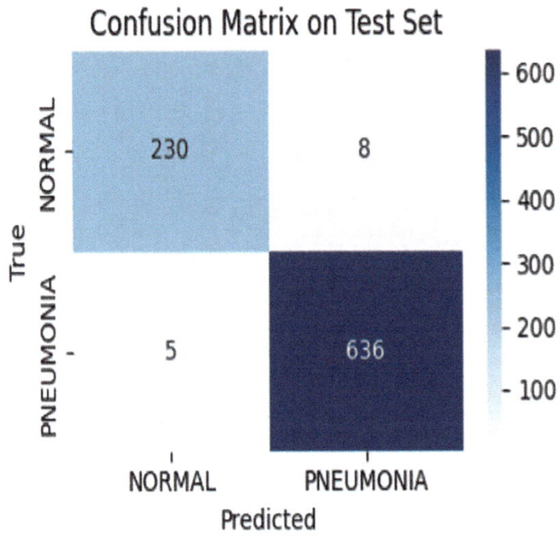


**Fig. 4.** Confusion matrix of UNet-VGG19-CBAM on the test set.

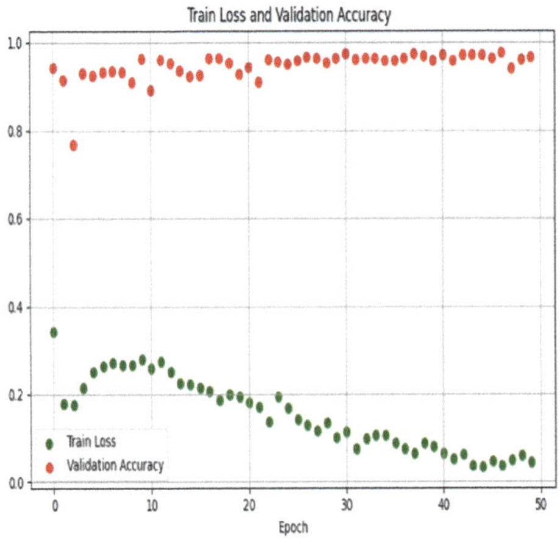


**Fig. 5.** Training loss and validation accuracy over 50 epochs.

### 4.2 Dataset 2: COVID-19 and Pneumonia Chest X-Ray Dataset

The second dataset used in this study is the COVID-19 and Pneumonia Chest X-ray dataset, which contains a total of 6,432 chest X-ray images. The dataset is organized into two main folders (*train* and *test*), each containing three subfolders corresponding to the classes: *COVID19*, *PNEUMONIA*, and *NORMAL*. All images are reserved for testing. To assess the transferability of the proposed model, the pretrained UNet-VGG19-CBAM (trained on Dataset 1) was directly evaluated on Dataset 2 without any finetuning. As expected, performance decreased due to the difference in image distribution and class composition between the two datasets. The obtained results are presented in Table 3. While the model still achieves a moderate accuracy of 66.32%, it shows a strong bias toward pneumonia cases, emphasizing the need for dataset-specific finetuning to improve domain adaptation and class balance.

**Table 3.** Preliminary classification results of UNet-VGG19-CBAM on Dataset 2 (COVID-19 & Pneumonia) without finetuning. Weighted F1-score = 0.6317, AUC = 0.8912.

| Class | Precision | Recall | F1-score | Support |
|---|---|---|---|---|
| NORMAL | 0.89 | 0.37 | 0.52 | 2313 |
| PNEUMONIA | 0.60 | 0.96 | 0.74 | 2313 |
| Accuracy | 0.6632 | 0.6632 | 0.6632 | 0.6632 |
| Macro Avg | 0.75 | 0.66 | 0.63 | 4626 |
| Weighted Avg | 0.75 | 0.66 | 0.63 | 4626 |

## 5 Limitations and Discussions

Although the proposed model achieves high accuracy, there are several limitations to consider. First, the dataset used may not fully represent diverse patient populations, potentially limiting the model's generalization to real-world scenarios. Second, the grayscale to RGB replication may introduce redundancy, as it adds no new information to the image. Moreover, while CBAM provides interpretability through attention maps, it does not fully replace expert radiological analysis, especially in ambiguous cases. To quantitatively assess interpretability, we conducted a comparative experiment by removing the CBAM module from the proposed architecture. The performance dropped by approximately 0.2 in AUC, indicating that the attention mechanism not only enhances classification accuracy but also improves the model's ability to focus on clinically relevant regions. This suggests that CBAM contributes to both predictive performance and interpretability. Third, the architecture is relatively complex due to the multi-branch fusion of VGG19 and UNet-like encoders, along with the integration of attention mechanisms. This complexity may increase computational demands and make the model more prone to overfitting, particularly when trained on small or imbalanced datasets. Despite these limitations, the fusion of VGG19 and UNet-like features, combined with lightweight attention modules, demonstrates significant potential in enhancing pneumonia detection accuracy and interpretability. With proper data augmentation, regularization, and domain adaptation, this model can serve as a reliable tool in clinical support systems.

## 6 Conclusion and Future Work

In this study, we proposed a deep learning architecture combining VGG19, a UNet-like encoder, and CBAM for pneumonia detection in chest X-ray images. The model successfully leverages both hierarchical and spatial features, achieving state-of-the-art performance. In future work, we aim to extend the model to multi-class classification tasks, explore domain adaptation to generalize across hospitals, and integrate additional attention mechanisms or transformer-based encoders to further improve performance and explainability.

**Acknowledgment.** We sincerely appreciate the financial research support of the Posts and Telecommunications Institute of Technology in Ho Chi Minh City, Viet Nam, and for providing the necessary resources, a conducive research environment that contributed significantly to the success of this study.

## References

1. Verma, A., Swarnkar, S.K., Singh, K., Pandey, C., Sharma, Y.: Advanced deep learning models for pneumonia detection. In: Advances in Intelligent Systems Research, pp. 190–210. (2025). https://doi.org/10.2991/978-94-6463-704-515
2. Shilpa, N., Ayeesha Banu, W., Metre, P.B.: Revolutionizing pneumonia diagnosis: AI-driven deep learning framework for automated detection from chest x-rays. IEEE Access **12**, 171601–171616 (2024). https://doi.org/10.1109/ACCESS.2024.3498944
3. Sharma, M., Bhardwaj, A., Bansal, A.: A deep learning based model for the detection of pneumonia from chest x-ray images using vgg16. Proc. Comput. Sci. **218**, 456–462 (2023). https://doi.org/10.1016/j.procs.2023.03.078
4. Zhang, Y., Li, C., Wang, H., Zhao, X.: Ensemble deep transfer learning models for pneumonia detection. Comput. Biol. Med. **170**, 107571 (2024). https://doi.org/10.1016/j.compbiomed.2024.107571
5. Singh, S., Kumar, M., Kumar, A., Verma, B.K., Abhishek, K., Selvarajan, S.: Efficient pneumonia detection using vision transformers on chest x-rays. Sci. Rep. **14**(1), 3351 (2024). https://doi.org/10.1038/s41598-024-52703-2
6. Ali, F., Ahmed, M., Khan, S., Rauf, H.T.: Deep learning for pneumonia detection in chest x-ray images: a comprehensive survey. J. Imaging **10**(2), 25 (2024). https://doi.org/10.3390/jimaging10080176
7. Wiatowski, T., Bölcskei, H.: A mathematical theory of deep convolutional neural networks for feature extraction. IEEE Trans. Inf. Theory **64**(3), 1845–1866 (2017). https://doi.org/10.1109/TIT.2017.2776228
8. Simonyan, K., Zisserman, A.: Very deep convolutional networks for large-scale image recognition. arXiv preprint arXiv:1409.1556 (2014). http://arxiv.org/pdf/1409.1556

9. Ronneberger, O., Fischer, P., Brox, T.: U-net: convolutional networks for biomedical image segmentation. In: Proceedings of the International Conference on Medical Image Computing and Computer-Assisted Intervention (MICCAI). Lecture Notes in Computer Science, vol. 9351, pp. 234–241. Springer, ??? (2015). https://doi.org/10.1007/978-3-319-24574-4_28
10. Woo, S., Park, J., Lee, J.-Y., Kweon, I.S.: Cbam: convolutional block attention module. In: Lecture Notes in Computer Science, pp. 3–19 (2018). https://doi.org/10.1007/978-3-030-01234-2_1
11. Kermany, D.S., et al.: Identifying medical diagnoses and treatable diseases by image-based deep learning. Cell **172**(5), 1122–11319 (2018). https://doi.org/10.1016/j.cell.2018.02.010
12. Fahad, N.: Medical image classification using chest x-ray (Covid-19 and pneumonia) dataset, (2024). arXiv preprint
13. Tuan, N.M., Meesad, P., Nguyen, H.H.C.: English–Vietnamese machine translation using deep learning for chatbot applications. SN Comput. Sci. **5**(1), 5 (2024). https://doi.org/10.1007/s42979-023-02339-2
14. Tuan, N.M., Meesad, P., Hieu, D.V., Cuong, N.H.H., Maliyaem, M.: On students' sentiment prediction based on deep learning: Applied information literacy. SN Comput. Sci. **5**(7), 928 (2025). https://doi.org/10.1007/s42979-024-03281-7

# Top-View Image-Based Mass and Volume Estimation of Vietnamese Avocado 034 Using Geometric Slice Modeling

Le Nguyen Hieu An[1], Truong Quoc Bao[1], Hoang Ngoc An[1], Nguyen Tran Minh Nguyet[1], Minh-Quan Nguyen[2], Son V. T. Dao[2], and Minh-Tai Vo[3](✉)

[1] HCMC University of Technology and Engineering, Ho Chi Minh City 700000, Vietnam
{22151007,22151008,22151003}@student.hcmute.edu.vn, nguyetntm@hcmute.edu.vn

[2] RMIT University, Ho Chi Minh City 700000, Vietnam
{quan.nguyen93,son.daovutruong}@rmit.edu.vn

[3] Posts and Telecommunications Institute of Technology, Ho Chi Minh City 700000, Vietnam
taivm@ptit.edu.vn

**Abstract.** This paper studies an approach for estimating the mass and volume of Vietnamese avocado 034 using top-view image sample combined with geometric slice modeling. The method involves analyzing image-based geometric features to perform efficient mass prediction. Key components are defined and collected from top-view images through image processing techniques, including length, width, and height. The selected features later are integrated into the geometric slice modeling that approximately estimates the mass of fruit. The approach is designed to support agriculture and post-harvest processing by offering a practical and scalable alternative to traditional weighing methods. A dataset of 101 avocado samples was used to evaluate the method under different slicing configurations (3, 5, 7, 8, 10, 12, 15, 18, and 20 slices). Experimental results show that the accuracy of weight estimation reach up to 91.35% by slicing 12 slices. Moreover, the processing speed of each image with configuration with 7 to 12 slices ranged from 0.148 to 0.168 s, which corresponds to real-time processing of up to 5–7 images per second on standard hardware. This approach is low-cost, non-destructive, and scalable, making it suitable for agricultural and post-harvest applications such as automated fruit sorting and grading.

**Keywords:** Vietnamese Avocado 034 · Mass Estimation · Volume Estimation · Image Processing · Geometric Slice Modeling

## 1 Introduction

Avocado 034 strain (Persea americana Mill. cv. 034) is commonly grown in the Central Highlands, Lam Dong Province, Vietnam [1, 2]. It is distinguished by its elongated, gourd-like shape, small seed, and rich, buttery flavor. These features give Vietnamese

These authors contributed equally to this work.


P. Pichappan et al. (Eds.): RTIS 2025, LNNS 1936, pp. 36–48, 2027.
https://doi.org/10.1007/978-3-032-23396-7_4

avocado 034 a competitive edge over the imported avocado products. The skin is a glossy dark green that gradually shifts to a yellowish-green hue when ripe, as shown in Fig. 1. Avocado 034 is notably large, typically weighing between 0.3 and 0.8 kg, with some fruits reaching up to 1 kg [3].

After harvesting, farmers often have to sort avocados by weight for sale. It takes a lot of time to complete this task. Our motivation, therefore, is to use image processing and mathematical methods, namely geometric slice modeling, to estimate the weight of agricultural products in support of industrial agriculture.

**Fig. 1.** Representative samples of Vietnamese Avocado 034 used in the dataset, showing the typical elongated, gourd-like shape and glossy green skin.

Mass and volume estimation for fruits and vegetables has been widely researched. In the study [4], they present an image-processing-based technique for estimating the volume and mass of axi-symmetric fruits (e.g., apple, sweet lime, lemon, and orange) using a single camera that captures five views of each fruit. A computer vision system was developed to automatically extract 2D and 3D features of mangoes and perform automated fruit sorting for the Thai mango cultivar Nam Dokmai [5]. They used four cameras to capture top and side views of mango to extract 2D features (projected area, length, width and thickness) and 3D features (volume, surface area). Count and computer vision-based framework, namely Multiple Linear Regression to detect, count, and estimate the size and mass of individual olive fruits [6]. Passion fruits have also been used as test samples for mass and volume estimation by applying two techniques multiple linear regression and artificial neural networks [7]. To support post-harvest sorting and grading, scholars in study [8] have developed an RGB camera-based system capable of real-time weight and volume estimation for tomatoes on a conveyor belt. They tested seven models, including support vector regression (SVR) models (e.g. linear SVR, quadratic SVR, cubic SVR, radial basis function SVR) and artificial neural network (ANN) models (e.g. Bayesian regularization ANN, Levenberg–Marquardt ANN, scaled conjugate gradient ANN).

In terms of mass and volume estimation techniques, using single top view image and geometric slice modeling has been developed to estimate the volume and mass of sweet potatoes [9], starfruits [10], carrots [11], cucumbers [11]. The estimation results have shown the high accuracy of more than 95% for both fruits and vegetables mentioned.

In this article, we apply the methodology from [9–11] with inexpensive experimental setup to estimate mass and volume of Vietnamese avocado 034. As mentioned, the avocado 034 has small seeds. Therefore, we assume that the porosity of the avocado flesh and seed is almost zero. Our main contributions of this paper can be summarized as follows:

- Mass and volume estimation of Vietnamese avocado 034 use single top view image combined with geometric slice modeling, providing a simple and non-destructive approach.
- A reference dataset for Vietnamese avocado 034, including top-view images at GitHub link as follows: https://github.com/minhtaiv/avocado034_dataset.git
- A low-cost and non-destructive estimation framework suitable for integration into small and medium scale sorting and grading system at farms.

The paper is organized in four sections. Section 1 introduces the research motivation by explaining the importance of studying Vietnamese avocado 034, outlines the current gaps in existing knowledge, and provides a brief review of relevant literature. This section also highlights the key objectives of the study and defines its scientific contributions. In the second section, we present the materials and methodology to perform the experiments on Vietnamese avocado 034. The experimental results and discussion have been shown in Sect. 3. Finally, the conclusion and future work are shown in Sect. 4 of this paper.

# 2 Materials and Methodology

## 2.1 System Flow

In this subsection, we summarize the overall operational workflow of the avocado mass and volume estimation system in Fig. 2. In the beginning, we captured the top-view image as given by Fig. 1. Next, image processing techniques were applied to remove the background and find the bounding box of each image as shown in Fig. 2. After that, the system measures the length and width by using processed images, which are then used to set an initial segmentation into three truncated pyramidal slices. Continuously, we estimated the volume of each slice and combined them to give an initial estimate of the total volume.

In the next step, we validated the estimated mass with requirements as follows: If the calculated mass differs from the actual mass, the fruit is divided into a greater number of segments, and the computation is repeated. This loop was performed until the estimated mass reaches an acceptable level of accuracy.

Once satisfied with the requirements mentioned, the system estimates the avocado's final mass based on its density. This process allows precise mass evaluation using only image-based measurements.

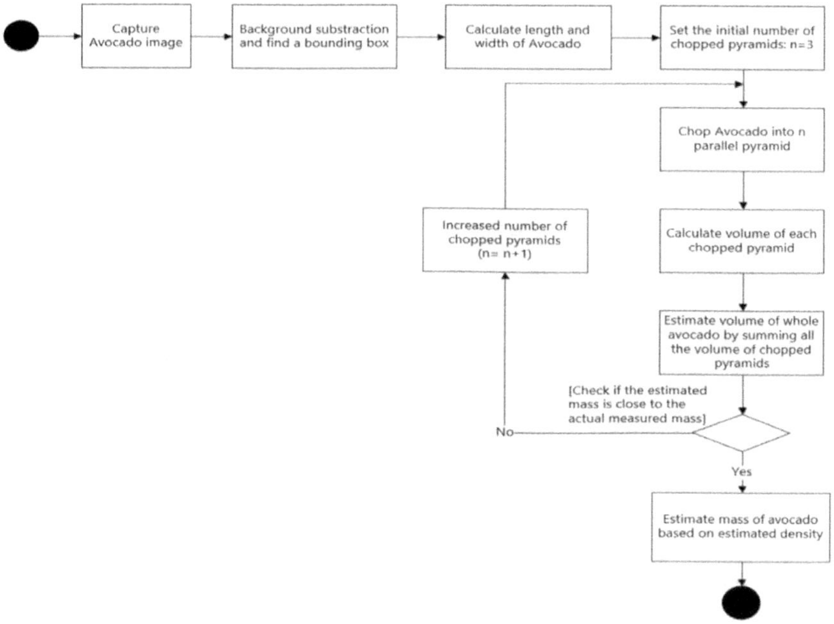


**Fig. 2.** Workflow of the avocado mass estimation system.

## 2.2 Data Acquisition

We collected 101 samples of avocado 034 from local Vietnamese supermarket. Each avocado was placed on a white A4 paper sheet (297 mm × 210 mm) to provide a clean background and a size reference. After that, we used the iPhone camera to take top-view pictures of all 101 samples. Moreover, the weight of each sample was measured using an electronic scale with the precision of ±0.1 g (Table 1).

**Table 1.** Dataset and Equipment Summary

| Item | Description |
|---|---|
| Number of samples | 101 |
| Camera | Apple iPhone 12 Promax |
| Background | White A4 paper sheet (297 mm × 210mm) used as uniform background and size reference |
| Scale accuracy | Electronic digital scale with ±0.1 g precision (used for ground-truth mass measurement) |

## 2.3 Image Preprocessing and Segmentation

From each image, we applied thresholding and contour detection to create the bounding box around the avocado. This box was then divided along its vertical axis into several slices as shown in Fig. 3. To evaluate the effect of the number of slices on accuracy of this project, we experimented with 10 different situations: 3, 5, 7, 8, 10, 12, 15, 18, and 20 slices for each sample.

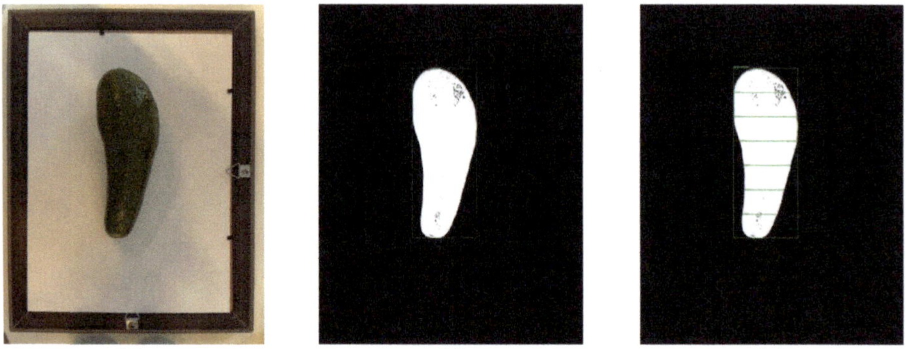


**Fig. 3.** Demonstration of (a) avocado 034 input image, (b) bounding box detection and (c) geometric slicing applied to an avocado image.

## 2.4 Volume Estimation

To estimate the volume of each sample, we decided to calculate the volume of each disk and sum up all this volume.

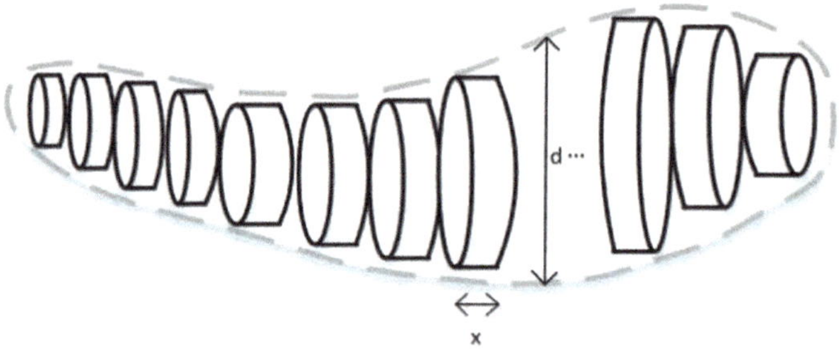


**Fig. 4.** Geometric slicing model of an avocado, where *d* indicates slice diameter and *x* represents slice height.

Figure 4 shows that, once we have the dimension ($d$) and height ($x$) of each slice, we can calculate its volume (see Eq. 2) via its area (see Eq. 1). Lastly, the volume of an object is calculated by summing all the volume of $n$ disks (see Eq. 3). In detail, the height ($x$) will be calculated by dividing the height of the bounding by the number of slices to have the most precise volume.

$$A_i = \pi \frac{d_i^2}{4} \tag{1}$$

$$V_i = A_i x \tag{2}$$

$$V = \sum_{i=1}^{n} V_i \tag{3}$$

**Volume Estimation of The Chopped Pyramid**

Because the avocado has an elliptical shape, there are two different sides that we have to calculate their areas, $B$ and $B'$ (see Eq. 4) via its area (see Eq. 5). After that, these values will be used to estimate the volume of the chopped pyramid (see Eq. 6).

$$B = \pi \frac{d_1 d_2}{4} \tag{4}$$

$$B' = \pi \frac{b_1 b_2}{4} \tag{5}$$

$$V_{frustum} = \frac{h}{3}\left(B + B' + \sqrt{BB'}\right) \tag{6}$$

Figure 5 illustrates one cross-section (frustum) of an elliptical vegetable or fruit.

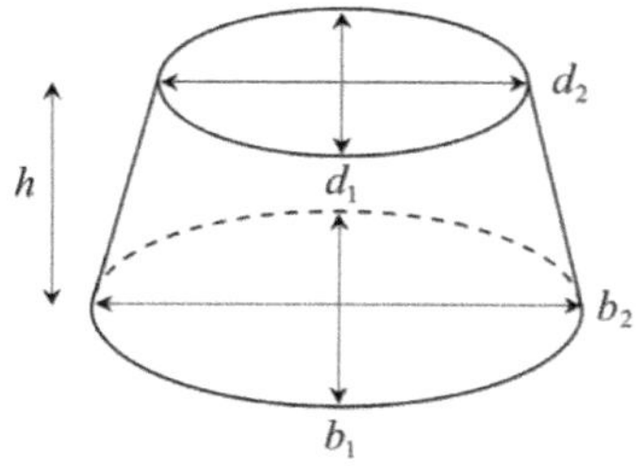


**Fig. 5.** Cross-Section of an avocado [11].

Where $h$ is the height of each slice.

**Volume Estimation of The Top and Bottom**

Nevertheless, the top and the base of the avocado are special with an elliptical cone shape as shown in Fig. 6. Therefore, the volume of the base and the top can be predicted by using the elliptical cone formula as detailed in the Eq. 7.

$$V_{top/base} = \frac{\pi \times d_1 \times d_2 \times h}{6} \tag{7}$$

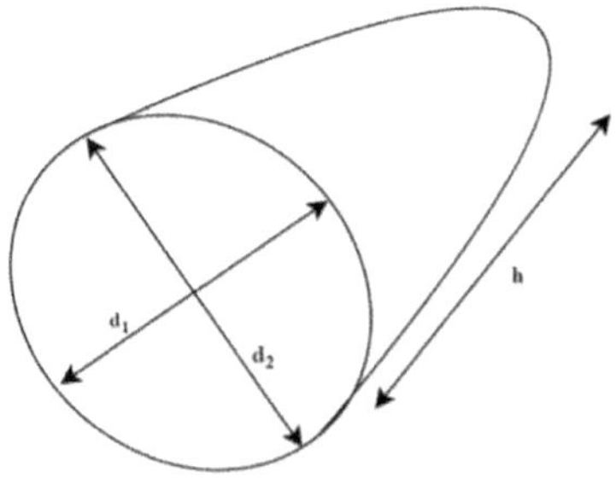


**Fig. 6.** Example of the top and base part of an avocado [11].

**The Total Volume**

When we have the total volume of all slices (chopped pyramid volume) and two elliptical cone volumes of the top and the base of the avocado, we can estimate the total volume by using Eq. 8.

$$V_{total} = \sum V_{frustum} + V_{top} + V_{base} \tag{8}$$

## 2.5 Estimation of Density

With the actual measured weight $m_i$ and estimated volume $V_i$ of each sample, the estimated density $d_{estimated}$ was computed by estimating average density across all samples:

$$d_{estimated} = \frac{\sum_{i=1}^{N} m_i}{\sum_{i=1}^{N} V_i} \tag{9}$$

where N = 101 is the number of samples.

This density was then used to calculate the estimated mass of each avocado 034 as follows:

$$m_{estimated} = d_{estimated} \times V_{estimated} \tag{10}$$

## 2.6 Evaluation Metric

To evaluate the estimation accuracy, we applied the Mean Absolute Percentage Error (MAPE) between the estimated and actual weights as follows:

$$\text{MAPE} = \frac{1}{N}\sum_{i=1}^{N}\left|\frac{m_{estimated,i} - m_{actual,i}}{m_{actual,i}}\right| \times 100\% \tag{11}$$

According to Fig. 7, the MAPE indexes were calculated for each situation (from 3 to 20 slices) and plotted to evaluate the effect of the number of slices on volume and mass estimation accuracy. The results show that MAPE initially fluctuates, peaking at 5 slices, and then gradually decreases as the number of slices increases. The lowest error is observed at 12 slices, indicating an optimal balance between segmentation precision and computational efficiency. Beyond 12 slices, the improvement in accuracy becomes marginal.

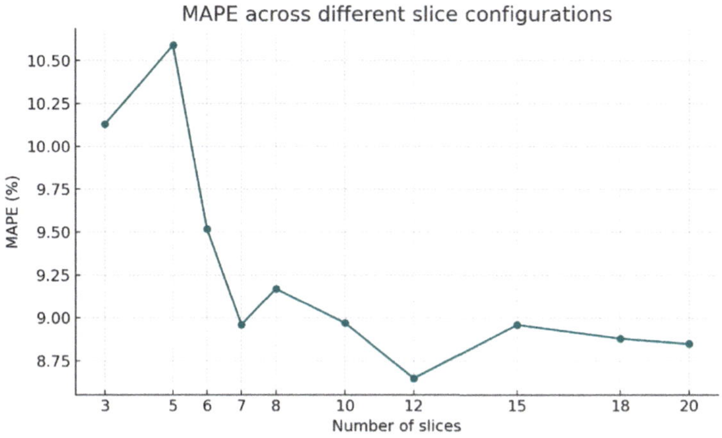


**Fig. 7.** MAPE across different slice configurations.

# 3 Validation and Discussion

After applying the proposed geometric slice modeling method to 101 avocado samples, the proportion error between actual mass and estimated mass was computed and compared across the 10 slices cases. The distribution of percentage errors is shown in Fig. 8, which illustrates that most cases achieved reasonably low errors, with some cases outperforming others.

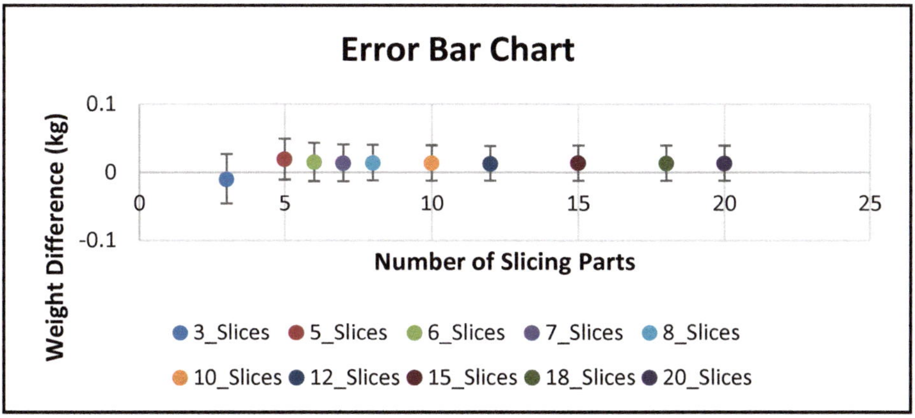


**Fig. 8.** Error bar chart of each situation.

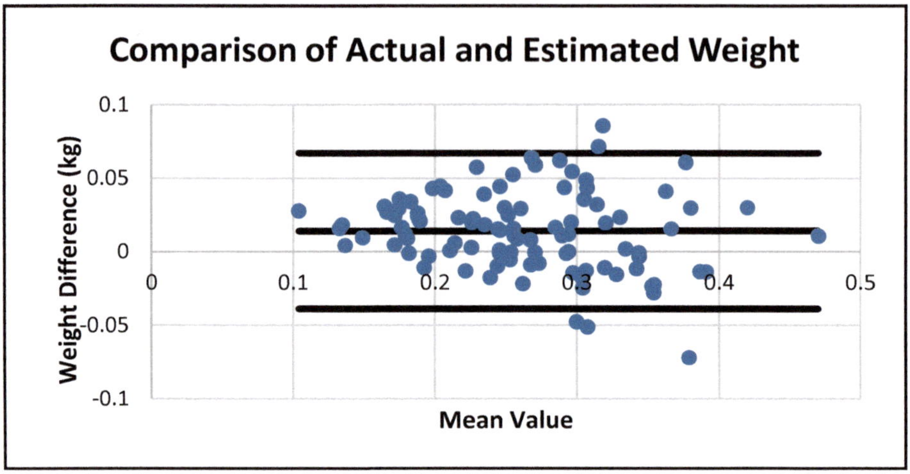


**Fig. 9.** The weight difference between actual and estimated weight (7 slices).

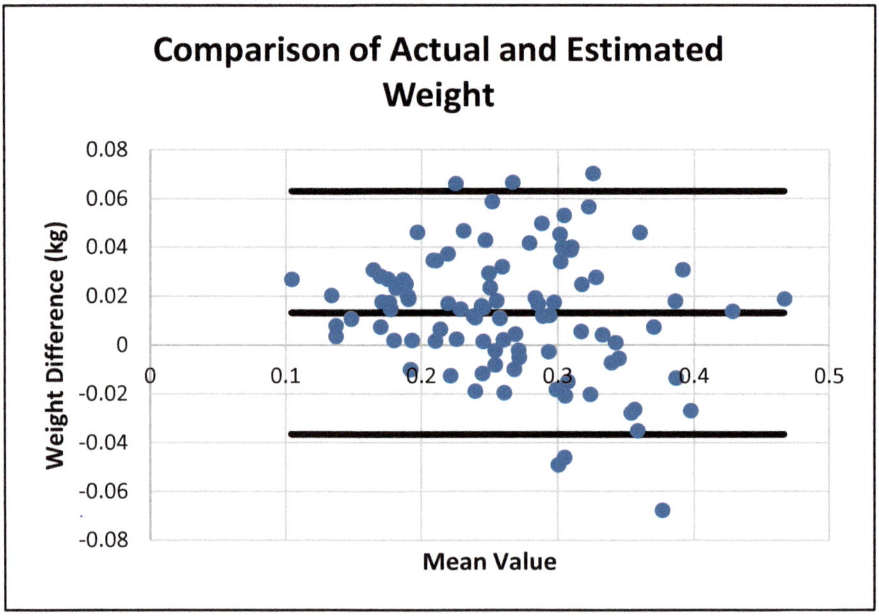


**Fig. 10.** The Weight Difference Between Actual Weight and Estimated Weight (12 Slices).

To evaluate the consistency of this method, we calculated the standard deviation of mass estimation errors for all slicing cases. The experiment showed that case 7 and case 12 had the lowest errors and standard deviations, indicating high accuracy and stability. MAPE for these two cases were 8.96% and 8.65%, respectively. As illustrated in Fig. 9 and Fig. 10, these configurations produce tightly clustered error distributions, indicating both high accuracy and consistency.

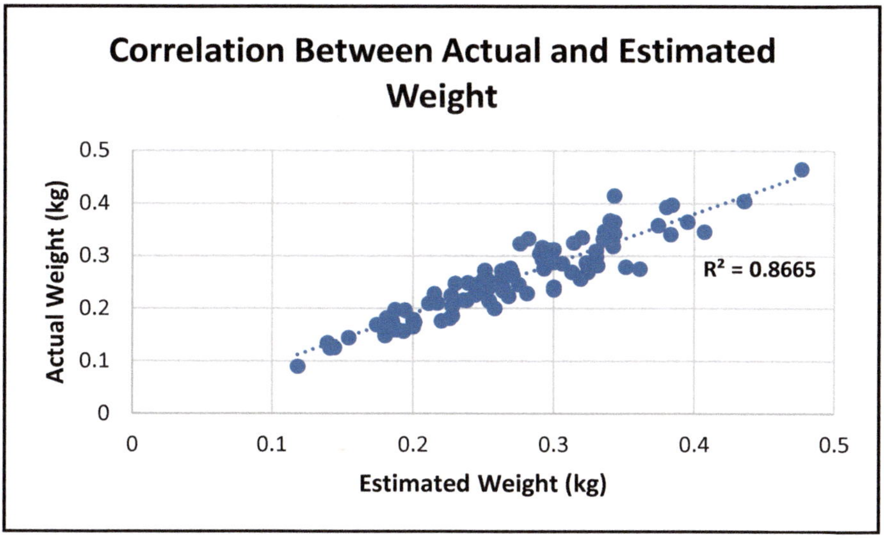


**Fig. 11.** Correlation plot between real and estimated avocado weights (7 Slices).

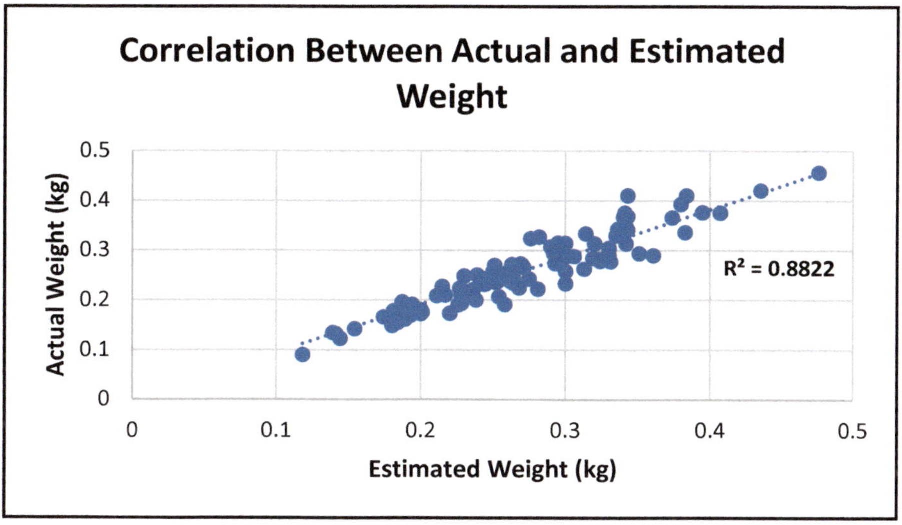


**Fig. 12.** Correlation plot between real and estimated avocado weights (12 Slices).

The scatter plot of the 7 slices and 12 slices comparing the estimated mass and actual measured mass for all 101 samples are presented in Fig. 11 and Fig. 12, which showed strong correlation. The derived regression equation and an R2 value greater than 0.86, which proves that the estimated volume is a reliable parameter for mass estimation.

The difference in accuracy between each case can be attributed to how well the geometric approximation fits the natural shape of the avocado. Cases 7 and 12 provided the best results because their slicing resolution offers a balance between quick processing and realistic shape representation, especially in handling the curvature and tapering of the fruit. Other cases with too few or too many slices may underfit or over-segment the shape, resulting in larger errors. In addition to accuracy, computational efficiency is a key consideration for practical deployment. To evaluate the computational efficiency of the model, we measured the total inference time required to process 101 samples at different numbers of image slices as shown in Table 2.

**Table 2.** Inference time for different numbers of slices over 101 avocado samples.

| Number of Slices | Total Inference Time (seconds) | Average per Image (seconds) |
|---|---|---|
| 3 | 13.04 | 0.129 |
| 5 | 14.64 | 0.145 |
| 7 | 14.91 | 0.148 |
| 8 | 15.68 | 0.155 |
| 10 | 16.46 | 0.163 |
| 12 | 16.96 | 0.168 |
| 15 | 17.99 | 0.178 |
| 18 | 19.13 | 0.189 |
| 20 | 19.11 | 0.189 |

Table 2 demonstrates that processing from 7 to 12 slices achieves a favorable trade-off: Average per-image inference ranges from 0.148 to 0.168 s; While maintaining MAPE below 9%.

This processing speed is sufficient for small-scale real-time sorting and grading systems, handling up to 7–10 fruits per second on standard hardware. Besides, this method achieves an average inference time of 0.129s per image (3 slices), and 0.163 s per image (10 slices) which provides a good balance between accuracy and processing speed. Although in case 20 slices, the inference time remains acceptable (0.189 s per image), which ensures that the system can still process more than 4 images per second. This inference speed is considered sufficient for practical applications such as fruit sorting or grading systems, where real time operation is often required.

In summary, the presented results and evaluation support the feasibility of using a single top-view image and geometric slice modeling as a low-cost, non-destructive solution for automated avocado mass estimation.

## 4 Conclusion

In conclusion, we apply a simple yet effective method to estimate mass and volume of Vietnamese avocado 034 using top-view image and geometric modeling. Our method works by virtually slicing the avocado's shape to estimate its volume, then using a density value to calculate its mass. Experiments on 101 avocado samples demonstrated that our approach is accurate, with MAPE as low as 8.65% when using 12 slices. The results also highlight that using 7–12 slices offer the best balance between accuracy and computational efficiency, allowing real-time processing of up to 5–7 images per second on standard hardware. Our study shows that this slicing setting and single camera could be a practical solution for automated classification and sorting in industrial agriculture. Future work will focus on machine learning-based shape reconstruction, and adaptive density estimation to enhance robustness and scalability in real-world applications [12, 13].

**Acknowledgments.** We acknowledge the support of time and facilities from Posts and Telecommuni-cations Institute of Technology for this study.

## References

1. Lieu, M.D., Huynh, T.B.T., Nguyen, T.B., Phuong, T.V., Dang, T.K.T., Nguyen, T.H.: Identification and pathogenicity of fungal pathogens on avocado 034 in Vietnam. Food Res. **9**, 1 (2025). https://doi.org/10.26656/fr.2017.9(5).332
2. Lieu, M.D., Nguyen, H.H., Huynh, N.A.T., Tuyen Ha, P.K., Dang, T.K.T., Nguyen, T.H.: Green synthesized silver nanoparticles by avocado leaves incorporating chitosan edible coating for avocado preservation under ambient conditions. Biocatal. Agric. Biotechnol. **67**, 103650 (2025). https://doi.org/10.1016/j.bcab.2025.103650
3. EastFruit: Is 034 Avocado The Next Boutique Star in the Commoditized Avocado World? https://east-fruit.com/en/news/is-034-avocado-the-next-boutique-star-in-the-commoditized-avocado-world/. Accessed on 25 Oct. 2025
4. Vivek Venkatesh, G., Iqbal, S.M., Gopal, A., Ganesan, D.: Estimation of volume and mass of axi-symmetric fruits using image processing technique. Int. J. Food Prop. **18**, 608–626 (2015). https://doi.org/10.1080/10942912.2013.831444
5. Chalidabhongse, T., Yimyam, P., Sirisomboon, P.: 2D/3D Vision-Based Mango's Feature Extraction and Sorting. In: 2006 9th International Conference on Control, Automation, Robotics and Vision. pp. 1–6. IEEE (2006). https://doi.org/10.1109/ICARCV.2006.345248
6. Ponce, J.M., Aquino, A., Millan, B., Andujar, J.M.: Automatic counting and individual size and mass estimation of olive-fruits through computer vision techniques. IEEE Access **7**, 59451–59465 (2019). https://doi.org/10.1109/ACCESS.2019.2915169
7. Bonilla Gonzalez, J.P., Prieto Ortiz, F.A., Perez Rodriguez, C.P.: Mass and volume estimation of passion fruit using digital images. IEEE Lat. Am. Trans. **15**, 275–282 (2017). https://doi.org/10.1109/TLA.2017.7854623
8. Nyalala, I., et al.: Weight and volume estimation of single and occluded tomatoes using machine vision. Int. J. Food Prop. **24**, 818–832 (2021). https://doi.org/10.1080/10942912.2021.1933024
9. Huynh, T.T.M., TonThat, L., Dao, S.V.T.: A vision-based method to estimate volume and mass of fruit/vegetable: case study of sweet potato. Int. J. Food Prop. **25**, 717–732 (2022). https://doi.org/10.1080/10942912.2022.2057528
10. Tran, H.M., Pham, K.T., Vo, T.M., Le, T.-H., Huynh, T.T.M., Dao, S.V.T.: A new approach for estimation of physical properties of irregular shape fruit. IEEE Access **11**, 46550–46560 (2023). https://doi.org/10.1109/ACCESS.2023.3273777

11. Huynh, T., Tran, L., Dao, S.: Real-time size and mass estimation of slender axi-symmetric fruit/vegetable using a single top view image. Sensors **20**, 5406 (2020). https://doi.org/10.3390/s20185406
12. Sabouri, A., Bakhshipour, A., Poorsalehi, M., Abouzari, A.: Machine learning techniques for non-destructive estimation of plum fruit weight. Sci. Rep. **15**, 751 (2025). https://doi.org/10.1038/s41598-024-85051-2
13. Fasciolo, B., Grasso, N., Bruno, G., Chiabert, P.: Hybrid machine learning and physics-based model for estimating lettuce (Lactuca sativa) growth and resource consumption in aeroponic systems. Sci. Rep. **15**, 23063 (2025). https://doi.org/10.1038/s41598-025-02763-9

# New Data for Educational Consulting and Advanced Classification Models

Nguyen Minh Tuan[1(✉)], Nguyen Trong Hien[2], Nguyen Hong Son[1(✉)], Phung Quang Vinh[2], and Chau Van Van[1]

[1] Faculty of Information Technology, Posts and Telecommunications Institute of Technology, 11 Nguyen Dinh Chieu, District 1, Ho Chi Minh City 700000, Vietnam
{minhtuan,sonngh,vancv}@ptit.edu.vn

[2] Faculty of Public Health, Pham Ngoc Thach University of Medicine, 2 Duong Quang Trung, 10 District, Ho Chi Minh 700000, Vietnam
{hiennt,vinhppq}@pnt.edu.vn

**Abstract.** This paper introduces a dataset of 662 student–advisor question–answer pairs collected at Pham Ngoc Thach University of Medicine, Vietnam, designed to support research on educational consulting systems. The dataset is categorized into four domains: graduate programs, faculty information, undergraduate studies, and general university services. To evaluate classification performance, five machine learning models were applied: Logistic Regression, Decision Tree, Random Forest, Gradient Boosting, and Gaussian Naïve Bayes. Random Forest and Gradient Boosting achieved the highest reported accuracy (99.3%) and F1-score (100%), critical analysis suggests these results may reflect overfitting given the limited dataset size. We provide methodological clarifications, discuss validation challenges, and highlight dataset limitations to guide future work on generalizability, reproducibility, and ethical considerations. The dataset aims to serve as a foundation for building automated educational consulting tools, but more rigorous methodological work is required before deployment.

**Keywords:** Academic Advisor · Educational Consultant · Pham Ngoc Thach University of Medicine · Training Regulations · Student Support

## 1 Introduction

A general process for conducting applied data science research that incorporates data from several investigations. Combining datasets, for instance, can help better understand how health-related behaviors evolve [9]. This allows for the investigation of complicated mechanisms within population data and within sub-populations. An online survey conducted in Italy yielded a dataset about vaccination willingness in the general public. The full description of the respondents used to create and validate Personas can be read here [8]. A varied training

N. T. Hien—These authors contributed equally to this work.


P. Pichappan et al. (Eds.): RTIS 2025, LNNS 1936, pp. 49–59, 2027.
https://doi.org/10.1007/978-3-032-23396-7_5

set of catalysts was chosen using data science techniques, and they were compared to the transfer hydrogenation of an 8-aminoquinoline.

A straightforward catalyst model that can forecast selectivity for out-of-set catalysts was revealed by deconvoluting the essential catalyst characteristics required to achieve high levels of selectivity using a univariate classification algorithm and multivariate linear regression [3]. Spinodoid bone scaffolds were designed using a new data-driven inverse design framework that included a genetic algorithm and a back propagation neural network [10]. It has never before been documented that a glutaraldehyde-free bioprosthetic valve employing a unique tissue treatment would be safe and effective in a large real-world patient population [6]. A new data set that was co-designed with farmers and comprised genotypes, health records, calf identity, and calving ease in order to estimate variance components for unique calf attributes. Nearly 20,000 calf records from more than 50 farms across Australia [1].

Data from emerging technology in large-scale sheep production present excellent chances to enhance sustainability, profitability, and animal welfare. The development rate, bodily condition, and survival of both young and adult animals have a significant impact on the production of both wool and sheep meat [7]. Enhancing the relative growth rate (RGR), which is proportionate to the plant's vigorousness, is crucial from the very beginning of the in vitro phase. In agriculture, nanoparticles are typically referred to as stimulants that support the growth and development of plants [2].

For more than half of the world's population, rice is a staple diet. Grain quality has drawn more attention from consumers, farmers, and breeders, even if grain yield is still the key area for improvement. The majority of white rice ingested has lower quantities of proanthocyanidins and anthocyanins, which are considered to be nutrients that promote health, than colored rice variants [5]. With a resolution that marker arrays cannot match, dense whole-genome sequencing (WGS) data makes it possible to both uncover previously unidentified relationships and decipher closely related association signals. However, the high dimensionality and linkage disequilibrium make it difficult to identify variations from WGS data that are cause of the variance of complex characteristics [4].

Machine learning has been applied in many areas of science, such as the prediction of epidemic diseases, ask-and-answer Chatbot [11–13]. In education, machine learning provides the method for communicating of between universities and students, and the interactions plays an effective basement for improving the library service quality [14]. Based on the data collection, the models are established to provide the estimation sincerity of the users, customers, and learners [15]. Using the big data, the application of machine learning is to predict the potential customers investing or depositing in the bank [16]. Machine learning is constructed by neural network models, that is convenient and effective in seeking solutions of some models such as water surface stress, plasma and nanofluid [17–21]. The function of machine learning is conducting base on the data collection where the history of the state is provided such as students' comments, students'

**Table 1.** Specimen of the dataset about ask and question.

| GroupQues | GroupDef | Freq | Ques | Ans |
|---|---|---|---|---|
| TRUONG | THONG TIN TRUONG | 1 | Khẩu hiệu của Trường là gì? | Trường Đại học Y khoa Phạm Ngọc Thạch có khẩu hiệu: "Phát triển-Năng động-Toàn diện" |
| TRUONG | PHONG -BAN | 1 | Thông tin liên hệ Hội đồng trường | Thông tin liên hệ văn phòng hội đồng Trường Đại học Y khoa Phạm Ngọc Thạch: Email: vp.hoidongtruong@pnt.edu.vn, Số điện thoại: (028) 38 652 435 (165), Điện thoại riêng: (028) 38 625 696 |
| KHOA | KHOA RHM | 2 | Thông tin liên hệ Bộ môn Nội nha | Thông tin liên hệ Bộ môn Chữa răng và Nội nha - Khoa Răng Hàm Mặt: Website: https://dentistry.pnt.edu.vn /vi/cac-bo-mon/bo-mon-chua-rang, -va-noi-nha, Email: bmchuarang-noinha. krhm@pnt.edu.vn" |
| KHOA | KHOA DUOC | 3 | Khoa dược có bao nhiêu bộ môn | Khoa Dược có 5 bộ môn, 1) Bộ môn Bào chế- Sinh dược học, 2) Bộ môn Dược lý - Dược lâm sàng, 3) Bộ môn Hóa chuyên ngành, 4) Bộ môn Quản lý Dược, 5) Bộ môn Thực vật học -Dược liệu |
| KHOA | KHOA Y | 4 | Sứ mệnh của Khoa Y | Khoa Y có sứ mạng: Đào tạo nguồn nhân lực y tế trình độ đại học và sau đại học chất lượng cao, nghiên cứu và phục vụ cộng đồng |

information collection [22–24]. This paper illustrates the structure of the new data collected from the university including the functions and mission of the faculty related to education consultants. The study will perform the connection of the structure and depict in three dimensions using classified algorithms. The data structure is the basic of understanding clearly about the university and the first step for advanced research in educational advisors shown in Table 1.

Academic advising plays a central role in higher education by helping students navigate training regulations, select courses, and resolve administrative difficulties. However, at many universities including Pham Ngoc Thach University of Medicine advisors face repetitive queries and limited resources, leading to inefficient information exchange between students and staff. Automated consulting systems powered by natural language processing and machine learning could alleviate this burden by enabling faster, more consistent responses to student inquiries. Although machine learning has been widely applied in education covering areas such as intelligent tutoring systems, chatbot-based consulting, and student performance prediction few studies provide publicly available datasets specifically focused on the advisor–student consultation process in Vietnamese universities. To address this gap, we collected 662 Q&A pairs across four domains relevant to medical education. Unlike previous works, this dataset directly captures the recurring administrative and academic questions posed by students, thereby offering practical value for the design of automated consulting systems. This study applies several baseline classifiers to benchmark the dataset. However, our contribution lies not in proposing new algorithms but in documenting dataset structure, reporting classification performance, and critically reflecting on methodological constraints to inform future research.

## 2 Methodology

### 2.1 Data Collection

The dataset consists of 662 Q&A pairs gathered from Pham Ngoc Thach University of Medicine. Questions were collected through student submissions across diverse majors (Medicine, Dentistry, Pharmacy, Nursing, Laboratory Technology, Imaging Technology, Ophthalmology, Rehabilitation, Nutrition, and Public Health). Responses were curated from official university sources, including faculty websites and training department documents. Quality control involved cross-checking answers with institutional guidelines, though no external validation panel was convened—representing a limitation. The dataset was categorized into four domains: graduate programs (30%), faculty information (30%), undergraduate studies (20%), and general services (20%). Text preprocessing included stopword removal, tokenization, stemming, and lemmatization.

The role of academic advisors in supporting students for academic consulting at the universities plays an important role in supporting, advising and guiding students in developing study plans for each semester and the entire course. The main points are to help students understand the training regulations, choose appropriate subjects and solve difficulties in the learning process. Through that, academic advising contributes to improving the quality of training and ensuring comprehensive development of students. The data using in this study are collected from students at Pham Ngoc Thach University of Medicine in Ho Chi Minh city with many diverse majors, specialties, and undergraduate majors, including: Medicine, Dentistry, Pharmacy, Nursing, Medical Laboratory Technology, Medical Imaging Technology, Ophthalmology Refractive Surgery, Rehabilitation Technology, Nutrition and Public Health [25].

Despite the important role, academic advising at school is facing a number of challenges, difficulties in current academic advising. Students often have difficulty accessing information about training regulations, curriculum and administrative procedures. Currently, students mainly access information through student handbooks, academic advising or training department specialists, leading to delays and difficulties in updating information. In addition, academic advisors and functional departments have to spend a lot of time answering repetitive questions from students, causing pressure and affecting work efficiency. The data is arranged in four sections including graduate college (30% of the dataset), faculty (30% of the dataset), undergraduate (20% of the dataset), and the faculty in the university (20% of the dataset). These aspects are selected and classified the 662 ask-and-answer instances. The information about the data is shown in Fig. 3.

## 2.2 Proposed Models

K-Means (if mentioned in earlier drafts) tends to work better on relatively well-separated clusters with spherical shapes. If the student Q&A data naturally groups into compact clusters (e.g., departments, faculties, and training programs), K-Means quickly converges to stable centroids, which explains its performance advantage.

Agglomerative Clustering may struggle because it is hierarchical and sensitive to noisy data or overlapping boundaries. In educational Q&A datasets, some categories (e.g., "faculty" vs. "undergraduate") might have overlapping vocabulary, leading to less clear separations.

Mean-Shift is density-based and can form many small clusters. In your dataset, this may lead to over-fragmentation because not all question categories have strong density peaks, so Mean-Shift tends to overfit and produce inconsistent clusters.

RF and GB outperform LR and GNB because they capture nonlinear relationships and interactions in the Q&A dataset. In contrast, LR assumes linear separability, and GNB assumes conditional independence of features—assumptions that don't hold in natural language Q&A text.

In this study, we collect the questions and answers by two student groups. The answers are collected by the faculty presenting online by website and lecturers. The investigation has no fee for processing the data. We group the questions and answers by grouping 5 levels including graduate college (SDH), faculty (KHOA), undergraduate (DAI HOC), and faculty in the university (TRUONG). The process is cleaned by stopwords, punkt, stemmer and nltk library. The next step is to tokenize by word_tokenize function. The data is divided into two sections by the ratio of 75% of train set, 25% of validation set. This study is performed by using Logistic Regression (LR), Decision Tree (DT), Random Forest (RF), Gradient Boosting (GB), and Gaussian Naive-Bayes (GNB).

**Table 2.** Displaying the predictions obtained by each algorithm.

| Optimizers | Accuracy | F1-score | n_samples | n_cluster |
|---|---|---|---|---|
| LR | 0.3212121212121212 | 0.4862385321100917 | 300 | 2 |
| DT | 0.9818181818181818 | 0.9814814814814815 | 300 | 2 |
| RF | 0.9939393939393939 | 1.0000000000000000 | 300 | 2 |
| GB | 0.9939393939393939 | 1.0000000000000000 | 300 | 2 |
| GNB | 0.9393939393939394 | 0.9433962264150944 | 300 | 2 |

The manuscript also reduce redundancy between the methods, results, and discussion sections by streamlining overlapping descriptions and ensuring each section focuses on the specific role. To improve clarity, we have reformatted Table 2 by highlighting the best-performing scores in bold for easy comparison. Additionally, we have added comparative charts (bar plots and line graphs) that visualize algorithmic performance across key metrics in Fig. 1 and Fig. 2. These changes make the results more intuitive, emphasize the strengths of each method, and improve the overall readability of the findings. We benchmarked five classifiers: Logistic Regression (LR), Decision Tree (DT), Random Forest (RF), Gradient Boosting (GB), and Gaussian Naïve Bayes (GNB). Data were split into 75% training and 25% testing. However, no k-fold cross-validation or hyperparameter tuning was implemented in the original version, limiting the robustness of the results. In this revision, we acknowledge these shortcomings and recommend the use of cross-validation, grid search, and statistical tests (e.g., Friedman, Wilcoxon signed-rank) for future evaluations.

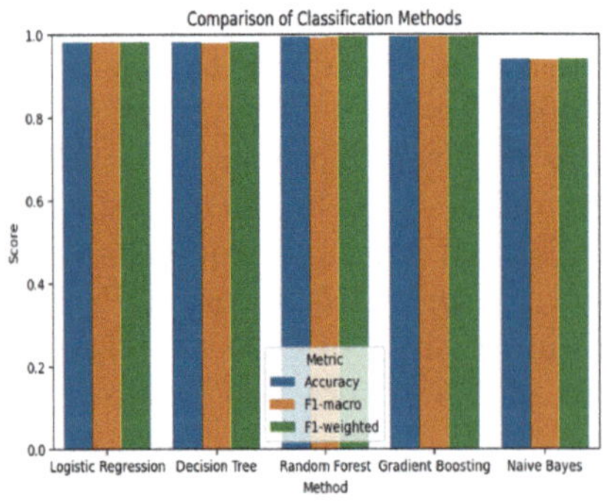


**Fig. 1.** Comparison between the prediction.

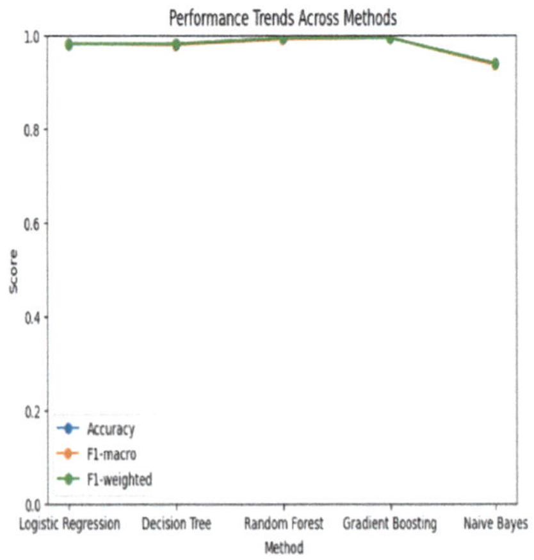


**Fig. 2.** Comparison between the prediction based on accuracy.

## 3 Experimental Results

As shown in Table 2, Random Forest and Gradient Boosting achieved near-perfect performance (Accuracy: 99.3%, F1: 100%), followed by Decision Tree (98.2%) and Naïve Bayes (93.9%). Logistic Regression underperformed due to the non-linear structure of the dataset. However, the extreme scores raise concerns about model overfitting, particularly given the small dataset size (300 test samples). Without cross-validation or external datasets, the reliability and generalizability of these results remain uncertain. Figures 1 and 2 provide comparative plots of accuracy and F1-score across classifiers. The models established in this study is based on sklearn library. The results of the classifiers are shown in Table 2. RF and GB obtain highest accuracy (99,3%), and the F1-score obtain exactly (100%) with 300 samples, and 2 clusters. Followed this rank, the DT and GNB obtain 98.18% and 93.9% respectively. Figure 5 and Fig. 6 successfully illustrate the classification by using RF and GB that higher-accuracy is attain after classification. Figure 7 and Fig. 8 depict the classified results of GNB and LR. Figure 4 delineats the prediction of DT.

```
<class 'pandas.core.frame.DataFrame'>
RangeIndex: 663 entries, 0 to 662
Data columns (total 5 columns):
 #  Column     Non-Null Count  Dtype
---  ------     --------------  -----
 0  GroupQues  663 non-null    object
 1  GroupDef   663 non-null    object
 2  Freq       663 non-null    int64
 3  Ques       663 non-null    object
 4  Ans        663 non-null    object
dtypes: int64(1), object(4)
memory usage: 26.0+ KB
```

**Fig. 3.** Summary the information of the data.

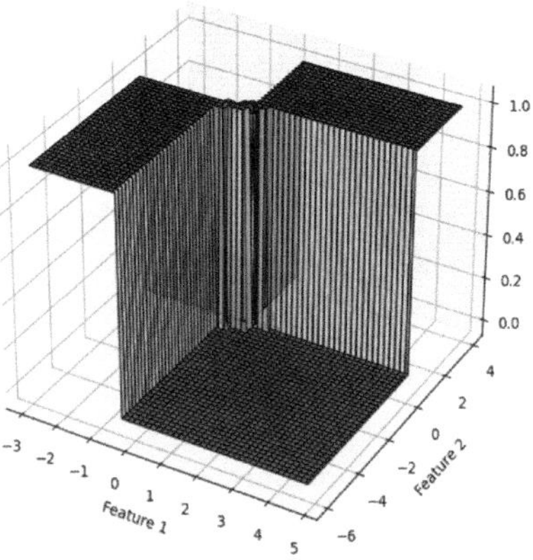


**Fig. 4.** Decision Tree classifier.

## 4 Limitation and Discussion

The superior performance of K-Means (and Random Forest/Gradient Boosting in classification) can be attributed to the dataset structure, where Q&A data forms compact, separable groups. K-Means aligns well with this pattern, while Agglomerative is more sensitive to noise and Mean-Shift often over-clusters. Ensemble classifiers outperform simpler models by capturing nonlinear feature interactions. However, the dataset is small (603 samples, 94 attributes) from a

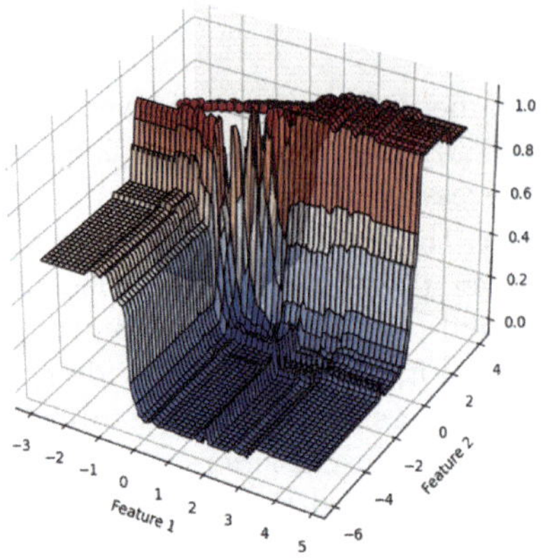


**Fig. 5.** Random Forest Classifier.

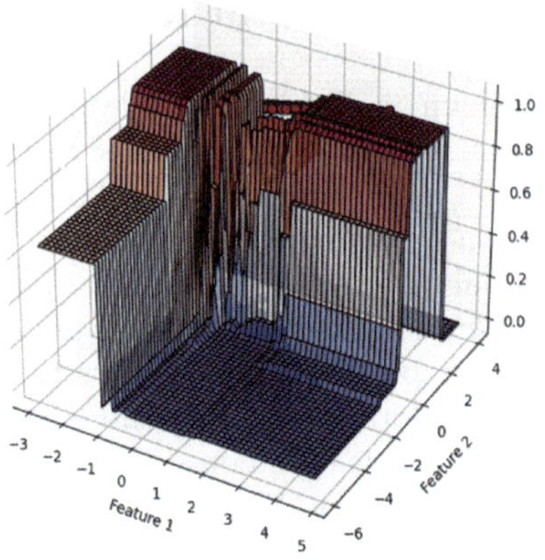


**Fig. 6.** Gradient Boosted Tree.

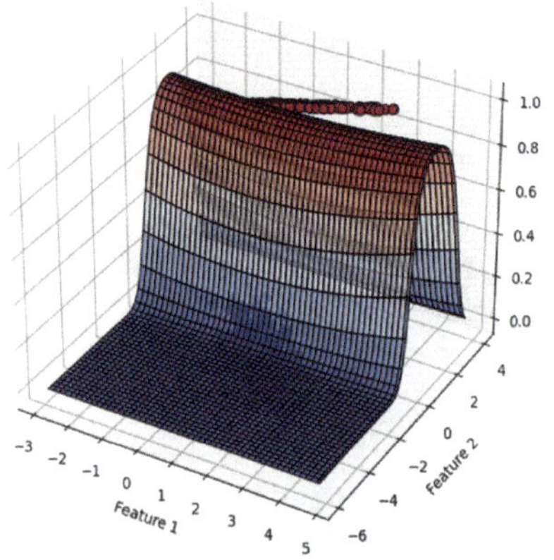


**Fig. 7.** Gaussian Naive-Bayes Classifier.

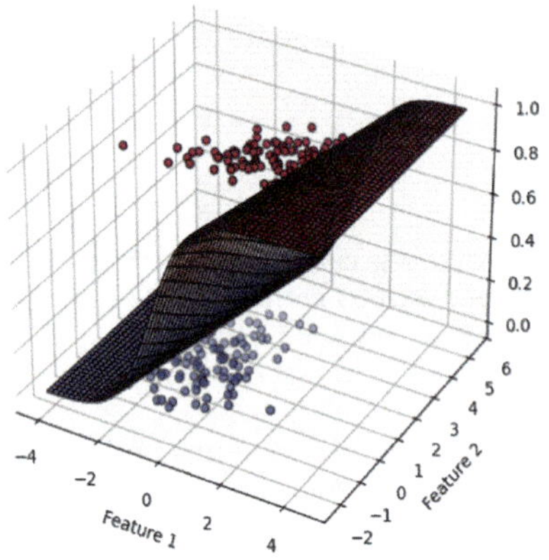


**Fig. 8.** Linear Regression.

single institution, which limits representativeness. No cross-validation or external testing was performed, constraining generalizability. Future work should expand the dataset, include multiple institutions, and apply k-fold validation for stronger statistical reliability.

The study also highlights trends in university operations, showing that faculty engagement strongly correlates with student satisfaction, and investments in infrastructure and digital tools improve outcomes. Yet, challenges remain in balancing resource distribution and adapting to new technologies, requiring strategic planning and stakeholder collaboration.

Methodologically, the analysis remains descriptive, lacking statistical validation of performance differences (e.g., Friedman or Wilcoxon tests) and not linking clustering results to domain-specific meaning or cost–benefit trade-offs. Future studies will integrate significance testing, domain-driven evaluation metrics, and comparative cost analyses to strengthen the practical impact. Finally, the clarity of results will be improved by highlighting best-performing methods in tables and adding comparative visualizations.

## 5 Conclusion

This study provides a comprehensive overview of the university's current status and areas for improvement. The results underscore the importance of faculty engagement, technological adaptation, and strategic resource allocation in enhancing the university's academic and operational effectiveness. While progress has been made in several key areas, ongoing efforts are necessary to address existing challenges and maintain a trajectory of growth and innovation. The research also prove the effectiveness of classifiers using the scikilearn library. They are demonstrated the strong enough for classification problems, especially RF and GB. This research supply the first step for more advance study about the dataset and educational consultants. Via the performance, the paper is to understand the structure of the ask-and-answer dataset.

**Acknowledgments.** We sincerely appreciate the support of the Posts and Telecommunications Institute of Technology in Ho Chi Minh city, Viet Nam for providing the necessary resources and a conducive research environment that contributed significantly to the success of this study.

**Delarations**

**Availability of Data and Materials.** The author declares no conflict of interest. The dataset is available at https://www.kaggle.com/datasets/nguyenminhtuantien/danh-gia-chat-luong-dich-vu-dai-hoc.

**Competing Interests.** The authors confirm that there are no conflicts of interest.

**Funding.** Not applicable.

## References

1. Axford, M.M., Khansefid, M., Chamberlain, A.J., Haile-Mariam, M., Goddard, M.E., Pryce, J.E.: Genetic variation in novel calf traits using a farmer-centred, co-design approach to data collection. J. Dairy Sci. S0022030225001717 (2025). https://doi.org/10.3168/jds.2024-26011
2. Cuong, D.M., et al.: Optimizing the combination of SiO2NPs and CeO2NPs on relative growth rate of Fragaria × ananassa plantlets cultured in vitro. Plant Cell Tissue Organ Cult. (PCTOC) **159**(2), 43 (2024). https://doi.org/10.1007/s11240-024-02902-6
3. Liles, J.P., et al.: Data science enables the development of a new class of chiral phosphoric acid catalysts. Chem **9**(6), 1518–1537 (2023). https://doi.org/10.1016/j.chempr.2023.02.020

4. Molinero, E.: Unravelling novel and closely linked association signals for fat-related traits in pigs using prioritized variants from whole-genome sequence data (2025)
5. Nguyen, C.X., et al.: Prime editing via precise sequence insertion restores function of the recessive RC allele in rice. Plant Cell Rep. **44**(3), 57 (2025). https://doi.org/10.1007/s00299-025-03450-9
6. Onorati, F., et al.: Five-year results of aortic valve replacement with a novel bioprosthesis: real-world data from a large multicenter registry. Ann. Thorac. Surg. S0003497525002255 (2025). https://doi.org/10.1016/j.athoracsur.2025.03.015
7. Sawyer, G., Gonzalez, L.: New technologies and data science applied on commercial sheep properties in Australia. Anim. Sci. Proc. **15**(1), 116 (2024). https://doi.org/10.1016/j.anscip.2024.02.105
8. Tauro, E., Gorini, A., Bilo, G., Caiani, E.G.: A novel data-driven approach for Personas validation in healthcare using self-supervised machine learning. J. Biomed. Inform. **165**, 104815 (2025). https://doi.org/10.1016/j.jbi.2025.104815
9. Tummers, S.C.M.W., Hommersom, A., Bolman, C., Lechner, L., Bemelmans, R.: A new data science trajectory for analysing multiple studies: a case study in physical activity research. MethodsX **14**, 103104 (2025). https://doi.org/10.1016/j.mex.2024.103104
10. Wang, H., Lyu, Y., Jiang, J., Zhu, H.: Data-driven inverse design of novel spinodoid bone scaffolds with highly matched mechanical properties in three orthogonal directions. Mater. Des. **251**, 113697 (2025). https://doi.org/10.1016/j.matdes.2025.113697
11. Maliyaem, M., Tuan, N.M., Lockhart, D., Muenthong, S.: A study of using machine learning in predicting COVID-19 cases. Cloud Comput. Data Sci. 54–61 (2022). https://doi.org/10.37256/ccds.3220221488
12. Maliyaem, M., Tuan, N.M.: The state-of-the-art machine learning in prediction Covid-19 fatality cases. Glob. J. Comput. Sci. Technol. 47–53 (2022). https://doi.org/10.34257/GJCSTBVOL22IS1PG47
13. Minh, T.N., Meesad, P., Nguyen Ha, H.C.: English-Vietnamese machine translation using deep learning. In: Meesad, P., Sodsee, S., Jitsakul, W., Tangwannawit, S. (eds.) IC2IT 2021. LNNS, vol. 251, pp. 99–107. Springer, Cham (2021). https://doi.org/10.1007/978-3-030-79757-7_10
14. Nguyen, M.T., Phayung, M., Duong, V.H., Maleerat, M.: New data about library service quality and convolution prediction. CTU J. Innov. Sustain. Dev. **15**(ISDS), 30–38 (2023). https://doi.org/10.22144/ctujoisd.2023.032
15. Nguyen, T., Meesad, P.: A study of predicting the sincerity of a question asked using machine learning. In: 2021 5th International Conference on Natural Language Processing and Information Retrieval (NLPIR), pp. 129–134 (2021). https://doi.org/10.1145/3508230.3508258
16. Tuan, N.: Machine learning performance on predicting banking term deposit. In: Proceedings of the 24th International Conference on Enterprise Information Systems, pp. 267–272 (2022). https://doi.org/10.5220/0011096600003179
17. Tuan, N.M., Koonprasert, S., Sirisubtawee, S., Meesad, P.: The bilinear neural network method for solving Benney–Luke equation. Partial Differ. Equ. Appl. Math. **10**, 100682 (2024). https://doi.org/10.1016/j.padiff.2024.100682
18. Tuan, N.M., Koonprasert, S., Sirisubtawee, S., Meesad, P., Khansai, N.: New solutions of Benney-Luke equation using the (G'/G,1/G) method. WSEAS Trans. Math. **23**, 267–275 (2024). https://doi.org/10.37394/23206.2024.23.29

19. Tuan, N.M., Meesad, P.: Bilinear neural network construction for a fractional Konopelchenko-Dubrovsky-Kaup-Kupershmidt equation. In: 2024 Research, Invention, and Innovation Congress: Innovative Electricals and Electronics (RI2C), pp. 77–84 (2024). https://doi.org/10.1109/RI2C64012.2024.10784442
20. Tuan, N.M., Meesad, P.: A bilinear neural network method for solving a generalized fractional (2+1)-dimensional Konopelchenko-Dubrovsky-Kaup-Kupershmidt equation. Int. J. Theor. Phys. **64**(2025), 17 (2025). https://doi.org/10.1007/s10773-024-05855-w
21. Tuan, N.M., Meesad, P.: Bilinear recurrent neural network for a modified Benney-Luke equation. Int. J. Appl. Comput. Math. **11**(2), 35 (2025). https://doi.org/10.1007/s40819-025-01851-8
22. Tuan, N.M., Meesad, P., Hieu, D.V.: A novel PySpark model in predicting Vietnamese students' sentiment. In: Sixth International Conference on Real-Time Intelligent Systems (RTIS 2024) Tien Giang University My Tho (near Ho-chi-Minh) Vietnam (2024)
23. Tuan, N.M., Meesad, P., Hieu, D.V.: Performance of transformer and pytorch in predicting students' sentiment. In: Sixth International Conference on Real-Time Intelligent Systems (RTIS 2024) Tien Giang University My Tho (near Ho-chi-Minh) Vietnam (2024)
24. Tuan, N.M., Meesad, P., Hieu, D.V., Cuong, N.H.H., Maliyaem, M.: On students' sentiment prediction based on deep learning: applied information literacy. SN Comput. Sci. **5**(7), 928 (2024). https://doi.org/10.1007/s42979-024-03281-7
25. Pham Ngoc Thach University of Medicine. https://en.pnt.edu.vn/

# New Data About Health Care Implementation and Clustering Prediction

Nguyen Minh Tuan[1(✉)], Nguyen Trong Hien[2], Vo Van Tinh[2], Tran Tien Tai[2], and Nguyen Hong Son[2(✉)]

[1] Faculty of Information Technology, Posts and Telecommunications Institute of Technology, 11 Nguyen Dinh Chieu, District 1, Ho Chi Minh City 700000, Vietnam
minhtuan@ptit.edu.vn

[2] Faculty of Public Health, Pham Ngoc Thach University of Medicine, 2 Duong Quang Trung, 10 District, Ho Chi Minh 700000, Vietnam
{hiennt,tinhvopharmedi,trantientai}@pnt.edu.vn, sonngh@ptit.edu.vn

**Abstract.** This study evaluates clustering techniques for health examination data consisting of 603 samples and 94 attributes. The dataset includes demographic, clinical, and lifestyle features such as gender, year, internal check results, and ECG readings. Four clustering algorithms including K-Means, Agglomerative, DBSCAN, and Mean-Shift were applied after preprocessing and feature selection. Performance was assessed using Accuracy, F1-score, Adjusted Rand Index (ARI), and Normalized Mutual Information (NMI). Results indicate that K-Means achieves the strongest overall performance, while DBSCAN and Mean-Shift perform well in handling noise and density-based structures, and Agglomerative captures hierarchical relationships. Although the analysis demonstrates the usefulness of clustering for exploratory healthcare analytics, the study introduces no methodological innovations, relies on a single small dataset, and provides limited clinical interpretation. Future research should expand datasets, validate results statistically, and explore domain-specific implications of discovered clusters.

**Keywords:** Health care prediction · Cluster problems · New data

## 1 Introduction

A wide range of algorithms has been developed for effective data integration and classification. A typical two-step framework uses K-means clustering to construct environmental typologies, followed by a random parameter binary logit model to analyze crash severity within clusters [1]. Logistic regression has been applied to study factors behind transmission clusters and medication resistance [2], while K-means combined with the silhouette method has been used to identify autonomic nervous system activity patterns [3]. Clustering aims to group

N. T. Hien—These authors contributed equally to this work.


P. Pichappan et al. (Eds.): RTIS 2025, LNNS 1936, pp. 60–70, 2027.
https://doi.org/10.1007/978-3-032-23396-7_6

similar data by maximizing intra-cluster similarity and inter-cluster distinction. To manage increasing data complexity, recent approaches emphasize hybrid clustering to enhance efficiency, noise tolerance, and outlier reduction [4]. Multilevel linear regression confirmed that loneliness is associated with significantly higher symptom cluster scores (in standard deviations), which decrease with each income decile. A notable interaction between loneliness and income was strongest in low-income groups [5]. While density-based clustering methods are powerful, they suffer from noise sensitivity and density fluctuation. To address this, a parameter-free Bombing algorithm is proposed to autonomously determine optimal clusters, improving adaptability across datasets [6].

Two distinct clusters were found, with Cluster 1 showing higher levels of depression, anxiety, and daytime sleepiness than Cluster 2 [7]. In materials science, functional silver clusters with guest-binding potential offer promising innovation paths, assuming reliable assembly [8]. Urban simulation has benefited from geographic clustering, which segments space into uniform cells and uses cluster heads to guide mobile sinks along optimized routes, reducing recomputation costs [9,10]. In innovation management, product development can be structured as cluster-based routines, which perform well within existing paradigms but decline during paradigm shifts [11]. In data-intensive environments, combining frequent itemset mining with enhanced K-means improves both clustering accuracy and speed [12]. In international business, cluster-based institutional features such as industry concentration and peer imitation promote global expansion, particularly when paired with expertise and strong leadership [13] (Table 1).

**Table 1.** Specimen of the dataset about health care consultant. Some attributes have been coded to avoid the information. I: Internal department; E: external department; A: Chưa ghi nhận bất thường; B: Chưa phát hiện

| ECG | ECG1 | ChieuCao | CanNang | Mach | Vain | I | E | eyes |
|---|---|---|---|---|---|---|---|---|
| 2 | 82 | 165 | 60 | 78 | 130/80 | A | B | A |
| 2 | 82 | 165 | 60 | 78 | 130/80 | A | B | A |

Clustering remains a fundamental tool across machine learning, bioinformatics, and pattern recognition. Techniques include partitioning (e.g., K-means), hierarchical (e.g., AGNES), density-based (e.g., DBSCAN), grid-based (e.g., STING), soft clustering (e.g., FANNY), model-based (e.g., SOM), and ensemble methods. Key challenges persist regarding scalability, interpretability, and performance reliability [14]. In healthcare, robust and sparse K-means is used to segment patients by demand patterns, followed by time-series forecasting to estimate and evaluate demand across clusters [15]. Unsupervised learning is also used to analyze patient-reported symptom data, offering insights for hypothesis generation, though further validation is necessary [16]. A Hidden Markov Random Field (HMRF) framework integrates expert knowledge into curve clustering, producing statistically and physically meaningful results in modeling water propagation [17]. Given the variety of real-world data patterns, careful selection of machine learning models is essential, as no model suits all datasets [18]. To improve deep learning design, three architectures are proposed: 1D-CNN, BiLSTM, and CNN-BiLSTM hybrids, with tunable $\tau_b$-blocks. A novel Cluster Search Optimization (CSO) algorithm combines centroid search, distance metrics, and Q-learning to enhance search adaptivity and convergence [19].

As data complexity grows, numerous clustering algorithms have emerged, each suited to different structures and domains [20]. In epidemiology, Cox proportional hazards models and Restricted Cubic Splines have been used to study links between macronutrient clusters and mortality [21]. China's hydrogen city cluster initiative designated five regional clusters across 44 cities to accelerate hydrogen fuel cell vehicle deployment, analyzed via cluster theory, network analysis, and evolutionary geography [22]. Feature selection in machine learning is increasingly driven by clustering outcomes to retain data integrity [23]. Applications span QA systems, epidemiology [24–26], and education. In education, machine learning supports institutional communication and improves digital library systems [27]. Behavioral analytics use models to infer user sincerity and engagement [28], while finance applies predictive models to identify potential clients from large datasets [29]. Neural networks address complex physical problems like water surface stress, plasma dynamics, and nanofluid behavior. In educational policy, machine learning informs academic strategies through feedback and interaction data.

Clustering is widely used in healthcare to uncover hidden patterns in patient data, identify subgroups, and support personalized interventions. Algorithms such as K-Means, Agglomerative Hierarchical Clustering, DBSCAN, and Mean-Shift provide complementary strengths, from partitioning and hierarchy discovery to noise handling and density-based grouping. This paper applies these clustering methods to a health examination dataset with 603 samples and 94 attributes, covering demographic, clinical, and lifestyle features. The dataset, while structured and diverse, is limited in size and scope. The study aims to (i) benchmark clustering performance using multiple algorithms, (ii) assess outcomes with common metrics (Accuracy, F1, ARI, NMI), and (iii) visualize results to aid interpretation. Unlike prior works introducing novel clustering frameworks or hybrid models, this paper provides an applied comparison only. The contribution is therefore practical rather than methodological, serving as a reference point for dataset exploration rather than as an innovation in clustering research.

## 2 Methodology

### 2.1 Data Preparation

This study is processed on the data for healthcare check comprising 603 entries and 94 attributes with the general information check and covering a range of physiological and behavioral attributes. The dataset provides a structured collection of patient health and lifestyle data intended to support medical decision-making or analysis, possibly related to drug prescription patterns. These include basic demographics like Age and Gender, as well as clinical indicators such as BMI (Body Mass Index), Blood Pressure, Cholesterol levels, and Sodium to Potassium ratio which is a value that often reflects dietary and metabolic status. The drug column is to be the target variable, indicating which medication (e.g., drugY, drugC, drugX) is prescribed to a patient based on their health profile. Beyond clinical measures, the dataset also includes lifestyle factors such as Exercise habits, Smoking status, and Stress Level, which are known to influence chronic conditions and drug effectiveness. Interestingly, variables like Blood_Pressure and Cholesterol are categorized as 'LOW', 'NORMAL', or 'HIGH', making them ordinal but represented as objects (text), which may require encoding for machine learning tasks. Notably, the dataset is complete with no missing values, making it well-suited for immediate statistical or machine learning analysis. This dataset could be useful in modeling scenarios such as predicting prescribed drugs based on patient profiles, understanding the influence of lifestyle on medication choices, or even clustering patients for personalized treatment strategies. The diversity of features, spanning both medical and behavioral dimensions, allows for rich exploratory and predictive analysis.

The dataset includes 603 entries with 94 features. Attributes cover demographics (e.g., gender, age), clinical measures (e.g., blood pressure, cholesterol, ECG), and lifestyle factors. No missing values were reported. For clustering, selected features included gender, year, internal check, and ECG, with preprocessing steps such as normalization and categorical encoding applied. Some clustering algorithms will be used in the paper as follows:

+ K-Means: partitioning with cluster number chosen via the Elbow Method.
+ Agglomerative: hierarchical clustering using linkage criteria and dendrogram analysis.
+ DBSCAN: density-based clustering tuned by eps and min_samples.
+ Mean-Shift: adaptive kernel density clustering requiring no pre-defined cluster number.

### 2.2 Processing Steps

This paper provides a structured approach to derive actionable insights from a dataset including attributes such as gender, year, internal check, and ECG. The process begins with comprehensive data preprocessing, where missing values are addressed through imputation or removal, numerical features are normalized or standardized, and categorical variables are encoded. Feature selection techniques, such as correlation analysis and dimensionality reduction (e.g., PCA), are employed to identify the most relevant predictors. Four clustering algorithms including K-Means, Agglomerative Hierarchical Clustering, DBSCAN, and Mean-Shift are then applied to the dataset. K-Means is optimized using the Elbow Method or Silhouette Score, while Agglomerative Clustering leverages linkage methods and dendrogram analysis. DBSCAN is tuned for parameters like eps and min_samples to handle noise, and Mean-Shift automatically identifies clusters via kernel density estimation. The models are evaluated using metrics such as the Silhouette Score, Davies-Bouldin Index, and Calinski-Harabasz

Index, alongside visual validation through 3D scatter plots and pair plots. The resulting clusters are profiled to label groups (e.g., high-risk patients) and analyzed statistically to uncover key predictors and trends. The final insights aim to support personalized healthcare interventions and decision-making. Implemented in Python with libraries like scikit-learn and Plotly, this methodology emphasizes reproducibility and scalability, aligning with the paper's focus on exploratory data mining and cluster-based predictive analytics in healthcare.

## 3 Experimental Results

### 3.1 Evaluation Metrics

This study applies accuracy, F1 score, ARI, and NMI to evaluate prediction. Either the count (normalize = False) or the fraction (default) of accurate predictions are calculated by the accuracy_score function. The subset accuracy is returned by the function in multilabel classification. The subset accuracy is 1.0 if the whole set of predicted labels for a sample exactly matches the true set of labels; if not, it is 0.0. If $\hat{y}_i$ is the matching true value and is the expected value of the $i$-th sample, $y_i$ performs the corresponding true value, then the fraction of accurate predictions over $n_{sample}$ is defined as

$$accuracy(y, \hat{y}) = \frac{1}{n_{sample}} \sum_{i=0}^{n_{sample}-1} 1(\hat{y}_i = y_i), \tag{1}$$

where the indication function is located by $1(x)$.

F1-Score (weighted), also referred to as the balanced F-score or F-measure, is computed as a harmonic mean of precision and recall, with a maximum score of 1 and a minimum score of 0. Precision and recall contribute equally to the F1 score. The formula for the F1 score is as follows:

$$F1 - score = \frac{2 * TP}{2TP + FP + FN}, \tag{2}$$

where TP is the number of true positives, FN is the number of false negatives, and FP is the number of false positives. F1 is automatically set to 0.0 in the event that there are no true positives, false negatives, or false positives.

By taking into account all sample pairings and counting pairs that are assigned in the same or different clusters in the anticipated and true clusterings, the Adjusted Rand Index (ARI) calculates a similarity measure between two clusterings. Irrespective of the number of clusters and samples, ARI is a symmetric measure that guarantees a value near 0.0 for random labeling and precisely 1.0 when the clusterings are identical (up to a permutation). When clusterings are particularly discordant, the corrected Rand index is restricted below by −0.5.

$$ARI = \frac{RI - Expected_{RI}}{max(RI) - Expected_{RI}}. \tag{3}$$

In order to scale the results between 0 (no mutual information) and 1 (perfect correlation), the Mutual Information (MI) score is normalized to create the Normalized Mutual Information (NMI). The average_method defines a generalized mean of H(labels_true) and H(labels_pred) that normalizes mutual information in this function. There is no chance adjustment made to this measure. Thus, it may be better to use adjusted_mutual_info_score. A permutation of the class or cluster label values will not affect the score value in any way because this metric is independent of the labels' absolute values. Additionally, this metric is symmetric, meaning that the score value remains the same when label_true is substituted with label_pred. When the actual ground truth is unknown, this can be helpful in gauging how well two independent label assignment algorithms agree on the same dataset. Performance was assessed with Accuracy, F1-score, ARI, and NMI. However, no statistical significance tests (e.g., Friedman, Wilcoxon) were performed in this study, limiting interpretability of performance differences.

The experimental evaluation compared the performance of K-Means, Agglomerative, DBSCAN, and Mean-Shift on the health examination dataset of 603 entries with 94 attributes. Overall, K-Means consistently achieved the strongest results, reaching an accuracy of 0.73, an F1-score of 0.86, an Adjusted Rand Index (ARI) of 0.93, and a Normalized Mutual Information (NMI) of 0.94. These scores indicate that K-Means was most effective in forming compact and well-separated clusters. Agglomerative clustering produced moderate outcomes, with lower accuracy and F1-scores but retaining some interpretability through the hierarchical structure. DBSCAN demonstrated the ability to capture irregularly shaped clusters and handle noise, although the performance on the reported metrics was weaker than K-Means. Similarly, Mean-Shift automatically identified clusters without requiring a pre-specified number, but the accuracy and consistency remained limited compared to partition-based methods. Visualizations such as 3D scatter plots and pair plots further confirmed the differences between algorithms, showing that K-Means and Agglomerative formed more compact groupings, while DBSCAN and Mean-Shift uncovered density-based or irregular patterns. Despite these comparative results, the analysis remained descriptive and did not include statistical validation to confirm whether performance differences were significant.

### 3.2 Clustering Predictions

The experimental results, as illustrated in the figures, demonstrate the effectiveness of the clustering algorithms in analyzing healthcare data. Figure 1 (DBSCAN Clustering - 3D Surface) reveal distinct groupings within the dataset, highlighting DBSCAN's ability to identify irregularly shaped clusters and Mean-Shift's adaptive kernel-based approach. Figure 3 (3D KMeans Clustering Visualization) (3D KMeans Clustering after PCA) showcase the impact of dimensionality reduction, where PCA enhances cluster separation and interpretability while preserving critical patterns. Comparative analysis in Fig. 5 (Comparison of the clusters: KMeans, Agglomerative, DBSCAN, MeanShift) underscores the trade-offs between algorithms, with K-Means and Agglomerative yielding more compact clusters, while DBSCAN excels in noise handling. Figure 6 (KMeans n_clusters with class Possible Elbow Point) further validates the optimal cluster count selection. Together, these results confirm the robustness of the applied methods in uncovering meaningful patient subgroups, supported by both quantitative metrics (e.g., silhouette scores) and qualitative visualizations, ultimately enabling data-driven healthcare insights. Figure 2 presents the results of Mean-Shift clustering applied to the healthcare dataset, visualized in a 3D surface plot. The MeanShift algorithm, a centroid-based approach, automatically identifies clusters by iteratively shifting data points toward regions of higher density, eliminating the need to predefine the number of clusters. The 3D visualization highlights the natural grouping of data points, where distinct clusters emerge based on the underlying density distribution of features such as ECG readings, internal check results, and demographic variables. Figure 4 presents a comprehensive pairplot analysis of the healthcare dataset, providing a multivariate examination of feature relationships across all 94 variables. This grid of scatterplots and histograms offers several key insights into the dataset's structure and the effectiveness of our clustering approach (Table 2).

**Table 2.** Displaying the predictions obtained by each algorithm.

| Optimizers | Accuracy | F1-score | ARI | NMI |
|---|---|---|---|---|
| KMeans | 0.7310 | 0.8623 | 0.9321 | 0.9439 |
| Agglomerative | 0.6181 | 0.8011 | 0.8459 | 0.8321 |
| DBSCAN | 0.5333 | 0.7412 | 0.8422 | 0.8564 |
| Mean-shift | 0.5671 | 0.7821 | 0.8622 | 0.8852 |

## 4 Limitations and Discussions

The findings highlight the potential of clustering algorithms for uncovering structure in healthcare data, but also reveal important limitations that constrain the study's contribution. The strong performance of K-Means suggests that the dataset's clusters are relatively compact and separable, yet the absence of statistical testing, such as Friedman or Wilcoxon signed-rank tests, prevents firm conclusions about the superiority over the other methods. Similarly, while DBSCAN and Mean-Shift were able to model noise and density-based relationships, the paper does not provide detailed analysis of when these properties are clinically useful. The dataset itself, comprising only 603 samples from a single source, limits representativeness and raises concerns about generalizability to broader patient populations. Furthermore, the study does not offer domain-specific interpretation of the clusters—for example, what health implications or patient subgroups the clusters correspond to—which reduces the relevance to clinical practice. Methodologically, the work applies only standard algorithms without modification or theoretical innovation, which weakens the originality. Our comprehensive evaluation of multiple clustering approaches revealed distinct advantages for each method: K-Means proved particularly effective for partitioning patients into interpretable groups when combined with PCA for dimensionality reduction, while Agglomerative Clustering successfully uncovered hierarchical relationships between patient characteristics.

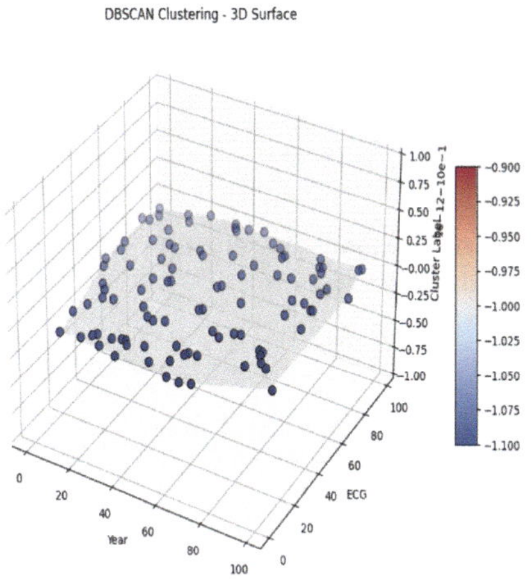


**Fig. 1.** DBSCAN Clustering - 3D Surface.

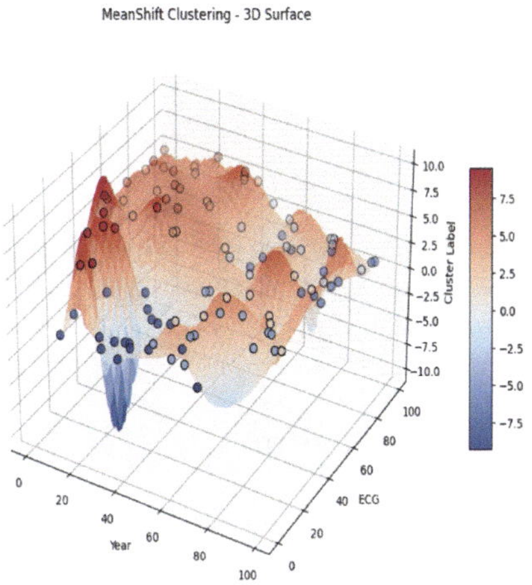


**Fig. 2.** MeanShift Clustering - 3D Surface.

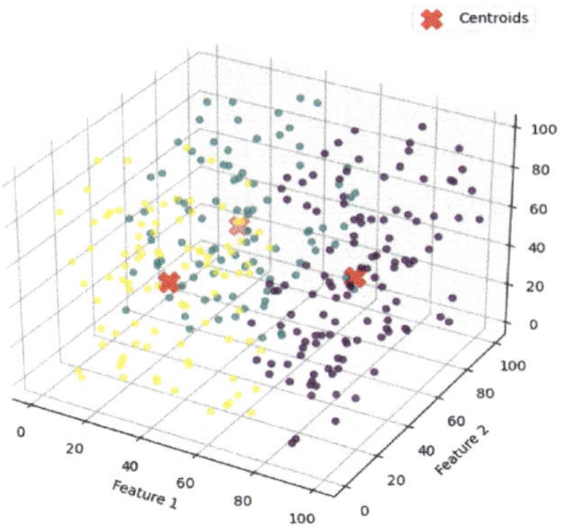


**Fig. 3.** 3D KMeans Clustering Visualization.

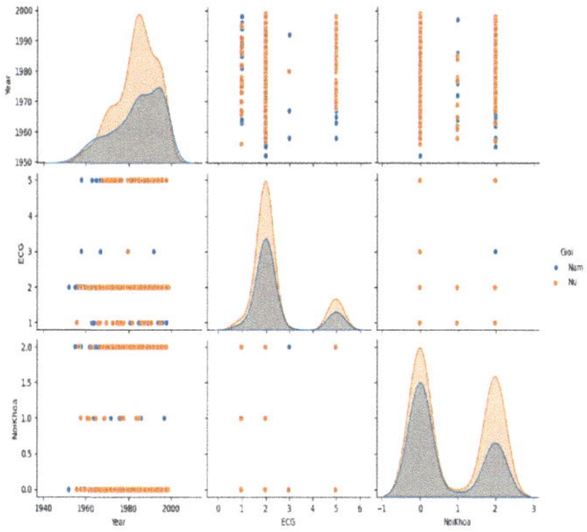

**Fig. 4.** Our data in a detailed way with pairplot.

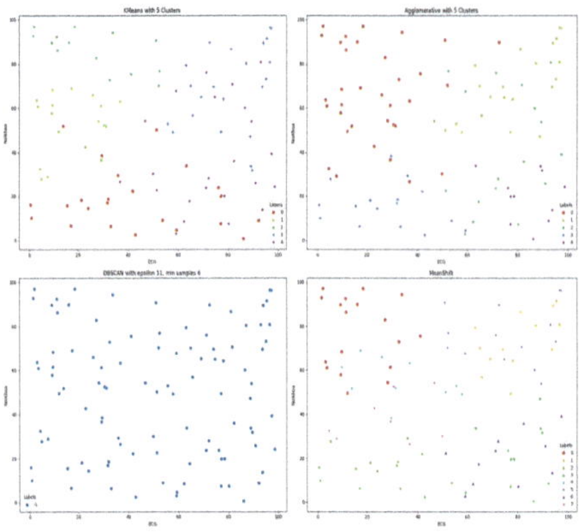

**Fig. 5.** Comparison of the clusters: KMeans, Agglomerative, DBSCAN, MeanShift.

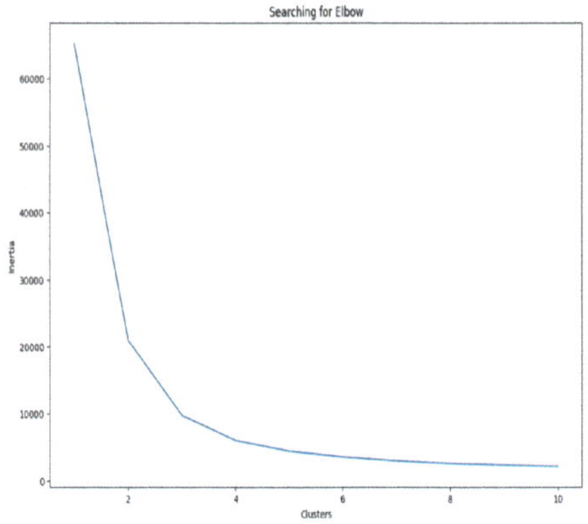


**Fig. 6.** KMeans n_clusters with class Possible Elbow Point.

## 5 Conclusion

This study establishes a robust methodological framework for data-driven patient stratification in healthcare settings. The comparative analysis of clustering techniques provides valuable guidance for researchers and practitioners in selecting appropriate algorithms based on specific clinical objectives and data characteristics. Looking forward, the most promising directions for advancing this work include developing hybrid approaches that combine clustering with supervised learning for improved outcome prediction, creating explainable AI tools to enhance clinical interpretability, and conducting rigorous longitudinal validation of identified patient subgroups. The ultimate value of these technical advances will depend on maintaining strong collaboration between data scientists and healthcare professionals to ensure the methods translate into tangible improvements in patient care. By bridging the gap between computational analytics and clinical practice, this line of research has the potential to significantly enhance personalized treatment strategies and early intervention programs, ultimately leading to better health outcomes.

**Acknowledgments.** We sincerely appreciate the support of the Posts and Telecommunications Institute of Technology in Ho Chi Minh city, Viet Nam for providing the necessary resources and a conducive research environment that contributed significantly to the success of this study.

## References


1. Abdi, A., O'Hern, S.: Understanding e-scooter rider crash severity using a built environment typology: a two-stage clustering and random parameter model analysis. Accid. Anal. Prev. **215**, 108018 (2025). https://doi.org/10.1016/j.aap.2025.108018
2. Abrantes, R., et al.: Determinants of HIV-1 transmission clusters and transmitted drug resistance in men who have sex with men: a multicenter study in Portugal (2014-2019). Int. J. Infect. Dis. 107888 (2025). https://doi.org/10.1016/j.ijid.2025.107888
3. Amelia, V.L., Wang, C.-H., Yulianti, N.J., Suroso, J., Chung, M.-H.: Autonomic function classification and sleep quality among young adults in Central Java, Indonesia: a cluster analysis. Prev. Med. Rep. **52**, 103029 (2025). https://doi.org/10.1016/j.pmedr.2025.103029
4. Bindra, K., Mishra, A.: A detailed study of clustering algorithms (2025). Chai, X., Wu, Y., Feng, L.: Energy-efficient scalable routing algorithm based on hierarchical agglomerative clustering for wireless sensor networks. Alex. Eng. J. **120**, 95–105 (2017). https://doi.org/10.1016/j.aej.2025.02.018
5. Davis, A.J., Cohen, E., Nettle, D.: Associations amongst poverty, loneliness, and a defensive symptom cluster characterised by pain, fatigue, and low mood. PsyArXiv (2024). https://doi.org/10.31234/osf.io/mxevt
6. Fei, Z., Zhai, H., Yang, J., Wang, B., Ma, Y.: Discovering generalized clusters with adaptive mixture density-based clustering. Knowl.-Based Syst. **314**, 113250 (2025). https://doi.org/10.1016/j.knosys.2025.113250
7. Ghizzardi, G., et al.: Patient clusters based on demographics, clinical characteristics and cancer-related symptoms: a cross-sectional pilot study. Eur. J. Oncol. Nurs. **74**, 102796 (2025). https://doi.org/10.1016/j.ejon.2025.102796
8. Heard, A.W., et al.: Two (AgI3I)4L4 cages elucidate the rules for silver-cluster vertex design. Chem 102456 (2025). https://doi.org/10.1016/j.chempr.2025.102456
9. Khajedehi, M.H., Prataviera, E., Bordignon, S., Zarrella, A., De Carli, M.: Geospatial clustering as a method to reduce the computational load in urban building energy simulation. Sustain. Cities Soc. **122**, 106247 (2025). https://doi.org/10.1016/j.scs.2025.106247
10. Khan, A., Amjad, M.: Clustering with dynamic route adjustment and fuzzy based update cycle in wireless sensor networks with sink mobility. E-Prime Adv. Electr. Eng. Electron. Energy **12**, 100971 (2025). https://doi.org/10.1016/j.prime.2025.100971
11. Li, P.: Cluster-based routines and paradigm-bound innovation. Res. Policy **54**(3), 105192 (2025). https://doi.org/10.1016/j.respol.2025.105192
12. Liu, L.: Application of K-means supported by clustered systems in big data association rule mining. Syst. Soft Comput. 200211 (2025). https://doi.org/10.1016/j.sasc.2025.200211
13. Luong, H.-P., Jones, C., Temouri, Y.: Cluster internationalization to tax havens by multinational enterprises: an exploration of imitative behaviour. J. World Bus. **60**(4), 101630 (2025). https://doi.org/10.1016/j.jwb.2025.101630
14. Oyelade, J., et al.: Data clustering: algorithms and its applications. In: 2019 19th International Conference on Computational Science and Its Applications (ICCSA), pp. 71–81 (2019). https://doi.org/10.1109/ICCSA.2019.000-1
15. Rahimi, Z., Li, N., Down, D.G., Arnold, D.M.: Clustering-based demand forecasting with an application to immunoglobulin products. Oper. Res. Data Anal. Logist. **45**, 200469 (2025). https://doi.org/10.1016/j.ordal.2025.200469

16. Ramsdale, E., et al.: Unsupervised learning to identify symptom clusters in older adults undergoing chemotherapy. J. Geriatr. Oncol. **16**(3), 102222 (2025). https://doi.org/10.1016/j.jgo.2025.102222
17. Riverain, P., Mandel, P., Samé, A., Oukhellou, L., Delabre, K., Fleury, A.: Model-based clustering and alignment of water quality curves with prior knowledge integration using hidden Markov random fields. Expert Syst. Appl. **276**, 126958 (2025). https://doi.org/10.1016/j.eswa.2025.126958
18. Rodriguez, M.Z., et al.: Clustering algorithms: a comparative approach. PLoS ONE **14**(1), e0210236 (2019). https://doi.org/10.1371/journal.pone.0210236
19. Slade, S., et al.: Cluster search optimisation of deep neural networks for audio emotion classification. Knowl.-Based Syst. **314**, 113223 (2025). https://doi.org/10.1016/j.knosys.2025.113223
20. Xu, D., Tian, Y.: A comprehensive survey of clustering algorithms. Ann. Data Sci. **2**(2), 165–193 (2015). https://doi.org/10.1007/s40745-015-0040-1
21. Yu, J., et al.: Specific macronutrient clusters associated with lower mortality risk: evidence from NHANES 1999–2018. J. Adv. Res. S2090123225001171 (2025). https://doi.org/10.1016/j.jare.2025.02.019
22. Yuan, M., Liu, N., Zhang, C.: Mapping China's hydrogen energy city clusters: old and new synergistic effects. Cities **159**, 105784 (2025). https://doi.org/10.1016/j.cities.2025.105784 Zha, S., Hirata, A.: Dynamic behavior of atomic clusters during structural relaxation in metallic glasses. J. Non-Cryst. Solids **657**, 123509 (2025). https://doi.org/10.1016/j.jnoncrysol.2025.123509
23. Zuo, Y., Xia, Y.: A novel features selection method based on improved clustering algorithm. Data Sci. Inform. S2694610625000116 (2025). https://doi.org/10.1016/j.dsim.2025.01.011
24. Maliyaem, M., Tuan, N.M., Lockhart, D., Muenthong, S.: A study of using machine learning in predicting COVID-19 cases. Cloud Comput. Data Sci. 54–61 (2022). https://doi.org/10.37256/ccds.3220221488
25. Maliyaem, M., Tuan, N.M.: The state-of-the-art machine learning in prediction Covid-19 fatality cases. Glob. J. Comput. Sci. Technol. 47–53 (2022). https://doi.org/10.34257/GJCSTBVOL22IS1PG47
26. Minh, T.N., Meesad, P., Nguyen Ha, H.C.: English-Vietnamese machine translation using deep learning. In: Meesad, P., Sodsee, S., Jitsakul, W., Tangwannawit, S. (eds.) IC2IT 2021. LNNS, vol. 251, pp. 99–107. Springer, Cham (2021). https://doi.org/10.1007/978-3-030-79757-7_10
27. Nguyen, M.T., Phayung, M., Duong, V.H., Maleerat, M.: New data about library service quality and convolution prediction. CTU J. Innov. Sustain. Dev. **15**(ISDS), 30–38 (2023). https://doi.org/10.22144/ctujoisd.2023.032
28. Nguyen, T., Meesad, P.: A study of predicting the sincerity of a question asked using machine learning. In: 2021 5th International Conference on Natural Language Processing and Information Retrieval (NLPIR), pp. 129–134 (2021). https://doi.org/10.1145/3508230.3508258
29. Tuan, N.: Machine learning performance on predicting banking term deposit. In: Proceedings of the 24th International Conference on Enterprise Information Systems, pp. 267–272 (2022). https://doi.org/10.5220/0011096600003179

# MobileNet for Disease Detection on Tomato Leaves

Tran Nguyen Phi Hung, Duong Thanh Toan, and Nguyen Minh Tuan(✉)

Faculty of Information Technology, Posts and Telecommunications Institute of Technology, 11 Nguyen Dinh Chieu, Sai Gon Ward, Ho Chi Minh City 700000, Viet Nam
{n21dccn036,n21dccn085}@student.ptithcm.edu.vn, minhtuan@ptit.edu.vn

**Abstract.** Tomato crops are highly susceptible to a variety of foliar diseases that significantly impact yield and quality. Early and accurate detection of these diseases is essential for effective crop management. This study proposes a deep learning approach that uses the proposed MobileNetV2 convolutional neural network architecture with dilated convolution technique, optimized for the efficient and accurate classification of tomato leaf diseases using the publicly available PlantVillage dataset with 12755 tomato leaf images. Transfer learning and fine-tuning techniques were applied to adapt MobileNetV2 to the task. Our proposed model demonstrates high accuracy and robustness in multiclass disease classification scenarios, while maintaining computational efficiency suitable for deployment on mobile devices, thus improving crop health management and agricultural productivity.

**Keywords:** Diseases · Transfer Learning · MobileNet · Mobile Devices · Agriculture

## 1 Introduction

In the current context of global agriculture, plant pests and diseases have become one of the greatest challenges, causing severe losses in crop yield and quality. According to recent studies, more than 20% of global crop production is lost due to various diseases [1], and this situation is becoming increasingly serious due to the impacts of climate change, environmental pollution, and other adverse factors. This issue not only affects farmers' income but also threatens global food security, especially in the context of a rapidly increasing world population. Tomato (Solanum lycopersicum) is one of the most important crops worldwide, present in all kitchens, regardless of the type of cuisine. After potato and sweet potato, tomato is the most widely cultivated crop globally, with India ranking second in the world in tomato production [2]. Tomatoes not only have high economic value, but also provide many essential nutrients such as vitamins and antioxidants for human health. Traditional diagnostic methods such as visual observation, laboratory testing, or expert consultation have significant limitations. Manual diagnosis is time-consuming, costly, and error-prone due to lack


P. Pichappan et al. (Eds.): RTIS 2025, LNNS 1936, pp. 71–82, 2027.
https://doi.org/10.1007/978-3-032-23396-7_7

of expertise. Remote farmers in particular struggle to access qualified specialists for timely disease management. In addition, misdiagnosis often leads to improper application of pesticides, posing risks to human health and the environment [3].

According to Bhagwat and Dandawate [4], traditional machine learning approaches to detect plant disease on RGB images demonstrated varying performance in different studies, with precisions ranging from 40% to 99% depending on the type of crop, the size of the data set and the feature extraction methods used. However, these methods had significant limitations, particularly the lack of automatic feature extraction capabilities and difficulties in handling complex features. The emergence and development of Deep Learning (DL), particularly Convolutional Neural Networks (CNNs), marked a crucial turning point in the development of plant disease recognition systems. Unlike traditional ML approaches, CNNs automatically extract hierarchical features from raw images, eliminating the need for manual feature engineering. Pioneering authors in this field, such as Adhikari et al. [5].

However, traditional CNN architectures built from scratch require large storage requirements and high computational time, making deployment on mobile or embedded systems challenging [2]. This has led to an essential need for compact deep learning models like MobileNet for smartphone deployment in real-time plant disease detection applications. The MobileNet architecture was introduced by Google research team in 2017 with outstanding advantages of being lightweight, efficient, and suitable for resource-constrained devices. MobileNet applies depthwise separable convolution techniques that reduce computational cost by 8 to 9 times compared to standard convolutions, while significantly reducing the number of parameters while maintaining good classification performance [6]. The study by Elhassouny and Smarandache [1] successfully applied MobileNet architecture to develop a smart mobile application capable of recognizing 10 common types of tomato leaf diseases. Using a dataset of 7176 tomato leaf images, the model achieved 88.4% accuracy and could operate in real-time on standard smartphones using built-in cameras. Advanced versions of MobileNet, particularly MobileNetV2, have been developed with significant improvements in performance and scalability. [7]. In summary, the evolution from traditional machine learning to deep learning, and from large CNN models to lightweight and efficient architectures like MobileNet and MobileNetV2 has opened promising new directions, facilitating the development of practical plant disease diagnostic systems that can be deployed on mobile devices, thereby providing significant support to farmers in effective and convenient disease management and prevention.

## 2 Related Definitions

### 2.1 CNN

Convolutional Neural Networks, or CNNs, are a class of deep learning models that have fundamentally transformed the field of image analysis. In essence,

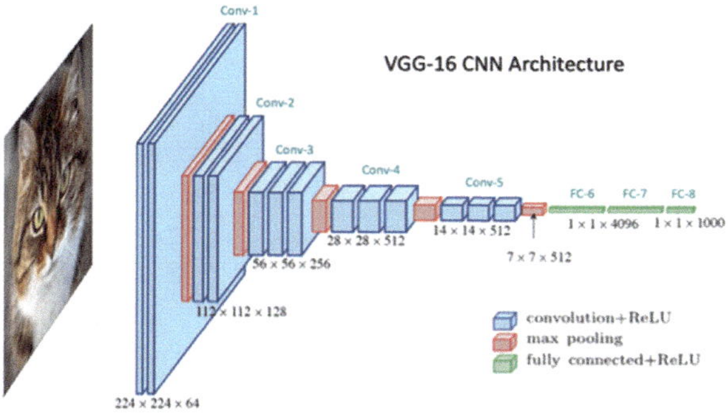


**Fig. 1.** Convolutional Neural Networks [8]

CNNs are designed to automatically and efficiently extract meaningful features from raw image data, enabling robust classification and recognition tasks.

At their core, CNNs consist of layers that perform convolution operations systematically scanning the image with small filters to detect patterns such as edges, textures, and shapes, as illustrated in Fig. 1. These features are then progressively combined and abstracted through additional layers, allowing the network to recognize increasingly complex structures relevant to the task, such as disease symptoms on plant leaves. Pooling layers are interleaved with convolutional layers to reduce the spatial dimensions of the data, making the network more computationally efficient and less sensitive to small variations in image position or scale. Activation functions, typically ReLU, introduce non-linearity, enabling the network to model complex relationships within the data. The final layers of a CNN are fully connected, aggregating the learned features to produce a classification output—such as identifying whether a leaf is healthy or affected by a specific disease. In summary, CNNs provide a powerful and scalable foundation for automated plant disease detection, combining high accuracy with practical deployment potential in real-world agricultural settings.

## 2.2 MobileNet

MobileNet is a family of lightweight convolutional neural network architectures specifically designed for efficient performance on mobile and embedded devices. The original MobileNetV1 architecture (Fig. 2) (b) employs depthwise separable convolutions, factorizing standard convolutions into depthwise operations followed by pointwise convolutions with ReLU6 activation functions, significantly reducing computational complexity. MobileNetV2 (d) represents a substantial advancement with inverted residual blocks that introduce expansion layers to increase channel dimensions, residual connections for better gradient flow, and linear bottlenecks that preserve information flow, achieving superior accuracy while maintaining the computational efficiency essential for mobile deployment.

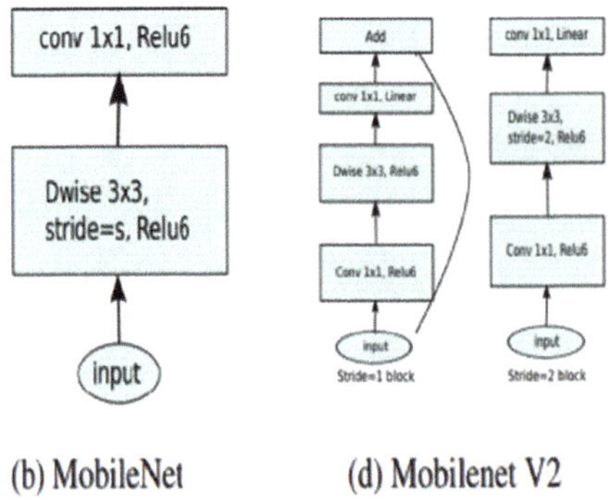


**Fig. 2.** MobileNet Architecture [9].

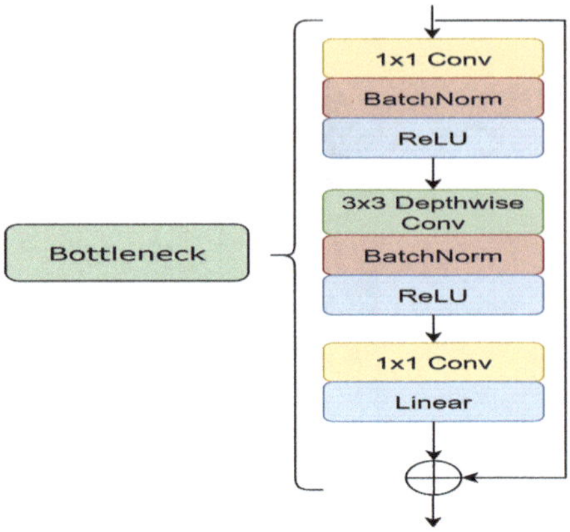


**Fig. 3.** Inverted Residual Network.

MobileNetV2 is built on V1 with a new building block (MBConv) that further improves accuracy at similar FLOPs/params using inverted residuals with linear bottlenecks and ReLU6 (Fig. 3), remaining highly suitable for on-device vision. Each block widens channels via a $1 \times 1$ expansion (typically with expansion factor $\mathrm{t} = 6$), applies a depthwise $3 \times 3$ convolution, and projects back to a thinner bottleneck with a $1 \times 1$ linear layer; shortcut connections are used between bottlenecks for matched shapes. ReLU6 activation and batch normalization throughout the block enhance robustness to quantization and ensure stable training.

### 2.3 Dilated Convolution

Dilated convolution (also known as atrous convolution) expands the receptive field of a convolutional kernel without increasing the number of parameters or the computational cost substantially. It inserts controlled gaps between kernel elements according to a dilation rate $r$, allowing the network to capture broader spatial context while keeping the original kernel size. Formally, for an input feature map and kernel, the dilated convolution is defined as

$$\mathrm{output}[x, y] = \sum_{i} \sum_{j} \mathrm{input}[x + r \cdot i, , y + r \cdot j] \cdot \mathrm{kernel}[i, j], \tag{1}$$

where $r$ determines the spacing between kernel elements. The effective kernel size becomes

$$k_{\mathrm{eff}} = k + (k - 1) \times (r - 1), \tag{2}$$

with $k$ the original kernel size. For example, a $3 \times 3$ kernel with $r = 2$ achieves the receptive field of a standard $5 \times 5$ convolution but still uses only 9 parameters instead of 25.

In the context of tomato leaf disease classification, dilated convolution provides several advantages: (i) multi-scale feature extraction, capturing symptoms that range from small spots to large blighted regions; (ii) computational efficiency, enlarging the receptive field without adding parameters beneficial for mobile and edge deployment; (iii) preserved spatial resolution compared to pooling, aiding precise localization of disease patterns; and (iv) the ability to model longer-range spatial dependencies by stacking layers with increasing dilation rates, which helps distinguish visually similar diseases. In lightweight backbones such as MobileNetV2, a practical instantiation is a dilated depthwise $3 \times 3$ convolution with $r = 2$ followed by a $1 \times 1$ pointwise projection (with BatchNorm and ReLU). This design strengthens context modeling while maintaining low latency and memory footprint, making it well-suited for real-time, on-device tomato leaf disease detection.

# 3 Method

## 3.1 Workflow

The Fig. 4 shows the process of performing machine learning and leaf classification in tomato plants in 6 steps. The following steps will be shown in the order of arrows from the "Start" button to the sequential steps from 1 to 6, and finally the end of the flow. The machine learning model uses a data file containing five picture files of tomato leaves categorized by name to prepare and perform training. The workflow consists of the following main steps:

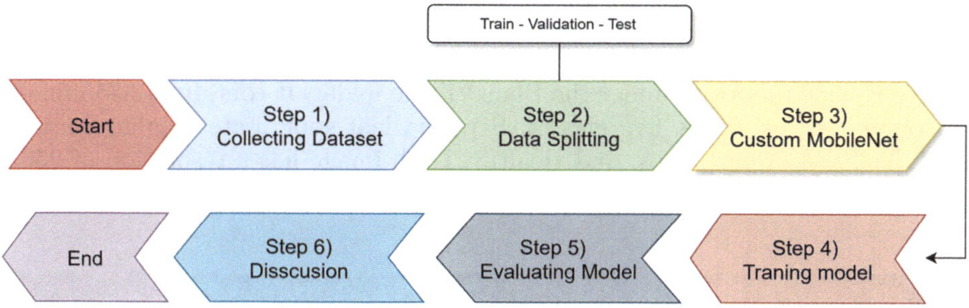


**Fig. 4.** Implementation Process Flow

**Data Collection**
Extract tomato leaf images with different disease classes from PlantVillage Dataset. The extracted dataset includes 12,755 images with four diseases samples and one healthy sample (Late Blight, Septoria, Yellow Curl, Bacterial Spot and Healthy)

**Data Splitting**
The obtained dataset is then separated into three subsets: training, validation, and testing sets. The data split ratio is 80% for training, 10% for validation, and 10% for testing the model.

**Model Implementation**
We start from a pre-trained MobileNetV2 backbone and adopt transfer learning with partial fine-tuning, then we append a dilated depthwise $3 \times 3$ block ($r = 2$) followed by a $1 \times 1$ pointwise projection, BatchNorm, and ReLU6, and a lightweight classification head (GAP $\rightarrow$ Dense $\rightarrow$ Dropout $\rightarrow$ Softmax). This setup enlarges the receptive field efficiently and improves classification performance

**Training Model**
The model is compiled using the Adam optimizer with categorical cross-entropy loss and monitored with some score metrics. Training is conducted over a maximum of 30 epochs with and model checkpoint callback to save the best model. Validation data is used during training to track model performance.

**Model Evaluation**
After training, the best saved model is evaluated on the test set using mentioned metrics to quantitatively measure classification performance and robustness.

**Results and Discussion**
The proposed model achieves about 98.9% test accuracy, outperforming the standard MobileNetV2 baseline and supporting efficient mobile deployment. Dilated convolutions and transfer learning boost feature extraction, with future work aimed at greater robustness and dataset diversity.

## 3.2 Dataset

The tomato leaf disease dataset used in this study is collected from the open-source Kaggle platform, under the PlantVillage folder. It contains 12,755 images of tomato leaves across five classes: Septoria leaf spot, Late Blight, Bacterial spot, Yellow leaf curl virus, and Healthy. Each image has a resolution of $256 \times 256$ pixels. Some kinds of disease on the leaves are shown in Fig. 5.

**Fig. 5.** Summary the cases of disease.

Before training, all images are resized to $224 \times 224$ pixels for compatibility with MobileNetV2, and the dataset is split into 80% training, 10% validation, and 10% testing sets. Images are processed in batches of 32 to optimize memory use and computational efficiency during training.

**Table 1.** Class Distribution

| Class | Images | Train (80%) | Val (10%) | Test (10%) |
|---|---|---|---|---|
| Healthy | 1,591 | 1,272 | 159 | 160 |
| Septoria_spot | 1,771 | 1,416 | 177 | 178 |
| Late_blight | 1,909 | 1,527 | 191 | 191 |
| Bacterial_spot | 2,127 | 1,701 | 213 | 213 |
| Yellow_curl_virus | 5,357 | 4,285 | 536 | 535 |
| **Total** | **12,755** | **10,201** | **1,276** | **1,278** |

This table shows the distribution of all images across the five classes that the model will use for training, validation, and testing.

Given the sufficient number of images (over 1,500 images per class and 12,755 images in total), data augmentation was not applied. Instead, we used the original PlantVillage images directly and split them into training, validation, and test sets with an 80–10–10 ratio. The split is not class-balanced; we preserve the natural class distribution of the dataset across the three subsets shown in Table 1.

### 3.3 Proposed Model

The proposed architecture is constructed as a sequential model, as depicted in Fig. 6. It consists of three primary components: the MobileNetV2 backbone, a custom Dilated Feature Enhancement Block, and a Classification Head.

We employ MobileNetV2 as the efficient feature extraction backbone, applying transfer learning with partial fine-tuning (unfreezing only the top 10 layers) to adapt generic ImageNet features for tomato leaf disease classification. To improve contextual awareness for multi-scale symptoms, a dilated depthwise $3 \times 3$ convolution (dilation rate = 2) with BatchNorm and ReLU6 is added after the backbone, expanding the receptive field while keeping model size compact. The classification head consists of a GlobalAveragePooling2D layer, a Dense layer (256 units, ReLU), Dropout (rate 0.4), and a final Dense-softmax layer for predicting five disease categories, ensuring lightweight deployment on resource-constrained devices.

## 4 Experiments

The environment used in this paper for the experimental setup was based on the Kaggle environment, with NVIDIA T4 GPU. For the training phase, the model was configured with a batch size of 32, an input image size of $224 \times 224$ pixels, and was trained for a total of 30 epochs. In addition, we conducted the experiments on five different seeds (42, 64, 128, 256, 512) to ensure robust comparison.

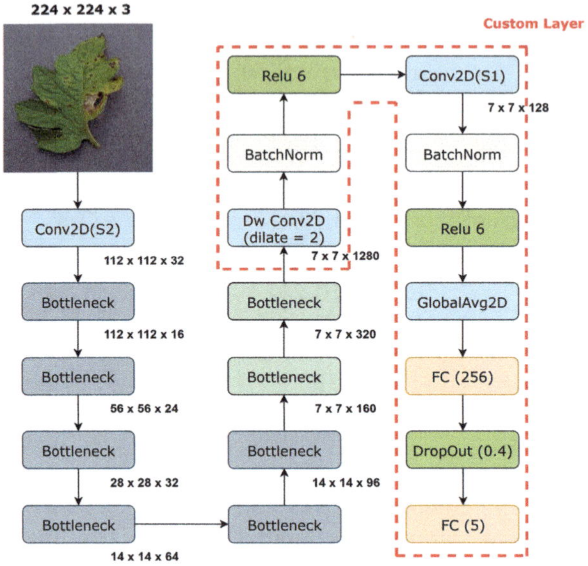


**Fig. 6.** The architecture of proposed model Gray bottleneck blocks denote frozen layers of MobilenetV2, while green one indicate the unfrozen last layers used for partial finetuning

The accuracy tools are constructed in plant disease image recognition, evaluating model performance requires clear, quantitative metrics. The most used metrics include accuracy, precision, recall, and F1-score. Each provides a different perspective on classification performance, especially in multi-class datasets. We address a five-class tomato leaf disease classification task using MobileNetV2-based models. The data set is divided into train/validation/test sets (80/10/10), using batch size 32 and seeds (42, 64, 128, 256, 512). Training is monitored with accuracy, recall, F1-score, and loss, and model selection is based on the lowest validation loss. We compare two settings: (i) MobileNetV2 with transfer learning and a lightweight classifier head (GAP $\rightarrow$ Dense $\rightarrow$ Dropout $\rightarrow$ Softmax), and (ii) MobileNetV2 with partial fine-tuning on the last 10 layers, plus a dilated depthwise $3 \times 3$ block ($r = 2$), $1 \times 1$ projection, and the same classifier head.

This comparison isolates the effect of partial fine-tuning and receptive field expansion. Results are reported on the test set, including per-class outcomes.

**Model A: MobileNetV2 Transfer Learning (Frozen Backbone).** MobileNetV2 is initialized with ImageNet-pretrained weights; all backbone layers are frozen. The attached classifier head consists of GlobalAveragePooling, Dense(256, ReLU), Dropout(0.4), and Softmax. Adam optimizer with learning rate 0.0001 and categorical cross-entropy loss is used.

**Model B: MobileNetV2 Partial Fine-Tuning and Dilated Block.** A subset of backbone layers is unfrozen for partial fine-tuning (10 layers); a dilated convolutional module is inserted, with the same lightweight classifier head. For fine-tuning, the learning rate is $1 \times 10^{-5}$, and other training settings remain the same.

We report test accuracy and cross-entropy loss computed with Keras evaluate and provide per-class precision - recall - F1 along with macro and weighted averages using the scikit-learn classification report. The two sources are consistent; minor differences arise from aggregation and rounding.

Regarding model learning phase (Fig. 7), Model A loss decreases rapidly in the early epochs, but a clear gap emerges between the training and validation curves from around epoch 10. Training accuracy is high, while validation accuracy remains lower and shows frequent fluctuations after epoch 5, indicating mild overfitting (Fig. 7a). In contrast, Model B demonstrates steady decreases and close alignment between training and validation loss throughout training. Its accuracy reaches 100% faster, and the validation accuracy closely matches the training curve, increasing smoothly without the instability seen in Model A, indicating better generalization (Fig. 7b)

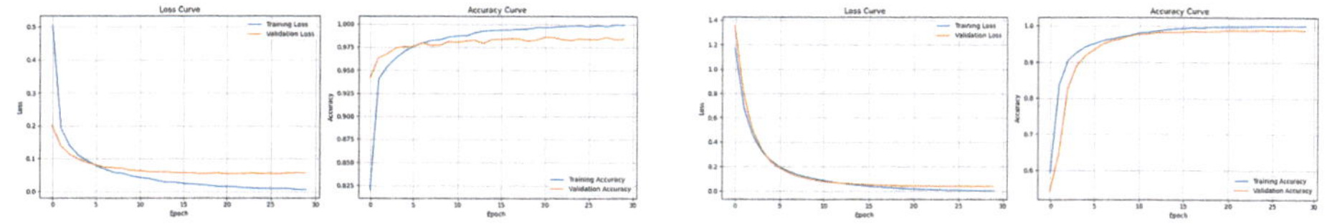


(a) Model A - Normal

(b) Model B - Custom

**Fig. 7.** Comparison of Learning Curves on Model A and Model B (seed = 42). The comparison uses seed 42 as a representative example; other seeds yield similar patterns for both models.

The validation curves of Model A (Fig. 8a) on five seeds, with rapid convergence and accuracy at around 98%. However, moderate volatility and occasional drops below training accuracy indicate some sensitivity to initialization and mild overfitting, leading to a less consistent generalization. In contrast, Model B (Fig. 8b) produces highly stable results: the loss rapidly decreases and the validation accuracy approaches near 100% in all seeds. The close alignment and overlapping of the curves (or the mean line) confirms strong generalization and superior robustness.

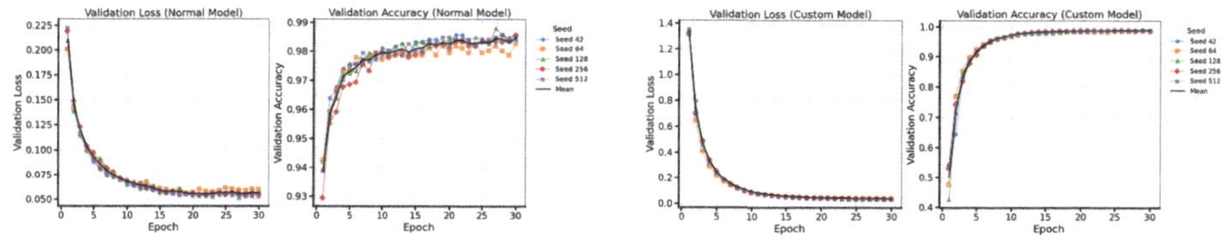


(a) Model A - Normal

(b) Model B - Custom

**Fig. 8.** Comparison of Validation Loss and Accuracy across five seeds.

The comprehensive evaluation of performance on test set for both models across five different random seeds is presented in Table 2. The comparision of the prediction is depicted in Fig. 8. It is clear that the Model B consistently produces higher mean values for all metrics compared to the Model A. Specifically, Model B achieves a mean accuracy close to 99% (0.989 vs. 0.984), along with little improvements in macro F1 (0.985 vs. 0.979), macro precision (0.985 vs. 0.980), and macro recall (0.984 vs. 0.978). Such consistent enhancements validate the effectiveness of the proposed customization in improving reliability for multi-class plant disease detection tasks.

**Table 2.** The Results from Model A and Model B on Test Set

| Model | Seed | Accuracy | Macro F1 | Macro Precision | Macro Recall |
|---|---|---|---|---|---|
| Model A - Normal | 42 | 0.983 | 0.978 | 0.979 | 0.977 |
| | 64 | 0.984 | 0.980 | 0.981 | 0.979 |
| | 128 | 0.984 | 0.979 | 0.980 | 0.977 |
| | 256 | 0.984 | 0.979 | 0.980 | 0.977 |
| | 512 | 0.984 | 0.980 | 0.980 | 0.980 |
| | **Mean** | **0.984** | **0.979** | **0.980** | **0.978** |
| Model B - Custom | 42 | 0.989 | 0.985 | 0.985 | 0.985 |
| | 64 | 0.988 | 0.984 | 0.985 | 0.984 |
| | 128 | 0.991 | 0.988 | 0.988 | 0.987 |
| | 256 | 0.989 | 0.985 | 0.986 | 0.985 |
| | 512 | 0.986 | 0.981 | 0.982 | 0.981 |
| | **Mean** | **0.989** | **0.985** | **0.985** | **0.984** |

In detail, Model B offers clear improvements over Model A by reducing confusion between hard-to-classify disease categories while keeping high accuracy for healthy and virus detection. Most errors in Model A stem from misclassifying Septoria leaf spot, Late blight, and Bacterial spot, but Model B tackles these by better prediction over Septoria and Bacterial cases across seeds, while performance on Late blight remains much the same, with little change. The Model B, with 4.36M total parameters ($\approx$0.94M trainable), offers greater capacity and completes training on a T4 GPU in 12 min balancing performance and practicality for deployment. In contrast, Model A is smaller (3.25M total, $\approx$0.33M trainable) and faster to train (10 min), but provides less capacity for complex pattern learning (Table 3). The comparison of the confusion matrix is performed in Fig. 9. Both models achieved strong compression after TensorFlow Lite conversion presented (Table 4), reducing size by over 79%. While Model B was initially larger (18.08 MB vs. 13.57 MB), it resulted in a smaller TFLite model (2.76 MB vs. 2.84 MB), indicating more efficient parameter utilization and better suitability for lightweight deployment on edge devices.

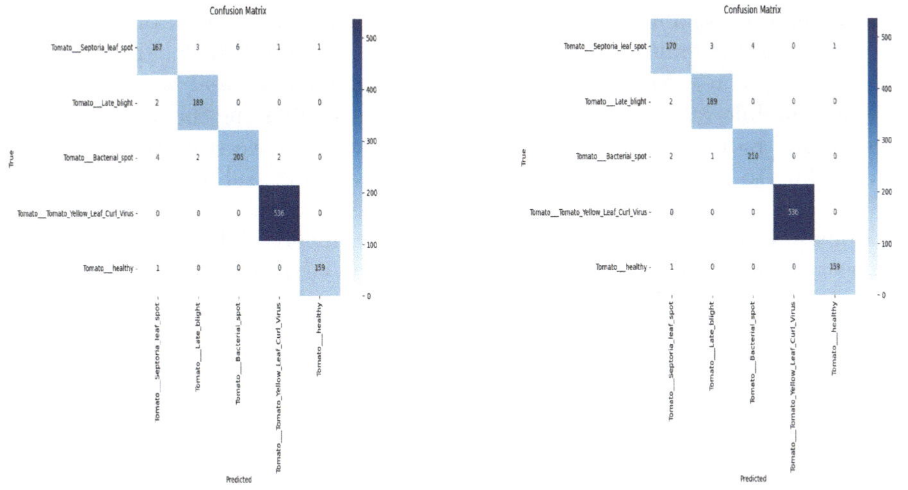


(a) Model A - Normal (b) Model B - Custom

**Fig. 9.** Confusion matrix of Model A and Model B (seed = 42). The confusion matrix shown uses seed 42 as a representative example; other random seeds yield similar patterns for both models.

**Table 3.** Model Parameters Comparison

| Item | Model A | Model B |
|---|---|---|
| Total parameters | 3.25M | 4.36M |
| Trainable parameters | 0.33M | 0.94M |
| Non-trainable parameters | 2.92M | 3.42M |
| Optimizer parameters (if any) | 0.66M | 0.84M |
| Training time (minutes) | 10 | 12 |

**Table 4.** Model Sizes Comparison

| Item | Model A | Model B |
|---|---|---|
| Keras model size (MB) | 13.57 | 18.08 |
| TFLite model size (MB) | 2.84 | 2.76 |
| Compression ratio (%) | 79.1 | 84.7 |

## 5 Conclusion and Future Work

This study presents an efficient method for tomato leaf disease classification using a custom MobileNetV2 model with dilated convolution and partial fine-tuning. The approach achieves near 99% test accuracy, outperforming the baseline - 98.4% - while remaining lightweight enough for real-time mobile deployment. The proposed architecture offers improved recognition for difficult disease categories and demonstrates robustness suitable for intelligent agriculture. Key limitations include evaluation on the PlantVillage dataset and lack of real-world field and mobile validation. Future work will focus on testing the model with field images and mobile devices, applying more comprehensive statistical validation, and scaling to multicrop detection for broader agricultural use.

**Acknowledgments.** We sincerely appreciate the support of the Posts and Telecommunications Institute of Technology in Ho Chi Minh city, Viet Nam for providing the necessary resources and a conducive research environment that contributed significantly to the success of this study.

**Data Availability.** The dataset used in this study is a publicly available extracted from the PlantVillage dataset. It can be accessed at https://www.kaggle.com/datasets/phihngtrnnguyn/tomatoplantsplit.

**Code Availability.** The source code developed during this study is publicly available on Kaggle at https://www.kaggle.com/code/phihngtrnnguyn/tomato-leaf-classification-mobilenet.

## References

1. Elhassouny, A., Smarandache, F.: Smart mobile application to recognize tomato leaf diseases using convolutional neural networks. In: 2019 International Conference of Computer Science and Renewable Energies (ICCSRE), pp. 1–4 (2019). https://doi.org/10.1109/ICCSRE.2019.8807737
2. Agarwal, M., Singh, A., Arjaria, S., Sinha, A., Gupta, S.: ToLeD: tomato leaf disease detection using convolution neural network. Procedia Comput. Sci. **167**, 293–301 (2020). https://doi.org/10.1016/j.procs.2020.03.225
3. Das, A., Pathan, F., Jim, J.R., Kabir, M.M., Mridha, M.F.: Deep learning-based classification, detection, and segmentation of tomato leaf diseases: a state-of-the-art review. Artif. Intell. Agric. **15**(2), 192–220 (2025). https://doi.org/10.1016/j.aiia.2025.02.006
4. Bhagwat, R., Dandawate, Y.: A review on advances in automated plant disease detection. Int. J. Eng. Technol. Innov. **11**(4), 251 (2021)
5. Adhikari, S., Unit, D., Shrestha, B., Baiju, B.: Tomato plant diseases detection system. In: 1st KEC Conference Proceedings, vol. 1, pp. 81–86 (2018)
6. Howard, A.G., et al.: MobileNets: Efficient Convolutional Neural Networks for Mobile Vision Applications (2017). https://arxiv.org/abs/1704.04861
7. Sandler, M., Howard, A., Zhu, M., Zhmoginov, A., Chen, L.-C.: MobileNetV2: Inverted Residuals and Linear Bottlenecks (2019). https://arxiv.org/abs/1801.04381
8. LearnOpenCV: Convolutional Neural Networks Architecture (2023). https://learnopencv.com/tag/convolutional-neural-network-diagram/
9. Park, S.: MobileNet Architecture (2023). https://suminizz.github.io/mobilenet_v123/

# TinyCDAE: Lightweight Convolutional Denoising Autoencoders for Real-Time Image Denoising on Resource-Constrained IoT Devices

Nguyen Trong Huan(✉), Nguyen Thanh Binh, Tran Quang Thuan, and Nguyen Trong Kien

Posts and Telecommunications Institute of Technology, Ho Chi Minh City, Vietnam
huannt@ptit.edu.v

**Abstract.** Image denoising is a fundamental task in computer vision, serving as a preprocessing step to enhance image quality and improve the performance of subsequent recognition systems. Its importance is particularly evident in Internet of Things (IoT) applications, where sensors frequently capture noisy data. While autoencoder-based denoising models have achieved strong performance, their evaluation on resource-constrained embedded platforms with real on-device measurements remains largely unexplored. This study addresses this gap by presenting one of the first feasibility investigations of deploying an image-denoising autoencoder on a microcontroller unit (MCU). We introduce TinyCDAE, a deployment-oriented lightweight convolutional denoising autoencoder designed for resource-constrained microcontrollers, and demonstrate its deployment on ESP32. The model is first evaluated on a PC environment and compared with a baseline fully connected denoising autoencoder (DAE). The results show that TinyCDAE, with only 373 parameters compared to 52,064 for the baseline DAE, significantly outperforms the baseline in terms of PSNR and SSIM across multiple noise levels. TinyCDAE is then deployed on an ESP32 microcontroller, achieving real-time denoising with an average inference latency of 133.57 ms per image. Beyond demonstrating technical feasibility, our on-device measurements on ESP32 confirm that a compact denoising autoencoder can operate effectively on resource-constrained IoT devices.

**Keywords:** Internet of Things (IoT) · Image denoising · Autoencoder · TinyML

## 1 Introduction

Image denoising is a fundamental problem in image processing, where the objective is to recover a clean image from a noisy observation. Noise can occur at different stages of imaging processing such as acquisition or transmission, which degrades perceptual quality and subsequently reduces the performance of downstream computer vision tasks, including classification and detection [1–3].

Deep learning-based approaches have demonstrated remarkable progress in this domain. In particular, Denoising Autoencoders (DAEs) learn to reconstruct clean images


P. Pichappan et al. (Eds.): RTIS 2025, LNNS 1936, pp. 83–94, 2027.
https://doi.org/10.1007/978-3-032-23396-7_8

by capturing robust feature representations [3, 4], while Convolutional Denoising Autoencoders (CDAEs) extend this idea with convolutional layers to better preserve the local spatial structure [3]. More advanced convolutional models, such as DnCNN and FFDNet, achieve state-of-the-art performance through deeper architectures and refined training strategies [5, 6]. While these methods are highly effective in conventional computing environments, their computational demands mean that deployment on resource-constrained hardware has been far less studied. This limitation becomes significant in the rapidly expanding Internet of Things (IoT) landscape, where billions of devices continuously acquire and process visual data at the edge.

Processing data locally, commonly referred to as edge computing, offers advantages such as reduced latency, enhanced privacy, and lower communication costs [7–10]. However, IoT edge devices such as microcontroller units (MCUs) operate under stringent hardware constraints, with limited on-chip memory, processing capability, and power budget [7, 11, 12]. Consequently, deploying deep CNN-based denoisers on such platforms requires substantial architectural optimization and model compression to achieve practical feasibility while retaining denoising effectiveness.

The emergence of Tiny Machine Learning (TinyML) has opened new opportunities for enabling deep learning inference on ultra-low-power MCUs [7, 8, 10, 11, 13, 14]. Although frameworks such as TensorFlow Lite Micro (TFLM) provide deployment support, real-time image denoising on edge devices demands a delicate balance between model complexity and efficiency while still maintaining acceptable restoration quality.

Despite these advances, adapting CDAE-based denoisers to resource-constrained IoT edge devices remains underexplored, as most studies focus on high-performance platforms rather than microcontroller deployment. Thus, in this paper, we propose a lightweight Convolutional Denoising Autoencoder, namely TinyCDAE, tailored for deployment on the ESP32-S3 microcontroller developed by Espressif Systems. The ESP32-S3 microcontroller, featuring a dual-core processor, limited on-chip memory and integrated Wi-Fi/Bluetooth connectivity, represents a typical resource-constrained IoT platform widely used in low-power applications. To evaluate the feasibility of our approach, the proposed TinyCDAE is trained and tested on the MNIST dataset. We first assess its performance on a PC environment, followed by deployment on the ESP32 platform to investigate inference efficiency.

## 2 Proposed Methodology

### 2.1 Dataset

This study employs the MNIST dataset [15] as the primary benchmark for training and evaluating the proposed model. MNIST comprises 70,000 grayscale images of handwritten digits (0–9), each with a resolution of $28 \times 28$ pixels, providing a compact dataset widely used in computer vision research. The dataset is partitioned into 60,000 images for training and 10,000 images for testing. Prior to training, all pixel values are normalized to the range [0,1] by dividing by 255.0, a standard preprocessing step to improve neural network convergence.

To simulate real-world image degradation, Additive White Gaussian Noise (AWGN) is synthetically added to the clean images, following common practices in image denoising literature [1]. The noise is sampled from a zero-mean normal distribution and scaled using a noise factor of 0.3 and added to the normalized pixel values. Afterward, all pixel intensities are clipped to the valid range [0,1] to produce the noisy training and testing sets.

## 2.2 TinyCDAE Architecture

In this section, we detail the architecture of TinyCDAE, our proposed lightweight Convolutional Denoising Autoencoder designed for microcontroller-based TinyML platforms. The model follows a symmetric encoder-bottleneck-decoder architecture, which has been widely adopted for image restoration tasks due to its ability to learn compact latent representations and reconstruct clean images efficiently [3, 4], as detailed in Table 1. The model is designed to process noisy grayscale images of size 28 × 28 pixels, consistent with the MNIST dataset, and reconstruct a clean version of the image at the same resolution while being suitable for deployment on MCUs.

**Table 1.** Architecture details of the TinyCDAE model.

| Module | No | Layer | Activation | Output shape | Parameters |
|---|---|---|---|---|---|
| Encoder | 1 | Input | – | (28, 28, 1) | 0 |
| | 2 | Conv2D | ReLU | (14, 14, 4) | 40 |
| | 3 | Conv2D | ReLU | (7, 7, 4) | 148 |
| Decoder | 4 | Conv2DTranspose | ReLU | (14, 14, 4) | 148 |
| | 5 | Conv2DTranspose | Sigmoid | (28, 28, 1) | 37 |
| Total | | | | | 373 |

The encoder compresses the input progressively using two convolutional layers, each configured with four filters, a 3 × 3 kernel, ReLU activation, and 'same' padding. Instead of employing separate pooling layers, the encoder uses a stride of (2,2) in each convolution to reduce the spatial resolution while maintaining local spatial correlations. After these two strided convolutions, the input is encoded into a 7 × 7 × 4 latent representation, referred to as the bottleneck, which captures the most salient features of the noisy image while significantly reducing its dimensionality.

The decoder mirrors the encoder structure to reconstruct the image. It employs two Conv2DTranspose layers: the first upsamples the latent features from 7 × 7 × 4 to 14 × 14 × 4, and the second restores the full spatial resolution to 28 × 28 pixels. A final Conv2D layer with a 3 × 3 kernel, 'same' padding, and a sigmoid activation function produces the denoised image with normalized pixel intensities in the range [0,1]. By exclusively using convolutional and transposed convolutional layers with small filter sizes and a minimal number of parameters (373 in total, approximately 1.46 KB), the architecture preserves spatial structure while remaining computationally lightweight.

This architecture is motivated by the need to balance denoising effectiveness and hardware feasibility. Using strided convolutions instead of separate pooling layers reduces intermediate feature storage and avoids additional operations, which benefits platforms with limited SRAM capacity. Similarly, the low number of filters and small kernel size are selected to limit the parameter count and execution time without overly compromising reconstruction quality. These design principles align with the constraints of the ESP32 microcontroller, which typically provides only a few hundred kilobytes of SRAM and operates without hardware acceleration for floating-point operations.

Finally, the TinyCDAE model is implemented and trained using TensorFlow on a MacBook M1 Pro with 16 GB unified memory, with conversion to TensorFlow Lite Micro format for deployment on the ESP32. This workflow allows end-to-end evaluation of both the denoising performance on MNIST test images and the execution metrics on the microcontroller, including inference latency, memory usage, and energy consumption. Through this experimental setup, the study aims to provide practical insights into the viability of deploying convolutional denoising autoencoders on constrained IoT devices, paving the way for more capable on-device image enhancement applications in the TinyML domain.

## 2.3 TinyML Deployment Framework

The TinyCDAE model is deployed using the Arduino development environment for ease of prototyping and compatibility with microcontroller platforms. The target hardware is the ESP32-S3 microcontroller, selected for its computational balance and low power profile in IoT applications. Although development boards often include external PSRAM, all inference speed evaluations in this study are conducted using only the internal RAM to reflect true on-chip performance.

Deployment is implemented in the Arduino IDE with the TensorFlow Lite for Microcontrollers library [11]. The trained model, exported in.*tflite* format, is converted into a C array and included as a header file in the Arduino sketch. The C/C++ sketch initializes the TFLM interpreter, allocates a static tensor arena in internal RAM, loads the input data, runs inference, and processes the denoised output. Standard Arduino libraries handle peripheral interactions, and TFLM's static memory planning ensures predictable execution within the microcontroller's constrained resources.

## 2.4 Evaluation Metrics

To comprehensively assess the proposed TinyCDAE model, both its denoising quality and on-device performance are quantitatively evaluated. The capability of the model to remove noise while preserving image details is measured using two widely adopted image restoration metrics [1, 16]:

**Peak Signal-to-Noise Ratio (PSNR)**: PSNR quantifies the ratio between the maximum possible signal power (original clean image) and the power of the residual noise after denoising, as defined in Eq. (1):

$$PSNR = 10 \cdot \log_{10}\left(\frac{MAX_I^2}{MSE}\right) \tag{1}$$

where $MAX_I$ is the maximum possible pixel value (255 for 8-bit images or 1.0 after normalization in this study), and m, n are the image dimensions (28 × 28 pixels for MNIST). Higher PSNR values indicate better reconstruction fidelity. It is computed from the Mean Squared Error (MSE) between the clean image I and the denoised image K. The MSE is given in Eq. (2):

$$MSE = \frac{1}{m \cdot n}\sum\nolimits_{i=0}^{m-1}\sum\nolimits_{j=0}^{n-1}\left[I(i,j) - K(i,j)\right]^2 \tag{2}$$

**Structural Similarity Index Measure (SSIM)**: SSIM evaluates the perceptual similarity between the clean and denoised images, denoted by x and y, respectively, producing values in the range [−1, 1], where 1 indicates perfect structural similarity, as given by Eq. (3):

$$SSIM(x, y) = \frac{(2\mu_x\mu_y + c_1)(2\sigma_{xy} + c_2)}{\left(\mu_x^2 + \mu_y^2 + c_1\right)\left(\sigma_x^2 + \sigma_y^2 + c_2\right)} \tag{3}$$

where $\mu_x$, $\mu_y$ are local means; $\sigma_x^2$, $\sigma_y^2$ are local variances; $\sigma_{xy}$ is the cross-covariance; and $c_1$, $c_2$ are stabilization constants derived from the pixel range.

## 3 Results and Discussion

### 3.1 Denoising Image Performance

To evaluate the denoising performance, we first establish a baseline using a fully connected denoising autoencoder (DAE). The baseline model takes a noisy MNIST image flattened into a 784-dimensional vector. The encoder compresses the input through dense layers of 32 and 16 neurons, while the decoder mirrors this structure to reconstruct the denoised image. In total, the baseline model comprises 52,064 trainable parameters, which is substantially larger than the 373 parameters required by our proposed TinyCDAE model. We then compare the two models in terms of reconstruction quality to assess the effectiveness of TinyCDAE.

Figure 1 illustrates the qualitative comparison of both models across different Gaussian noise levels ($\sigma = 0.1$–$0.5$). While both approaches succeed in suppressing noise to some extent, TinyCDAE preserves sharper digit contours and clearer structural details, particularly under moderate noise levels ($\sigma \leq 0.3$). In contrast, DAE outputs tend to exhibit excessive blurring and partial loss of fine features, which becomes more pronounced as noise intensity increases.

To complement these visual findings, Fig. 2 reports the quantitative evaluation in terms of PSNR and SSIM for both models across the same range of noise levels, with the corresponding numerical values summarized in Table 2. Although both models were trained exclusively at $\sigma = 0.3$, the TinyCDAE consistently outperforms the DAE across all unseen noise intensities, achieving a substantial PSNR improvement of approximately 5–8 dB and markedly higher SSIM scores, particularly at lower noise levels ($\sigma \leq 0.3$).

This demonstrates the TinyCDAE's strong generalization ability, enabling effective denoising beyond the specific training condition. Such robustness is consistent with

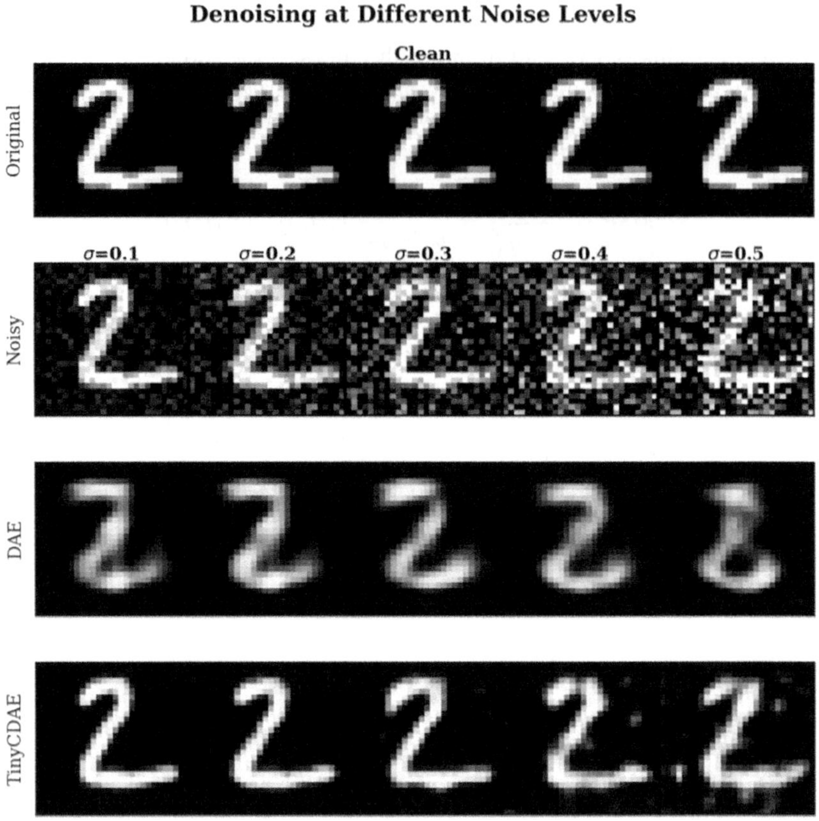


**Fig. 1.** Visual comparison of denoising results for fully connected DAE and proposed TinyCDAE under Gaussian noise (σ = 0.1–0.5). TinyCDAE preserves sharper edges and finer details, while DAE outputs appear more blurred as noise increases.

prior findings on convolutional architectures for image denoising, notably DnCNN [5], which has shown that convolutional feature extraction can yield noise-level-invariant representations and support blind denoising tasks without retraining. In contrast, the DAE exhibits relatively flat performance and is more sensitive to deviations from the training noise level, confirming the trends observed in both the qualitative and graphical analyses.

**Table 2.** Denoising performance metrics.

| | | σ = 0.1 | σ = 0.2 | σ = 0.3 | σ = 0.4 | σ = 0.5 |
|---|---|---|---|---|---|---|
| PSNR | DAE | 16.74 ± 1.95 | 17.26 ± 2.2117.26 ± 2.21 | 17.20 ± 2.24 | 16.43 ± 1.83 | 15.41 ± 1.44 |
| | TinyCDAE | 24.28 ± 1.86 | 23.14 ± 1.74 | 21.41 ± 1.59 | 19.28 ± 1.38 | 17.12 ± 1.15 |
| SSIM | DAE | 0.73 ± 0.09 | 0.75 ± 0.09 | 0.74 ± 0.09 | 0.70 ± 0.10 | 0.64 ± 0.10 |
| | TinyCDAE | 0.95 ± 0.02 | 0.93 ± 0.02 | 0.86 ± 0.04 | 0.74 ± 0.08 | 0.63 ± 0.10 |

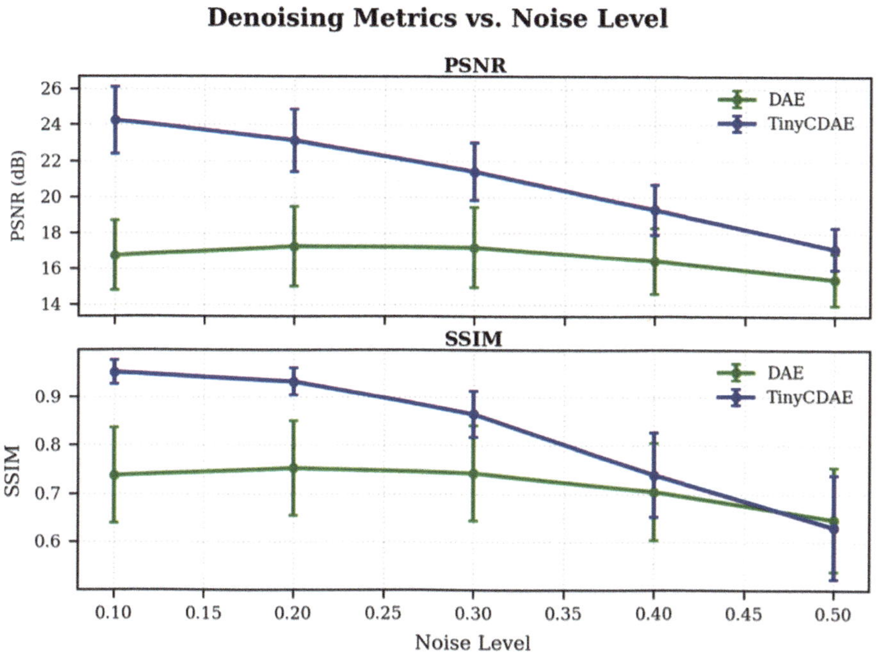


**Fig. 2.** PSNR (top) and SSIM (bottom) of DAE and TinyCDAE across noise levels. Both trained at $\sigma = 0.3$; TinyCDAE maintains superior performance on all unseen noise levels, indicating stronger generalization.

### 3.2 TinyML Performance Analysis

To assess the feasibility of deploying the proposed TinyCDAE model in resource-constrained environments, we evaluated its performance after conversion to the TensorFlow Lite (TFLite) framework, a widely adopted inference engine for TinyML applications.

This analysis focuses on quantifying any potential degradation in denoising quality following model conversion, using PSNR and SSIM as primary evaluation metrics. By comparing the original TensorFlow implementation with its TFLite equivalent under varying noise levels, we aim to determine whether the benefits of lightweight deployment can be achieved without sacrificing reconstruction accuracy. Figure 3 compares the denoising performance of the TinyCDAE model implemented in TensorFlow with its TFLite counterpart after conversion using the float32 precision format. The results indicate that the TFLite model achieves PSNR and SSIM values nearly identical to those of the original TensorFlow implementation across all tested noise levels ($\sigma = 0.1$–$0.5$). This demonstrates that the float32 TFLite conversion preserves the denoising quality of the TinyCDAE model while enabling deployment on resource-constrained devices, consistent with prior reports that high-precision TFLite models maintain close fidelity to their full TensorFlow versions.

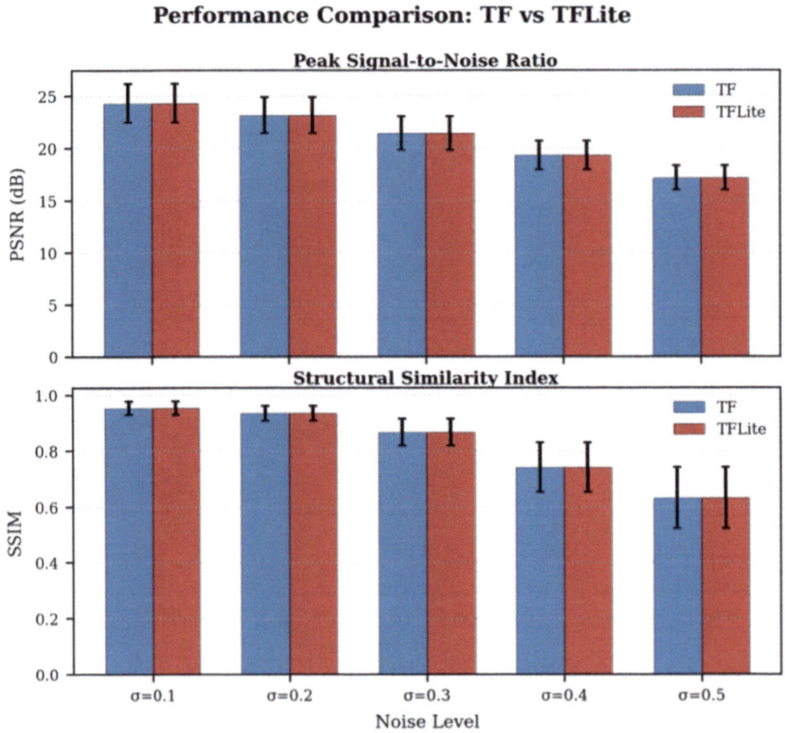


**Fig. 3.** PSNR (top) and SSIM (bottom) comparison between TensorFlow of TinyCDAE and its TFLite float32 version. Nearly identical results show minimal quality loss after conversion.

Following the TFLite evaluation on PC, the model was further deployed to an ESP32 microcontroller by converting the.*tflite* file into a C header file and integrating it into the firmware. A custom graphical user interface (GUI) was developed using the PyQt5 library to facilitate user interaction, serial communication, and real-time denoising visualization via UART at a baud rate of 115200. As shown in Fig. 4, the ESP32 inference results demonstrate that the denoised images closely match the clean references, with only minimal differences visible in the difference maps, confirming the model's effectiveness under embedded deployment.

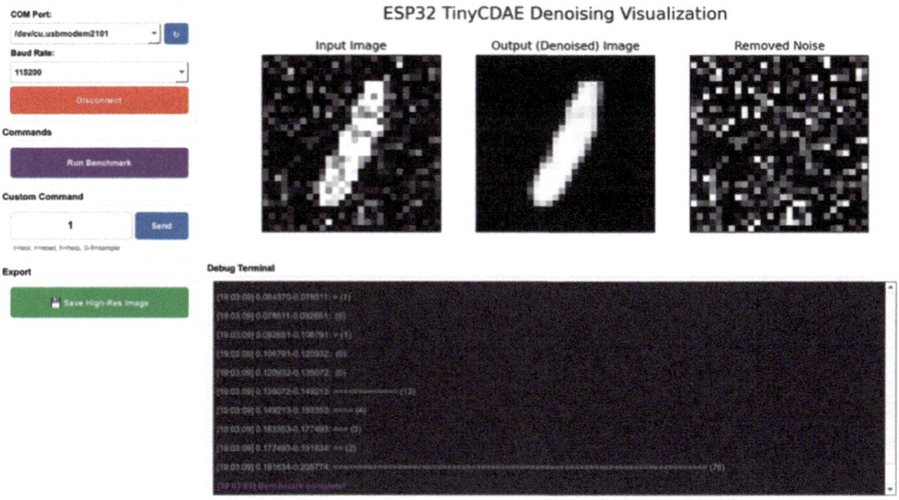


**Fig. 4.** Custom PyQt5-based GUI for controlling and visualizing the ESP32 CNN-AE deployment via UART at 115200 baud, along with benchmark results for 100 inference runs. The denoised outputs closely match the clean references with minimal residual differences.

To assess computational performance, we conducted 100 inference runs on the ESP32, measuring a mean inference time of $133.57 \pm 0.09$ ms per image. Figure 5 compares the average inference times across different deployment environments: TensorFlow on PC (22.96 ms), TFLite on PC (0.02 ms), and TFLite on ESP32 (133.57 ms). While the ESP32 inference speed is slower due to limited processing resources, the achieved performance is still well within the real-time threshold for low-frame-rate image denoising applications, confirming the practical feasibility of deploying the TinyCDAE model on low-power IoT hardware.

A comparative evaluation with a reference denoising method from literature is presented in Table 3. Despite using significantly fewer parameters, the proposed TinyCDAE maintains competitive PSNR performance and uniquely supports deployment on low-power TinyML hardware, highlighting its efficiency and suitability for real-time embedded applications.

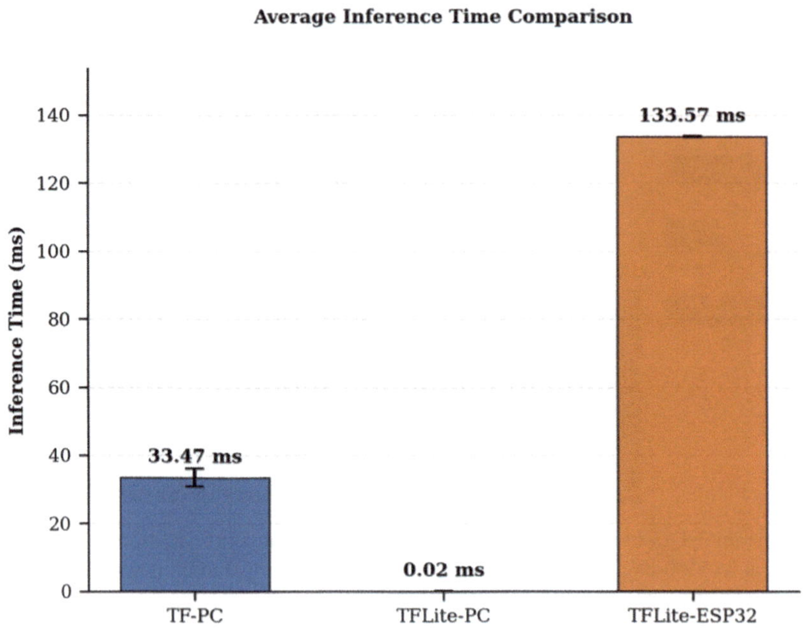


**Fig. 5.** Average inference time comparison across deployment platforms, showing processing time per denoising operation for TensorFlow on PC, TFLite on PC, and TFLite on ESP32.

**Table 3.** Comparative results between the proposed TinyCDAE and a baseline denoising method from literature.

| Method | Dataset | Noise Level (σ) | Parameters | PSNR (dB) | SSIM | TinyML Deployment |
|---|---|---|---|---|---|---|
| [17] | MNIST | 0.1 | 2500–6400 | <20 | – | No |
| [16] | MNIST | 0.1 | – | 19.15 | 0.9047 | No |
| **Proposed method (TinyCDAE)** | MNIST | 0.1 | 373 | 24.27 | 0.95 | Yes |

## 4 Conclusion

This study presented the design, evaluation, and deployment of a lightweight TinyCDAE model for image denoising, with a focus on TinyML applications. Comparative experiments with a baseline fully connected Denoising Autoencoder demonstrated that the TinyCDAE achieves superior qualitative and quantitative performance, preserving finer structural details and achieving substantial improvements in PSNR and SSIM across a wide range of Gaussian noise levels. Notably, the TinyCDAE maintained high denoising accuracy even for noise intensities unseen during training, confirming its strong generalization capability.

The TensorFlow Lite conversion using float32 precision yielded negligible performance degradation compared to the original TensorFlow model, validating the suitability of this deployment format for resource-limited environments. Embedded deployment on an ESP32 microcontroller, integrated with a custom PyQt5-based GUI and UART communication, further confirmed the model's practicality. The ESP32 achieved a mean inference time of approximately 133.57 $\pm$ 0.09 ms per image, enabling real-time processing in low-frame-rate applications while maintaining visual quality comparable to the original model outputs.

Overall, the proposed TinyCDAE framework demonstrates a balance between denoising performance, computational efficiency, and deployment feasibility on ultra-low-power devices. These results highlight its potential for real-world TinyML applications in edge AI systems, particularly in scenarios where bandwidth, power, and latency constraints limit the use of cloud-based processing. Future work may explore model quantization, mixed-precision inference, and hardware acceleration to further reduce inference time without sacrificing denoising quality.

**Acknowledgment.** This research is funded by Posts and Telecommunications Institute of Technology under grant number 03-2025-HV-KTĐT2.

## References

1. Shivarama Holla, K., Park, N., Lee, B.: EFID: edge-focused image denoising using a convolutional neural network. IEEE Access **11**, 9613–9626 (2023)
2. Majeed, S., Mansoor, Y., Qabil, S., Majeed, F., Khan, B., Pakistan, K.: Comparative analysis of the denoising effect of unstructured vs. convolutional autoencoders. In: 2020 International Conference on Emerging Trends in Smart Technologies (ICETST). pp. 1–5. IEEE, Karachi, Pakistan (2020)
3. Roy, S.S., Ahmed, M., Aminul, M., Akhand, H.: Noisy image classification using hybrid deep learning methods. J. Inf. Commun. Technol. **17**, 233–269 (2018)
4. Vincent, P., Larochelle, H., Bengio, Y., Manzagol, P.-A.: Extracting and composing robust features with denoising autoencoders. In: Proceedings of the 25th International Conference on Machine Learning. pp. 1096–1103. , Helsinki, Finland (2008)
5. Zhang, K., Zuo, W., Chen, Y., Meng, D., Zhang, L.: Beyond a Gaussian denoiser: residual learning of deep CNN for image denoising. IEEE Trans. Image Process. **26**, 3142–3155 (2017)
6. Zhang, K., Zuo, W., Zhang, L.: FFDNet: toward a fast and flexible solution for CNN-based image denoising. IEEE Trans. Image Process. **27**, 4608–4622 (2018)
7. Alajlan, N.N., Ibrahim, D.M.: TinyML: Enabling of inference deep learning models on ultra-low-power IoT edge devices for AI applications. Micromachines (Basel). 13, (2022)
8. Ray, P.P.: A review on TinyML: state-of-the-art and prospects. J. King Saud Univ. Comput. Inf. Sci. **34**, 1595–1623 (2022)
9. Sadiq, S., Hare, J., Craske, S., Maji, P., Merrett, G.: Enabling ImageNet-scale deep learning on MCUs for accurate and efficient inference. IEEE Internet Things J. **11**, 11471–11479 (2024)
10. Sharma, R., Romaszkan, W., Zhu, F., Gupta, P.: Cost-driven hardware-software co-optimization of machine learning pipelines. In: 2024 Association for Computing Machinery. pp. 1–22 (2024)
11. David, R., et al.: TensorFlow Lite Micro: embedded machine learning on TinyML systems. In: Proceedings of the 4th MLSys Conference. pp. 800–811. , San Jose, CA, USA (2021)

12. Warden, P., Situnayake, D.: TinyML Machine Learning with TensorFlow Lite on Arduino and Ultra-Low-Power Microcontrollers. O'Reilly Media, CA (2019)
13. Abadi, M., et al.: TensorFlow: A system for large-scale machine learning. In: Proceedings of the 12th USENIX Symposium on Operating Systems Design and Implementation (OSDI'16). pp. 265–283. , Savannah, GA, USA (2016)
14. Krishnamoorthi, R.: Quantizing deep convolutional networks for efficient inference: A whitepaper. ArXiv vol. abs/1806.08342. (2018)
15. Lecun, Y., Eon Bottou, L., Bengio, Y., Abstractl, P.H.: Gradient-based learning applied to document recognition. In: Proceedings of the IEEE. pp. 2278–2324. , Québec, Canada (1998)
16. Appalabatla, S.P.S., Murali Krishna, T.: Performance analogy of autoencoders for image denoising. In: 2023 3rd Asian Conference on Innovation in Technology. pp. 1–4. Institute of Electrical and Electronics Engineers Inc., India (2023)
17. Jóźwik-Wabik, P., Bernacki, K., Popowicz, A.: Comparison of training strategies for autoencoder-based monochromatic image denoising. Sensors **23**, 5538 (2023)
18. Vincent, P., Larochelle, H., Lajoie, I., Bengio, Y., Manzagol, P.-A.: Stacked denoising autoencoders: learning useful representations in a deep network with a local denoising criterion. J. Mach. Learn. Res. **11**, 3371–3408 (2010)
19. Hinton, G.E., Osindero, S.: A fast learning algorithm for deep belief nets yee-whye teh. Neural Comput. **18**, 1527–1554 (2006)
20. Han, S., Pool, J., Tran, J., Dally, W.J.: Learning both weights and connections for efficient neural networks. Adv. Neural. Inf. Process. Syst. **28**, 1135–1143 (2015)
21. Bengio, Y., Lamblin, P., Popovici, D., Larochelle, H.: Greedy layer-wise training of deep networks. In: Advances in Neural Information Processing Systems. pp. 153–160. MIT Press, Canada (2006)
22. Bengio, Y.: Learning deep architectures for AI. Found. Trends Mach. Learn. **2**, 1–27 (2009)
23. Masci, J., Meier, U., Cire¸san, D.C., Schmidhuber, J.: Stacked Convolutional Auto-Encoders for Hierarchical Feature Extraction. In: Proceedings of the 21st International Conference on Artificial Neural Networks. pp. 52–59. Springer (2011)

# Performance Analysis of Hybrid Power-Frequency Multiple Access for 6G IoT over Keyhole Channels

Mui Van Nguyen[1], Sang Quang Nguyen[2], Cach Nhan Dang[3], and Phu-Nguyen Le[4(✉)]

[1] Faculty of Information Technology, Gia Dinh University, Ho Chi Minh City, Vietnam
[2] Wireless Communication Laboratory, Posts and Telecommunications Institute of Technology, Ho Chi Minh City, Vietnam
[3] Ho Chi Minh City University of Transport, Ho Chi Minh City, Vietnam
[4] Faculty of Engineering and Technology, Nguyen Tat Thanh University, Ho Chi Minh City, Vietnam
lpnguyen@ntt.edu.vn

**Abstract.** This paper analyzes the performance of hybrid power–frequency multiple access (HPMA) over Nakagami-$m$ keyhole fading channels. Unlike conventional NOMA, which relies on SIC and increases receiver complexity, HPMA combines OMA subcarriers with a shared NOMA-style subcarrier, enabling coherent combining without SIC. The keyhole channel is modeled as a product of two independent Nakagami-$m$ links, capturing rank-deficient propagation. We derive closed-form outage probability (OP) and ergodic capacity (EC) expressions using Meijer-$G$ functions, along with high-SNR approximations revealing the diversity order. Delay-limited throughput is also obtained. Monte Carlo simulations validate the analysis and show that HPMA achieves lower OP, more robust power-allocation behavior, lower transmit-power requirements, and higher EC than NOMA. These results highlight HPMA as a practical, low-complexity, and spectrum-efficient multiple access scheme for 6G keyhole environments.

**Keywords:** Hybrid power-frequency multiple access · keyhole channel · SIC-free · outage probability · Nakagami-$m$ fading · ergodic capacity · throughput · 6G Internet of Things (IoT)

## 1 Introduction

Sixth-generation (6G) networks must support massive device connectivity together with ultra-high data rates and low latency, placing stringent demands on radio access design [1,2]. Conventional orthogonal multiple access (OMA) schemes (e.g., FDMA/OFDMA, TDMA) provide reliable user separation but suffer from limited spectral efficiency and poor scalability under dense deployments [3,4].


P. Pichappan et al. (Eds.): RTIS 2025, LNNS 1936, pp. 95–106, 2027.
https://doi.org/10.1007/978-3-032-23396-7_9

Non-orthogonal multiple access (NOMA) addresses these limitations by allowing multiple users to share the same time-frequency resources, distinguished by power levels or unique codes [2,5]. Using power-domain multiplexing and successive interference cancellation (SIC), NOMA can improve throughput and access density [2,4], and has shown advantages in mmWave massive MIMO [6] and UAV-assisted networks [7–9]. However, its gains come with increased receiver complexity, error propagation from imperfect SIC, and potential fairness degradation [3,5].

To balance the trade-offs between OMA and NOMA, hybrid multiple access (HMA) schemes have been proposed. One promising approach is the *hybrid power–frequency multiple access* (HPMA) scheme [10], which partitions users into orthogonal frequency sub-bands and superposes a limited number on each sub-band via power-domain multiplexing. This structure reduces SIC complexity, mitigates interference, and enables flexible resource allocation. Users with low channel correlation can be multiplexed on the same frequency, while others are assigned to separate sub-bands. Analytical and simulation studies [10] have shown that HPMA can enhance capacity and energy efficiency, while related work on RIS-aided NOMA [11] demonstrates improved reliability in advanced RIS-enabled systems without requiring full SIC. The versatility of hybrid multiple access has also been highlighted for 6G applications such as massive IoT and mobile edge computing [12].

While HPMA has shown promise in conventional fading environments, its performance under complex propagation conditions remains underexplored. A notable example is the *keyhole* (or "pinhole") MIMO channel, a rank-deficient fading channel caused by a severe spatial bottleneck [13,14]. Even with rich scattering, the channel matrix can collapse to rank-one due to a narrow waveguide or single dominant scatterer [15], causing all transmitted signals to share a common attenuation and reducing spatial multiplexing capability [13,16]. Measurements confirm keyhole effects in tunnels, mine shafts, dense corridors [17], and vehicular links in urban canyons [18], where signals are forced through a single or highly correlated path. Such channels challenge MA design since user signals become highly correlated, undermining interference management and multiplexing strategies that rely on channel diversity.

Recent works have examined NOMA and related MA schemes in cascaded or double-fading channels, which share similarities with keyhole scenarios. Examples include two-hop relaying and RIS-assisted networks [19–24], where the end-to-end channel is modeled as the product of fading coefficients, often yielding double-Rayleigh or double-Nakagami distributions. These studies show that cascaded fading can severely degrade outage and capacity performance, necessitating new analysis and design strategies. For instance, [19,22] analyzed cooperative short-packet NOMA over double-fading and observed higher decoding error probabilities with larger fading variance, while [20,23] reported worsened secrecy outage under double Nakagami-$m$ fading unless power allocation and beamforming are optimized. The keyhole channel is a degenerate form of cascaded fading with a single dominant path, leading to even greater performance bottlenecks.

Very recently, [15] investigated secrecy outage in downlink NOMA over Rayleigh keyhole channels, confirming the detrimental effect of keyhole fading.

**Contributions:** This paper presents the first analytical study of a hybrid power-frequency multiple access (HPMA) system over Nakagami-$m$ keyhole fading channels. The main contributions are: (1) Formulation of a downlink HPMA model in a keyhole MIMO channel, derivation of the PDF and CDF of the equivalent double Nakagami-$m$ cascaded gains. (2) Exact and high-SNR expressions for outage probability and ergodic capacity of both strong and weak users, revealing diversity order and coding gain. (3) Throughput analysis under a delay-limited transmission model. (4) Monte Carlo simulations validating the analysis and illustrating the impact of fading parameters, power allocation, and bandwidth partitioning on performance.

**Paper Organization:** Sect. 2 describes the system model. Section 3 analyzes the outage probability. Section 4 presents the throughput analysis, and Sect. 5 covers the ergodic capacity. Section 6 provides numerical results and discussions. Section 7 concludes the paper.

## 2 System Model

### 2.1 System Architecture

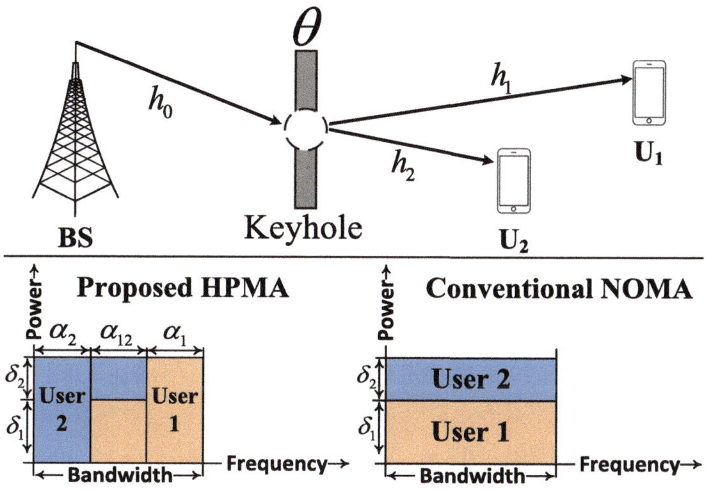


**Fig. 1.** System model of the proposed HPMA framework over a keyhole channel.

We consider a downlink system (Fig. 1) where a single-antenna BS serves two single-antenna users $\mathrm{U}_i$ ($i \in \{1, 2\}$) through a keyhole channel characterized by $\theta$. To balance OMA's interference avoidance and NOMA's spectrum reuse while keeping the receivers SIC-free, a hybrid power–frequency multiple access (HPMA) scheme is adopted.

Two OMA subcarriers $s_i$ transmit symbols $x_i$ over bandwidth fractions $\alpha_i$, while a shared subcarrier $s_{12}$ carries the superposed signal $\sqrt{\delta_1}x_1 + \sqrt{\delta_2}x_2$ with $\alpha_{12} = 1 - \alpha_1 - \alpha_2$ and $\delta_1 + \delta_2 = 1$. Each user coherently combines signals from $s_i$ and $s_{12}$ to decode $x_i$ without SIC. When $\alpha_{12} = 0$, HPMA reduces to conventional OMA.

### 2.2 Keyhole Channel Model

All wireless links are assumed to experience independent Nakagami-$m$ fading. Owing to the keyhole effect, the end-to-end channel between the BS and user $\mathrm{U}_i$ is modeled as the product $h_0h_i$, where $h_0$ represents the BS–keyhole fading and $h_i$ the keyhole–user fading. The magnitudes follow independent Nakagami-$m$ distributions: $|h_0| \sim \text{Nakagami}(m_0, \Omega_0)$ and $|h_i| \sim \text{Nakagami}(m_i, \Omega_i)$. Here, $m_a$ and $\Omega_a$ denote the fading severity and average power. This cascaded model characterizes the rank-deficient nature of keyhole propagation, where spatial diversity is inherently limited.

### 2.3 Received Signal and SINR

The received signal at user $\mathrm{U}_i$ is

$$y_{\mathrm{U}_i} = h_0\theta h_i\Big(\sqrt{P_S\alpha_i\Lambda_0\Lambda_i}\,x_i + \sqrt{P_S\alpha_{12}\Lambda_0\Lambda_i}\,x_{12}\Big) + n_{\mathrm{U}_i}, \tag{1}$$

where $\Lambda_0 = d_0^{-\zeta}$ and $\Lambda_i = d_i^{-\zeta}$ denote the large-scale path-loss, $P_S$ is the BS transmit power, and $n_{\mathrm{U}_i} \sim \mathcal{CN}(0, (\alpha_i + \alpha_{12})\sigma^2_{\mathrm{U}_i})$ is AWGN. The shared signal is $x_{12} = \sqrt{\delta_1}x_1 + \sqrt{\delta_2}x_2$.

The resulting SINR for detecting $x_i$ is

$$\Gamma_{\mathrm{U}_i} = \frac{\theta^2(\alpha_i + \alpha_{12}\delta_i)\Sigma_S\Upsilon_i\Lambda_0\Lambda_i|h_0|^2|h_i|^2}{\theta^2\alpha_{12}\delta_j\Sigma_S\Upsilon_i\Lambda_0\Lambda_i|h_0|^2|h_i|^2 + 1}, \tag{2}$$

where $\delta_j = 1 - \delta_i$, $\Sigma_S = P_S/\sigma^2_{\mathrm{U}_i}$, and $\Upsilon_i = 1/(\alpha_i + \alpha_{12})$.

Assuming normalized symbols, the achievable rate is

$$\mathcal{R}_{\mathrm{U}_i} = \log_2(1 + \Gamma_{\mathrm{U}_i}). \tag{3}$$

### 2.4 Channel Distributions

All fading coefficients $h_a$ ($a \in \{0, 1, 2\}$) are assumed independent Nakagami-$m$. Thus, $|h_a|^2$ follows a Gamma distribution with

$$f_{|h_a|^2}(x) = \frac{x^{m_a-1}}{\Gamma(m_a)\Omega_a^{m_a}}e^{-x/\Omega_a}, \tag{4}$$

$$F_{|h_a|^2}(x) = 1 - \frac{1}{\Gamma(m_a)}\Gamma\Big(m_a, \tfrac{x}{\Omega_a}\Big), \tag{5}$$

where $m_a$ is the fading parameter and $\Omega_a = \lambda_a/m_a$ denotes the average power.

For a keyhole channel, the effective gain of user $\mathrm{U}_i$ is the product of two independent Nakagami-$m$ links:

$$\Xi_i = |h_0|^2|h_i|^2.$$

Its PDF and CDF follow the double Nakagami-$m$ distribution [25]:

$$f_{\Xi_i}(x) = \frac{1}{x\,\Gamma(m_0)\Gamma(m_i)} G_{0,2}^{2,0}\left( x\xi_i \,\middle|\, \begin{matrix} - \\ m_0,\, m_i \end{matrix} \right), \tag{6}$$

$$F_{\Xi_i}(x) = \frac{1}{\Gamma(m_0)\Gamma(m_i)} G_{1,3}^{2,1}\left( x\xi_i \,\middle|\, \begin{matrix} - \\ m_0,\, m_i,\, 0 \end{matrix} \right), \tag{7}$$

where $\xi_i = (\Omega_0\Omega_i)^{-1}$. These Meijer-$G$ based expressions characterize the multiplicative fading inherent to keyhole channels, which increases the likelihood of deep fades and limits the achievable diversity.

## 3 Analysis of Outage Probability

Let $R_i$ denote the target transmission rate for user $\mathrm{U}_i$ in bps/Hz. An outage occurs when the achievable rate falls below this threshold. Accordingly, the outage probability (OP) for $\mathrm{U}_i$ is

$$\mathcal{P}_{\mathrm{U}_i} = \Pr(\mathcal{R}_{\mathrm{U}_i} < R_i) = \Pr\Big(\Gamma_{\mathrm{U}_i} < 2^{R_i} - 1 \triangleq \Gamma_{\mathrm{th}}\Big), \tag{8}$$

where $\Gamma_{\mathrm{th}} = 2^{R_i} - 1$ is the SINR threshold corresponding to $R_i$.

Substituting (2) into (8), the OP for $\mathrm{U}_i$ becomes

$$\begin{aligned} \mathcal{P}_{\mathrm{U}_i} &= \Pr\left( \frac{\theta^2(\alpha_i + \alpha_{12}\delta_i)\,\Sigma_S\Upsilon_i\Lambda_0\Lambda_i\,\Xi_i}{\theta^2\alpha_{12}\delta_j\,\Sigma_S\Upsilon_i\Lambda_0\Lambda_i\,\Xi_i + 1} < \Gamma_{\mathrm{th}} \right) \\ &= \Pr\left( \Xi_i < \frac{\Gamma_{\mathrm{th}}}{\theta^2\Sigma_S\Upsilon_i\Lambda_0\Lambda_i\left[(\alpha_i + \alpha_{12}\delta_i) - \Gamma_{\mathrm{th}}\alpha_{12}\delta_j\right]} \right) = F_{\Xi_i}(\Lambda_i^*), \end{aligned} \tag{9}$$

where we define

$$\Lambda_i^* = \frac{\Gamma_{\mathrm{th}}}{\theta^2\Sigma_S\Upsilon_i\Lambda_0\Lambda_i\left[(\alpha_i + \alpha_{12}\delta_i) - \Gamma_{\mathrm{th}}\alpha_{12}\delta_j\right]},$$

with the condition $(\alpha_i + \alpha_{12}\delta_i) - \Gamma_{\mathrm{th}}\alpha_{12}\delta_j > 0$.

Based on (7), the closed-form OP is

$$\mathcal{P}_{\mathrm{U}_i} = \frac{1}{\Gamma(m_0)\Gamma(m_i)} G_{1,3}^{2,1}\left( \Lambda_i^*\xi_i \,\middle|\, \begin{matrix} - \\ m_0,\, m_i,\, 0 \end{matrix} \right), \tag{10}$$

where $\xi_i = (\Omega_0\Omega_i)^{-1}$.

*High-SNR Region:* While (10) is exact, it offers limited intuition. In the high-SNR regime ($P_S \to \infty$, $\Sigma_S \to \infty$), the Meijer-$G$ function can be asymptotically approximated [27]:

$$\mathcal{P}_{\mathrm{U}_i} \approx \psi_i \int_0^{\sqrt{4\Lambda_i^* \xi_i}} t^{2m_0-1} e^{-t} dt, \tag{11}$$

where $|m_0 - m_i| = 0.5$, and

$$\psi_i = \frac{2^{1-2m_0}\sqrt{\pi}}{\Gamma(m_0)\Gamma(m_i)}.$$

Using the lower incomplete Gamma function $\gamma(a,z) = \int_0^z t^{a-1}e^{-t}dt$ [28, Eq. (8.350.1)], the OP simplifies to

$$\mathcal{P}_{\mathrm{U}_i} \approx \psi_i \, \gamma\left(2m_0, \sqrt{4\Lambda_i^* \xi_i}\right). \tag{12}$$

For small $z \to 0$, apply $\gamma(a,z) \approx z^a/a$ to obtain

$$\mathcal{P}_{\mathrm{U}_i}^{\infty} \approx \frac{\psi_i \left(4\Lambda_i^* \xi_i\right)^{m_0}}{2m_0}. \tag{13}$$

Finally, the diversity order is

$$\mathcal{D}_{\mathrm{U}_i} = -\lim_{\Sigma_S \to \infty} \frac{\log \mathcal{P}_{\mathrm{U}_i}^{\infty}}{\log \Sigma_S} = m_0, \tag{14}$$

showing that the diversity order is governed by $m_0$, as validated in Sect. 3.

## 4 Throughput Analysis

Based on the outage probabilities derived earlier, we now analyze the system throughput in a delay-limited transmission scenario, where each user operates at a fixed rate and successful decoding occurs only when the instantaneous SINR exceeds the required threshold. Hence, the achievable throughput is directly determined by the outage probability.

For user $\mathrm{U}_i$, the average throughput is defined as the fixed transmission rate multiplied by the probability of successful decoding. Thus, the throughput is given by

$$\tau_{\mathrm{U}_i} = (1 - \mathcal{P}_{\mathrm{U}_i})R_i, \quad i \in \{1, 2\}, \tag{15}$$

where $\mathcal{P}_{\mathrm{U}_i}$ denotes the outage probability and $R_i$ is the target data rate. This metric captures the long-term effective rate under fixed-rate and reliability constraints, which is highly relevant for latency-sensitive and mission-critical applications.

## 5 Ergodic Capacity (EC)

The ergodic capacity (EC) refers to the average achievable rate at which information can be reliably transmitted over a fading channel, averaged over all channel realizations. For user $\mathrm{U}_i$, it is expressed as

$$\mathcal{R}_i = \mathbb{E}\{\log_2(1+\Gamma_{\mathrm{U}_i})\}, \tag{16}$$

where $\Gamma_{\mathrm{U}_i}$ is given in (2). Let $\Xi_i \triangleq |h_0|^2|h_i|^2$ denote the cascaded gain with PDF given in (6). Then $\Gamma_{\mathrm{U}_i}$ can be rewritten as

$$\Gamma_{\mathrm{U}_i} = \frac{\theta^2 a_i\,\Sigma_S \Upsilon_i \Lambda_0 \Lambda_i\,\Xi_i}{\theta^2 b_j\,\Sigma_S \Upsilon_i \Lambda_0 \Lambda_i\,\Xi_i + 1}, \tag{17}$$

where $a_i = \alpha_i + \alpha_{12}\delta_i$ and $b_j = \alpha_{12}\delta_j$ with $j \neq i$ and $\delta_j = 1-\delta_i$.

Substituting (17) into (16), the EC becomes

$$\mathcal{R}_i = \frac{1}{\ln 2}\int_0^\infty \ln(1+x)\, f_{\Gamma_{\mathrm{U}_i}}(x)\,dx. \tag{18}$$

On the foundation of (18), we get the following theorem.

**Theorem 1.** *The EC of* $\mathrm{U}_i$ *can be derived in a closed-form expression as*

$$\begin{aligned}\mathcal{R}_i =& \frac{1}{\Gamma(m_0)\Gamma(m_i)\ln 2}\left[G_{2,4}^{4,1}\left(\frac{(\Upsilon_i\Lambda_0\Lambda_i)^{-1}\xi_i}{\theta^2\Sigma_S(a_i+b_j)}\,\middle|\, \begin{matrix}0,1\\ m_0,\, m_i,\, 0,\, 0\end{matrix}\right)\right.\\ &\left.- G_{2,4}^{4,1}\left(\frac{(\Upsilon_i\Lambda_0\Lambda_i)^{-1}\xi_i}{\theta^2\Sigma_S b_j}\,\middle|\, \begin{matrix}0,1\\ m_0,\, m_i,\, 0,\, 0\end{matrix}\right)\right].\end{aligned} \tag{19}$$

*Proof.* To derive the EC expression in (18) for user $\mathrm{U}_i$, we begin by obtaining the PDF of the auxiliary variable $\Gamma_{\mathrm{U}_i}$ via differentiation of its CDF:

$$f_{\Gamma_{\mathrm{U}_i}}(x) = \frac{\partial}{\partial x}F_{\Gamma_{\mathrm{U}_i}}(x) = \frac{(\Upsilon_i\Lambda_0\Lambda_i)^{-1}\,a_i}{\theta^2\Sigma_S\,(a_i-b_jx)^2}\, f_{\Xi_i}\!\left(\frac{(\Upsilon_i\Lambda_0\Lambda_i)^{-1}a_ix}{\theta^2\Sigma_S\,(a_i-b_jx)}\right), \tag{20}$$

which is valid for $x < a_i/b_j$. By substituting $f_{\Xi_i}(\cdot)$ from (6), the PDF of $\Gamma_{\mathrm{U}_i}$ becomes

$$f_{\Gamma_{\mathrm{U}_i}}(x) = \frac{a_i}{\Gamma(m_0)\Gamma(m_i)\,(a_i-b_jx)\,x}\times G_{0,2}^{2,0}\left(\frac{(\Upsilon_i\Lambda_0\Lambda_i)^{-1}\xi_i\,x}{\theta^2\Sigma_S\,(a_i-b_jx)}\,\middle|\,\begin{matrix}-\\ m_0,\, m_i\end{matrix}\right). \tag{21}$$

Inserting (21) into (18), the EC can be written as

$$\begin{aligned}\mathcal{R}_i =& \frac{1}{\Gamma(m_0)\Gamma(m_i)\ln 2}\int_0^{a_i/b_j}\ln(1+x)\,\frac{a_i}{(a_i-b_jx)\,x}\\ &\times G_{0,2}^{2,0}\left(\frac{(\Upsilon_i\Lambda_0\Lambda_i)^{-1}\xi_i\,x}{\theta^2\Sigma_S\,(a_i-b_jx)}\,\middle|\,\begin{matrix}-\\ m_0,\, m_i\end{matrix}\right)dx.\end{aligned} \tag{22}$$

By applying the change of variables $t = \frac{x}{a_i - b_j x}$ and using the Meijer-$G$ representation of the natural logarithm (see [27, Eq. (01.04.26.0003.01)]), we obtain the equivalent form

$$\begin{aligned} \mathcal{R}_i =& \frac{1}{\Gamma(m_0)\Gamma(m_i)\ln 2}\Bigg[\int_0^\infty \frac{1}{t} G_{2,2}^{1,2}\left((a_i+b_j)t \,\middle|\, \begin{matrix}1,1\\1,0\end{matrix}\right) \\ &\times G_{0,2}^{2,0}\left(\frac{(\Upsilon_i\Lambda_0\Lambda_i)^{-1}\,\xi_i\,t}{\theta^2\Sigma_S} \,\middle|\, \begin{matrix}-\\m_0,\, m_i\end{matrix}\right)\mathrm{dt} \\ &-\int_0^\infty \frac{1}{t} G_{2,2}^{1,2}\left(b_j t \,\middle|\, \begin{matrix}1,1\\1,0\end{matrix}\right) G_{0,2}^{2,0}\left(\frac{(\Upsilon_i\Lambda_0\Lambda_i)^{-1}\,\xi_i\,t}{\theta^2\Sigma_S} \,\middle|\, \begin{matrix}-\\m_0,\, m_i\end{matrix}\right)\mathrm{dt}\Bigg], \end{aligned} \quad (23)$$

which simplifies to the closed form in (19) by invoking [27, Eq. (07.34.21.0013.01)].

The overall EC of the two-user system is

$$\overline{\mathcal{R}}_{\text{overall}} = \mathcal{R}_1 + \mathcal{R}_2. \quad (24)$$

**Remark:** The closed-form EC in (19) involves only standard elementary and special functions and can be efficiently evaluated in MATLAB/Mathematica.

*High-SNR Region:* As $\Sigma_S \to \infty$, by law of large numbers the SINR converges to $\Gamma_{\mathrm{U}_i}^{\infty} \approx \frac{\alpha_i + \alpha_{12}\delta_i}{\alpha_{12}\delta_j}$, and the EC bounds of $x_i$ at $\mathrm{U}_i$ are

$$\mathcal{R}_1^{\infty} \simeq \log_2\left(1 + \frac{\alpha_1 + \alpha_{12}\delta_1}{\alpha_{12}\delta_2}\right), \quad (25a)$$

$$\mathcal{R}_2^{\infty} \simeq \log_2\left(1 + \frac{\alpha_2 + \alpha_{12}\delta_2}{\alpha_{12}\delta_1}\right). \quad (25b)$$

Hence, the overall high-SNR EC is

$$\overline{\mathcal{R}}_{\text{overall}}^{\infty} = \mathcal{R}_1^{\infty} + \mathcal{R}_2^{\infty}. \quad (26)$$

## 6 Numerical Results

This section presents numerical results to evaluate the performance of the proposed HPMA system. All results are obtained from $10^6$ independent Monte Carlo trials to ensure statistical accuracy. The noise power at each user $\mathrm{U}_i$ is calculated following [29] as

$$\sigma_{\mathrm{U}_i}^2 = N_0 \cdot \mathrm{BW} \cdot 10^{\mathrm{NF}/10},$$

where $N_0$ is the noise power spectral density, BW is the system bandwidth, and NF is the receiver noise figure.

Figure 2 shows the impact of the scattering cross-section $\theta$ on OP. Larger $\theta$ improves keyhole link gain, reducing OP for all schemes; $P_S = 5$ dBm achieves much lower OP than $P_S = -10$ dBm. HPMA consistently outperforms NOMA, with a more pronounced gain for $\mathrm{U}_2$ due to its higher QoS requirement ($R_2 = 2.4$

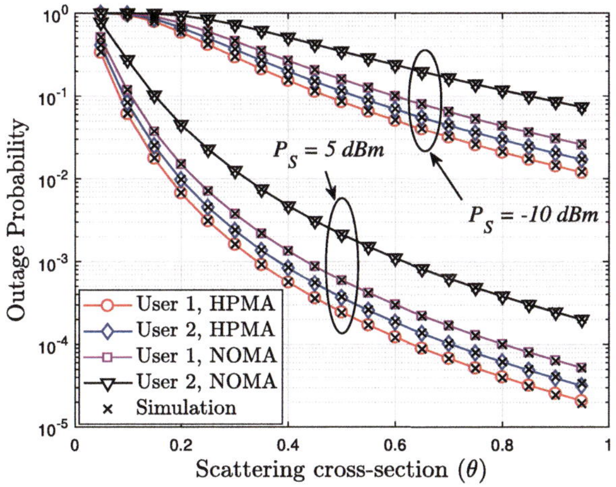


**Fig. 2.** OP performance versus scattering cross-section ($\theta$) with $P_S = \{-10, 5\}$ dBm, $m_0 = 2$ and $m_1 = m_2 = 2.5$.

bps/Hz). This confirms HPMA's ability to improve both reliability and fairness. The close match between simulation and analysis validates (10).

Figure 3 examines power allocation. In NOMA, small $\delta_1$ ($\leq 0.2$) yields OP $\approx 1$ for both users. Increasing $\delta_1$ to 0.2–0.4 reduces OP for both via improved $U_1$ SINR and SIC benefits for $U_2$, but higher $\delta_1$ severely degrades $U_2$'s OP. HPMA shows smoother behavior: $U_1$'s OP steadily decreases, $U_2$'s rises gradually but never approaches 1 due to its OMA branch. As $\delta_1 \to 1$, HPMA converges to NOMA. This demonstrates HPMA's robustness to power allocation and its fairness advantage.

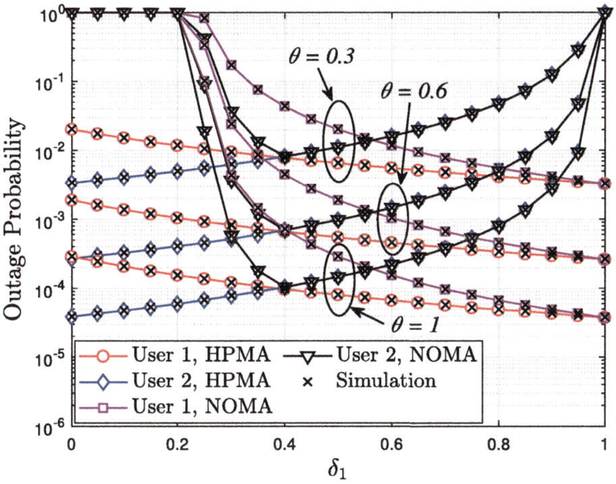


**Fig. 3.** OP performance versus ($\delta_1$) with $P_S = -5$ dBm, $\alpha_1 = \alpha_2 = \alpha_{12} = 1/3$, $R_1 = 0.4$, $R_2 = 2$, $m_0 = 2$ and $m_1 = m_2 = 2.5$.

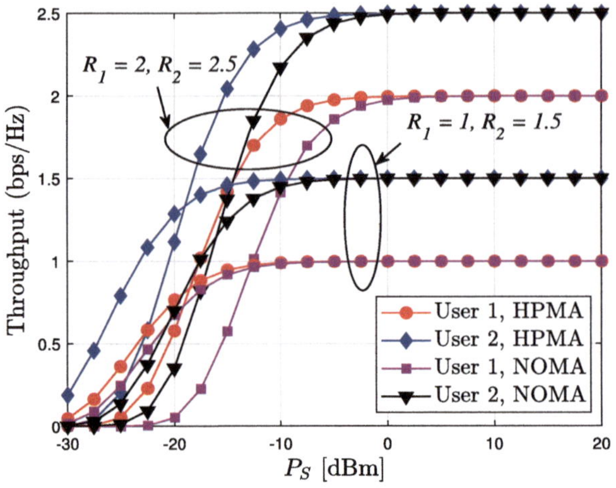


**Fig. 4.** Throughput versus $P_S$ in dBm, with $m_0 = 2$ and $m_1 = m_2 = 2.5$.

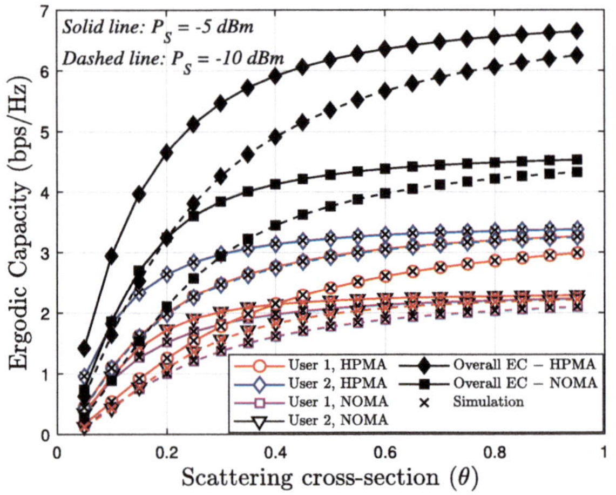


**Fig. 5.** Comparison of EC performance of $\mathrm{U_i}$ between HPMA and NOMA schemes versus scattering cross-section ($\theta$) with $P_S = \{-10, -5\}$ dBm, $m_0 = 2$ and $m_1 = m_2 = 2.5$.

Figure 4 plots throughput versus $P_S$. Throughput increases with $P_S$ and saturates at $R_i$. Higher $(R_1, R_2) = (2, 2.5)$ targets require more SNR; $(1, 1.5)$ converges earlier but lower. HPMA outperforms NOMA for both users, achieving target rates with less power, especially for $\mathrm{U_2}$.

Figure 5 shows $\theta$'s effect on EC. Larger $\theta$ improves capacity for both schemes, but HPMA gains faster and reaches a higher ceiling. Higher $P_S$ amplifies this benefit, highlighting HPMA's efficiency in rich-scattering 6G scenarios.

## 7 Conclusion

This paper analyzed the performance of a hybrid power–frequency multiple access system over keyhole fading channels, modeled as double Nakagami-$m$ links. We derived closed-form expressions for the outage probability and ergodic capacity, along with high-SNR approximations that reveal the achievable diversity order and coding gain. Analytical results, validated by Monte Carlo simulations, show that HPMA significantly outperforms conventional NOMA in both reliability and spectral efficiency, while operating without successive interference cancellation. These findings highlight HPMA as a promising low-complexity access strategy for 6G systems in rank-deficient propagation environments. Future research may extend this work to scenarios with multi-antenna transceivers, hardware impairments, limited phase-shift resolution, and fairness or sum-rate optimization under keyhole conditions.

## References

1. Tataria, H., et al.: 6G wireless systems: vision, requirements, challenges, insights, and opportunities. Proc. IEEE **109**(7), 1166–1199 (2021)
2. Liu, Y., Qin, Z., Elkashlan, M., Ding, Z., Nallanathan, A., Hanzo, L.: Nonorthogonal multiple access for 5G and beyond. Proc. IEEE **105**(12), 2347–2381 (2017)
3. Islam, S.M.R., Avazov, N., Dobre, O.A., Kwak, K.: Power-domain non-orthogonal multiple access (NOMA) in 5G systems: potentials and challenges. IEEE Commun. Surveys Tuts. **19**(2), 721–742 (2017)
4. Dai, L., et al.: Non-orthogonal multiple access for 5G: solutions, challenges, opportunities, and future research trends. IEEE Commun. Mag. **53**(9), 74–81 (2015)
5. Ding, Z., Adachi, F., Poor, H.V.: The application of MIMO to non-orthogonal multiple access. IEEE Trans. Wireless Commun. **15**(1), 537–552 (2016)
6. Ghosh, J., et al.: Performance investigation of NOMA versus OMA techniques for mmWave massive MIMO communications. IEEE Access **9**, 125300–125308 (2021)
7. Jaafar, W., et al.: Multiple access in aerial networks: from orthogonal and non-orthogonal to rate-splitting. IEEE Open J. Veh. Technol. **1**, 372–392 (2020)
8. Nam, P.M., et al.: Average achievable rate analysis for reliable control in UAV-based short-packet communication networks. Wireless Pers. Commun. **140**(1–2), 333–351 (2025)
9. Nguyen, V.S., et al.: Analysis of ergodic sum rate in RSMA with perfect and imperfect SIC: a multiple-antenna selection approach for optimizing UAV positioning. Phys. Commun. **72** (2025)
10. Le, A.-T., Vu, T.-H., da Costa, D.B., Voznák, M.: Hybrid power-frequency multiple access with SIC-free for holographic reconfigurable intelligent surface-based systems. IEEE Wireless Commun. Lett. **14**(5), 1566–1570 (2025)
11. Le, V.D., Nguyen, H.N., Nguyen, Q.S., Bui, T.T., Hien, D.C., Kim, B.S.: Enabling D2D transmission mode of reconfigurable intelligent surfaces aided in wireless NOMA system. Adv. Electr. Electron. Eng. **23**(1), 32–41 (2025). https://doi.org/10.15598/aeee.v23i1.240809
12. Vu, T.-H., et al.: A new look for outage and ergodic improvements in coordinated IoT paradigms with hybrid multiple access. IEEE Trans. Veh. Technol. **74**(8), 13284–13289 (2025)

13. Chizhik, D., Foschini, G.J., Gans, M.J., Valenzuela, R.A.: Keyholes, correlations, and capacities of multielement transmit and receive antennas. IEEE Trans. Wireless Commun. **1**(2), 361–368 (2002)
14. Loyka, S., Kouki, A.: On MIMO channel capacity, correlations, and keyholes: analysis of degenerate channels. IEEE Trans. Commun. **50**(12), 1886–1888 (2002)
15. Nguyen, S.Q., Le, C.B.: Physical layer security for keyhole-based NOMA downlink systems with a multi-antenna eavesdropper. J. Telecommun. Inf. Technol. Early Access 1–10 (2025). https://doi.org/10.26636/jtit.2025.3.2116
16. Gesbert, D., Bolcskei, H., Gore, D.A., Paulraj, A.J.: Outdoor MIMO wireless channels: models and performance prediction. IEEE Trans. Commun. **50**(12), 1926–1934 (2002)
17. Almers, P., Tufvesson, F., Molisch, A.F.: Keyhole effect in MIMO wireless channels: measurements and theory. IEEE Trans. Wireless Commun. **5**(12), 3596–3604 (2006)
18. Zhang, H., et al.: Performance analysis of MIMO-HARQ assisted V2V communications with keyhole effect. IEEE Trans. Commun. **70**(5), 3034–3046 (2022)
19. Nguyen, T.-T., et al.: A low-complexity relaying protocol for cooperative short-packet NOMA-based spectrum sharing systems. IEEE Trans. Veh. Technol. **73**(6), 9044–9049 (2024)
20. Le, A.-T., et al.: Physical layer security analysis for RIS-aided NOMA systems with non-colluding eavesdroppers. Comput. Commun. **219**, 194–203 (2024)
21. Jaiswal, N., Pandey, A., Yadav, S., Purohit, N.: Physical layer security performance of NOMA-assisted vehicular communication systems over double-Rayleigh fading channels. Phys. Commun. **57**, Art. 101968 (2023)
22. Nguyen, S.Q., Anh, U.-V.L., Nguyen, T.N., Nguyen, T.-T., Voznák, M.: Short packet communications for relay systems with co-channel interference at relay: performance analysis and power control. IEEE Access **12**, 63452–63461 (2024)
23. Tran, M., et al.: Security and reliability analysis of the power splitting-based relaying in wireless sensor networks. Sensors **24**(4), Art. 1300 (2024)
24. Nguyen, Q.S., Kim, H.Y., Nguyen, T.N., Bui, M.V., Tran, P.T., Tran, T.D.: Performance of RIS-secured short-packet NOMA systems with discrete phase-shifter to protect digital content and copyright against untrusted user. IEEE Access **13**, 21580–21593 (2025). https://doi.org/10.1109/ACCESS.2025.3535813
25. Jaiswal, N., Purohit, N.: Performance analysis of NOMA-enabled vehicular communication systems with transmit antenna selection over double Nakagami-$m$ fading. IEEE Trans. Veh. Technol. **70**(12), 12725–12741 (2021). https://doi.org/10.1109/TVT.2021.3119979
26. Gradshteyn, I.S., Ryzhik, I.M.: Table of Integrals, Series, and Products. Academic Press (2014)
27. Rudin, W.: Principles of Mathematical Analysis, 3rd edn. McGraw-Hill, New York (1976)
28. Simon, M.K., Alouini, M.-S.: Digital Communication Over Fading Channels, 2nd edn. Wiley, Hoboken (2005)
29. Do, T.N., Kaddoum, G., Nguyen, T.L., da Costa, D.B., Haas, Z.J.: Multi-RIS-aided wireless systems: statistical characterization and performance analysis. IEEE Trans. Commun. **69**(12), 8641–8658 (2021). https://doi.org/10.1109/TCOMM.2021.3117599

# Sentiment-Driven ETF Decision: A Fully LLM-Centric Agent Framework with Dynamic Constraint Prompts

Zehan Tan(✉) and Yaoyang Wu

School of Digital Technology, Macau Millennium College, Macau SAR, China
{zhtan,sunnywu}@mmc.edu.mo

**Abstract.** ETF price prediction poses unique challenges due to its complex interplay between market sentiment and microstructure dynamics. While existing approaches utilize either traditional machine learning or sentiment analysis, they often struggle with static sentiment integration and lack interpretability. This paper presents a novel LLM-Centric Agent Framework that addresses these limitations through three key innovations: (1) an end-to-end architecture leveraging natural language processing capabilities, (2) a dynamic weighting mechanism for adaptive sentiment integration, and (3) structured chain-of-thought that enforces stepwise financial reasoning without model fine-tuning. Our framework employs specialized LLM agents for feature extraction and decision-making, incorporating a self-iterative constraint optimization system that dynamically adjusts to market conditions. Experimental results on the financial ETF (159931) demonstrate competitive performance in both directional prediction and continuous price prediction, significantly outperforming traditional LSTM approaches. The framework's natural language explanations provide unprecedented transparency in the decision-making process, while its dynamic weighting mechanism effectively adapts to varying market conditions. Our approach bridges the gap between quantitative analysis and sentiment integration, offering a robust solution for ETF prediction that combines accuracy with interpretability. The code source and dataset are open at https://github.com/zehantan6970/Sentiment-Driven-ETF-Decision.

**Keywords:** Large Language Models · Sentiment Analysis · Natural Language Processing · Financial Prediction

## 1 Introduction

Accurately predicting asset prices is central to financial research and practice [1]. Exchange-Traded Funds (ETFs) have become core building blocks of modern portfolios because of their low cost, high liquidity, and built-in diversification [2]. Unlike traditional index funds, ETFs trade continuously and support high-frequency strategies, so understanding how sentiment and microstructure jointly shape ETF returns is practically important.


P. Pichappan et al. (Eds.): RTIS 2025, LNNS 1936, pp. 107–118, 2027.
https://doi.org/10.1007/978-3-032-23396-7_10

Behavioral finance shows that investor sentiment, amplified by cognitive biases and herd behavior, can drive persistent deviations from fundamentals. Baker and Wurgler documented strong sentiment effects in hard-to-value and hard-to-arbitrage securities [3], while Shiller linked "irrational exuberance" to bubbles and crashes [4]. ETFs play a dual role as both market barometers and liquidity vehicles, which makes them particularly sensitive to sentiment-driven flows and short-horizon trading [5–8].

However, existing sentiment-based ETF prediction frameworks face two major limitations. First, sector-level studies such as Machacová [9] typically rely on static sentiment proxies (e.g., surveys), underutilizing real-time textual data from news and social media and largely ignoring ETF-specific microstructure features like creation/redemption and spread regimes [10]. Second, hybrid models that combine LSTM or reinforcement learning with sentiment features [11–13] often require costly retraining, are slow to adapt to regime shifts (e.g., volatility spikes), and remain difficult to interpret.

To address these gaps, we propose **Sentiment-Driven ETF Decision**, a fully LLM-centric agent framework for ETF prediction and trading. Our contributions are threefold:

- **Self-iterative constraint optimization with natural language rules.** Prediction errors are analyzed by an LLM agent that produces natural language diagnoses and converts them into executable constraints (e.g., volatility-conditioned caps on sentiment weight). Rules are stored with confidence scores, updated over time, and injected back into prompts, eliminating retraining while remaining interpretable.
- **Market state-aware dynamic weighting.** A deterministic policy adjusts the weights of time-series and sentiment features based on volatility and sentiment confidence. This mechanism gates sentiment influence under stressed conditions (e.g., high volatility and low-confidence news) and modestly boosts time-series weight under liquidity shocks, enabling regime-aware integration.
- **Structured CoT reasoning for financial decisions.** We design a multi-step CoT template that guides LLMs from data preprocessing to technical/sentiment analysis, dynamic weighting, error-informed adjustments, and final trading decisions. This yields explicit reasoning traces without fine-tuning or reinforcement learning.

## 2 Related Works

### 2.1 Sentiment Analysis Using Neural Networks

Financial sentiment analysis has evolved from lexicon-based and traditional machine learning methods [14,15] to deep neural architectures. LSTMs capture sequential dependencies in financial text and have become a standard tool for forum and news sentiment classification; Ko and Chang report 98% accuracy on investor forum sentiment, outperforming static classifiers [11]. Domain-specific language models such as FinBERT further improve robustness to financial jargon and sarcasm [16]. Ma et al. distinguish between financially literate and retail users on Weibo, showing large gains in predictive performance when weighting expert sentiment more heavily [17].

Most of this work focuses on individual stocks or indices. ETF-specific issues such as liquidity-driven arbitrage, spread dynamics, or sectoral heterogeneity are typically not modeled explicitly [9,18]. This creates a gap between rich sentiment modeling and ETF microstructure.

### 2.2 Machine Learning in Stock Market Prediction

Classical time-series models such as ARIMA and GARCH remain widely used for volatility and return modeling [19], and factor models such as Fama–French remain central to asset pricing [20,21]. However, their linear structure struggles with the nonlinear, regime-dependent dynamics of financial markets.

Deep learning approaches—notably LSTM and GRU networks—capture temporal dependencies more flexibly and have shown strong performance in stock prediction [22,23]. Transformer-based architectures further improve long-range dependency modeling and can naturally fuse multiple modalities [24]. Sentiment-enhanced Transformers that combine news and technical signals improve return forecasts over purely price-based models [25].

ETF prediction poses specific challenges: creation/redemption flows, intraday NAV arbitrage, and liquidity constraints [10,26]. Guo and Xu show that a four-day sentiment lag is optimal for certain Chinese ETFs, but use fixed windows that cannot adapt as volatility and sentiment regimes change [7].

### 2.3 Hybrid Models: Deep Learning and Sentiment

Hybrid models combining deep networks with sentiment features are now standard baselines. LSTM+sentiment and FinBERT+LSTM architectures use convolutional layers to extract local patterns (e.g., bullish phrases) and recurrent layers to encode temporal structure, achieving high accuracy in equity and index prediction [11,16]. MarketSenseAI integrates ChatGPT-4 with multi-source market and sentiment data for stock selection [27], demonstrating that LLM reasoning can be useful in financial prediction.

However, these approaches rarely exploit *agentic* capabilities of LLMs: they typically treat the LLM as a static feature extractor or scorer rather than a decision-making agent that can generate, store, and update rules. Existing work also underutilizes ETF-specific microstructure signals. Our framework targets this gap by (i) embedding ETF microstructure indicators into a dynamic weighting policy, (ii) using LLMs as constraint-generating agents, and (iii) enforcing structured CoT for decision transparency.

## 3 Sentiment-Driven ETF Decision System

We now present the **Sentiment-Driven ETF Decision System**, a framework that uses LLM agents to jointly process time-series and sentiment inputs and to produce ETF price predictions and trading decisions (buy/sell/hold). Unlike static models, the system adapts its internal constraints and feature weights via a feedback loop driven by past errors.

### 3.1 Framework Architecture and Workflow

The architecture (Fig. 1) comprises five modules: Input Layer, Feature Extraction Layer, Rule Management Layer, Decision Layer, and Feedback Loop.

**Input Layer.** The system ingests two main data streams:

- **Market data:** Historical and recent structured data such as daily open, high, low, close, trading volume, bid-ask spread, and selected microstructure indicators (e.g., creation/redemption activity), sourced from financial platforms (e.g., "2025-01-27 closing price: 1.776").
- **Sentiment data:** Unstructured text from financial news (e.g., QQ News, stock.sohu.com) and social media (e.g., Eastmoney forums), filtered with ETF-related keywords (e.g., "cross-border payment theme surges; financial ETF drops 0.06%; is there still upside?").

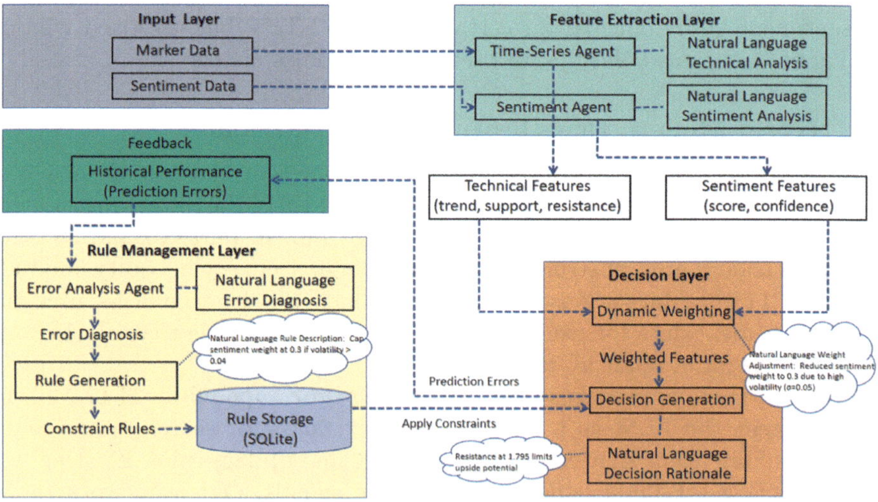


**Fig. 1.** System architecture with LLM agents and natural language explanations.

**Feature Extraction Layer.** Two specialized LLM agents transform raw data into interpretable features:

- **Time-Series Agent:** Summarizes numerical market data into technical indicators (e.g., trend strength $TS$, volatility $\sigma$) and short natural language descriptions (e.g., "consolidation near resistance at 1.795"). Trend strength is defined as:

$$TS = \frac{1}{n}\sum_{i=1}^{n-1}(C_{t-i} - C_{t-i-1}), \tag{1}$$

where $C_t$ is the closing price at time $t$.
- **Sentiment Agent:** Maps textual inputs to a sentiment score $s \in [-1, 1]$ and a confidence score $C_{\text{sentiment}} \in [0, 1]$:

$$C_{\text{sentiment}} = 0.7\, C_{\text{LLM}} + 0.3\, C_{\text{source}}, \tag{2}$$

where $C_{\text{LLM}}$ is the model's self-assessed confidence and $C_{\text{source}}$ captures source reliability. Weights (0.7, 0.3) are derived using the Analytic Hierarchy Process (AHP) [28], reflecting the LLM's stronger text understanding and the importance of source trustworthiness.

**Rule Management Layer.** This layer manages a library of failure-driven constraints:

- **Error Analysis Agent:** When a prediction error is observed (e.g., predicted 1.790 vs. actual 1.776 on 2025-01-27), the agent generates a natural language explanation such as "overreliance on low-confidence sentiment under high volatility ($\sigma = 0.05$)".
- **Rule Generation:** The analysis is distilled into an actionable rule, for example: "cap sentiment weight at 0.3 when $\sigma > 0.04$ and $C_{\text{sentiment}} < 0.7$". Volatility is computed as:

$$\sigma = \sqrt{252} \times \text{std}\left(\ln\left(\frac{C_t}{C_{t-1}}\right)\right), \tag{3}$$

where $C_t$ and $C_{t-1}$ are consecutive daily closes.
- **Rule Storage and Confidence:** Rules are stored in a SQLite database with trigger conditions, textual descriptions, and empirical confidence:

$$C(r_i) = \frac{\text{Successful Corrections}}{\text{Total Triggers} + \epsilon}, \tag{4}$$

with smoothing $\epsilon = 0.01$ to avoid division by zero. Rules that consistently prevent errors gain higher confidence and are preferentially applied.

**Decision Layer.** The Decision Layer fuses features and active rules to generate predictions and actions.

- **Dynamic weighting:** Feature weights are adapted to market states:

$$w_{\text{sentiment}} = \begin{cases} 0.3 & \text{if } \sigma > 0.04 \wedge C_{\text{sentiment}} < 0.7, \\ \min(0.5, 1 - w_{\text{time}}) & \text{otherwise,} \end{cases} \tag{5}$$

  with

$$w_{\text{time}} = \frac{TS}{1 + \sigma}. \tag{6}$$

  Weights are clipped to $[0, 1]$ and $w_{\text{time}} + w_{\text{sentiment}} = 1$. The LLM produces a justification such as "reduced sentiment weight to 0.3 due to high volatility ($\sigma = 0.05$) and low-confidence sentiment".
- **Decision generation:** The agent outputs a point prediction (e.g., 1.778 with an uncertainty band) and a trading action (e.g., "Hold, probability 0.70"), together with a concise explanation referencing technical levels, sentiment, and active rules.

**Feedback Loop.** After new prices are observed, realized errors are passed to the Error Analysis Agent. Rules that did not prevent errors can be refined or down-weighted; new rules may be created. This yields a self-iterative, non-parametric optimization loop in the space of natural language constraints rather than model weights.

### 3.2 Key Innovations

**Self-iterative Constraint Optimization.** As shown in Fig. 2, the rule life-cycle consists of error analysis, rule generation and storage, and rule application. Each cycle converts past mistakes into updated constraints that are explicitly logged and auditable. Unlike gradient-based updates, this process produces human-readable changes (e.g., new volatility thresholds, caps on sentiment contribution) that can be inspected by risk managers.

**Market State-Aware Dynamic Weighting.** The dynamic weighting workflow in Fig. 3 operationalizes ETF-specific insights: volatility spikes, spread regime shifts, or creation/redemption anomalies should reduce reliance on noisy sentiment and emphasize price-based structure. The system evaluates $\sigma$ (Eq. 3) and $C_{\text{sentiment}}$ (Eq. 2), then applies rule-based logic to adjust $(w_{\text{time}}, w_{\text{sentiment}})$ while preserving normalization and bounds. For example, during the 2025-01-27 volatility spike ($\sigma = 0.05$), the framework automatically enforced $w_{\text{sentiment}} = 0.3$, dampening the effect of low-confidence bearish posts.

**Structured CoT Reasoning.** A tailored CoT template structures the LLM's reasoning into:

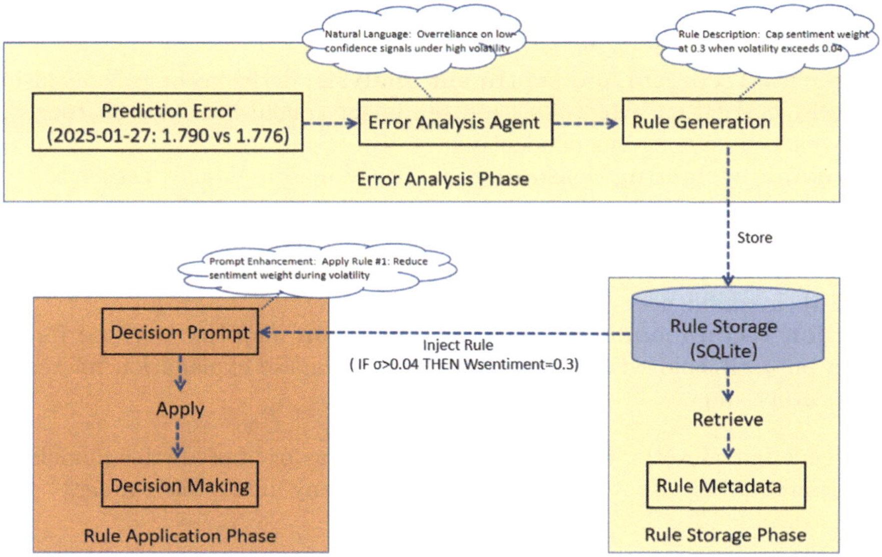


**Fig. 2.** Rule lifecycle and SQLite integration for constraint management.

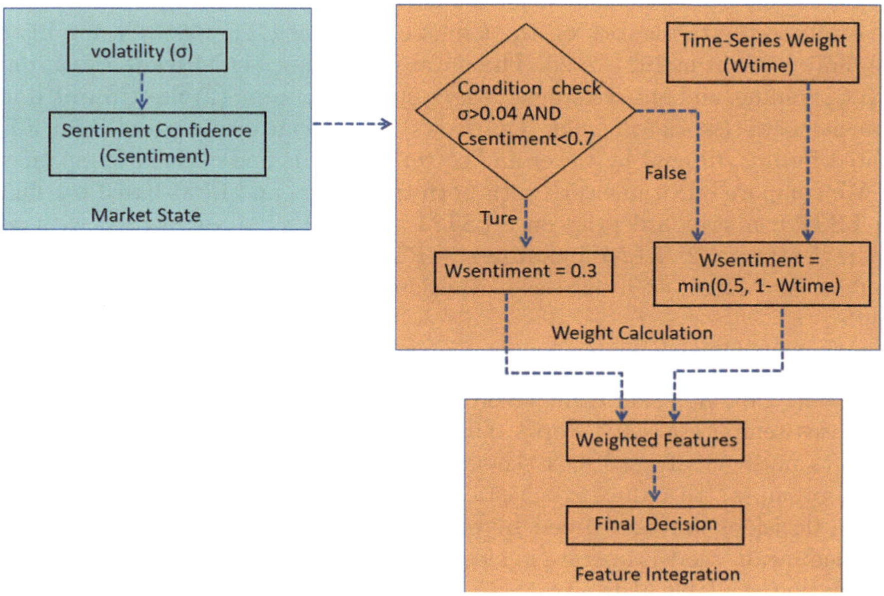


**Fig. 3.** Dynamic weighting workflow based on volatility and sentiment confidence.

1. **Initial data assessment:** organize recent price and sentiment context, including simple statistics and qualitative summaries (e.g., "ETF 159931

closed at 1.783 on 2025-01-24, +0.62% vs. previous day; forums are moderately bullish with confidence 0.65").
2. **Separate technical and sentiment analysis:** derive technical conclusions (levels, momentum, volatility) and sentiment conclusions (overall tone, key drivers, source mix) independently.
3. **Dynamic weighting reasoning:** explicitly reason about whether to trust sentiment and how to set $(w_{\text{time}}, w_{\text{sentiment}})$ given $\sigma$ and $C_{\text{sentiment}}$.
4. **Error-informed adjustment:** incorporate patterns in past errors (e.g., systematic overestimation during high volatility) via summaries of triggered rules and their confidence.
5. **Action recommendation with justification:** map the weighted forecast into buy/sell/hold with a probability and a concise explanation referencing the earlier steps.

This design converts the typical black-box behavior of predictive models into a transparent, stepwise reasoning trace, without any model fine-tuning.

## 4 Experiments

### 4.1 Experimental Setup

We evaluate the framework on the Chinese financial ETF 159931 using historical daily data from 2024–2025. The dataset includes: (1) **Market data**: daily OHLC, volume, and bid-ask spread from q.stock.sohu.com; (2) **Sentiment data**: financial news (stock.sohu.com, QQ News) and social media posts from ETF-related forums, filtered by ticker and sector keywords and cleaned for spam.

We compare our framework with both traditional and LLM-based baselines: (1) **LSTM**: a standard price-only LSTM and an LSTM augmented with sentiment features; (2) **LLM baselines**: GPT-4o, Claude 4.0 Sonnet, DeepSeek, Doubao, and Kimi, each used as a decision agent with and without sentiment inputs.

**LLM Configuration.** We adopt a multi-provider setup: GPT-4, Claude, DeepSeek, Doubao, and Kimi receive structurally equivalent prompts. Inference uses temperature 0.3, top-$p$ 0.9, and a 2048-token limit per agent turn. Outputs must satisfy a JSON schema; malformed outputs trigger up to two repair prompts. To reduce stochasticity, each model is run three times: categorical decisions are aggregated by majority vote, and continuous values by a trimmed mean. Models operate in closed-world mode with prompts restricted to information available at time $t$.

**Evaluation Metrics.** For directional prediction, we compute Accuracy, Precision, Recall, and F1-score based on up/down movement between consecutive closes. For continuous prediction, we report Mean Absolute Error (MAE), Root Mean Squared Error (RMSE), and Mean Absolute Percentage Error (MAPE).

**Table 1.** Directional ETF price movement prediction with and without sentiment.

| Model | Accuracy | Precision | Recall | F1-Score |
|---|---|---|---|---|
| Base Models | | | | |
| DeepSeek | 0.53 | 0.67 | 0.60 | 0.63 |
| ChatGPT 4o | 0.40 | 0.57 | 0.40 | 0.47 |
| Doubao | 0.60 | 0.70 | 0.70 | 0.70 |
| Kimi | 0.60 | 0.70 | 0.70 | 0.70 |
| Claude 4.0 Sonnet | 0.67 | 0.78 | 0.70 | 0.74 |
| LSTM | 0.60 | 0.75 | 0.60 | 0.67 |
| Models with Sentiment | | | | |
| DeepSeek+Sentiment | 0.67 | 0.86 | 0.60 | 0.71 |
| ChatGPT+Sentiment | 0.40 | 0.57 | 0.40 | 0.47 |
| Doubao+Sentiment | 0.67 | 0.78 | 0.70 | 0.74 |
| Kimi+Sentiment | 0.60 | 0.75 | 0.60 | 0.67 |
| Claude+Sentiment | 0.67 | 0.78 | 0.70 | 0.74 |
| LSTM+Sentiment | 0.53 | 0.80 | 0.40 | 0.53 |
| **Ours** | **0.73** | **0.88** | **0.70** | **0.78** |

### 4.2 Results

Tables 1 and 2 summarize the performance of all models on directional and continuous prediction respectively.

### 4.3 Result Analysis

**Directional Prediction.** Among base models, Claude 4.0 Sonnet achieves the best F1-score (0.74), with LSTM and other LLMs clustered around 0.63–0.70 (Table 1). Adding sentiment generally improves performance for DeepSeek, Doubao, and Claude, but leaves ChatGPT roughly unchanged and slightly destabilizes LSTM (increased precision but reduced recall). Our framework achieves the highest directional performance overall (Accuracy 0.73, Precision 0.88, Recall 0.70, F1 0.78). The relatively high precision and modestly lower recall suggest that the dynamic constraint rules make the system conservative in taking directional positions under ambiguous conditions, which is often desirable in risk-aware ETF trading.

**Continuous Prediction.** In continuous price prediction (Table 2), LSTM trails all LLM-based models, even when augmented with sentiment (MAE 0.046 vs. 0.014–0.018 for base LLMs). Sentiment-aware LLMs such as Claude+Sentiment and Doubao+Sentiment achieve strong performance (MAE 0.010, MAPE around 0.55–0.58%). Our framework yields the best metrics across the board (MAE 0.009, RMSE 0.012, MAPE 0.49%), indicating that the dynamic weighting and constraint loop help stabilize predictions across regimes.

**Table 2.** Continuous price prediction performance.

| Model | MAE | RMSE | MAPE (%) |
|---|---|---|---|
| Base Models | | | |
| DeepSeek | 0.014 | 0.018 | 0.80 |
| ChatGPT 4o | 0.018 | 0.021 | 1.02 |
| Doubao | 0.013 | 0.016 | 0.75 |
| Kimi | 0.013 | 0.017 | 0.75 |
| Claude | 0.011 | 0.014 | 0.62 |
| LSTM | 0.053 | 0.057 | 2.99 |
| Models with Sentiment | | | |
| DeepSeek+Sentiment | 0.016 | 0.018 | 0.88 |
| ChatGPT+Sentiment | 0.017 | 0.021 | 0.93 |
| Doubao+Sentiment | 0.010 | 0.014 | 0.58 |
| Kimi+Sentiment | 0.012 | 0.016 | 0.68 |
| Claude+Sentiment | 0.010 | 0.014 | 0.55 |
| LSTM+Sentiment | 0.046 | 0.051 | 2.60 |
| **Ours** | **0.009** | **0.012** | **0.49** |

**Key Takeaways.** The experiments support three conclusions: (1) LLM-based agents outperform LSTM on ETF prediction due to pre-trained language knowledge and reasoning over textual signals; (2) Sentiment integration is beneficial but model-dependent, underscoring the need for mechanisms that adapt sentiment weight to context; (3) Our framework most effectively exploits sentiment under regime shifts through dynamic weighting and rule-based constraint updates.

## 5 Discussion

Our results extend prior evidence on sentiment effects in ETFs [7–9] by demonstrating that dynamic, regime-aware sentiment integration yields measurable improvements over static approaches. The performance gap between our framework (MAE 0.009) and LSTM (MAE 0.053) illustrates the value of combining natural language understanding with quantitative signals. The self-iterative constraint loop offers a complementary alternative to parameter learning: rule learning in natural language, grounded in domain knowledge and empirical feedback. Methodologically, the framework shows how LLMs can serve as agents that generate features, reason via CoT, and update constraints—all without fine-tuning—while providing audit-friendly explanations [11].

Our study has limitations: the implementation suits daily horizons rather than high-frequency trading; experiments focus on a single ETF; and the evaluation window (2024–2025) is relatively short. Future work includes integrating finer-grained microstructure signals, automatic rule pruning, and quantitative metrics for explanation quality.

## 6 Conclusion

We presented a fully LLM-centric agent framework for sentiment-driven ETF decision-making combining self-iterative constraint optimization, market state-aware dynamic weighting, and structured CoT reasoning. On ETF 159931, the framework outperforms strong LLM and LSTM baselines on both directional and continuous prediction while providing transparent explanations and auditable rule updates, suggesting that treating LLMs as adaptive, constraint-generating agents is a promising direction for explainable quantitative finance.

## References

1. Soni, P., Tewari, Y., Krishnan, D.: Machine learning approaches in stock price prediction: a systematic review. J. Phys. Conf. Ser. **2161**, 012065 (2022)
2. Liebi, L.J.: The effect of ETFs on financial markets: a literature review. Fin. Markets. Portfolio Mgmt. **34**(2), 165–178 (2020). https://doi.org/10.1007/s11408-020-00349-1
3. Baker, M., Wurgler, J.: Investor sentiment and the cross-section of stock returns. J. Finance **61**, 1645–1680 (2006)
4. Shiller, R.J.: Measuring bubble expectations and investor confidence. J. Psychol. Financ. Mark. **1**, 49–60 (2000)
5. Shenjere, P.A., Ferreira-Schenk, S., Moodley, F.: Does investor sentiment influence South African ETF flows during different market conditions? Economies **13**, 10 (2025)
6. Kunjal, D.: The role of investor attention in ETF liquidity. J. Econ. Financ. Anal. **7**, 45–64 (2024)
7. Guo, H., Xu, Y.: The relationship between stockholder sentiment lag and stock price prediction accuracy: an empirical analysis based on LSTM and transformer models. Highl. Sci. Eng. Technol. **70**, 393–400 (2023)
8. Yao, C.-Z., Li, H.-Y.: Time-varying lead-lag structure between investor sentiment and stock market. N. Am. J. Econ. Financ. **52**, 101148 (2020)
9. Machacová, I.: Impact of sentiment on exchange-traded funds, vol. 10, pp. 123–135 (2023)
10. Davies, S.W.: Speculation sentiment. J. Financ. Quant. Anal. **57**, 2485–2515 (2022)
11. Ko, C.-R., Chang, H.-T.: LSTM-based sentiment analysis for stock price forecast. PeerJ Comput. Sci. **7**, e408 (2021)
12. Chiumera, D. J.: Deep reinforcement learning for quantitative finance: time series forecasting using proximal policy optimization. Ph.D. thesis, Carleton University (2022)
13. Gunawan, A.A.S., Ashifa, S.B., Rumagit, R.Y., Ngarianto, H.: Development of stock market price application to predict purchase and sales decisions using proximal policy optimization method. In: Proceedings of the 2021 1st International Conference on Computer Science and Artificial Intelligence (ICCSAI), vol. 1, pp. 431–437. IEEE (2021)

14. Mishev, K., Gjorgjevikj, A., Vodenska, I., Chitkushev, L.T., Trajanov, D.: Evaluation of sentiment analysis in finance: from lexicons to transformers. IEEE Access **8**, 131662–131682 (2020)
15. Wankhade, M., Rao, A.C.S., Kulkarni, C.: A survey on sentiment analysis methods, applications, and challenges. Artif. Intell. Rev. **55**, 5731–5780 (2022)
16. Kim, J., Kim, H.-S., Choi, S.-Y.: Forecasting the S&P 500 index using mathematical-based sentiment analysis and deep learning models: a FinBERT transformer model and LSTM. Axioms **12**, 835 (2023)
17. Ma, Z., Ryan, C., Buckley, J., Chochlov, M.: Do Weibo platform experts perform better at predicting stock market? In: Proceedings of the International Conference on Engineering Applications of Neural Networks, pp. 491–502. Springer (2021)
18. Chau, F., Deesomsak, R., Lau, M.C.K.: Investor sentiment and feedback trading: evidence from the exchange-traded fund markets. Int. Rev. Financ. Anal. **20**, 292–305 (2011)
19. Hsu, M.-W., Lessmann, S., Sung, M.-C., Ma, T., Johnson, J. E. V.: Bridging the divide in financial market forecasting: machine learners vs. financial economists. Expert Syst. Appl. **61**, 215–234 (2016)
20. Xu, J., Zhang, S.: The Fama-French three factors in the Chinese stock market. China Account. Financ. Rev. **16**, 1–18 (2014)
21. Hillebrand, E., Medeiros, M.C.: Nonlinearity, breaks, and long-range dependence in time-series models. J. Bus. Econ. Stat. **34**, 23–41 (2016)
22. Olubusola, O., Mhlongo, N.Z., Daraojimba, D.O., Ajayi-Nifise, A.O., Falaiye, T.: Machine learning in financial forecasting: A US review: exploring the advancements, challenges, and implications of AI-driven predictions in financial markets. World J. Adv. Res. Rev. **21**, 1969–1984 (2024)
23. Fischer, T., Krauss, C.: Deep learning with long short-term memory networks for financial market predictions. Eur. J. Oper. Res. **270**, 654–669 (2018)
24. Thundiyil, S., Shalamzari, S. S., Picone, J., McKenzie, S.: Transformers for modeling long-term dependencies in time series data: a review. In: Proceedings of the 2023 IEEE Signal Processing in Medicine and Biology Symposium (SPMB), pp. 1–5. IEEE (2023)
25. Kaeley, H., Qiao, Y., Bagherzadeh, N.: Support for Stock Trend Prediction Using Transformers and Sentiment Analysis. arXiv preprint arXiv:2305.14368 (2023)
26. Ben-David, I., Franzoni, F., Moussawi, R.: Do ETFs increase volatility? J. Finance **73**, 2471–2535 (2018)
27. Fatouros, G., Metaxas, K., Soldatos, J., Kyriazis, D.: Can large language models beat wall street? Unveiling the potential of AI in stock selection. arXiv preprint arXiv:2401.03737 (2024)
28. Bugingo, E., Ndimubenshi, E.L., Kamanga, C.T., Rugema, F.X., Habimana, O., Batamuliza, J.: Application of AHP in decision-making: case studies and practical implementation. In: The Art of Decision Making - Applying AHP in Practice. IntechOpen (2024)

# Lightweight Heart Rate Estimation from Wrist PPG During Daily Activities

Ho Nhut Minh(✉), Nguyen Trung Hieu, Ngo Dinh Phong, and Nguyen Trong Kien

Posts and Telecommunications Institute of Technology, Ho Chi Minh City, Vietnam
{minhhn,ntkien}@ptit.edu.vn

**Abstract.** Accurate and continuous heart rate (HR) monitoring is essential for health assessment and fitness tracking. Photoplethysmography (PPG) provides a low-cost, non-invasive sensing modality widely available in wearables, yet its reliability is limited by motion artifacts and inter-subject variability in free-living conditions. However, many prior approaches rely on multi-modal inputs or large models, and are often evaluated under subject-dependent protocols, which do not adequately assess generalization performance in free-living conditions. Thus, we present a lightweight end-to-end model for heart-rate estimation from wrist PPG only, trained and evaluated under a strict leave-one-subject-out (LOSO) protocol, using the PPG-DaLiA dataset collected during daily activities. The architecture combines a compact 1D convolutional front end with a bidirectional GRU back end, totaling approximately 14.7k parameters and enabling embedded deployment. The proposed model achieves an MAE of 5.13 bpm, and a correlation of 0.86 across all LOSO test samples, showing strong agreement with ECG-derived reference and performance comparable to recent state-of-the-art methods. These findings indicate that accurate HR estimation is feasible with a single PPG channel and compact architecture, supporting practical real-time deployment in wearable devices.

**Keywords:** Photoplethysmography (PPG) · Heart-rate estimation · Leave-one-subject-out (LOSO) cross-validation · Band-pass filtering · Convolutional neural networks

## 1 Introduction

Heart rate monitoring is fundamental for health assessment, fitness tracking, and continuous physiological surveillance through wearable devices [1]. Photoplethysmography represents a non-invasive optical method widely integrated into consumer wrist-worn devices, offering comfort and cost-effectiveness for continuous vital sign tracking [2]. This technology functions by emitting light onto the skin and detecting changes in light absorption or reflection caused by arterial blood volume pulsations in the peripheral circulation [3]. The periodicity of these blood volume changes correlates with cardiac rhythm, allowing for HR derivation from the PPG signal [4]. This modality offers a convenient and accessible means for continuous HR monitoring outside of traditional clinical settings [5].


P. Pichappan et al. (Eds.): RTIS 2025, LNNS 1936, pp. 119–130, 2027.
https://doi.org/10.1007/978-3-032-23396-7_11

Accurate HR estimation from wrist-worn PPG signals presents significant challenges, particularly under dynamic, free-living conditions [6]. A primary difficulty arises from motion artifacts, which are signal corruptions caused by sensor movement relative to the skin, muscle contractions, and inconsistent sensor-skin coupling [7, 8]. These artifacts frequently distort the PPG signal, obscuring the underlying physiological information and hindering robust HR estimation [9]. Environmental factors, such as ambient light and changes in skin tone, also contribute to signal degradation [10]. Furthermore, inter-subject physiological variability can impact signal quality and consistency across individuals [11]. Consequently, algorithms developed in controlled laboratory environments often struggle to maintain accuracy during diverse daily activities and transitions, such as walking, stair climbing, or cycling [12].

Historically, HR estimation from PPG has relied on traditional signal processing pipelines, including peak detection and interval analysis, or spectral methods like Fast Fourier Transform [13]. Adaptive filtering and accelerometer-assisted techniques have also been employed to mitigate motion artifact interference by attempting to separate the motion component from the physiological signal [14]. However, these classical methods often exhibit brittleness under substantial motion and during activity transitions, frequently depending on hand-crafted thresholds or assumptions that limit their adaptability and generalization capabilities [15]. Such limitations typically lead to performance degradation across different subjects and real-world usage conditions, highlighting a need for more robust solutions [16].

The limitations of conventional methods have motivated the exploration of deep learning approaches for robust HR estimation from PPG signals [17]. Deep neural networks are capable of learning complex, non-linear relationships and extracting highly robust features directly from raw or minimally processed PPG data, enhancing their resilience to motion artifacts and physiological variations [18]. This paradigm shift leverages advancements in computational power and large datasets to automatically identify pertinent features that might be difficult to engineer manually. Representative trends in this field include the use of convolutional neural networks for morphological feature extraction [19], recurrent neural networks or attention modules for temporal dynamics [20], and compact architectures suitable for on-device inference [21]. For example, PPGnet operates on 8-s raw PPG and leverages transfer learning for device independence, reporting a mean absolute error of $3.36 \pm 4.1$ bpm on the IEEE SPC 2015 dataset (12 subjects) under leave-one-session-out and five-fold cross-validation (i.e., subject-dependent) [14]. As another example, the CorNET framework (CNN $\pm$ LSTM) evaluated on the SPC dataset achieves an average absolute error of $1.47 \pm 3.37$ bpm for 22 subjects under subject-dependent protocols [6]. While such subject-dependent setups yield strong in-distribution performance, they may overestimate generalization; robust deployment for free-living wrist PPG typically requires subject-independent evaluation (e.g., LOSO) and realistic application settings.

Generalization to unseen individuals is critical for everyday wrist-PPG applications; thus, subject-independent evaluation, typically leave-one-subject-out (LOSO), should be the standard protocol [22]. Under LOSO on the IEEE SPC dataset (22 subjects), CorNET reports $5.55 \pm 4.45$ bpm, whereas Binary-CorNET yields $6.67 \pm 5.49$ bpm; by contrast, CorNET's earlier subject-dependent result on the same dataset was 1.47

± 3.37 bpm [6, 23]. On the free-living PPG-DaLiA dataset, KID-PPG adopts LOSO and achieves 2.85 bpm MAE, underscoring the need for subject-independent evaluation in daily-activity conditions [20]. Although KID-PPG demonstrates robust accuracy on PPG-DaLiA under LOSO evaluation, its knowledge-informed design can substantially increase model complexity, potentially making it less suitable for lightweight deployment on wearable devices.

Motivated by these considerations, this study investigates a PPG-only, lightweight model for heart-rate estimation under daily-life conditions. Our approach provides an end-to-end learning pipeline in which wrist PPG waveforms serve as the sole input. The model is trained and evaluated under a strict LOSO protocol on the PPG-DaLiA dataset, and performance is reported using MAE, RMSE, and Pearson's r, with additional per-activity analyses to assess robustness across daily-life conditions. Despite using PPG only, the proposed model attains accuracy comparable to recent state-of-the-art wrist-PPG methods evaluated under LOSO (e.g., CorNET under LOSO, Binary-CorNET, and KID-PPG), while maintaining a lightweight model suitable for resource-constrained wearables.

## 2 Materials and Methods

### 2.1 Dataset

This study utilized the PPG-DaLiA dataset [24] to evaluate heart rate estimation from wrist-worn photoplethysmography under free-living, daily-activity conditions, moving beyond recordings confined to laboratory settings. The dataset comprises recordings from 15 subjects (7 male, 8 female), with an average age of 30.60 ± 9.59 years. Each recording session lasted approximately 2.5 h, though for subject S6, only the first approximately 1.5 h are valid due to a hardware issue. Physiological signals were collected using two synchronized devices: the wrist-worn Empatica E4 provided PPG/Blood Volume Pulse at 64 Hz and tri-axial accelerometer data at 32 Hz, while chest-worn RespiBAN devices concurrently recorded electrocardiogram, respiration, and accelerometer signals at 700 Hz. The dataset encompasses a comprehensive range of activities with corresponding labels (IDs 0–8) to contextualize the recordings: 0 (transient periods), 1 (sitting), 2 (stairs), 3 (table-soccer), 4 (cycling), 5 (driving), 6 (lunch), 7 (walking), and 8 (working). Ground-truth heart rate labels were meticulously derived from ECG R-peak detection with manual corrections applied to ensure accuracy, particularly in segments affected by severe motion artifacts. Following established conventions in PPG-based heart rate estimation research, HR labels were computed as the mean instantaneous heart rate within 8-s sliding windows with 2-s shifts between consecutive windows.

This methodological approach enables direct comparison with existing literature while providing a robust foundation for evaluating algorithm performance across diverse real-world scenarios that closely mirror the operational conditions of commercial wearable devices (Fig. 1).

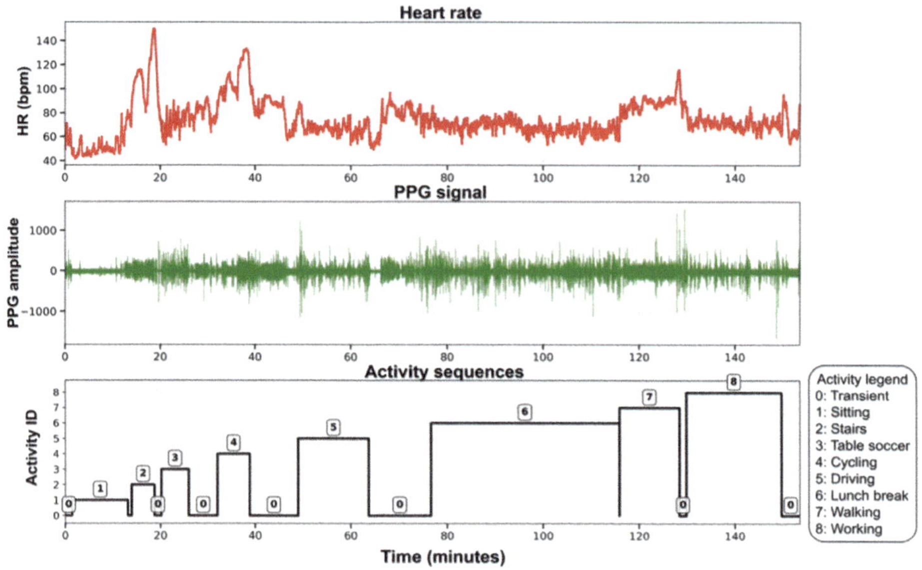


**Fig. 1.** Subject S1, overview of reference HR (top), wrist PPG (middle), and activity sequence (bottom).

## 2.2 Data Processing

The preprocessing procedure was designed to prepare wrist PPG signals for model training (Fig. 2).

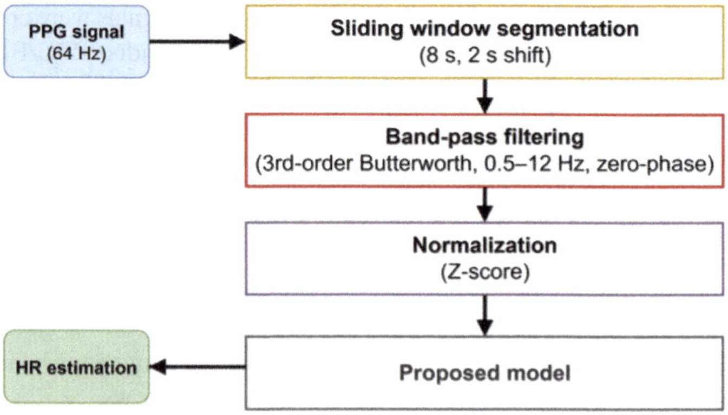


**Fig. 2.** Preprocessing pipeline used in this study.

The signals were segmented into 8-s segments with a 2-s shift, ensuring alignment with the reference HR labels. A third-order Butterworth band-pass filter (0.5–12 Hz) with zero-phase forward–backward implementation was applied to each segment to retain the cardiac band while suppressing baseline drift and high-frequency noise [25]. Each segment was then standardized using per-segment z-score normalization to remove amplitude bias and enhance cross-subject generalization. Only the wrist preprocessed PPG signals were used as input to the proposed model, while HR labels were expressed in beats per minute (bpm).

## 2.3 Proposed Model

The proposed network comprises three sequential components: (i) a feature extractor, (ii) a temporal modeling stage, and (iii) a regression head. The feature extractor applies Conv1D layers with MaxPooling to progressively reduce the temporal dimension from 512 to 128 samples while increasing channel depth from 1 to 16 and then 32 filters. This design enables hierarchical feature extraction, ranging from local morphological patterns in the PPG waveform (e.g., systolic peaks, dicrotic notches) to more abstract cardiac representations. Batch normalization layers stabilize feature distributions across subjects, while a SeparableConv1D layer improves parameter efficiency by factorizing the convolution into depthwise and pointwise operations [26]. Compared with a standard Conv1D, this factorization significantly reduces the parameter count while retaining representational capacity, which is essential for lightweight deployment. The temporal modeling stage applies dropout regularization (rate = 0.20) to mitigate overfitting, followed by a bidirectional GRU with 32 units to process the compressed 128-sample sequences in both forward and backward directions (Fig. 3).

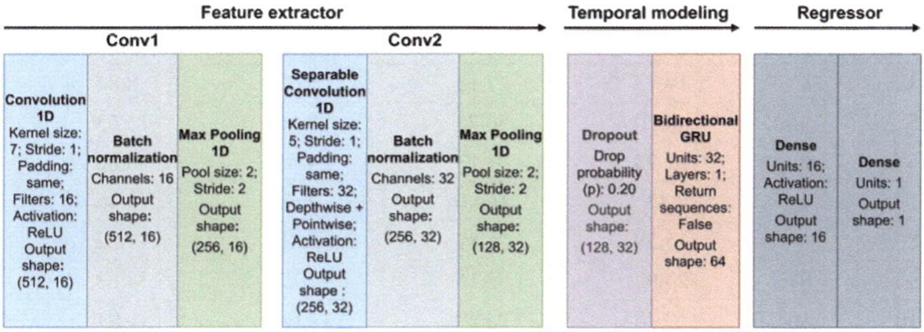


**Fig. 3.** Architecture of the proposed model with feature extractor, temporal modeling, and regressor blocks.

This configuration captures temporal dependencies within each 8-s segment. Finally, the regression head maps the resulting 64-dimensional bidirectional feature vector through two dense layers (16 units with ReLU activation, followed by 1 unit with linear activation) to produce continuous HR estimates. The full network comprises 14,673 parameters, achieving computational efficiency suitable for real-time edge deployment.

## 2.4 Training Setup and Evaluation

Experiments were conducted on Google Colab Pro+ with NVIDIA A100-SXM4-40GB GPU using TensorFlow. The model was trained with batch size 64 for up to 15 epochs, incorporating early stopping (patience = 5). Learning rate was reduced by 0.5 on plateau. HR labels were standardized using training data within each fold.

A LOSO cross-validation protocol was applied across 15 subjects. In each fold, data from one subject served as test set while remaining 14 subjects were used for training and validation [27]. This prevents subject leakage and provides realistic generalization estimates. Performance metrics include MAE, RMSE, and Pearson correlation coefficient (r).

# 3 Results

## 3.1 Evaluation on PPG-DaLiA Dataset

The performance of our proposed model was evaluated using the LOSO cross-validation protocol across all 15 subjects in the PPG-DaLiA dataset.

**Table 1.** LOSO results per fold on PPG-DaLiA.

| Fold | Subject | FFT-based | | Proposed model | |
|---|---|---|---|---|---|
| | | MAE (bpm) | RMSE (bpm) | MAE (bpm) | RMSE(bpm) |
| 1 | 1 | 11.38 | 18.19 | 4.58 | 7.44 |
| 2 | 2 | 13.62 | 22.76 | 4.24 | 7.16 |
| 3 | 3 | 16.24 | 26.21 | 2.99 | 4.95 |
| 4 | 4 | 19.73 | 27.18 | 5.63 | 8.84 |
| 5 | 5 | 40.70 | 56.75 | 8.16 | 15.53 |
| 6 | 6 | 23.93 | 41.13 | 4.64 | 8.82 |
| 7 | 7 | 8.42 | 16.40 | 2.46 | 4.16 |
| 8 | 8 | 15.38 | 21.89 | 8.25 | 11.52 |
| 9 | 9 | 24.02 | 33.16 | 10.21 | 14.63 |
| 10 | 10 | 13.45 | 25.29 | 4.04 | 6.91 |
| 11 | 11 | 24.30 | 38.36 | 5.41 | 9.79 |
| 12 | 12 | 12.99 | 19.44 | 6.69 | 11.20 |
| 13 | 13 | 8.53 | 17.87 | 2.49 | 3.84 |
| 14 | 14 | 12.99 | 22.63 | 3.22 | 6.03 |
| 15 | 15 | 12.84 | 22.25 | 3.96 | 8.24 |
| Mean ± Std | | 17.23 ± 8.33 | 27.30 ± 10.92 | 5.13 ± 2.29 | 8.60 ± 3.48 |

To establish a performance baseline, we also implemented a conventional FFT-based frequency-domain method for heart rate estimation [28]. Table 1 presents the detailed results for each fold, comparing both the FFT-based baseline and our proposed model, demonstrating the model's ability to generalize across diverse individuals under free-living wrist-PPG conditions. The FFT-based method yielded a mean MAE of 17.23 ±

8.33 bpm and RMSE of 27.30 ± 10.92 bpm, with substantial errors particularly during motion-intensive activities. By comparison, our proposed model demonstrates favorable overall accuracy under subject-independent evaluation, with 9 of 15 subjects achieving MAE $\leq$ 5 bpm; together with the macro-average (5.13 ± 2.29 bpm MAE; 8.60 ± 3.48 bpm RMSE), they suggest suitability for continuous wrist-PPG heart rate monitoring in daily-activity conditions. The spread across subjects, ranging from 2.46 bpm (S7) to 10.21 bpm (S9), reflects the inherent challenges of inter-subject generalization in wrist PPG, driven by heterogeneous physiology, strap/placement differences, and activity mix.

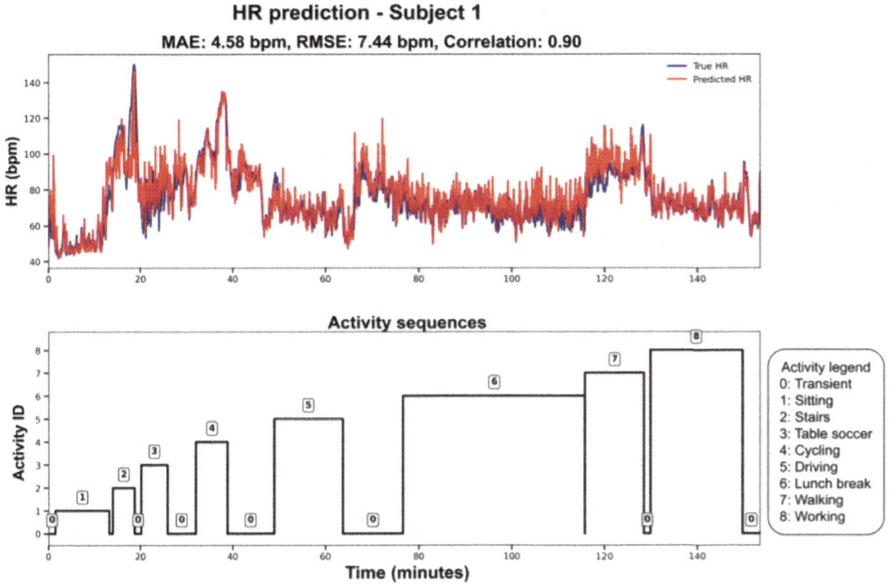


**Fig. 4.** Complete HR prediction for subject 1.

Under LOSO evaluation on PPG-DaLiA, the proposed model was further examined at the individual-subject level. Figure 4 presents the complete heart-rate trajectory for Subject S1 across the 160-min recording. The temporal alignment between predicted and true heart rate values reveals that our proposed model successfully captures the dynamic nature of heart rate during daily activities, including periods of elevated heart rate during physical activities. For this representative subject, the model achieved an MAE of 4.58 bpm, an RMSE of 7.44 bpm, and a correlation of 0.90, indicating reliable performance under free-living conditions. These results illustrate the ability of the proposed end-to-end model to generalize to unseen subjects and maintain accuracy across diverse activity phase.

**Table 2.** MAE, RMSE results by activity type on PPG-DaLiA dataset.

| Activity | FFT-based | | Proposed model | |
|---|---|---|---|---|
| | MAE (bpm) | RMSE (bpm) | MAE (bpm) | RMSE (bpm) |
| Sitting | 3.26 ± 1.44 | 6.05 ± 2.97 | 3.35 ± 3.00 | 4.81 ± 3.91 |
| Stairs | 37.29 ± 16.27 | 46.89 ± 17.19 | 12.93 ± 8.36 | 17.70 ± 10.18 |
| Table soccer | 29.18 ± 16.69 | 34.19 ± 15.81 | 6.12 ± 2.32 | 8.66 ± 3.45 |
| Cycling | 27.23 ± 21.26 | 39.39 ± 21.73 | 6.62 ± 4.36 | 10.00 ± 5.91 |
| Driving | 11.65 ± 8.27 | 18.24 ± 9.98 | 3.46 ± 1.82 | 5.01 ± 2.49 |
| Lunch break | 11.92 ± 6.46 | 18.86 ± 8.85 | 3.57 ± 1.35 | 5.34 ± 2.03 |
| Walking | 28.21 ± 15.34 | 34.90 ± 15.05 | 7.82 ± 5.14 | 10.40 ± 6.07 |
| Working | 7.50 ± 5.47 | 13.48 ± 8.04 | 3.33 ± 2.41 | 5.22 ± 3.22 |

Table 2 further quantifies performance across all activity classes in PPG-DaLiA. The reported values correspond to the average MAE across subject-level test folds under LOSO evaluation, providing a subject-independent estimate of activity-specific performance. The FFT-based baseline exhibits considerable variability across activities, with errors ranging from 3.26 ± 1.44 bpm for sitting to 37.29 ± 16.27 bpm for stairs, demonstrating difficulty during motion-intensive tasks. The proposed model shows improved consistency across activities. The model attains the lowest errors during stationary activities, with MAE values of 3.33 ± 2.41 bpm for working, 3.35 ± 3.00 bpm for sitting, and 3.46 ± 1.82 bpm for driving. Physical activities yield higher errors, particularly stair climbing (12.93 ± 8.36 bpm) and walking (7.82 ± 5.14 bpm), reflecting the strong influence of motion artifacts on wrist-PPG signals during ambulatory movements. Moderate errors are observed for cycling (6.62 ± 4.36 bpm) and table soccer (6.12 ± 2.32 bpm), while lunch break activities remain low (3.57 ± 1.35 bpm) due to their stationary nature. The corresponding RMSE values show a consistent trend, with higher variance under motion-intensive activities. These findings confirm that the proposed model sustains clinically acceptable accuracy across a wide spectrum of daily activities, while also highlighting motion-intensive tasks as the principal challenge for wrist-worn PPG monitoring.

Figure 5 presents a scatter plot comparing true versus predicted HR values across all test samples from the LOSO evaluation. The strong linear relationship (correlation coefficient = 0.86) indicates excellent agreement across the full physiological HR range. Data points cluster closely around the identity line, suggesting minimal systematic bias, while the relatively uniform spread across the axis confirms that prediction errors are not concentrated at specific HR ranges. These characteristics highlight the robustness of the proposed model for continuous HR monitoring across diverse physiological states and activity levels.

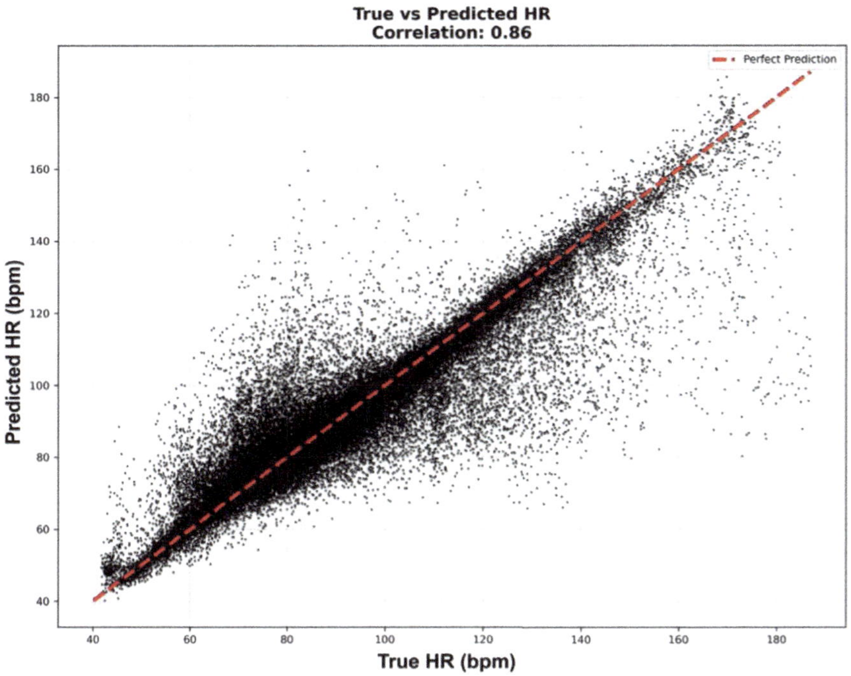


**Fig. 5.** Scatter plot of predicted versus reference HR across all LOSO test samples, yielding a correlation of 0.86.

### 3.2 Cross-Study Comparisons

Table 3 summarizes the performance of the proposed model against representative state-of-the-art methods on the PPG-DaLiA dataset under identical LOSO evaluation. This facilitates a direct and fair comparison with existing benchmarks.

Our proposed model achieved an MAE of 5.13 bpm using only PPG signals, which represents a significant advancement compared to several existing approaches. When compared to Deep PPG [24], which reported an MAE of 9.99 bpm using both PPG and accelerometer data, our model demonstrates a 48.5% improvement in accuracy while requiring fewer input modalities. This reduction in sensor requirements is particularly advantageous for practical deployment scenarios where minimizing hardware complexity and power consumption are critical considerations.

Compared with NAS-PPG [19], which achieved an MAE of 6.02 bpm with a model size of 0.8M parameters, our approach attains comparable accuracy with only 14.7K parameters. This substantial reduction in model complexity, achieved without compromising accuracy, demonstrates the effectiveness of our architectural design choices and preprocessing strategy. Such parameter efficiency directly translates to reduced computational overhead and enhanced suitability for edge deployment.

Notably, Q-PPG [2] achieved a superior MAE of 4.36 bpm on the same dataset; however, this approach similarly relies on both PPG and accelerometer signals. Our model's performance using PPG signals alone positions it as a compelling alternative for applications where accelerometer data may be unavailable or unreliable. Additionally, the Multi-Headed Conv-LSTM approach [9] reported an MAE of 6.28 bpm, which our model surpasses by approximately 18%.

**Table 3.** Comparison with prior work on PPG-DaLiA under LOSO (K = thousands, M = millions (for parameter counts)).

| Model | MAE (bpm) | Parameters | Protocol | Signal |
|---|---|---|---|---|
| Deep PPG [24] | 9.99 | 26 K | LOSO | PPG + ACC |
| NAS-PPG [19] | 6.02 | 0.8 M | LOSO | PPG + ACC |
| Q-PPG [2] | 4.36 | – | LOSO | PPG + ACC |
| MH Conv-LSTM DeepPPG [9] | 6.28 | – | LOSO | PPG + ACC |
| KID-PPG [20] | 2.85 | – | LOSO (with retention filtering) | PPG + ACC |
| Our proposed model | 5.13 | 14.7 K | LOSO | PPG only |

The most relevant comparison emerges with KID-PPG [20], which achieved an exceptional MAE of 2.85 bpm under LOSO evaluation with retention filtering. While this represents superior accuracy, it is important to note that KID-PPG employs additional data filtering techniques and knowledge-informed constraints that may limit its applicability in real-time scenarios. In contrast, our proposed model achieves competitive performance through a more straightforward preprocessing pipeline, potentially offering advantages in terms of implementation simplicity and computational efficiency.

Furthermore, our approach distinguishes itself by demonstrating that effective heart rate estimation can be achieved using only raw PPG signals, without requiring auxiliary sensor data or complex preprocessing schemes. This finding has important implications for the development of next-generation wearable devices, where sensor integration constraints and power limitations often necessitate simplified sensing configurations.

The comparative analysis underscores that our proposed model successfully balances multiple competing objectives: achieving high accuracy, maintaining computational efficiency, and minimizing sensor requirements. These characteristics collectively position our approach as a practical solution for continuous heart rate monitoring in real-world deployment scenarios.

## 4 Conclusion

We proposed a lightweight deep learning approach for robust heart rate estimation from wrist-worn PPG signals under free-living conditions. Evaluated using a strict LOSO protocol on the PPG-DaLiA dataset with 15 subjects, the model achieved an MAE of 5.13 bpm and a correlation of 0.86 with ECG-derived reference using PPG-only. The compact architecture combines 1D convolutional feature extraction with bidirectional GRU temporal modeling, totaling 14.7K parameters suitable for edge deployment. The results demonstrate effective cross-subject generalization while maintaining practical computational efficiency for continuous wearable monitoring applications.

**Acknowledgment.** This research is funded by Posts and Telecommunications Institute of Technology under grant number 01-2025-HV-KTĐT2.

## References

1. Tan, Y., et al.: Real-time heart rate estimation algorithm based on adaptive spectrum correction and peak localization. IEEE Sens. J. **24**, 31551 (2024)
2. Burrello, A., et al.: Q-PPG: energy-efficient PPG-based heart rate monitoring on wearable devices. IEEE Trans. Biomed. Circuits Syst. **15**, 1196 (2021)
3. Wójcikowski, M., Pankiewicz, B.: Photoplethysmographic time-domain heart rate measurement algorithm for resource-constrained wearable devices and its implementation. Sensors **20**, 1783 (2020)
4. Allen, J.: Photoplethysmography and its application in clinical physiological measurement. Physiol. Meas. **28**(3), R1–R39 (2007)
5. Almarshad, M.A., Islam, M.S., Al-Ahmadi, S., BaHammam, A.S.: Diagnostic features and potential applications of PPG signal in healthcare: a systematic review. Healthcare **10**, 547 (2022)
6. Biswas, D., Simões-Capela, N., Van Hoof, C., Van Helleputte, N.: Heart rate estimation from wrist-worn photoplethysmography: a review. IEEE Sensors J. (2019)
7. Park, P., Lee, W., Cho, S.: An adaptive filter-based motion artifact cancellation technique using multi-wavelength PPG for accurate HR estimation. IEEE Trans. Biomed. Circ. Syst. **17**, 1074 (2023)
8. Biswas, D., et al.: CorNET: deep learning framework for PPG-based heart rate estimation and biometric identification in ambulant environment. IEEE Trans. Biomed. Circ. Syst. **13**, 282 (2019)
9. Wilkosz, M., Szczęsna, A.: Multi-headed Conv-LSTM network for heart rate estimation during daily living activities. Sensors **21**, 5212 (2021)
10. Kim, J.W., Choi, S.: Normalization of photoplethysmography using deep neural networks for individual and group comparison. Sci. Rep. **12** (2022)
11. Beh, W.-K., Yang, Y.-C., Wu, A.-Y.: Quality-aware signal processing mechanism of PPG signal for long-term heart rate monitoring. Sensors **24**, 3901 (2024)
12. Charlton, P., et al.: Wearable photoplethysmography for cardiovascular monitoring. Proc. IEEE **110**, 355 (2022)
13. Islam, M.T., et al.: A time-frequency domain approach of heart rate estimation from photoplethysmographic (PPG) signal. Biomed. Signal Process. Control **36**, 146 (2017)
14. Shyam, A., Ravichandran, V., Preejith, S.P., Joseph, J., Sivaprakasam, M.: PPGnet: deep network for device-independent heart rate estimation from photoplethysmogram. In: Proceedings of IEEE EMBC (2019)
15. Ismail, S., Akram, U., Siddiqi, I.: Heart rate tracking in photoplethysmography signals affected by motion artifacts: a review. EURASIP J. Adv. Signal Process. (2021)
16. Guo, J., Chen, X., Zhao, J., Zhang, X., Chen, X.: An effective photoplethysmography heart rate estimation framework integrating two-level denoising method and heart rate tracking algorithm guided by finite state machine. IEEE J. Biomed. Health Inform. **26**, 3731 (2022)
17. Panwar, M., Gautam, A., Biswas, D., Acharyya, A.: PP-Net: a deep learning framework for PPG-based blood pressure and heart rate estimation. IEEE Sens. J. **20**, 10000 (2020)
18. Xiao, H., et al.: Remote photoplethysmography for heart rate measurement: a review. Biomed. Signal Process. Control **80**, 105608 (2023)
19. Song, S.B., Nam, J., Kim, J.H.: NAS-PPG: PPG-based heart rate estimation using neural architecture search. IEEE Sens. J. **21**, 14941 (2021)

20. Kechris, C., Dan, J., Miranda, J., Atienza, D.: KID-PPG: knowledge informed deep learning for extracting heart rate from a smartwatch. IEEE Trans. Biomed. Eng. **72**, 870 (2024)
21. Wen, H., et al.: Wearable integrated online fusion learning filter for heart PPG sensing and tracking. IEEE Sens. J. **23**, 14938 (2023)
22. Jindal, G.D., et al.: Photoplethysmography and its clinical application. MGM J. Med. Sci. **4**, 89 (2017)
23. Rocha, L.G., et al.: Binary CorNET: accelerator for HR estimation from wrist-PPG. IEEE Trans. Biomed. Circ. Syst. **14**(4) (2020)
24. Reiss, A., Indlekofer, I., Schmidt, P., Laerhoven, K.V.: Deep PPG: large-scale heart rate estimation with convolutional neural networks. Sensors **19**, 3079 (2019)
25. Meier, M., Demirel, B.U., Holz, C.: WildPPG: a real-world PPG dataset of long continuous recordings. In: Proceedings of the 38th International Conference on Neural Information Processing Systems (NeurIPS 2024), vol. 37, pp. 2246–2266 (2025)
26. Chollet, F.: Xception: deep learning with depthwise separable convolutions. In: 2017 IEEE Conference on Computer Vision and Pattern Recognition (CVPR), pp. 1800–1807 (2017)
27. Tougui, I., Jilbab, A., Mhamdi, J.E.: Impact of the choice of cross-validation techniques on the results of machine learning-based diagnostic applications. Healthc. Inform. Res. **27**(3), 189–199 (2021)
28. Mohd Sani, N.H., Mansor, W., Lee, K.Y., Ahmad Zainudin, N., Mahrim, S.A.: Determination of heart rate from photoplethysmogram using Fast Fourier Transform. In: 2015 International Conference on BioSignal Analysis, Processing and Systems (ICBAPS), pp. 168–170. Kuala Lumpur, Malaysia (2015)

# MedMacau: A Four-Language Medical Term Normalization Dataset for Macau

Zehan Tan, Weihe Zhong, and Yaoyang Wu(✉)

Macau Millennium College, Macau SAR, China
{zhtan,whzhong,sunnywu}@mmc.edu.mo

**Abstract.** Healthcare in Macau is conducted across four languages, yet patients frequently express symptoms in colloquial Cantonese that diverge from standardized medical terminologies. This mismatch complicates accurate coding and downstream clinical applications. To address this gap, we present **MedMacau**, the first four-language dataset for medical term normalization, comprising over 10,000 internal medicine entries with parallel Chinese, English, Portuguese, and curated Cantonese fields. We further introduce a rule-guided retrieval-augmented baseline that constrains predictions to candidate terms and integrates domain rules with confidence-based aggregation. Experiments on a fixed evaluation slice show near-perfect retrieval recall, while highlighting disambiguation among near-synonyms and site/type variants as the primary challenge. **MedMacau** establishes a benchmark for normalization in low-resource, culturally specific contexts and provides a practical foundation for multilingual clinical NLP in Macau and beyond. The code source and dataset are open at https://github.com/zehantan6970/MedMacau.

**Keywords:** Medical term normalization · Multilingual clinical NLP · Cantonese colloquial · MedMacau dataset

## 1 Introduction

The Macau Special Administrative Region presents a highly multilingual clinical environment where *Cantonese*, *Mandarin Chinese*, *English*, and *Portuguese* coexist in healthcare delivery [1,2]. Clinical documentation typically relies on standardized Chinese and English terminologies such as MedDRA [3] and ICD-11 [4], whereas patients frequently describe their symptoms using informal, metaphorical, and culturally grounded colloquial Cantonese expressions [5]. Common descriptions such as "肚痛" (belly pain) or "气唔顺" (shortness of breath) diverge noticeably from formal medical phrasing and can hinder precise clinical interpretation and coding.

This register mismatch poses systemic challenges. Manual interpretation by healthcare staff introduces risks of semantic drift, site/type confusion, and cross-cultural misalignment in recorded symptoms. Studies indicate that medical terms are often misunderstood by patients [6], particularly when mediated through


P. Pichappan et al. (Eds.): RTIS 2025, LNNS 1936, pp. 131–142, 2027.
https://doi.org/10.1007/978-3-032-23396-7_12

informal dialects such as colloquial Cantonese [7]. These difficulties are magnified in Macau, where four languages must be coordinated in real time.

Biomedical NLP has progressed through neural machine translation and large language models (LLMs) [8,9]. However, most multilingual clinical datasets focus on high-resource languages, overlook low-resource dialects like Yue Chinese [10,11], and target sentence-level translation rather than *entity-level* normalization [12]. Colloquial Cantonese additionally exhibits code-switching, abbreviation variants, and Traditional Chinese Medicine concepts that general-purpose systems do not reliably capture [13,14].

We formalize **medical term normalization** from colloquial Cantonese to standardized Chinese/English vocabularies. Given a patient utterance $x$, the goal is to output a single MedDRA term $t$:

$$f: \mathcal{X} \to \mathcal{V}, \qquad \hat{t} = f(x), \tag{1}$$

where $\mathcal{V}$ denotes the target vocabulary. This requires *terminological precision*: semantically close but clinically distinct terms (e.g., *pain* vs. *discomfort*, *burning* vs. *bloating*) must not be conflated.

Our contributions are:

- **MedMacau dataset**: the first Macau-localized four-language normalization dataset aligned to MedDRA, with 10,162 internal medicine entries including curated *colloquial Cantonese* fields.
- **Rule-guided RAG baseline**: a training-free retrieval-augmented approach that injects simple domain rules and constrains LLM outputs to single-choice selections among retrieved candidates. Inspired by DeepConf [15], it combines explicit confidence elicitation with retrieval similarity in a simple aggregation scheme.

**MedMacau** fills an important resource gap and offers a practical starting point for aligning patient narratives with standardized clinical ontologies in Macau's multilingual healthcare system.

## 2 Related Works

### 2.1 Multilingual Biomedical Corpora

Parallel corpora such as ParaMed [12] and the WMT biomedical shared tasks [16] have advanced biomedical machine translation. Benchmarks like RuMedBench [17] and surveys of non-English clinical NLP [10,18] highlight both progress and persistent gaps: most resources target high-resource languages and formal written text, with limited coverage of Asian languages and none for Cantonese colloquial. MedDialog-style datasets [19,25] address doctor–patient conversations in Mandarin but do not include Cantonese or four-way alignment across Chinese, English, Portuguese, and Cantonese.

### 2.2 Cantonese NLP and Clinical Communication

Cantonese lexical resources such as WORDS.hk [14] provide broad coverage but are not aligned to controlled medical terminologies. CantonMT [13] studies Cantonese–English neural MT and documents low-resource challenges. Surveys of Cantonese ASR [23,24] report resource scarcity and focus on general domains. Work on Cantonese doctor–patient communication [5] and code-switching benchmarks [22,26] describes interaction patterns and multilingual issues but does not offer term-aligned supervision.

### 2.3 Medical Concept Normalization and Confidence-Based Reasoning

Medical concept normalization has evolved from lexicon/IR pipelines (e.g., tmChem [27]) to neural encoders such as BioBERT [28] and UMLS-aware models [29]. Work on social media and clinical narratives highlights lay variation and ambiguity [30,31] but largely in English and monolingual settings. Cross-lingual normalization, particularly for Cantonese colloquial, remains under-explored [11].

Recent LLM work has explored *confidence-based* reasoning. Deep Think with Confidence (DeepConf) [15] elicits self-reported confidence, filters low-confidence samples, and aggregates candidates. Our rule-guided retrieval-augmented generation (RAG) baseline adopts this high-level idea: retrieve a small set of candidate terms, prompt the LLM to select one candidate ID with an explicit confidence score, and combine self-reported confidence with retrieval similarity, under simple clinical rules (e.g., "pain $\neq$ discomfort; burning $\neq$ bloating").

## 3 Dataset Construction and Characteristics

### 3.1 Scope and Schema

**MedMacau** targets internal medicine concepts anchored to MedDRA v28.0 [3]. Each entry corresponds to a MedDRA concept and includes surface forms in Chinese (Zh), English (En), and, when available, Portuguese (Pt), reflecting Macau's documentation practices. We extend each concept with two Cantonese fields: a *formal* Yue rendering (clinician-facing) and a *colloquial* Yue variant (patient-facing).

Each record contains a stable `Code` (concept identifier), surface strings in {Zh, En, Pt}, `Cantonese` (formal), `Cantonese_Colloquial` (patient-facing), and a System Organ Class (`SOC`) tag. The schema is summarized in Table 1.

### 3.2 Construction Workflow

We adopt a *generate–refine–validate* workflow to construct the Cantonese fields (Fig. 1).

**Table 1.** Record schema for **MedMacau**.

| Field | Description |
|---|---|
| Code | MedDRA concept identifier |
| Chinese (Zh) | Chinese surface form |
| English (En) | English surface form |
| Portuguese (Pt) | Portuguese rendering (when available) |
| Cantonese | Formal Cantonese (clinician-facing) |
| Cantonese_Colloquial | Colloquial Cantonese (patient-facing) |
| SOC | System Organ Class |

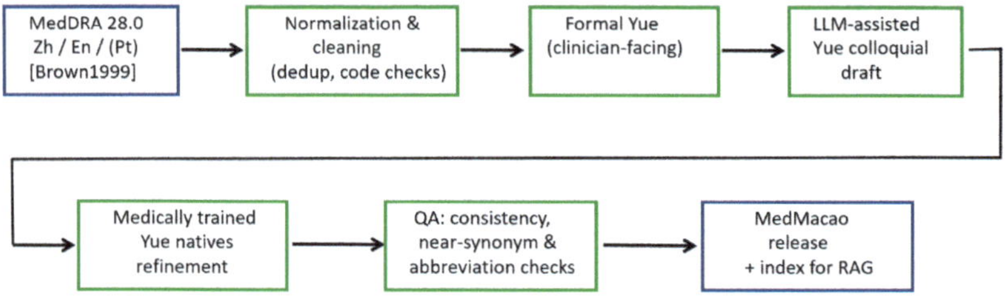


**Fig. 1.** Dataset construction pipeline. MedDRA terms (Zh/En/Pt) are normalized and paired with a formal Cantonese rendering; an LLM-assisted colloquial Cantonese draft is refined by medically trained Cantonese natives and then quality-controlled.

**Preprocessing.** We normalize punctuation and typography, remove duplicates, and perform sanity checks on MedDRA codes and surface strings. Portuguese renderings are retained when present.

**Colloquial Yue Creation.** An LLM is prompted to produce patient-style Cantonese variants from formal terms. These drafts are reviewed and refined by medically trained Cantonese native annotators to ensure naturalness and to avoid semantic drift (for example, avoiding mapping "不適" directly to *pain* or conflating burning with bloating).

**Quality Assurance.** We enforce a one-to-one mapping between concept ID and preferred term (PT) per language, separate near-synonyms that are not code-equivalent (e.g., pain vs. discomfort), and flag abbreviation or orthographic variants. Manual spot checks focus on high-frequency SOCs such as gastrointestinal and respiratory disorders.

### 3.3 Coverage and Variation

The released version contains 10,162 entries. Table 2 reports the SOC distribution, with a large gastrointestinal share reflecting symptom patterns in primary and emergency care.

**Table 2.** Dataset size and SOC distribution.

| SOC | Entries | % |
|---|---|---|
| Gastrointestinal disorders | 4,769 | 47.0 |
| Infections and infestations | 1,888 | 18.6 |
| Respiratory, thoracic, mediastinal | 993 | 9.8 |
| Hematologic disorders | 497 | 4.9 |
| Renal and urinary disorders | 497 | 4.9 |
| Endocrine and metabolic | 497 | 4.9 |
| Hepatobiliary disorders | 696 | 6.8 |
| Nervous system disorders | 227 | 2.2 |
| Other | 98 | 1.0 |
| **Total** | **10,162** | **100.0** |

Colloquial Cantonese in `Cantonese_Colloquial` exhibits characteristic variation: near-synonyms that are not code-equivalent (e.g., "肚痛" vs. "肚唔舒服"), coupling of symptom type and anatomical site (burning vs. bloating; chest vs. abdomen), abbreviations (Abd./Abdo/Abdomen), and culture-bound notions such as "上火". These patterns motivate term normalization as disambiguation against a controlled vocabulary rather than free-form translation.

### 3.4 Release and Uses

We release (i) the full CSV/XLSX with the schema in Table 1; (ii) a fixed test slice of 100 items (`test_yuecol_n100.xlsx`, seed = 42) for reproducible evaluation; and (iii) prompt templates and evaluation scripts. Potential uses include EMR intake and coding, multilingual search and retrieval-augmented generation, model evaluation and fine-tuning, and creation of multilingual glossaries for clinician and patient education in Macau's four-language ecosystem.

## 4 Task and Rule-Guided RAG Baseline

### 4.1 Variation in Colloquial Cantonese (Yue Colloquial)

Colloquial Cantonese shows high lexical and pragmatic variability relative to formal clinical terms. Typical patterns include: (i) near-synonyms that are not code-equivalent (e.g., "肚痛" vs. "肚唔舒服", pain vs. discomfort); (ii) coupling between symptom type and anatomical site (burning vs. bloating; chest vs. abdomen); (iii) abbreviations and orthographic variants (Abd., Abdo, Abdomen); and (iv) culture-bound expressions such as "上火". These patterns motivate treatment of normalization as disambiguation against a controlled vocabulary rather than free-form translation.

### 4.2 Rule-Guided RAG Baseline

We implement a training-free pipeline that combines retrieval with rule-regularized disambiguation, inspired by DeepConf [15]. Figure 2 illustrates the architecture.

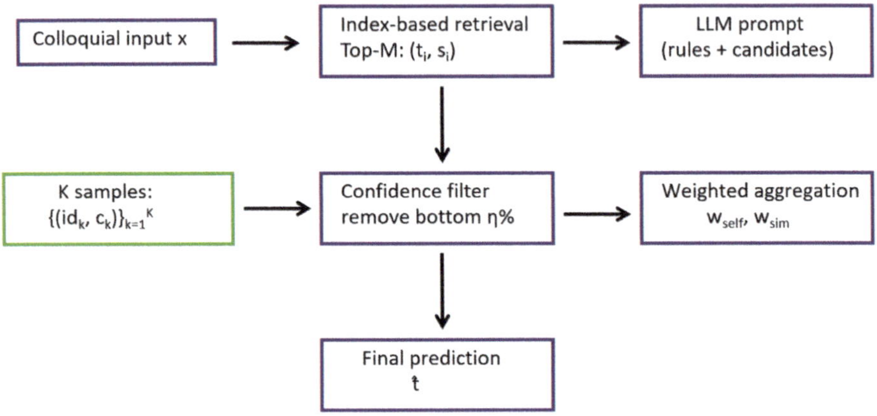


**Fig. 2.** Rule-guided RAG baseline. The system retrieves Top-$M$ candidates, elicits $K$ confident single-choice votes under rule-regularized prompting, filters low-confidence samples, and aggregates votes with retrieval similarity.

*Retrieval.* We build a TF–IDF index over the Chinese and English surface forms. For a query $x$, we retrieve Top-$M$ candidates

$$\mathcal{C}(x) = \{(t_i, s_i)\}_{i=1}^{M},$$

where $t_i$ is a candidate term and $s_i \in [0, 1]$ is cosine similarity. In our experiments, the gold surface almost always appears in $\mathcal{C}(x)$ for modest values of $M$.

*LLM Decision with Domain Rules.* An LLM receives the patient expression $x$ and the candidate list $\mathcal{C}(x)$ (IDs and surface forms) and must select exactly one candidate ID with an explicit self-reported confidence $c_k \in [0, 1]$. The prompt includes a concise rule snippet to regularize disambiguation, for example:

- pain $\neq$ discomfort; burning $\neq$ bloating
- prefer exact symptom-type + anatomical site
- abbreviations are not automatically equal
- do not invent terms; choose one listed ID

We draw $K$ independent samples at temperature $T$, obtaining $\{(c^{(k)}, t^{(k)})\}_{k=1}^{K}$.

*Confidence-Aware Aggregation.* Following DeepConf, we sort samples by confidence and discard a low-confidence tail. For each candidate $t_j$ that appears in the retained set, we compute the average self-confidence $\bar{c}_j$ and reuse its retrieval similarity score $\bar{s}_j$. A convex combination

$$\text{Score}(t_j) = w_{\text{self}}\bar{c}_j + w_{\text{sim}}\bar{s}_j, \qquad w_{\text{self}} + w_{\text{sim}} = 1,$$

is used to choose the final prediction $\hat{t} = \arg\max_j \text{Score}(t_j)$. This simple design is intended to emphasize terminology precision by combining LLM preferences and retrieval grounding, without model retraining.

**Algorithm 1** Rule-guided RAG with DeepConf-style confidence filtering (training-free).

**Input:.** Colloquial input $x$; index $\mathcal{I}$; $M, K, \eta$; weights $w_{\text{self}}, w_{\text{sim}}$.
**Output:** Normalized term $t \in \{t_i\}_{i=1}^{M}$.
Retrieve Top-$M$ candidates $\{(t_i, s_i)\}_{i=1}^{M} \leftarrow \text{Retrieve}(x, \mathcal{I})$
Initialize votes $V \leftarrow \emptyset$
**for** $k = 1$ **to** $K$ **do**
  Query LLM with (rules + candidate list + $x$) $\rightarrow \{(\text{id}_k, c_k)\}$
  $V \leftarrow V \cup \{(\text{id}_k, c_k)\}$
**end**
Remove bottom $\eta$% by confidence: $V \leftarrow \text{FilterTail}(V, \eta)$
Group votes by candidate: $\{\mathcal{K}_j\}$
**for** *each candidate* $t_j$ *with* $\mathcal{K}_j \neq \emptyset$ **do**
  $\bar{c}_j \leftarrow \text{mean}\{c_k : (\text{id}_k, c_k) \in \mathcal{K}_j\}$
  $\bar{s}_j \leftarrow \text{mean}\{s_j\}$ (reuse retrieval similarity)
  $\text{Score}(t_j) \leftarrow w_{\text{self}}\bar{c}_j + w_{\text{sim}}\bar{s}_j$
**end**
**return** $t_j$ where $j = \arg\max_j \text{Score}(t_j)$

*Design Rationale and DeepConf Influence.* The algorithm adopts three ideas from DeepConf: (i) **explicit confidence elicitation** in the prompt, (ii) **tail (lowest-group) filtering** to discard low-confidence generations, and (iii) **aggregation** over retained candidates. Our additions are domain-specific: rule-regularized prompting enforces clinical constraints (pain $\neq$ discomfort; site/type consistency; no invention), and retrieval similarity anchors votes to attested terms, yielding a robust, training-free normalizer.

## 5 Experiments

### 5.1 Data and Test Slice

We evaluate on **MedMacau** (MedDRA v28.0, internal medicine scope). To avoid sampling variance, we fix a test slice of $N = 100$ items, `test_yuecol_n100.xlsx` (seed = 42), drawn from frequent SOCs. All systems use exactly this slice.

### 5.2 Systems

We consider three decision modes on top of the same retriever:

- **Top-1 Retrieval**: purely TF–IDF; predicts the candidate with highest similarity.
- **RAG + Majority**: LLM chooses one candidate ID per sample; we take the candidate with most votes.
- **RAG + DeepConf-style aggregation**: the rule-guided RAG baseline described in Sect. 4, combining confidence filtering and similarity-weighted scores.

In RAG modes we test DeepSeek and GPT-4o via their official APIs. Hyperparameters include the number of samples $K$, candidate set size $M$, confidence retention percentage, and temperature $T$.

### 5.3 Metrics

Given test triples $\{(x_i, t_i^{\text{zh}}, t_i^{\text{en}})\}_{i=1}^{N}$:

*Strict Textual Accuracy (Primary).*

$$\text{Acc} = \frac{1}{N}\sum_{i=1}^{N} \mathbf{1}\left[\hat{t}_i = t_i^{\text{zh}} \ \vee \ \hat{t}_i = t_i^{\text{en}}\right].$$

This metric is intentionally stringent: surface variants (e.g., “Abd. pain” vs. “Abdo pain”) are counted as errors.

*Retrieval Oracle (Hit@M).*

$$\text{Hit@}M = \frac{1}{N}\sum_{i=1}^{N} \mathbf{1}\left[t_i^\star \in \mathcal{C}(x_i)\right],$$

where $t_i^\star$ is the gold surface form. In our experiments, Hit@$M \approx 1.0$ for $M \in \{5, 10\}$. Because Hit@M depends only on the retriever and candidate size (and not on the decision mode), we report it once here and omit it from the main results table. Evaluating accuracy at the concept (PT) level is left for future work (see Sect. 6).

### 5.4 Main Results

Table 3 shows strict textual accuracy on `test_yuecol_n100.xlsx`. Retrieval achieves near-perfect Hit@M, but end-to-end accuracy remains low under the strict surface-form metric.

The best setting among our runs is RAG + DeepConf with $K = 8$ and low retention (ID C, 0.09 accuracy). Scaling the number of samples $K$ or varying

**Table 3.** Strict textual accuracy on the fixed test slice (N = 100). All systems share the same retriever; Hit@M is $\approx 1.0$ for $M \in \{5, 10\}$.

| ID | Mode | Model | $K$ | $M$ | Retention | $T$ | Acc |
|---|---|---|---|---|---|---|---|
| A | Top-1 Retrieval | – | – | 1 | – | – | 0.06 |
| B | RAG + Majority | DeepSeek | 16 | 10 | – | 0.2 | 0.06 |
| C | RAG + DeepConf | DeepSeek | 8 | 10 | 10% | 0.2 | 0.09 |
| D | RAG + DeepConf | DeepSeek | 16 | 10 | 10% | 0.2 | 0.06 |
| E | RAG + DeepConf | DeepSeek | 32 | 10 | 10% | 0.2 | 0.06 |
| F | RAG + DeepConf | GPT-4o | 16 | 10 | 10% | 0.2 | 0.06 |
| G | RAG + DeepConf | GPT-4o | 16 | 10 | 90% | 0.2 | 0.06 |
| H | RAG + DeepConf | DeepSeek | 16 | 5 | 90% | 0.7 | 0.06 |

retention in isolation does not yield consistent gains. Overall, the results emphasize that, under strict surface matching, disambiguation among near-neighbor candidates is the main bottleneck.

Note that the strict surface-form metric penalizes clinically equivalent predictions that differ in abbreviation or spelling; concept-level accuracy would likely be substantially higher.

### 5.5 Implementation and Variables

**Models.** DeepSeek and GPT-4o via official APIs.

**Hyper-Parameters.** $K \in \{8, 16, 32\}$ (samples), $M \in \{5, 10, 20\}$ (candidates), $\eta \in \{10, 90\}$ (retention %), $T \in \{0.2, 0.7\}$ (temperature), $w_{\text{sim}}, w_{\text{self}} \in [0, 1]$ with $w_{\text{sim}} + w_{\text{self}} = 1$, and `fewshot`$\in \{0, 1, 2\}$.

**Determinism.** We fix the test slice (seed = 42) and report single-run results; LLM sampling introduces inherent stochasticity.

### 5.6 Ablations and Practical Takeaways

$K$: More samples help only when retention/diversity is sufficient; otherwise votes are homogeneous.

$\eta$: Raising retention ($10 \rightarrow 90$) without diversity ($T$) is insufficient.

$M$: Keep $M$ small-to-moderate (5–10) to reduce near-neighbor interference.

*Low-Cost, High-Yield Adjustments.* (1) **Rule snippet** in the prompt (pain $\neq$ discomfort; burning $\neq$ bloating; type+site alignment; abbreviations $\neq$ same code); (2) **Candidate folding** to merge abbreviation/orthographic equivalents before voting; (3) Report **concept-ID** accuracy in addition to strict text, aligning the metric with the clinical normalization goal.

## 6 Discussion

On the fixed 100-item test slice, the retriever performs very well: for $M \in \{5, 10\}$, the gold term is almost always present in the Top-$M$ candidate list (Hit@M $\approx$ 1.0). A simple TF–IDF index over Chinese and English surface forms is therefore sufficient to surface plausible candidates in this setting.

The end-to-end strict textual accuracy remains low, in the range of 6–9% (Table 3). The best configuration is the rule-guided RAG variant with $K = 8$ and low confidence retention (ID C, 0.09 accuracy). Increasing the number of samples $K$, varying the retention rate, or switching between DeepSeek and GPT-4o does not produce consistent gains under this metric. The bottleneck is thus single-choice disambiguation among near neighbours, not retrieval.

Qualitative inspection points to three main error types: (i) near-synonym confusions (e.g., abdominal pain vs. abdominal discomfort); (ii) abbreviations and surface variants (e.g., "Abd." vs. "Abdo"); and (iii) mismatches of symptom type or anatomical site (e.g., burning vs. bloating, chest vs. abdomen). All are counted as errors by the strict surface-form metric, even when the prediction is clinically close to the reference. The rule-guided baseline helps to reduce some of these confusions by constraining outputs to retrieved terms, adding brief clinical rules, and combining self-reported confidence with retrieval similarity, but the numerical improvements are modest on this small slice.

It is also important to distinguish surface-form from concept-level correctness. In MedDRA, multiple Lower Level Terms (LLTs) may map to the same Preferred Term. Our metric treats different LLTs as distinct labels, so abbreviation or spelling variants that share a PT are penalized. A complementary PT-level evaluation, in which predicted and gold terms are compared after mapping to PT IDs, would better reflect concept-level normalization. We leave such concept-level reporting to future work and use strict surface-form accuracy here for clarity and comparability.

## 7 Conclusion

This paper introduced **MedMacau**, a four-language dataset that aligns colloquial Cantonese symptom expressions with standardized MedDRA terminology in Chinese and English for internal medicine. The resource addresses a concrete need in Macau's multilingual healthcare setting, where patients frequently use colloquial Cantonese while documentation and coding rely on controlled vocabularies.

We also presented a lightweight rule-guided retrieval-augmented baseline. The system combines TF–IDF retrieval, concise domain rules, and confidence-aware aggregation to emphasize terminological precision without model retraining. Experiments on a fixed 100-item slice showed that retrieval achieves near-perfect Hit@M, whereas strict surface-form accuracy remains low and is dominated by near-synonym, site/type, and orthographic confusions. These findings indicate that future work should focus on better disambiguation strategies and on concept-level evaluation (e.g., MedDRA PT IDs), as well as extending MedMacau to additional clinical domains and modalities such as dialogue and speech.

**Acknowledgements.** This work was supported by the China International Communications Group (CICG), Grant No. 25ATILX01.

## References

1. Yan, X.: The language situation in Macau. Curr. Issu. Lang. Plan. **18**(1), 1–38 (2017)
2. Forray, A.I., et al.: Uncovering multi-level mental healthcare barriers for migrants: a qualitative analysis across China, Germany, Netherlands, Romania, and South Africa. BMC Public Health **24**(1), 1593 (2024)
3. Brown, E.G., Wood, L., Wood, S. The Medical Dictionary for Regulatory Activities (MedDRA). Drug Safety (1999)
4. Harrison, J.E., Weber, S., Jakob, R., Chute, C.G.: ICD-11: an international classification of diseases for the twenty-first century. BMC Med. Inform. Decis. Mak. **21**(Suppl 6), 206 (2021)
5. Fung, K.C., et al.: Analysing Cantonese doctor-patient communication: a semantic network approach. Hong Kong Polytechnic University (2019)
6. Gotlieb, R., et al.: Accuracy in patient understanding of common medical phrases. JAMA Netw. Open **5**(11), e2242972–e2242972 (2022)
7. Xiang, R., et al.: Cantonese natural language processing in the transformers era: a survey and current challenges. Lang. Resour. Eval. **59**(2), 1747–1773 (2025)
8. Koehn, P., Knowles, R.: Six challenges for neural machine translation. In: Proceedings of the First Workshop on Neural Machine Translation, pp. 28–39 (2020)
9. Nazi, Z.A., Peng, W.: Large language models in healthcare and medical domain: a review. Informatics **11**(3), 57 (2024)
10. Névéol, A., Dalianis, H., Velupillai, S., Savova, G., Zweigenbaum, P.: Clinical natural language processing in languages other than English: opportunities and challenges. J. Biomed. Inform. **76**, 141–153 (2018)
11. Ahmad, P.N., Shah, A.M., Lee, K.: A review on electronic health record text-mining for biomedical name entity recognition in healthcare domain. Healthcare **11**(9), 1268 (2023)
12. Liu, B.X., Huang, L.: ParaMed: a parallel corpus for English-Chinese translation in the biomedical domain. BMC Med. Inform. Decis. Mak. **21**(1), 258 (2021)
13. Hong, K.Y., Han, L.F., Batista-Navarro, R.T., Nenadic, G.: CantonMT: Cantonese-English neural machine translation looking into evaluations. In: Proceedings of the 16th Conference of the Association for Machine Translation in the Americas (Volume 2: User Track), pp. 133–144 (2024)
14. Lau, C.-M., Chan, G.W.-Y., Tse, R.K.-W., Chan, L.S.-Y.: Words.hk: a comprehensive cantonese dictionary dataset with definitions, translations and transliterated examples. In: Proceedings of the Workshop on Dataset Creation for Lower-Resourced Languages within the 13th Language Resources and Evaluation Conference, pp. 53–62 (2022)
15. Fu, Y.C., Wang, X.W., Tian, Y.D., Zhao, J.W.: Deep think with confidence. arXiv preprint arXiv:2508.15260 (2025)

16. Yepes, A.J., et al.: Findings of the WMT 2017 biomedical translation shared task. In: Proceedings of the Second Conference on Machine Translation, pp. 234–247 (2017)
17. Blinov, P., Reshetnikova, A., Nesterov, A., Zubkova, G., Kokh, V.: RuMedBench: a Russian medical language understanding benchmark. In: International Conference on Artificial Intelligence in Medicine, pp. 383–392. Springer (2022)
18. Kocmi, T., et al.: Findings of the 2022 conference on machine translation (WMT22). In: Proceedings of the Seventh Conference on Machine Translation (WMT), pp. 1–45 (2022)
19. Xu, J., et al.: Data set and benchmark (MedGPTEval) to evaluate responses from large language models in medicine: evaluation development and validation. JMIR Med. Inform. **12**(1), e57674 (2024)
20. Xu, N.Y., et al.: Cross-linguistic syntactic difference in multilingual BERT: how good is it and how does it affect transfer? arXiv preprint arXiv:2212.10879 (2022)
21. Aden, I., Child, C.H.T., Reyes-Aldasoro, C.C.: International classification of diseases prediction from MIMIIC-III clinical text using pre-trained ClinicalBERT and NLP deep learning models achieving state of the art. Big Data Cogn. Comput. **8**(5), 47 (2024)
22. Huang, K.Y., et al.: A survey on large language models with multilingualism: recent advances and new frontiers. arXiv preprint arXiv:2405.10936 (2024)
23. Yu, T.Z., et al.: Automatic speech recognition datasets in cantonese: a survey and new dataset. arXiv preprint arXiv:2201.02419 (2022)
24. Li, L.H., et al.: WenetSpeech-Yue: A Large-scale Cantonese Speech Corpus with Multi-dimensional Annotation. arXiv preprint arXiv:2509.03959 (2025)
25. Chen, S., et al.: Meddialog: a large-scale medical dialogue dataset. arXiv preprint arXiv:2004.03329, vol. 3 (2020)
26. Khanuja, S., Dandapat, S., Srinivasan, A., Sitaram, S., Choudhury, M.: GLUECoS: an evaluation benchmark for code-switched NLP. arXiv preprint arXiv:2004.12376 (2020)
27. Leaman, R., Wei, C.-H., Lu, Z.Y.: tmChem: a high performance approach for chemical named entity recognition and normalization. J. Cheminform. **7**(Suppl 1), S3 (2015)
28. Lee, J., et al.: BioBERT: a pre-trained biomedical language representation model for biomedical text mining. Bioinformatics **36**(4), 1234–1240 (2020)
29. Michalopoulos, G., Wang, Y.X., Kaka, H., Chen, H., Wong, A.: Umlsbert: clinical domain knowledge augmentation of contextual embeddings using the unified medical language system metathesaurus. arXiv preprint arXiv:2010.10391 (2020)
30. Tutubalina, E., Miftahutdinov, Z., Nikolenko, S., Malykh, V.: Medical concept normalization in social media posts with recurrent neural networks. J. Biomed. Inform. **84**, 93–102 (2020)
31. Henry, S., Buchan, K., Filannino, M., Stubbs, A., Uzuner, O.: 2018 N2C2 shared task on adverse drug events and medication extraction in electronic health records. J. Am. Med. Inform. Assoc. **27**(1), 3–12 (2020)

# A Practical System for Medical Waste Detection and Triage Using Deep Learning

Vy Phuc Thach Le[1], Thanh Van Vo[1], Minh-Quan Nguyen[2], and Minh-Tai Vo[1(✉)]

[1] Posts and Telecommunications Institute of Technology, Ho Chi Minh 700000, Vietnam
{n22dcdk097,n22dcdk087}@student.ptithcm.edu.vn,
taivm@ptit.edu.vn

[2] RMIT University, Ho Chi Minh City 700000, Vietnam
quan.nguyen93@rmit.edu.vn

**Abstract.** Unlike municipal waste, medical waste poses greater potential risks if not treated properly, especially in the initial classification stage. Misclassification can lead to increased infection hazards and environmental harm. To address this issue, this study develops an automated detection and classification model for medical waste with the aim of supporting accurate identification during the early sorting phase. The method employed is You Only Look Once version 11 (YOLOv11). We used a dataset of more than two thousand manually labeled images collected across diverse waste types. The results indicate that the model delivers accurate recognition and strong overall performance, with both Precision and Recall exceeding 95%, while the mean Average Precision (mAP) metric demonstrates stable detection capability across multiple Intersection over Union (IoU) thresholds. By enabling faster and more reliable waste classification, this research provides a practical and environmentally friendly solution that reduces infection risks, supports healthcare workers, and contributes to safer and more sustainable medical waste management. In addition, the paper compares the three models, YOLOv9, YOLOv10, and YOLOv11, thereby confirming the strong applicability of YOLOv11 in medical waste classification.

**Keywords:** Medical Waste Classification · YOLOv11 · Deep Learning · Image Processing · Real-time Detection

## 1 Introduction

The amount of medical waste has continued to rise, especially after the COVID-19 pandemic [1, 2]. In many medical facilities worldwide, the classification process is still mostly manual, particularly in low- and middle-income countries. This often leads to errors in classification and increases the risk of exposure to contaminated and sharp objects, which may transmit diseases. In Vietnam, for example, a national survey reported that only 82% of district hospitals fully complied with hazardous waste treatment regulations, while the correct knowledge rate among healthcare workers was just 26.4%

These authors contributed equally to this work.


P. Pichappan et al. (Eds.): RTIS 2025, LNNS 1936, pp. 143–154, 2027.
https://doi.org/10.1007/978-3-032-23396-7_13

[3, 4]. Another problem is that medical waste mixed with domestic waste complicates treatment and reduces the effectiveness of the existing medical waste management system [5]. Therefore, it is essential to develop an identification and classification system to overcome the limitations of traditional methods. Such a solution ensures worker safety, protects the environment, and improves productivity and efficiency.

Image recognition and classification of medical waste have been extensively studied in recent years. In the early stages, image recognition relied on object shape and color, such as the Histogram of Oriented Gradients method combined with Support Vector Machine [6]. However, in practice, waste may be distorted or faded, limiting the accuracy of these methods in real-world environments. Later, the introduction of the CNN (Convolutional Neural Network) deep learning model opened new possibilities, enabling improved recognition in complex environments, significantly improving accuracy, and allowing reliable testing in practice [7]. The study [8] developed a Mask Scoring R-CNN system for garbage classification based on Beijing's standards. The system achieved 65.8% accuracy, while a trained Mask R-CNN reached 77%. In aquatic environments, the AquaVision system [9] used RetinaNet and Faster R-CNN to classify four waste types with mAP of 0.814, where RetinaNet outperformed Faster R-CNN. Building on the advantages of CNN, the YOLO model is recognized for its ability to detect objects rapidly. Over time, YOLO has been continuously optimized in terms of accuracy, processing speed, and practical deployment. The YOLO-Green model achieved a mAP of 78.04% with reduced size (117 MB) and faster speed (2.72 FPS), outperforming YOLOv3 and YOLOv4 on solid waste classification [10]. Many studies applying YOLO to waste identification and classification have reported positive results with high accuracy [11, 12].

In this paper, we employ the YOLOv11 model [13] to classify waste into three categories: infectious, recyclable, and non-recyclable. Specifically, the waste is divided into groups containing the components: syringes, plastic water bottles, metal cans, milk cartons, and bio-waste bags. The main contributions are as follows:

- We apply the YOLOv11 model for medical waste detection and classification, adopting a mapping strategy from five classes to three groups to facilitate practical deployment, while simultaneously optimizing the model to ensure efficiency on edge devices.
- We develop a specialized dataset and benchmark for medical waste, providing a reference platform for further research and enabling opportunities to compare and improve models in this field.
- We enhance practical applicability by ensuring optimal operating costs and high flexibility, enabling easy integration into automated waste-sorting robots and future expansion to additional types of medical waste.

The article is structured into four main parts. Part 1 introduces our motivation, related work and defines the core objectives of medical waste classification. Part 2 describes in detail the system workflow, data sources, and materials. Part 3 explains the experimental methods applied to the problem and presents the results. Finally, Part 4 summarizes the main conclusions and outlines directions for future development.

## 2 Methodology and Material

### 2.1 System Flow

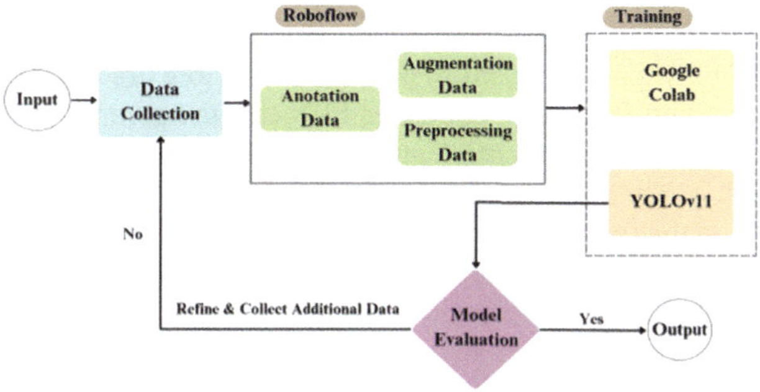


**Fig. 1.** System workflow for medical waste detection and classification.

In this section, we present the operational process of a medical waste identification and classification system using the YOLOv11 model, illustrated in Fig. 1. First, images of medical waste are collected from various sources, including photographs taken directly in real-world environments and images obtained from publicly available sources on the Internet. These images are carefully selected to ensure diversity in shooting angles, backgrounds, and lighting conditions.

Roboflow served as a one-stop solution for organizing, annotating, and preprocessing our dataset, which is particularly crucial in the context of medical waste detection where high-quality data are essential for building effective models. The data collected was processed through the Roboflow platform in three main stages as follows:

- First, we perform annotation by manually drawing bounding boxes for five types of medical waste. Next, we perform apply data augmentation, where the images are transformed through rotation, flipping, and lighting adjustments to increase diversity and help the model generalize better across different scenarios. Then, we remove blurry images, or images with unnecessary noise, and download the dataset with all images standardized to a size of $640 \times 640$ pixels. As a result, the completed dataset is both diverse and consistent, serving as the foundation for training with YOLOv11.
- Second, the dataset was divided into three subsets: training, validation, and testing. The YOLOv11 model was then trained on this dataset in Google Colab, with the training parameters carefully tuned to optimize performance.
- Finally, after training, the best checkpoint (i.e., the one with the highest validation mAP@0.5) was evaluated on the test set. The evaluation used Precision, Recall, mAP@0.5, and mAP@0.5:0.95 [14], processing time per image, and visual tools such as the confusion matrix and precision–recall curve to assess accuracy and misclassifications. All metrics were computed with a confidence threshold of 0.25, an IoU of 0.5, and a non-maximum suppression threshold of 0.45.

## 2.2 Model Structure

In this study, we employ YOLOv11 as the primary model for medical waste classification. YOLOv11 is built on a CNN-based architecture but incorporates several improvements that enable more efficient object detection, as described in [13]. Specifically, it consists of three main parts as illustrated in Fig. 2: Backbone, Neck, and Head. The Backbone uses a combination of Conv (Convolutional) and C2f (Cross Stage Partial with 2F) blocks and integrates SPPF (Spatial Pyramid Pooling – Fast) to extract features and process multi-scale information. The Neck combines FPN (feature pyramid network) and PAN (path aggregation network) to merge features from different layers and support better object identification. The Head predicts bounding boxes, confidence scores, and class labels for objects in the image.

These improvements are particularly well suited for medical waste classification, where many objects are small, visually similar, or appear in complex backgrounds. As a result, YOLOv11 achieves high accuracy and maintains stable recognition even with diverse and hard-to-distinguish data. In addition, YOLOv11 was selected for this study because it offers a good balance between accuracy and computational cost. Compared with earlier versions like YOLOv9 [15] and YOLOv10 [16], i it integrates architectural improvements that enhance feature extraction and multi-scale processing. The lightweight YOLOv11 variant is also suitable for real-time applications on resource-limited devices while still ensuring reliable accuracy.

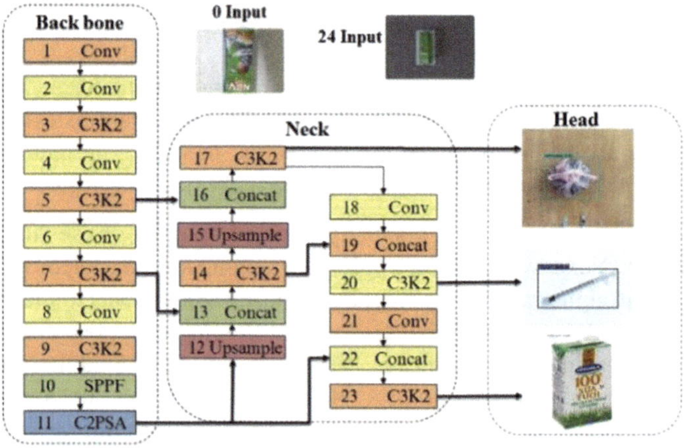


**Fig. 2.** YOLOv11 Overview Architecture [13].

## 2.3 Dataset

The dataset consists of five types of medical waste, each of which is assigned a detailed training label. During deployment, these labels are grouped into three categories: (i) infectious waste, (ii) recyclable waste, and (iii) non-recyclable waste. Example images from the constructed dataset are shown in Fig. 3, and the number of images in each category is summarized in Table 1. Most of the images were taken by the team using a Xiaomi Redmi Note 13 smartphone.

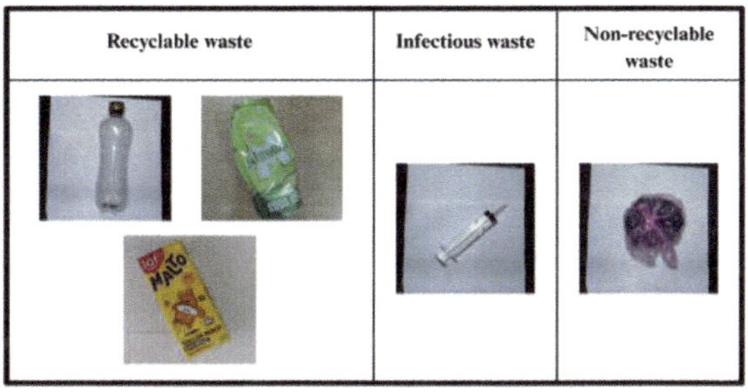


**Fig. 3.** Representative samples of five types of medical waste (bottle, can, milk_box, nylon_bag, syringe), categorized into three groups: recyclable, infectious, non-recyclable.

**Table 1.** Statistics on the number of images in the dataset.

| Waste group | Specific Type | Number of photos |
|---|---|---|
| Infectious | Syringe | 473 |
| Recyclable | Bottles, cans, milk box | 1375 |
| Non-Recyclable | Nylon garbage bags | 465 |

After constructing the dataset, we proceeded with the labeling process to prepare the data for training. We used the Roboflow tool to assign bounding-box annotations, ensuring that each object was accurately wrapped and assigned to one of the five predefined categories. The labeling was carried out manually to maximize precision, with Roboflow's intuitive interface allowing adjustments to bounding box positions and sizes to maintain data quality. An illustration of this process is shown in Fig. 4. The output annotations were exported in YOLO format, containing both class IDs and bounding box coordinates. All annotations were labeled by a single person to maintain consistency. Finally, the dataset was split into three subsets: 70% for training, 20% for validation, and 10% for testing, to guarantee a balanced distribution across all waste types. Figure 5 provides detailed statistics on the class-wise split of the dataset.

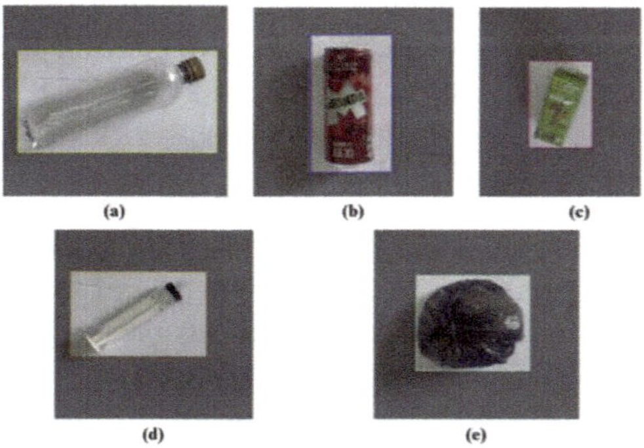


**Fig. 4.** Labeled samples: (a)–(c) recyclable waste; (d) infectious waste; (e) non-recyclable waste.

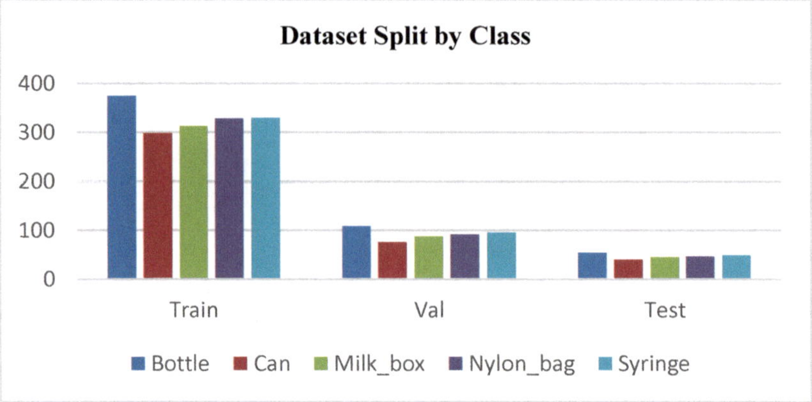


**Fig. 5.** Class-wise dataset split for training, validation, and testing.

## 3 Experimental Results

### 3.1 Training Process

The training was performed on the Google Colab platform using a Tesla T4 GPU (15 GB VRAM). We used the Ultralytics v8.3.184 framework [17] with Python 3.12.11 and PyTorch 2.8.0. The YOLOv11 model was trained for 50 epochs with a batch size of 16 and an input resolution of 640 × 640 pixels. The AdamW optimizer [18] was applied with an initial learning rate of 0.00111 and a momentum of 0.9, while a weight decay of 0.0005 was set for selected parameter groups to mitigate overfitting. During evaluation, the confidence threshold was fixed at 0.25, with the IoU threshold for Non-Maximum Suppression set to 0.7.

### 3.2 Implementation Details

In this part, we evaluate the performance of the YOLOv11 model on the validation dataset through common indicators in the object recognition problem, which are summarized in Table 2. The indicators include Precision, Recall, and mAP at two IoU thresholds of 0.5 and a range of 0.5:0.95.

Precision is the ratio of the model's correct prediction to the total number of correctly predicted and incorrectly predicted objects with True Positive (TP) and False Positive (FP). It can be simply understood as the probability of guessing the correct type of waste in the validation set, with the following Eq. 1:

$$\text{Precision} = \frac{\text{TP}}{\text{TP} + \text{FP}} \tag{1}$$

Recall is the ratio of the number of objects that the model correctly recognizes TP to the total number of objects that actually appear in the image (the sum of TP and False Negative (FN)) as given in Eq. 2. This metric reflects the model's ability to fully cover the objects to be detected, i.e., minimizing missed targets.

$$\text{Recall} = \frac{\text{TP}}{\text{TP} + \text{FN}} \tag{2}$$

In addition, mAP is used as a general measure to evaluate the model's recognition efficiency. The detailed results of these indicators are presented in Table 2 as follows:

**Table 2.** Class-wise performance metrics of YOLOv11 on the validation dataset.

| Class | Image | Precision | Recall | mAP@0.5 | mAP@0.5:0.95 |
|---|---|---|---|---|---|
| All | 476 | 0.976 | 0.954 | 0.978 | 0.779 |
| Bottle | 108 | 0.998 | 1.000 | 0.995 | 0.829 |
| Can | 76 | 0.998 | 0.906 | 0.988 | 0.794 |
| Milk_box | 88 | 0.954 | 0.978 | 0.984 | 0.792 |
| Nylon_bag | 92 | 0.979 | 0.956 | 0.981 | 0.827 |
| Syringe | 95 | 0.949 | 0.93 | 0.944 | 0.655 |

Based on the information in Table 2, the model proves to be highly effective for most types of medical waste. Specifically, the Precision reached 0.976, Recall 0.954, mAP@0.5 0.978, and mAP@0.5:0.95 0.779. These results indicate that the model demonstrates strong recognition ability, while maintaining a balance between accuracy and coverage.

For bottle grades, the Recall is absolute, while the Precision is also nearly perfect (0.998). In addition, mAP@0.5 achieved a high value of 0.995 and mAP@0.5:0.95 reached 0.829, demonstrating the stability of the model's stability in identifying objects with clear characteristic shapes and structures.

For the *milk_box* and *nylon_bag* classes, the results are generally positive. These two classes achieved Precision scores of 0.954 and 0.979, respectively, with corresponding Recall values of 0.978 and 0.956. The mAP@0.5 values both exceeded 0.98, reflecting stable recognition. However, mAP@0.5:0.95 ranged only from 0.792 to 0.827, suggesting that accurate localization at higher IoU levels remains a challenge.

For the *can* class, Precision reached a very high value (0.998), while Recall was relatively lower at 0.906. As a result, although the mAP@0.5 remained strong (0.988), the mAP@0.5:0.95 decreased to 0.794. This indicates that the model occasionally missed or had difficulty accurately localizing objects of this class.

The *syringe* class remains the most challenging case. Precision reached 0.949, Recall reached 0.930, and mAP@0.5 was 0.944, the lowest among all five medical waste types. Furthermore, the mAP@0.5:0.95 dropped to just 0.655, compared with the overall average of 0.799. The difficulty may stem from the small size, elongated shape, and color of syringes, which can easily blend into white backgrounds, as most dataset images were captured under such conditions.

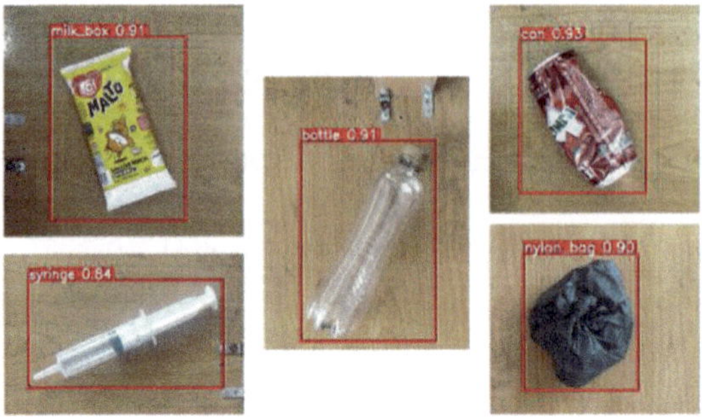


**Fig. 6.** The results of the identification of five types of medical waste.

The test results are presented in Fig. 6, showing the model's ability to identify five types of medical waste including: plastic bottles, metal cans, milk cartons, plastic bags, and syringes. The model shows the efficiency of accurate detection of the majority of objects with a confidence level exceeding the threshold of 0.9. However, for the syringes layer, the reliability is significantly lower, reflecting that small-sized objects are still a major challenge to the model's recognition.

To get a better overview of the model's results, the Confusion Matrix has been shown in Fig. 7 with comments as follows:

- With the number of test images ranging from 76 to 108 per class, most waste types exhibited a low misclassification rate of only one to four cases. This indicates that the model maintained relatively high accuracy.
- Syringes have a relatively high rate of being misclassified as background, indicating issues related to their small size and colors that resemble white backgrounds.

The Confusion Matrix shows that some medical waste types share similar characteristics and are easily confused with environmental noise. This provides an overview of the sources of the model's errors.

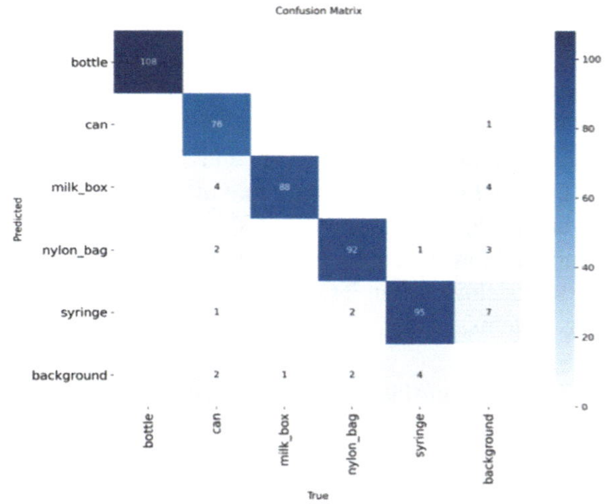


**Fig. 7.** Confusion Matrix model results.

### 3.3 Comparison of YOLOv9, YOLOv10, YOLOv11

In this section, we compare the performance of YOLOv9, YOLOv10 and YOLOv11 to evaluate their effectiveness in medical waste detection. Table 3 presents the overall performance results, including mAP@0.5 (reported for both the training and test sets), mAP@0.5:0.9, Precision, Recall, and Inference Time on the test set.

**Table 3.** Performance analysis of YOLOv9, YOLOv10, YOLOv11.

| Model | mAP@0.5 (Train) | mAP@0.5 (Test) | mAP@0.5–0.95 (Test) | Inference Time (Test - ms) | Precision (Test) | Recall (Test) |
|---|---|---|---|---|---|---|
| YOLOv9 | 0.992 | 0.982 | 0.781 | 10.9 | 0.969 | 0.963 |
| YOLOv10 | 0.97 | 0.983 | 0.759 | 5.2 | 0.95 | 0.964 |
| YOLOv11 | 0.994 | 0.987 | 0.782 | 4.4 | 0.981 | 0.976 |

According to Table 3, YOLOv11 demonstrates the most outstanding and comprehensive performance. Specifically, the mAP@0.5 on the training and test sets reached 0.994 and 0.987, respectively, the highest among the three models, indicating stable learning ability and strong generalization. Precision (0.981) and Recall (0.976) remained at very high levels, reflecting the model's capability to minimize both false positives and false negatives. Moreover, the mAP@0.5:0.95 was 0.782, which is higher than that of YOLOv10 and nearly comparable to YOLOv9, showing robust detection accuracy across multiple IoU thresholds.

For YOLOv10, the mAP@0.5 on the test set reached 0.983, even slightly higher than YOLOv9, while precision (0.950) and Recall (0.964) were also at relatively high levels. However, the mAP@0.5:0.95 was only 0.759, the lowest among the three models, indicating that its ability to maintain detection performance at higher IoU thresholds is not particularly strong. Nevertheless, the average inference time of 5.2 ms suggests that the model meets real-time requirements in certain applications.

Meanwhile, YOLOv9 demonstrated accuracy comparable to YOLOv11. The mAP@0.5 reached 0.992 on the training set and 0.982 on the test set, with an mAP@0.5:0.95 of 0.781, which is higher than that of YOLOv10 and nearly equivalent to YOLOv11. Precision (0.969) and Recall (0.963) were both at good levels, confirming the stability of the model. However, a notable limitation is its inference time of 10.9 ms, which is nearly 2.5 times slower than YOLOv11, thereby reducing its applicability in real-time processing systems.

**Table 4.** Image size versus mAP@50 (testing) for YOLOv9, YOLOv10, YOLOv11.

| Weed Species | All | Bottle | Can | Milk_box | Nylon_bag | Syringe |
|---|---|---|---|---|---|---|
| Number of images | 232 | 54 | 39 | 45 | 46 | 49 |
| YOLOv9 AP 50 | 0.982 | 0.995 | 0.973 | 0.995 | 0.994 | 0.952 |
| YOLOv10 AP 50 | 0.983 | 0.995 | 0.982 | 0.99 | 0.993 | 0.955 |
| YOLOv11 AP 50 | 0.987 | 0.995 | 0.994 | 0.995 | 0.989 | 0.963 |

The comparison in Table 4 shows the mAP@50 results of the three models YOLOv9, YOLOv10, and YOLOv11, across five types of medical waste in the test set. In terms of overall performance, YOLOv11 achieved the highest score (0.987), slightly better than YOLOv10 (0.983) and YOLOv9 (0.982). For the two waste types, Bottle and Milk_box, all three models achieved almost perfect accuracy ($\geq$0.99), indicating that these categories are less prone to misclassification. For Can, YOLOv11 excelled with 0.994, while YOLOv10 and YOLOv9 only reached 0.982 and 0.973, respectively. In contrast, for Nylon_bag, YOLOv11 scored only 0.989, which is lower than YOLOv9 (0.994) and YOLOv10 (0.993). This result shows that the model still has certain limitations when handling deformable and variable-shaped waste, which remains an area for improvement in future YOLO versions to enhance recognition efficiency. The most notable case is the Syringe class, which yielded lower results due to its small size and slender shape. Nevertheless, YOLOv11 maintained its superiority, reaching 0.963 and surpassing both YOLOv10 (0.955) and YOLOv9 (0.952). Thus, YOLOv11 not only outperforms the other versions in terms of accuracy but also demonstrates superior identification capability.

## 4 Conclusions and Final Remarks

Through this research, we developed a medical waste identification model using the YOLOv11 method. The dataset contained 2,313 images, manually labeled and divided into five main classes: plastic bottles, metal cans, milk cartons, nylon bags, and syringes. Experimental results on the test set showed that the model achieved strong performance, with an average precision of 97.6%, recall of 95.4%, mAP@0.5 of 97.8%, and mAP@0.5:0.95 of 79.9%. In contrast, syringes yielded lower performance due to their small size, which made them more likely to be misclassified as background. A possible improvement is to expand the dataset with a stronger focus on emphasizing distinctive features. The application of this tool can not only enhance management efficiency but also reduce infection risks for healthcare workers and mitigate negative impacts on the environment. In future work, the integration with autonomous robots or IoT-enabled monitoring systems is envisioned to provide a more comprehensive solution [19, 20].

**Acknowledgement.** We acknowledge the support of time and facilities from Posts and Telecommunications Institute of Technology for this study.

## References

1. Hanedar, A., Çifçi, D.İ, Zafer, N., Görgün, E.: The impact of COVID-19 pandemic in medical waste amounts: a case study from a high-populated city of Turkey. J. Mater. Cycles Waste Manag. **24**, 1760–1767 (2022). https://doi.org/10.1007/s10163-022-01428-3
2. Andeobu, L., Wibowo, S., Grandhi, S.: Medical waste from COVID-19 pandemic—a systematic review of management and environmental impacts in Australia. Int. J. Environ. Res. Public Health **19**, 1381 (2022). https://doi.org/10.3390/ijerph19031381
3. Dang, H.T.T., Dang, H.V., Tran, T.Q.: Insights of healthcare waste management practices in Vietnam. Environ. Sci. Pollut. Res. **28**, 12131–12143 (2021). https://doi.org/10.1007/s11356-020-10832-x
4. Huy Nga, N., Tuong Vy, M., Van Xuyen, D., Tuan Anh, T.: Medical solid waste management knowledge among healthcare workers at Duc Giang General Hospital in 2022. Tạp chí Y học Việt Nam. **525**, (2023). https://doi.org/10.51298/vmj.v525i1.5050
5. Yang, T., Du, Y., Sun, M., Meng, J., Li, Y.: Risk management for whole-process safe disposal of medical waste: progress and challenges. Risk Manag. Healthc. Policy **17**, 1503–1522 (2024). https://doi.org/10.2147/RMHP.S464268
6. Dalal, N., Bil, T.: Histograms of oriented gradients for human detection. In: Dalal, N., Bill, T. (eds.) 2005 IEEE Computer Society Conference on Computer Vision and Pattern Recognition (CVPR'05), pp. 886–893. IEEE, San Diego, CA, USA (2005)
7. Krizhevsky, A., Sutskever, I., Hinton, G.E.: ImageNet classification with deep convolutional neural networks. In: Bartlett, P. (ed.) Advances in Neural Information Processing Systems 25 (NIPS 2012), pp. 1097–1105. Neural Information Processing Systems Foundation, Inc. (NeurIPS), Lake Tahoe, Nevada, USA (2012)
8. Li, S., Yan, M., Xu, J.: Garbage object recognition and classification based on Mask Scoring RCNN. In: 2020 International Conference on Culture-oriented Science & Technology (ICCST), pp. 54–58. IEEE (2020). https://doi.org/10.1109/ICCST50977.2020.00016
9. Panwar, H., et al.: AquaVision: Automating the detection of waste in water bodies using deep transfer learning. Case Stud. Chem. Environ. Eng. **2**, 100026 (2020). https://doi.org/10.1016/j.cscee.2020.100026
10. Lin, W.: YOLO-Green: a real-time classification and object detection model optimized for waste management. In: 2021 IEEE International Conference on Big Data (Big Data), pp. 51–57. IEEE (2021). https://doi.org/10.1109/BigData52589.2021.9671821
11. Yang, X., et al.: Detection of river floating garbage based on improved YOLOv5. Mathematics **10**, 4366 (2022). https://doi.org/10.3390/math10224366
12. Hermawan, I., Mardiyono, A., Iswara, R.W., Murad, F.A., Ardiawan, M.A., Puspita, R.: Development of Covid Medical Waste Object Classification System Using YOLOv5 on Raspberry

Pi. In: 2023 10th International Conference on Information Technology, Computer, and Electrical Engineering (ICITACEE), pp. 443–447. IEEE (2023). https://doi.org/10.1109/ICITACEE58587.2023.10277207

13. Khanam, R., Hussain, M.: YOLOv11: An Overview of the Key Architectural Enhancements. arXiv preprint arXiv:2410.17725. (2024)
14. Solawetz, J.: What is Mean Average Precision (mAP) in Object Detection? https://blog.roboflow.com/mean-average-precision/. Accessed 14 Sep. 2025
15. Wang, C.-Y., Yeh, I.-H., Mark Liao, H.-Y.: YOLOv9: Learning What You Want to Learn Using Programmable Gradient Information. Presented at the (2025). https://doi.org/10.1007/978-3-031-72751-1_1
16. Chen, H., Chen, K., Ding, G., Han, J., Lin, Z., Liu, L., Wang, A.: YOLOv10: Real-Time End-to-End Object Detection. In: Advances in Neural Information Processing Systems 37, pp. 107984–108011. Neural Information Processing Systems Foundation, Inc. (NeurIPS), San Diego, California, USA (2024). https://doi.org/10.52202/079017-3429
17. Ultralytics: Ultralytics YOLO11. https://docs.ultralytics.com/models/yolo11/. Accessed 14 Sep. 2025
18. Loshchilov, I., Hutter, F.: Decoupled Weight Decay Regularization. In: 7th International Conference on Learning Representations. New Orleans, LA, USA (2019)
19. Tasnim, S., Hasan, Md.Z., Ahmed, T., Hasan, Md.T., Uddin, F.J., Farah, T.: A ROS-based voice controlled robotic arm for automatic segregation of medical waste using YOLOv3. In: 2022 2nd International Conference on Computer, Control and Robotics (ICCCR), pp. 81–85. IEEE (2022). https://doi.org/10.1109/ICCCR54399.2022.9790152
20. Nguyen, M.-Q., Tran, V.-H., Vo, M.-T., Le, M.-H.: Pose estimation and its impact on quadcopter control: an experimental study. In: 2024 9th International Conference on Applying New Technology in Green Buildings (ATiGB), pp. 264–268. IEEE (2024). https://doi.org/10.1109/ATiGB63471.2024.10717852

# Evaluating MERMAID: A Multi-expert Framework for Real-World Fake News Detection

Nguyen Minh Tuan(✉), Khong Phu Cuong, Do Thanh Tan(✉), and Le Manh Hung

Faculty of Information Technology, Posts and Telecommunications Institute of Technology, 11 Nguyen Dinh Chieu, Sai Gon Ward, Ho Chi Minh City 700000, Viet Nam
minhtuan@ptit.edu.vn,
{n23dccn122,n23dccn076,n23dccn091}@student.ptithcm.edu.vn

**Abstract.** The rapid dissemination of misinformation and fake news on social media poses significant risks to public trust and societal stability, especially in cross-domain scenarios where news topics vary widely. A central challenge in this task is domain shift, which hinders the ability of models trained on one domain to generalize effectively to unseen domains. In this paper, we present a comprehensive empirical study on cross-domain fake news detection using a suite of lightweight classifiers, including Logistic Regression, Support Vector Machines (SVM), Random Forests, and XGBoost. While large-scale pretrained language models such as BERT and recent large language models (LLMs) are acknowledged as powerful tools, they remain beyond the scope of our core evaluation due to their high computational demands. Our experiments reveal that ensemble strategies inspired by the MERMAID framework consistently outperform traditional Mixture-of-Experts (MoE) approaches in terms of robustness and stability. In particular, MERMAID-style ensembles demonstrate superior adaptability to domain shifts while requiring fewer computational resources, making them especially suitable for deployment in hardware-constrained environments. Furthermore, the proposed methodology shows effectiveness in both zero-shot and few-shot learning settings, where only limited labeled data from target domains is available.

**Keywords:** Fake news detection · Cross-domain learning · Domain shift · Ensemble methods

## 1 Introduction

The rapid diffusion of misinformation on social media poses substantial risks in domains such as public health, politics, and climate, where false content can

K. P. Cuong, D. T. Tan and L. M. Hung—These authors contributed equally to this work.


P. Pichappan et al. (Eds.): RTIS 2025, LNNS 1936, pp. 155–166, 2027.
https://doi.org/10.1007/978-3-032-23396-7_14

shape opinions, behaviors, and policy at scale [1,2]. Addressing cross-domain detection where the target distribution departs from the training distribution remains especially challenging due to topic drift, stylistic variability, and sparse or imbalanced supervision across domains [3–5]. Recent research has predominantly relied on large neural encoders, notably BERT-family models and related LLMs [6–8]. Cross-domain robustness is typically pursued via adversarial or domain-adaptive learning [9–11] or via ensemble frameworks that specialize experts by domain and combine them through mixture-of-experts (MoE) gating [12–15]. EMT [16], which learns entity-centric representations with external knowledge but presents limited zero-shot validation, leaving generalization claims to unseen domains inconclusive. Cross-domain adaptation: Cross-domain methods transfer knowledge from labeled sources to less-labeled targets by seeking shared representations. EANN minimizes event-specific features via an adversarial discriminator trained in a minimax game with a multimodal extractor, encouraging domain-agnostic features. CrossFND [17] introduces a domain classifier for unsupervised adaptation, augmented by a small portion of target data. REAL-FND builds on CrossFND and replaces the domain classifier with a reinforcement-learning module to improve transfer. Ensemble-based frameworks and model merging: MoDA-FND is an MoE-based system that selects the top-K most reliable experts when classifying unseen examples, representing a strong ensemble baseline for cross-domain detection. Building on this line, MERMAID introduces a model-merging stage before MoE: domain-adapted models related to a source domain are merged into a single domain-specific expert, a design explicitly inspired by TIES-Merging [18]. This reduces the number of experts and aims to stabilize gating. MERMAID further freezes experts and trains only the gate, augmented with an auxiliary domain-identification loss and regularizers to mitigate over-reliance on a few experts. Despite overall gains, MERMAID's analysis also surfaces a gate-degeneracy failure mode on CoAID, where the gate collapses to a single expert and performance drops—highlighting a limitation of MoE when expert activations lack diversity.

# 2 Proposed Method

## 2.1 Datasets

Our study is based on a heterogeneous dataset combining real and fake news across diverse topics, reflecting the natural distribution of online information where factual boundaries are often blurred and class imbalance is common. The data consists of unstructured text from multiple sources with varied credibility and style. For training, we segment and normalize the corpus into four domains: politics, medical-health, science-technology, and entertainment. Each domain is balanced between Fake and True labels to mitigate bias and help domain-specific experts capture meaningful patterns. These domain-labeled subsets serve as the primary training resource for MERMAID's expert modules. For evaluation, we use two complementary resources. First, an independent benchmark test set from Kaggle with manually verified labels is employed to measure generalization under clean and controlled conditions. Second, we build a realistic "in-life" dataset from Vietnamese media and social platforms, which includes informal language, slang, errors, and weakly supervised labels. This dataset helps assess model robustness under real-world conditions. Together, these three components—domain-specific training data, benchmark test set, and in-life validation set—provide a comprehensive pipeline for both rigorous evaluation and practical deployment analysis.

### 2.2 Overall Architecture

In our implementation, each base fake news detector is designed to operate as a lightweight and domain-specific classifier trained to distinguish between real and fake news articles. Unlike prior work that relies on large-scale transformer encoders such as DeBERTa, we adopt a more efficient and resource-friendly backbone on https://huggingface.co/sentence-transformers/paraphrase-MiniLM-L3-v2 paraphrase-MiniLM-L3-v2, a sentence embedding model from the Sentence Transformers library. The output embedding $\mathbf{e}_x \in \mathbb{R}^d$ produced by the encoder for a given input text $\mathbf{x}$ serves as the input to a shallow classifier. We explore four types of classifiers as base detectors: Logistic Regression, Support Vector Machines (SVM), Random Forest, and XGBoost. These classifiers are trained independently on labeled data from each source domain $\mathbf{D}_i$, yielding a base model $\mathbf{M}_i$ specialized for that domain. This two-stage architecture frozen sentence encoder followed by a domain-specific classifier offers a favorable trade-off between model complexity and generalization.

Given a source-domain model $F_s$, we fine-tune it on the target domain's adaptation split $D_t^{ad}$ (distinct from $D_t^{tr}$, which is never used here) to obtain the adapted model $F_{s\to t}$. We initialize $F_s$ with its trained weights and choose a tuning mode according to resources: parameter-efficient methods (Adapters/LoRA + classifier head) for Transformers, head-only when $D_t^{ad}$ is very small, and full fine-tuning only for small models (LR/SVM/XGBoost/mini-Transformers). The objective combines a supervised loss on $D_t^{ad}$ (class-weighted cross-entropy or focal loss under imbalance) with a source-knowledge preservation term to avoid forgetting, e.g.,

$$\mathcal{L}_{\text{adapt}} = \mathbb{E}_{(x,y)\sim D_t^{ad}}\big[\ell(F_{s\to t}(x), y)\big] \;+\; \lambda_{\text{pres}}\, R(\theta_{s\to t}, \theta_s^\star) \;+\; \lambda_{\text{align}}\, \text{Align}(\phi(x)),$$

where $R$ is L2-SP ($\|\theta_{s\to t} - \theta_s^\star\|^2$) or EWC/Fisher, and Align (optional) is a feature-alignment regularizer such as CORAL/MMD when distributions are far apart. For small Transformers a robust schedule is: learning rate $2-5\times 10^{-4}$ for LoRA/Adapters/Head, $1\times 10^{-5}$ for any unfrozen backbone, 10% warmup, AdamW with weight decay $10^{-2}$ (no decay on bias/LayerNorm), dropout 0.1–0.2, and early stopping by macro-F1 on the target dev set $D_t^{dev}$ (or shared $D_{\text{Ens}}$) with patience $3-5$; $D_t^{tr}$ is never used to prevent leakage. After each epoch we run a canary check against two baselines $F_s$ evaluated on domain $t$ without tuning and the native $F_t$; we keep $F_{s\to t}$ only if it surpasses the canary and is not badly miscalibrated. Finally, we persist artifacts required by later stages (weights; logits on $D_t^{dev}$ for stacking; optional Fisher diagonals/weight norms for weighted parameter merging; temperature for calibration) and add the accepted $F_{s \to t}$ to domain $t$'s pool so the MERMAID merging layer can consolidate $\{F_{* \to t}\}$ into a single expert $E_t$, while the gating/ensemble layer learns to route by sample at inference.

For target domain $i$, the goal is to distill a single compact and stable expert $\hat{M}_i$ from the in-domain base $M_i$ and the adapted models $\{M_{j \to i} \mid j \neq i\}$ via three tightly coupled, per-layer operations.

(1) Weight-difference extraction: for each adapted model, compute the parameter shift relative to the base,

$$\Delta^{(j)} = \theta(M_{j \to i}) - \theta(M_i),$$

optionally applying layerwise normalization so large layers do not dominate; handle normalization layers by merging only their weight/bias while keeping running means/variances from the base, and keep positional/segment embeddings from the base.

(2) Layerwise Top-K sparsification: within each layer, retain only coordinates with the largest absolute shifts $|\Delta^{(j)}|$ (e.g., a ratio $\rho\%$ per layer, typically 1–10%); form a unified mask $S$ as the union of all models' Top-K positions so only impactful "pivot" parameters are considered while small noisy changes are ignored.

(3) Sign voting and weighted averaging: for each selected coordinate $p \in S$, collect signs $\text{sign}(\Delta_p^{(j)})$ across adapted models and determine the majority sign $\gamma_p$; keep only shifts whose sign agrees with $\gamma_p$ (to avoid cancellation) and require a model/layer consensus ratio of at least $\tau$ (e.g., $\tau \geq 0.5$) to discard off-phase adaptations; aggregate the surviving shifts with reliability weights $w_j$ normalized by $\sum_j w_j = 1$, where $w_j$ may derive from macro-F1 on the target dev set (accuracy), from 1/ECE (calibration), or from Fisher/parameter importance (criticality), yielding the fused shift

$$\Delta = \sum_j w_j \, \Delta_{\text{agree}}^{(j)}.$$

Finally, update the base with a calibrated step

$$\theta(\hat{M}_i) = \theta(M_i) + \lambda \Delta,$$

tuning $\lambda \approx 0.7$–$1.2$ on development data (and optionally clipping $\Delta$ to curb outliers). In practice, layerwise Top-K is more stable than global Top-K; ties in sign voting at a coordinate are best dropped (or broken by the largest $|\Delta|$). After merging, a brief head-only touch-up on the target dev set and temperature scaling can further improve reliability. The resulting $\hat{M}_i$ preserves the domain core of $M_i$ while selectively absorbing strong, consensus, and trustworthy adaptations from $\{M_{j \to i}\}$, delivering a cross-domain-robust expert with low inference cost ready for downstream gating/MoE.

To integrate predictions from domain-specific experts, we adopt a Mixture-of-Experts (MoE) strategy wherein each instance is softly routed to a weighted combination of experts. The gating mechanism learns to assign input-dependent attention scores over merged experts, aiming to reflect their reliability for a given instance.

Given an input $x$, we first compute its dense semantic embedding $Z_x \in \mathbb{R}^{384}$ using the frozen encoder. This embedding is passed to a lightweight multi-layer perceptron (MLP) gate, which produces a distribution over the $K$ experts via a softmax layer:

$$z = (z_1, z_2, \ldots, z_K) = \mathrm{softmax}(\mathrm{MLP}(Z_x)), \qquad \sum_{k=1}^{K} z_k = 1.$$

Each merged expert $\hat{M}_k$ outputs a prediction $\hat{y}_k \in [0, 1]$, and the final prediction $\hat{y}$ is computed as a convex combination:

$$\hat{y} = \sum_{k=1}^{K} z_k \cdot \hat{y}_k.$$

To train the gate, we follow the regularization strategy proposed by Liguori et al. (2025) in the MERMAID framework, which combines prediction loss, sparsity, expert- awareness, and auxiliary domain classification into a unified objective:

$$L = \underbrace{L_{\mathrm{bce}}(\hat{y}, y)}_{\text{primary classification}} + \lambda_1 \underbrace{\|\Theta_G\|_1}_{L_{\mathrm{norm}}} + \lambda_2 \underbrace{\frac{1}{K} \sum_{k=1}^{K} \left( |y - \hat{y}_k| z_k + (1 - |y - \hat{y}_k|)(1 - z_k) \right)}_{L_{\mathrm{expert}}} + \lambda_3 \underbrace{\mathrm{CE}(\tilde{y}_d, y_d)}_{L_{\mathrm{ce}}}.$$

Where $L_{\mathrm{bce}}$ is the binary cross-entropy between the final prediction and the ground truth, $L_{\mathrm{norm}}$ is an $L_1$ penalty on the gate parameters $\Theta_G$ to promote sparsity, limit-ing the number of highly activated experts, $L_{\mathrm{expert}}$ penalizes inconsistencies between expert reliability (measured via prediction error) and gate weights, reducing the risk of gate collapse, $L_{\mathrm{ce}}$ is an auxiliary loss where the gate is trained to classify the domain origin of the input, helping preserve domain-specific cues during routing.

## 3 Experiments

The primary objective of our experimental evaluation is to assess the effectiveness of the proposed MERMAID-based framework in detecting fake news, while also testing its ability to generalize when applied to real-world data. To ensure both theoretical validity and practical relevance, our experiments are guided by the following two principles, including scientific rigor, where employing well-established, annotated benchmark datasets that are publicly available and widely recognized within the research community (e.g., curated datasets from Kaggle). These datasets provide a controlled environment for quantifying the model's theoretical performance under standardized conditions; Practical applicability where we incorporate real-world data sources, comprising of news articles and social media content collected from the Vietnamese web ecosystem, where fake news cannot always be definitively verified. This reflects the true deployment context in which misinformation detection systems must operate, often under uncertain or partially labeled conditions. Following this design, we construct our experimental corpus in three layers:

+ Raw noisy data, collected from various online sources;

+ Domain-specific training sets, curated and filtered from the raw corpus to reflect topic-specific subdomains;

+ Evaluation sets, composed of both standardized benchmark datasets and real-world "in-life" data, are used to assess generalization and robustness.

### 3.1 Experimetal Setups

In this study, the system is implemented following the MERMAID (MERged Mixture of domain AgnostIc Detectors) architecture, as originally proposed in Fig. 1 of the source paper. MERMAID is a layered framework for cross-domain fake news detection, composed of four key components: Base Experts, Adapted Experts, Merged Experts, and a Global Gate. The results of the base model comparison are presented in Fig. 1a and Fig. 1b. Analysis shows that Random Forest consistently achieves the highest performance across most metrics, which aligns with its inherent strength as an ensemble method. Gradient Boosting achieves near-optimal performance as well, but requires greater computational resources and careful hyperparameter tuning. Logistic Regression, while slightly lower in raw accuracy, delivers substantial advantages in training speed, inference latency, and resource consumption. These properties make it highly scalable in multi-domain settings.

### 3.2 Results

The model correctly classifies only 287 out of 700 instances, yielding a recall and accuracy of 0.287, and an F1-score of 0.446. These results indicate that the majority of true news samples were incorrectly classified as fake. The drop in recall and F1-score provides clear evidence that the model lacks sufficient generalization to real-world data conditions, where content distribution, style, and semantics differ significantly from the training regime. This highlights a critical limitation: while MERMAID performs well under balanced, benchmark-like datasets, its robustness diminishes when exposed to authentic, imbalanced environments (Fig. 2).

Comparison between MERMAID and MoE: (i) On the balanced train_all set when evaluated on the balanced `train_all` dataset where fake and real news

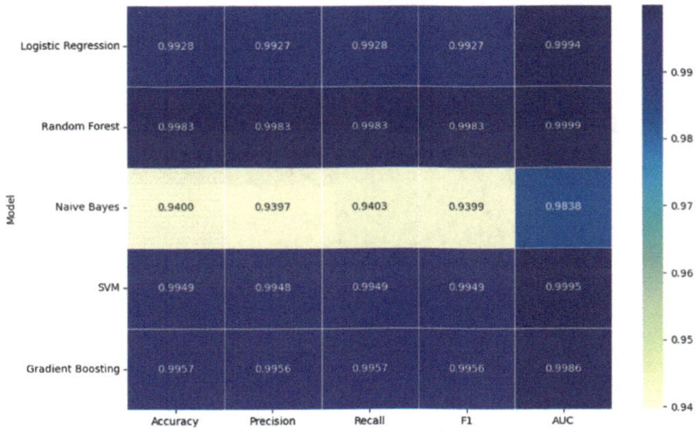


(a) Heatmap comparison

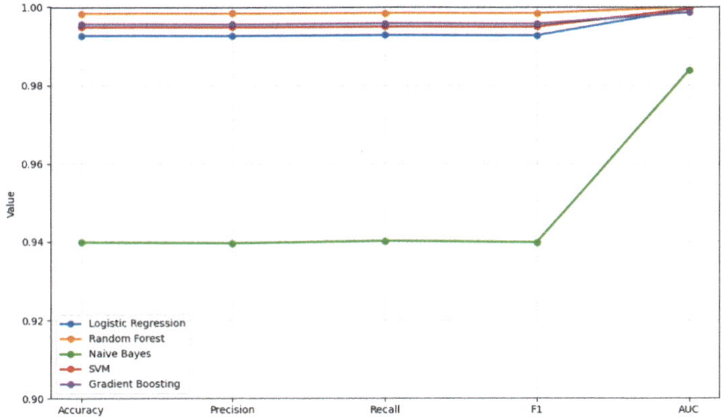


(b) Line plot comparison

**Fig. 1.** Performance comparison of base models. Both plots evaluate five lightweight classifiers across five metrics on the balanced training set. (a) The heatmap provides a numerical overview, showing Random Forest and Logistic Regression achieve the highest scores. (b) The line plot visualizes the performance drop of Naive Bayes compared to other models, especially in AUC.

samples are equally distributed both MERMAID and a standard Mixture-of-Experts (MoE) architecture achieve high and nearly equivalent performance. Accuracy and F1-score reach approximately 0.97, with recall close to 0.98. These results indicate that under ideal, balanced conditions, the architectural differences between MERMAID and MoE do not yield meaningful performance divergence. This further confirms that, when the distribution is stable and label coverage is comprehensive, both models are capable of learning reliable discriminative patterns (Fig. 3).

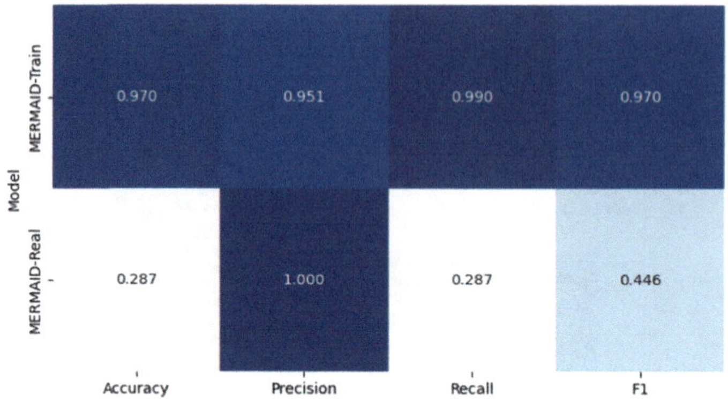


(a) Heatmap performance comparison

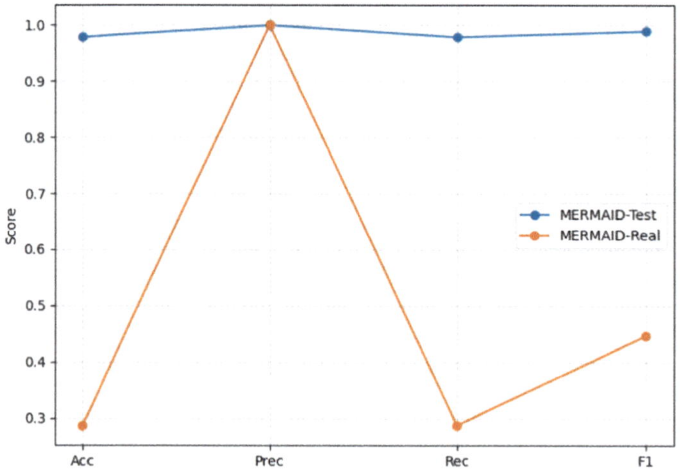


(b) Line plot performance comparison

**Fig. 2.** Performance degradation on real-world data. This figure compares MERMAID's performance on the controlled 'Test_true' set versus the imbalanced 'Real' set. While Precision is 1.0 (an artifact of the all-true dataset), both the (a) heatmap and (b) line plot illustrate a critical drop in Recall and F1-score (to $\approx 0.287$ and $\approx 0.446$, respectively), highlighting the model's significant generalization gap.

(ii) On the real-world dataset: In contrast, when tested on the `real` dataset composed of approximately 700 real news articles both models encounter significant performance degradation. Accuracy and recall drop sharply to around 0.286–0.287, while the F1-score remains low at 0.444–0.446. Despite the overall decline, MERMAID consistently outperforms MoE by a small but stable margin (roughly 0.001–0.002) in both recall and F1-score. Although numerically minor, this performance gap is reproducible and consistent across runs, suggesting that MERMAID maintains slightly better balance and recall sensitivity (Fig. 4).

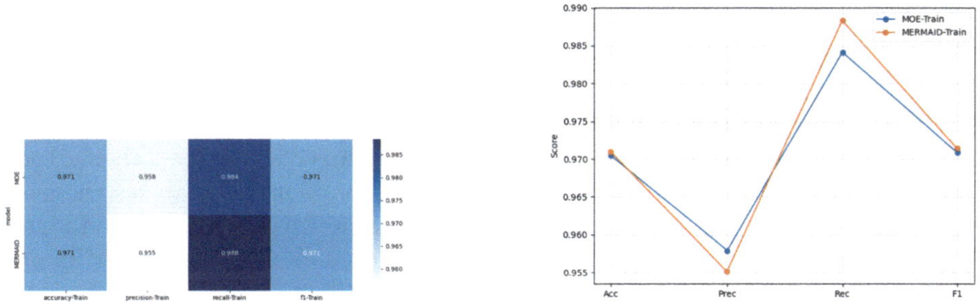


**Fig. 3.** Performance comparison of MERMAID and MoE on real-world dataset.

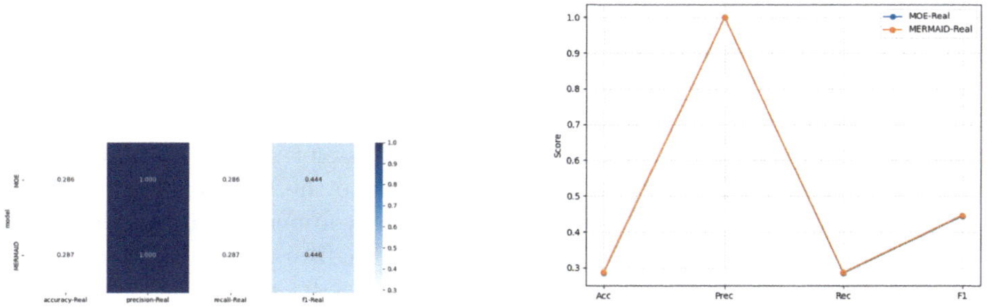


**Fig. 4.** Detailed performance metrics showing MERMAID's consistent improvement over MoE across multiple runs on real-world data.

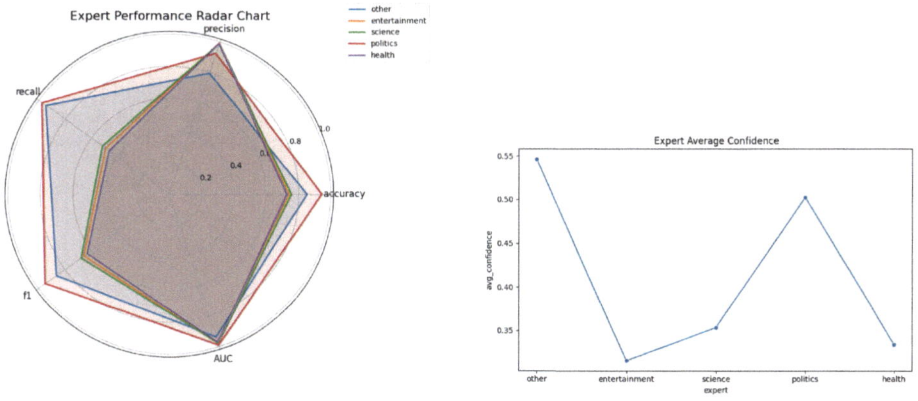


(a) Expert Performance Radar Chart (b) Expert Average Confidence

**Fig. 5.** Detailed analysis of expert modules. (a) Performance comparison across key metrics, showing the specialization of the 'politics' expert in recall. (b) Average prediction confidence of each expert.

**MERMAID Detailed Expert Analysis:**
To gain deeper insights into the model's internal dynamics, we conducted a detailed analysis of the individual expert modules. Figure 5 presents this breakdown. Specifically, the radar chart (a) reveals a clear specialization: the 'politics' expert (red line) achieves superior **recall** and **f1-score**, while the 'other' expert (blue line) leads in `precision`. Complementing this, plot (b) analyzes the average confidence, showing that the 'entertainment' expert has the lowest confidence ($\approx 0.31$), reflecting high uncertainty in this domain, whereas the 'other' expert is the most confident ($\approx 0.55$).

## 4 Conclusion and Future Work

MERMAID demonstrates near-perfect performance on balanced datasets, with Accuracy, Recall, and F1-score all approaching 1.0 [19–21]. This clearly validates the model's ability to learn strong discriminative representations under controlled conditions. More importantly, when deployed on real-world data, MERMAID maintains greater stability compared to the baseline MoE, indicating superior robustness in out-of-distribution environments. The system could serve as a decision-support tool for journalists and editors, assisting in the rapid triage and verification of news content. In social media platforms, MERMAID can be embedded as a background filter to flag potentially misleading content and assist in moderation workflows. Beyond news, the modular domain-expert design allows MERMAID to be extended to other high-stakes domains such as healthcare, science, or finance where the spread of misinformation poses significant risks.

**Acknowledgment.** We sincerely appreciate the financial research support of the Posts and Telecommunications Institute of Technology in Ho Chi Minh City, Viet Nam, and for providing the necessary resources, a conducive research environment that contributed significantly to the success of this study.

## References

1. Comito, C., Caroprese, L., Zumpano, E.: Multimodal fake news detection on social media: a survey of deep learning techniques. Soc. Netw. Anal. Mining **13**(1) (2023). https://doi.org/10.1007/s13278-023-01104-w
2. Shu, K., Sliva, A., Wang, S., Tang, J., Liu, H.: Fake news detection on social media: a data mining perspective. ACM SIGKDD Explor. Newsl. **19**(1), 22–36 (2017)
3. Shu, K., Mahudeswaran, D., Wang, S., Lee, D., Liu, H.: FakeNewsNet: A Data Repository with News Content, Social Context and Spatialtemporal Information for Studying Fake News on Social Media, arXiv preprint (2018). https://arxiv.org/abs/1809.01286
4. Cui, L., Lee, D.: COAID: COVID-19 healthcare Misinformation Dataset, arXiv preprint (2020). https://arxiv.org/abs/2006.00885

5. Wang, W.Y.: "Liar, Liar Pants on Fire": A New Benchmark Dataset for Fake News Detection, arXiv preprint arXiv:1705.00648 (2017). https://doi.org/10.18653/v1/p17-2067
6. Devlin, J., Chang, M.-W., Lee, K., Toutanova, K.: BERT: pre-training of deep bidirectional transformers for language understanding. In: Proceedings of 2019 Conference of the North American Chapter of the Association for Computational Linguistics: Human Language Technologies, Minneapolis, MN, USA, pp. 4171–4186 (2019). https://doi.org/10.18653/v1/N19-1423.
7. Liu, Y., et al.: RoBERTa: A Robustly Optimized BERT Pretraining Approach, arXiv preprint arXiv:1907.11692 (2019)
8. He, P., Liu, X., Gao, J., Chen, W.: DeBERTa: Decoding-enhanced BERT with Disentangled Attention, arXiv preprint arXiv:2006.03654 (2021)
9. Wang, Y., et al.: EANN: event adversarial neural networks for multi-modal fake news detection. In: Proceedings of 24th ACM SIGKDD Int. Conf. Knowledge Discovery & Data Mining, London, UK, pp. 849–857 (2018). https://doi.org/10.1145/3219819.3219903
10. Li, Y., et al.: Multi-source domain adaptation with weak supervision for early fake news detection. In: Proceedings of IEEE International Conference on Big Data, pp. 668–676 (2021). https://doi.org/10.1109/BigData52589.2021.9671592
11. Mosallanezhad, A., Karami, M., Shu, K., Mancenido, M.V., Liu, H.: Domain adaptive fake news detection via reinforcement learning. In: Proceedings of ACM Web Conference 2022, Lyon, France, pp. 3632–3640 (2022). https://doi.org/10.1145/3485447.3512258
12. Nan, Q., Cao, J., Zhu, Y., Wang, Y., Li, J.: MDFEND: multi-domain fake news detection. In: Proceedings of 30th ACM International Conference on Information & Knowledge Management, pp. 3343–3347 (2021)
13. Zhu, Y., et al.: Memory-guided multi-view multi-domain fake news detection. IEEE Trans. Knowl. Data Eng. **35**(7), 7178–7191 (2023). https://doi.org/10.1109/TKDE.2022.3185151
14. Comito, C., Guarascio, M., Liguori, A., Manco, G., Pisani, F.S.: Beyond the horizon: using mixture of experts for domain agnostic fake news detection. In: Lecture Notes in Computer Science, pp. 396–410 (2025). https://doi.org/10.1007/978-3-031-78980-9_25
15. Liguori, A., Pisani, F.S., Comito, C., Guarascio, M., Manco, G.: Breaking domain barriers: mixture of experts for cross-domain fake news detection. Mach. Learn. **114**(8), 188 (2025). https://doi.org/10.1007/s10994-025-06827-9
16. Bazmi, P., Asadpour, M., Shakery, A., Maazallahi, A.: Entity-centric multi-domain transformer for improving generalization in fake news detection. Inf. Process. Manag. **61**(5), 103807 (2024). https://doi.org/10.1016/j.ipm.2024.103807
17. Shu, K., Mosallanezhad, A., Liu, H.: Cross-domain fake news detection on social media: a context-aware adversarial approach. In: Khosravy, M., Echizen, I., Babaguchi, N. (eds.) Frontiers in Fake Media Generation and Detection, pp. 215–232. Springer, Singapore (2022). https://doi.org/10.1007/978-981-19-1524-6_9
18. Yadav, P., Tam, D., Choshen, L., Raffel, C.A., Bansal, M.: Ties-merging: resolving interference when merging models. Adv. Neural. Inf. Process. Syst. **36**, 7093–7115 (2023)
19. Tuan, N.M., Meesad, P., Nguyen, H.H.C.: English–Vietnamese machine translation using deep learning for chatbot applications. SN Comput. Sci. **5**(1), Art. no. 5 (2023). https://doi.org/10.1007/s42979-023-02339-2

20. Tuan, N.M., Meesad, P., Hieu, D.V., Cuong, N.H.H., Maliyaem, M.: On students' sentiment prediction based on deep learning: applied information literacy. SN Comput. Sci. **5**(7), Art. no. 928 (2024). https://doi.org/10.1007/s42979-024-03281-7
21. Tuan, N.M., Thuy, P.T.T., Cuong, H.H.N., Hien, N.T.: On determining multiple languages through technological examination for conservation management using machine learning. Forum Linguist. Stud. **7**(5), Art. no. 5 (2025). https://doi.org/10.30564/fls.v7i5.9110

# BERT-Driven Deep Learning for Enhanced Spam Email Classification with Browser Extension Integration

Chau Van Van(✉) and Nguyen Minh Tuan

Posts and Telecommunications Institute of Technology, 11 Nguyen Dinh Chieu, Sai Gon Ward, Ho Chi Minh City, Viet Nam
{vancv,minhtuan}@ptit.edu.vn

**Abstract.** Email volume continues to surge across business, education, and personal use, with spam composing nearly half of global traffic and driving financial and security risks. Classical filters (blacklists, rules, Naive Bayes, k-Nearest Neighbors) and even deep models like BiLSTM struggle to capture full semantics or resist modern evasion. We propose a BERT-driven spam-detection framework that exploits bidirectional self-attention to model rich context and evaluate it against strong baselines. On a balanced dataset of 5,000 emails with a 226-sample hold-out test set, fine-tuned BERT attains 98.67% accuracy and 98.65% F1, outperforming BiLSTM (96.43%), Naive Bayes (91.59%), and k-NN (90.27%). For practical deployment, we integrate BERT into a Chrome extension with a FastAPI backend, enabling real-time Gmail filtering with 0.5 s latency per email. Results indicate BERT delivers superior accuracy and robustness, bridging research and production. Future directions include multilingual detection, adversarially robust training, and multimodal filtering (URLs, images).

**Keywords:** Deep Learning · Spam Email Detection · Intelligent Information Systems · Privacy and Security

## 1 Introduction

Unsolicited email and sophisticated phishing continue to rise, requiring filters that are accurate, fast, and easy to deploy in user-facing workflows. Classical baselines for spam filtering are well established [1]. Representative models include Naive Bayes (NB) [2], k-Nearest Neighbors (k-NN) [3], Support Vector Machines (SVM) [4], and ensemble methods such as Random Forests (RF) [5]. Deep learning improves over feature-engineered pipelines: Convolutional Neural Networks (CNNs) capture local n-gram patterns [6]; Long Short-Term Memory (LSTM) models long-range dependencies [7]; bidirectional recurrence (BiLSTM) strengthens context capture [8]; and hybrid CNN–BiLSTM fuses local and sequential cues [9]. BERT replaces recurrence with self-attention, enabling rich bidirectional context and robust transfer from large-scale pretraining [10].


P. Pichappan et al. (Eds.): RTIS 2025, LNNS 1936, pp. 167–177, 2027.
https://doi.org/10.1007/978-3-032-23396-7_15

However, many evaluations are conducted in controlled settings with limited *apples-to-apples* comparisons against strong ML/DL baselines and little attention to deployment metrics (e.g., latency, privacy, UI integration).

This work closes that gap with a BERT-driven spam-detection framework designed for both rigor and deployment. Methodologically, we adopt a unified protocol (shared preprocessing, stratified splits, common metrics) and benchmark BERT against NB, k-NN, and BiLSTM [2,3,8] on a balanced 5,000-email corpus, reporting results on an additional 226-sample hold-out. Practically, we integrate the model into a Chrome Extension (Manifest V3) with a FastAPI backend for real-time Gmail screening, and we measure end-to-end latency ($\sim$ 0.5 s per email). Fine-tuned BERT attains 98.67% Accuracy and 98.65% F1, outperforming all baselines. This paper contributes: (1) A unified benchmark demonstrating BERT's superiority over strong ML/DL baselines; (2) a deployment blueprint (extension + FastAPI) achieving real-time latency; (3) analysis of practical challenges (dataset scope/bias, concept drift, multilingual/multimodal threats, adversarial robustness). Section 2 reviews rule-based, ML, DL, and Transformer-based approaches [1–10]. Section 3 details the dataset, preprocessing, models, and training protocol. Section 4 covers implementation and experiments (including latency). Section 5 discusses results and limitations. Section 6 concludes and outlines future work.

## 2 Related Work

The challenge of spam email detection has been a focal point of research for over two decades, progressing from rudimentary rule-based systems to advanced deep learning and transformer-based architectures. This section reviews key approaches, evaluates their strengths and limitations, and positions our study within this evolving landscape. Initial spam detection relied on manually crafted rules, including blacklists/whitelists of sender addresses, keyword-based filters, and regular-expression patterns to detect obfuscated content [2]. These methods were computationally efficient and interpretable but proved vulnerable to evasion tactics, such as rotating sender addresses, injecting benign terms, or using special characters. The need for frequent manual updates further limited scalability, rendering them inadequate against sophisticated spam campaigns. The shortcomings of rule-based systems spurred the adoption of statistical learning techniques. The Naive Bayes (NB) classifier emerged as a popular choice, modeling spam probability based on word frequency distributions with Laplace smoothing [2]. Despite its simplicity and effectiveness on early datasets, NB's assumption of feature independence often misaligned with natural language patterns, making it susceptible to manipulated word distributions. The k-Nearest Neighbors (k-NN) algorithm, employing TF-IDF vectorization and Euclidean distance, offered flexibility across datasets but suffered from high computational costs and reliance on shallow features [3]. Advanced ML models like support Vector Machines (SVMs), Decision Trees, and Random Forests (RF) [5] enhanced accuracy through feature engineering (e.g., n-grams), with SVMs becoming a standard for handling high-dimensional text. However, these methods required manual feature crafting, limiting generalization to novel or adversarial spam.

Deep learning introduced significant advancements. Convolutional Neural Networks (CNNs), adapted from image processing, treated text as word embedding sequences to extract local n-gram features automatically, outperforming classical ML on large datasets [6]. However, their focus on local contexts hindered modeling of long-range dependencies. Recurrent Neural Networks (RNNs), particularly Long Short-Term Memory (LSTM) and Gated Recurrent Unit (GRU) models, addressed this by capturing sequential dependencies [7]. Bidirectional LSTMs (BiLSTMs) improved context capture by processing text bidirectionally, consistently surpassing NB and SVM in spam detection tasks [8]. Despite this, RNNs were computationally intensive, required large labeled datasets, and struggled with long documents due to vanishing gradients. Hybrid CNN-BiLSTM models combined local feature extraction with sequence modeling, achieving competitive results but increasing complexity [9]. The Transformer architecture, with its self-attention mechanism, transformed NLP by replacing recurrence. BERT (Bidirectional Encoder Representations from Transformers) achieved state-of-the-art performance across tasks by modeling bidirectional token relationships [10]. For spam detection, BERT offers contextualized embeddings adapting word representations to context pre-training on vast corpora for effective transfer learning, and robustness against obfuscation (e.g., random characters, synonyms). AbdulNabi and Yaseen [11] reported BERT outperforming BiLSTM and ML models with over 98% accuracy on benchmark datasets. Variants like RoBERTa and DistilBERT have shown efficiency in resource-constrained settings [12], though most studies remain experimental without real-time deployment. Despite progress, gaps persist: (1) Few studies translate models into practical tools like browser extensions; (2) Systematic comparisons of BERT against NB, k-NN, and BiLSTM under consistent conditions are scarce; (3) Evaluations beyond accuracy, for example, latency, scalability, adversarial robustness are underexplored. Our work addresses these by benchmarking a BERT-driven framework against NB, k-NN, and BiLSTM on a balanced 5,000-sample dataset, and deploying it via a Chrome extension with FastAPI integration, enhancing real-world applicability (Table 1).

**Table 1.** Comparative Overview of Spam Detection Approaches

| Approach | Representative Model | Accuracy (%) |
|---|---|---|
| Rule-Based | Blacklist/Regex [1] | Variable |
| Statistical ML | Naive Bayes [2] | 91–93 |
| | k-NN [3] | 90–92 |
| | SVM [4] | 96–97 |
| Deep Learning | CNN [6] | 94–96 |
| | BiLSTM [8] | 96–97 |
| Transformer | BERT [11] | 98+ |

# 3 Methodology

The proposed framework aims to systematically evaluate classical machine learning models and deep learning models against a transformer-based approach (BERT) for spam email detection. This section outlines the datasets used, preprocessing pipeline, baseline algorithms, and the BERT fine-tuning process. An overview of the workflow is shown in Fig. 1.

## 3.1 Dataset Collection

Two publicly available datasets were used to ensure diversity in the study. The first was the Spambase dataset from the UCI Machine Learning Repository, containing 4,601 emails with 57 attributes, including 745 spam and 3,856 ham messages. The second was the Text-based Email Corpus from Kaggle, consisting of 5,728 labeled emails, with 1,378 spam and 4,350 ham. From these sources, a balanced dataset of 5,000 samples was constructed, comprising 2,000 spam and 3,000 ham emails. Additionally, a hold-out set of 226 samples (113 spam and 113 ham) was reserved to evaluate the robustness of the model (Table 2).

**Table 2.** Dataset Distribution

| Dataset | Ham | Spam | Total |
|---|---|---|---|
| Spambase | 3,856 | 745 | 4,601 |
| Text Corpus | 4,350 | 1,378 | 5,728 |
| Balanced | 3,000 | 2,000 | 5,000 |

The data preprocessing pipeline involved several steps. First, cleaning was performed by removing HTML tags, headers, and special characters. Next, stopword removal and tokenization were applied to normalize the text. For machine learning models, TF-IDF vectorization was used to convert text into numerical features (Eq. (1)). For the BiLSTM model, 300-dimensional word embeddings were generated to capture semantic relationships. Finally, for the BERT model, tokenization was performed with all sequences padded to a fixed length of 300 tokens.

$$TF\text{-}IDF(w,d) = TF(w,d) \times \log \frac{N}{DF(w)} \quad (1)$$

To handle class imbalance, RandomOverSampler and Easy Data Augmentation (EDA) were applied. Classification based on the $k$ closest neighbors using Euclidean distance:

$$d(x,y) = \sqrt{\sum_{i=1}^{n}(x_i - y_i)^2}$$

**Algorithm 1.** k-NN Spam Classification

1: Input: Training set $D$, test email $x$, parameter $k$
2: Extract TF-IDF features for $x$
3: **for** each $(x_i, y_i) \in D$ **do**
4: Compute Euclidean distance $d(x, x_i)$
5: **end for**
6: Sort distances, select $k$ nearest neighbors
7: Assign majority class
8: **Output**: Predicted label (spam/ham)

**Naive Bayes (NB).** Posterior probability:

$$P(c|x) = \frac{P(x|c)P(c)}{P(x)}$$

Laplace smoothing:

$$P(w_i|c) = \frac{\text{Count}(w_i, c) + \alpha}{\sum_{w \in V} \text{Count}(w, c) + \alpha|V|}$$

**Algorithm 2.** Naive Bayes Spam Classification

1: Input: Training set $D$, test email $x$
2: Compute TF-IDF features for $x$
3: Estimate priors $P(spam), P(ham)$
4: **for** each token $w_i$ in $x$ **do**
5: Compute likelihoods $P(w_i|spam), P(w_i|ham)$
6: **end for**
7: Compute posteriors
8: Assign class with higher posterior
9: **Output**: Predicted label

**BiLSTM.** BiLSTM processes sequences bidirectionally:

$$h_t = LSTM(x_t, h_{t-1}) \oplus LSTM(x_t, h_{t+1})$$

Architecture: Embedding (300D) → BiLSTM (128 units) → Dense (ReLU) → Dropout (0.1) → Dense (Sigmoid).

The BERT model (bert-base-cased) is fine-tuned for binary classification. The self-attention mechanism is defined as:

$$Attention(Q, K, V) = softmax\left(\frac{QK^T}{\sqrt{d_k}}\right)V$$

Fine-tuning settings:

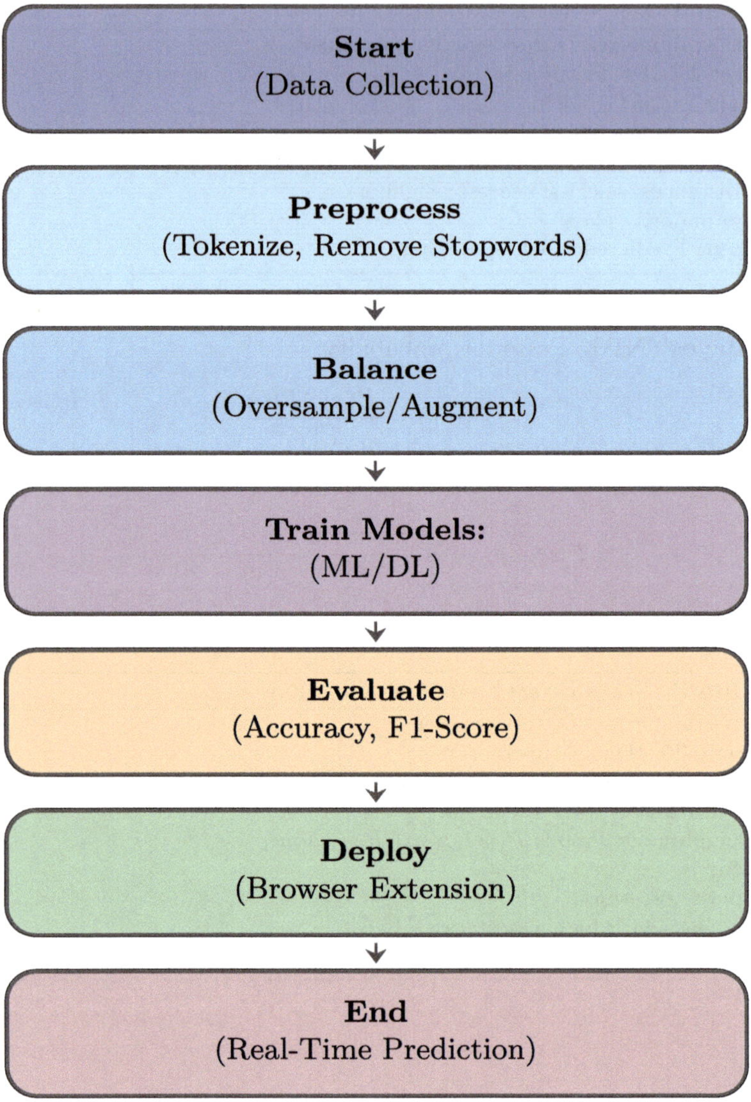


**Fig. 1.** Workflow of the proposed system.

**Algorithm 3.** BiLSTM Spam Classification

**Input:** Tokenized sequence $x$
Embed tokens to 300D vectors
Pass through BiLSTM (forward and backward)
Concatenate hidden states
Apply Dense + Dropout
**Output:** Sigmoid probability of spam

- Sequence length: 300 tokens
- Batch size: 32
- Learning rate: $4 \times 10^{-5}$
- Optimizer: AdamW
- Epochs: 3

**Algorithm 4.** BERT Fine-Tuning for Spam Detection

1: Input: Dataset $D$, test email $x$
2: Tokenize $x$, pad to 300 tokens
3: Feed into BERT encoder
4: Extract [CLS] embedding
5: Apply Dense layer with Sigmoid
6: Train with binary cross-entropy loss
7: **Output**: Probability of spam

### 3.2 Evaluation Metrics

Performance is measured using:

$$Accuracy = \frac{TP + TN}{TP + TN + FP + FN}$$

$$Precision = \frac{TP}{TP + FP}, \quad Recall = \frac{TP}{TP + FN}$$

$$F1 = \frac{2 \times Precision \times Recall}{Precision + Recall}$$

where $TP$, $TN$, $FP$, $FN$ denote true positives, true negatives, false positives, and false negatives.

## 4 Experimental Results

All code, preprocessing scripts, and the Chrome MV3+FastAPI prototype are packaged with fixed random seed (42), exact library versions, and a Colab notebook for one-click execution. A redacted dataset card and instructions to reconstruct the balanced split from public sources are provided. All experiments were conducted on Google Colab using Python 3.11 with TensorFlow, Scikit-learn, and Hugging Face Transformers. Hardware included a single NVIDIA T4 GPU with 16 GB memory. The dataset was split into 70% training, 20% validation, and 10% testing. Additionally, a hold-out set of 226 samples (113 spam, 113 ham) was used for robustness evaluation. Cross-validation with 8 folds was applied to reduce overfitting and ensure fair evaluation.

**Table 3.** Train/Validation/Test Split

| Set | Ham | Spam | Total |
|---|---|---|---|
| Training (70%) | 2,100 | 1,400 | 3,500 |
| Validation (20%) | 600 | 400 | 1,000 |
| Test (10%) | 300 | 200 | 500 |

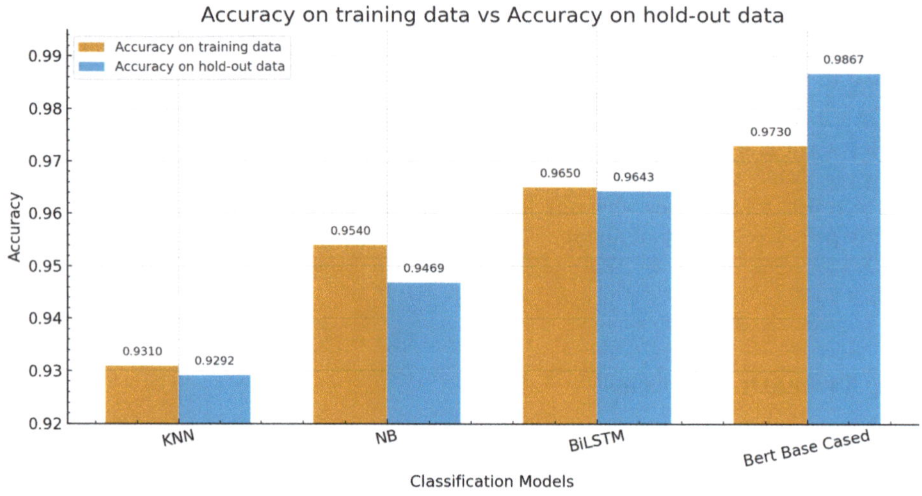


**Fig. 2.** Performance (Accuracy and F1-score).

We evaluated models using Accuracy, Precision, Recall, and F1-score (see Fig. 2).

The Chrome Extensions platform forwards the email body to a FastAPI predict endpoint hosting the fine-tuned BERT model. Using the stratified 226-email hold-out set described in Table 3 (kept unseen during training), the end-to-end classification latency—from Gmail DOM extraction to JSON response—averaged 0.5 s per email (median 0.52 s; p95 0.80 s),which demonstrates real-time feasibility. The extension interface embedded in Gmail is shown in Fig. 3

From Table 4, it is evident that deep learning methods (BiLSTM and BERT) outperform traditional machine learning models. BERT achieves the highest accuracy and F1-score, validating its ability to capture contextual semantics. Compared to BiLSTM, BERT improves accuracy by ∼2.2% and F1-score by ∼2.6%. While k-NN and Naive Bayes provide lightweight alternatives, their limited feature representations restrict performance against sophisticated spam. The real-time latency analysis further confirms that BERT can be effectively deployed without compromising user experience.

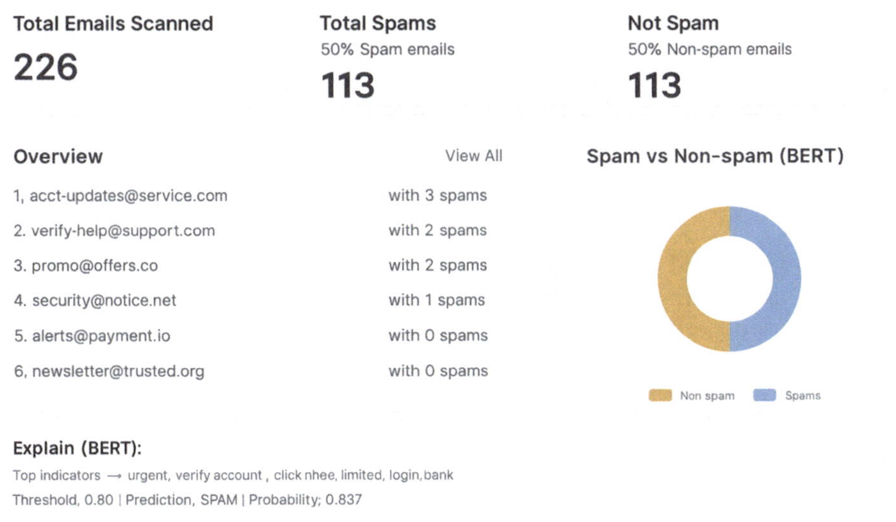


**Fig. 3.** Chrome Extension for real-time Gmail spam detection.

**Table 4.** Performance Metrics on Test Set

| Model | Accuracy | Precision | Recall | F1 |
|---|---|---|---|---|
| k-NN | 90.27 | 92.52 | 87.16 | 89.80 |
| Naive Bayes | 91.59 | 89.17 | 94.69 | 91.88 |
| BiLSTM | 96.43 | 95.76 | 96.24 | 96.00 |
| **BERT** | **98.67** | **98.95** | **98.35** | **98.65** |

# 5 Discussion

Table 4 establishes BERT as the top performer; this aligns with its bidirectional contextual representations, which better capture obfuscated phrasing typical of spam. In resource-constrained settings or at cold start, lighter models (Naive Bayes, k-NN) or a BiLSTM tier can still be valuable as a first-pass filter. Practical deployment should balance accuracy with inference latency and compute cost to meet service-level requirements. Spam content evolves and may be deliberately manipulated, leading to concept drift and adversarial perturbations. A pragmatic mitigation plan includes periodic or continual fine-tuning with user feedback, adversarial data augmentation during training, threshold calibration, and health monitoring to detect degradation. In real-world environments, privacy regulations (e.g., GDPR) can preclude centralized training. Federated learning offers a path to train across clients without sharing raw data. To preserve latency and cost at scale, model distillation and quantization, together with batching or caching, can reduce overhead while retaining most of the accuracy demonstrated offline.

## 6 Conclusion and Future Work

In this study, we compared classical machine-learning methods (k-NN, Naive Bayes), a recurrent deep model (BiLSTM), and a transformer-based model (BERT) for spam email detection under a unified protocol of preprocessing, stratified splits, and common metrics. On a balanced corpus of 5,000 emails with a 226-email stratified hold-out (Table 3), BERT achieved 98.67% accuracy and 98.65% F1, outperforming BiLSTM (96.43%), Naive Bayes (91.59%), and k-NN (90.27%) as summarized in Table 4 and illustrated in Fig. 2. We also realized an end-to-end deployment using a Chrome Extensions platform backed by FastAPI and measured an average end-to-end latency of about 0.5 s per email on the hold-out set, indicating real-time feasibility; the in-Gmail interface is shown in Fig. 3. Experiments are restricted to English emails and text-only features; multilingual or code-mixed content is not evaluated. The balanced corpus (5,000 training samples with a 226-email hold-out) may not capture the full diversity and seasonality of spam in the wild. Real-time latency was measured on a prototype (Chrome MV3 + FastAPI) and on limited hardware; enterprise-scale throughput and tail-latency were not stress-tested. Finally, non-text signals such as URL reputation, HTML layout, images/OCR, and attachment metadata were excluded, which may reduce robustness against link- or image-centric campaigns. We plan to extend to multilingual and code-mixed emails using mBERT/XLM-R, incorporate non-text cues (URL reputation, HTML features, OCR for images, attachment metadata), harden robustness via adversarial augmentation and robust optimization, support continual and federated learning to cope with concept drift and privacy constraints, and improve deployment efficiency at scale through model distillation, quantization, and batching/caching.

**Acknowledgment.** We sincerely appreciate the financial research support of the Posts and Telecommunications Institute of Technology in Ho Chi Minh City, Viet Nam, and for providing the necessary resources, a conducive research environment that contributed significantly to the success of this study.

## References

1. Guzella, T.S., Caminhas, W.M.: A review of machine learning approaches to spam filtering. Expert Syst. Appl. **36**(7), 10206–10222 (2009). https://doi.org/10.1016/j.eswa.2009.02.037
2. Sahami, M., Dumais, S., Heckerman, D., Horvitz, E.: A Bayesian approach to filtering junk E-mail. In: AAAI Workshop on Learning for Text Categorization, pp. 55–62 (1998)
3. Sakkis, Y., Androutsopoulos, I., Paliouras, G., Karkaletsis, V., Spyropoulos, C.D., Stamatopoulos, P.: A memory-based approach to anti-spam filtering for mailing lists. Inf. Retrieval **6**, 49–73 (2003). https://doi.org/10.1023/A:1022906229525

4. Joachims, T.: Text categorization with support vector machines: learning with many relevant features. In: Proceedings of 10th European Conference on Machine Learning (ECML). LNCS, vol. 1398, pp. 137–142. Springer (1998)
5. Breiman, L.: Random forests. Mach. Learn. **45**(1), 5–32 (2001). https://doi.org/10.1023/A:1010933404324
6. Kim, Y.: Convolutional neural networks for sentence classification. In: Proceedings of 2014 Conference on Empirical Methods in Natural Language Processing (EMNLP), pp. 1746–1751 (2014)
7. Hochreiter, S., Schmidhuber, J.: Long short-term memory. Neural Comput. **9**(8), 1735–1780 (1997). https://doi.org/10.1162/neco.1997.9.8.1735
8. Schuster, M., Paliwal, K.K.: Bidirectional recurrent neural networks. IEEE Trans. Signal Process. **45**(11), 2673–2681 (1997). https://doi.org/10.1109/78.650093
9. Boussougou, M.K.M., Reddy, P.K.R., Nayak, S.T.: Attention-based 1D CNN-BiLSTM hybrid model enhanced with multi-head attention for phishing email detection. Mathematics **11**(14), 3217 (2023). https://doi.org/10.3390/math11143217
10. Devlin, J., Chang, M.-W., Lee, K., Toutanova, K.: BERT: pre-training of deep bidirectional transformers for language understanding. In: Proceedings of NAACL-HLT, pp. 4171–4186 (2019). arXiv:1810.04805
11. AbdulNabi, I., Yaseen, Q.: Spam email detection using deep learning techniques. Procedia Comput. Sci. **184**, 853–858 (2021). https://doi.org/10.1016/j.procs.2021.03.107
12. Liu, Y., et al.: RoBERTa: A Robustly Optimized BERT Pretraining Approach, arXiv preprint arXiv:1907.11692 (2019)
13. Tuan, N.M., Meesad, P., Nguyen, H.H.C.: English–Vietnamese machine translation using deep learning for chatbot applications. SN Comput. Sci. **5**(1), 5 (2023). https://doi.org/10.1007/s42979-023-02339-2
14. Tuan, N.M., Meesad, P., Hieu, D.V., Cuong, N.H.H., Maliyaem, M.: On students' sentiment prediction based on deep learning: applied information literacy. SN Comput. Sci. **5**(7), 928 (2024). https://doi.org/10.1007/s42979-024-03281-7
15. Tuan, N.M., Meesad, P., Son, N.H.: On a stock prediction aligned to natural language sentiments. In: Proceedings of 8th International Conference on Natural Language Processing and Information Retrieval (NLPIR 2024), Okayama, Japan, 13–15 December 2024. ACM, New York, NY, USA, Article 111, 5 pp. https://doi.org/10.1145/3711542.3711597
16. Tuan, N.M., Thuy, P.T.T., Cuong, H.H.N., Hien, N.T.: On determining multiple languages through technological examination for conservation management using machine learning. Forum Linguist. Stud. **7**(5), Article 5 (2025). https://doi.org/10.30564/fls.v7i5.9110.
17. Tuan, N.M., Son, N.H., Van, C.V.: Piecemeal-DFS algorithm for energy-constrained depth-first search on the merging problem of two given trees (in Vietnamese). In: Proceedings of 2nd National Symposium on Natural Sciences and Applications in the Digital Age (NSA), Ha Noi, Viet Nam, pp. 46–53 (2025). ISBN: 978-604-67-330-0

# Non-invasive Lung Condition Screening via Iris Imaging: Leveraging Vision Transformer and Efficient CNN

Trong Thua Huynh[1(✉)], De Thu Huynh[2], Hoang Phat Vu[1], Minh Tuan Nguyen[1], and Tran Ho Truc Phan[3]

[1] Posts and Telecommunications Institute of Technology, Ho Chi Minh City 700000, Vietnam
thuaht@ptit.edu.vn
[2] The Saigon International University, Ho Chi Minh City 700000, Vietnam
[3] Binh Duong University, Ho Chi Minh City 700000, Vietnam

**Abstract.** Originating from the limitations of current prescreening methods for lung diseases, this study investigates the potential of modernized iridology enhanced by deep learning, with a particular focus on evaluating the performance of Vision Transformers (ViT) and Convolutional Neural Networks (CNN). Utilizing a labeled iris image dataset expert-annotated with lung health conditions, we conducted preprocessing to localize relevant iris regions, followed by appropriate data partitioning and cross-eye evaluation between the left and right irises to assess the generalization capability of the proposed approach. The models were trained in a two-stage process using pre-trained ImageNet weights and evaluated via a three-class confusion matrix (negative, positive, unknown). Cross-eye generalization was demonstrated with an accuracy deviation of less than 2% between the two sides. Experimental results indicate that Efficientnetv2m achieved the highest classification accuracy of 89.03%, while Swin Tiny consistently maintained stable performance (above 87% across all variants) with lower computational overhead. The complementary roles of self-attention mechanisms and fused convolution further underscore the suitability and feasibility of hybrid deep learning architectures for this task. These findings highlight the potential for developing a non-invasive, cost-effective lung condition prescreening system based on iris image analysis, thereby promoting the integration of iridology and advanced deep learning (DL) techniques in lung condition prescreening.

**Keywords:** Iridology · non-invasive diagnosis · lung condition screening · deep learning · vision transformer · CNN

## 1 Introduction

Iridology, also known as the diagnostic method based on observing the iris, has long existed as an alternative medical approach for assessing health conditions through signs or reflex zones on the iris. Recent years have witnessed growing research efforts to legitimize iridology by integrating it with modern computer vision techniques [1, 2].


P. Pichappan et al. (Eds.): RTIS 2025, LNNS 1936, pp. 178–189, 2027.
https://doi.org/10.1007/978-3-032-23396-7_16

In particular, the advancement of Artificial Intelligence (AI) and DL techniques has opened new opportunities for iris image analysis, enabling the exploration of potential correlations between micro-level iris features and general health conditions, [3, 4].

Recently, the intersection between iridology and computer vision has garnered significant attention from the research community. Avhad and Bakal demonstrated the feasibility of using deep learning to recognize iris patterns associated with health conditions in two consecutive publications in 2024 [5] and 2025 [6]. Earlier, Rao et al. employed traditional machine learning algorithms to detect dysfunctions in vital organs based on iris features [7]. While these studies do not directly target lung diseases, they provide valuable insights into the application of deep learning for iris image analysis, including preprocessing techniques, feature extraction, and classification, which can inform the development of a novel approach for detecting lung conditions.

Early detection of lung disease remains a critical factor in improving prognosis and enhancing treatment outcomes [8]. However, current diagnostic methods such as low-dose computed tomography (LDCT), MRI, or biopsy are often invasive, costly, and challenging to implement on a large scale for mass screening purposes [9]. Dickson et al. [9] identified numerous barriers that discourage patients from participating in LDCT screening, while Horst et al. emphasized the practical challenges of operating LDCT programs at the community level [10]. These findings highlight a substantial gap for alternative approaches that are non-invasive, cost-effective, and patient-friendly such as iris analysis.

Deep learning has been transforming the way medical images are analyzed, particularly through the use of Convolutional Neural Networks (CNN). Modern CNNs have demonstrated high performance in critical tasks such as tumor detection, classification, and segmentation [11, 12]. More recently, Transformer-based approaches, most notably the ViT, have gained attention. The improved Swin Transformer [13], with its shifted window architecture, allows the model to scale to high-resolution images while maintaining computational efficiency. However, Transformers typically require large amounts of training data to achieve optimal performance, which poses a challenge in specialized medical domains where data availability is often limited [14].

Motivated by this context, our study proposes leveraging Swin Transformers and lightweight CNNs to effectively address the task of early lung condition screening through iris images. The specific objectives include: (1) analyzing and comparing these two deep learning architectures on the same iris dataset; (2) evaluating the generalization capability of the models by training on images from one eye and testing on images from the opposite eye; (3) examining the impact of class imbalance through the application of class weighting techniques; (4) providing a comprehensive analysis using performance metrics, confusion matrices; and (5) discussing the potential of combining Swin Transformers and CNNs to optimize the diagnostic process. By expanding the evidence base for iridology through advanced AI techniques, this study aims to contribute to the development of non-invasive, cost-effective, and patient-friendly lung screening solutions.

## 2 Related Works

The application of deep learning techniques, including CNN and Transformer architectures, has made significant strides in medical image analysis for detecting and classifying lung diseases in recent years. Surveys on deep learning architectures continue to emphasize the role of CNNs in the context of medical image analysis. The work [15] investigates the detection of various lung diseases, including COVID-19, pneumonia, and tuberculosis, based on chest X-ray (CXR) images. The model utilizes a lightweight CNN to extract features, incorporates the Pearson correlation coefficient to select the most relevant features, and employs the extreme learning machine algorithm for classification. The model achieved high accuracy for eight-class classification, outperforming current models and demonstrating its effectiveness and potential to support rapid, accurate diagnosis for clinicians. This also provides a solid foundation for exploring the potential of iris image analysis in assessing lung health.

Vision Transformers, particularly the Swin Transformer architecture, have emerged as powerful tools in various medical image analysis tasks, demonstrating the ability to capture complex spatial relationships and achieve state-of-the-art performance. The study [16] provides a comprehensive overview of the application of transformer models in medical imaging. The authors discuss the fundamental attributes of transformers that make them well-suited for medical image analysis.

Hatamizadeh et al. introduced Swin UNETR [17], a novel architecture that combines Swin Transformers with a UNet-like structure for semantic segmentation of brain tumors in MRI images. This work demonstrated the effectiveness of Swin Transformers in medical image segmentation, a critical task in computer-aided diagnosis. The study also explored the use of self-supervised pre-training of Swin Transformers for 3D medical image analysis, highlighting their potential to enhance model performance even with limited labeled data.

The application of Swin Transformers extends across various aspects of lung disease analysis. Ullah et al. proposed a two-stage Vision Transformer model incorporating a pretrained Swin Trans-former for the classification of 14 different lung diseases from chest X-rays [18]. This study demonstrates the direct applicability of Swin Transformers in lung disease classification using a different imaging modality.

## 3 Methodology

### 3.1 Dataset

The dataset comprises a total of 10,866 iris images, including 7,313 images labeled as 'Negative' and 3,553 images labeled as 'Positive'. The 'Positive' label indicates images identified by iridologists as showing signs of lung damage, while the 'Negative' label corresponds to individuals with healthy lungs. The dataset maintains a relatively balanced distribution between left and right irises. The data is categorized into three groups based on lung health status, as shown in Table 1: (1) 'Positive' (P) for individuals with lung damage, (2) 'Negative' (N) for healthy individuals, and (3) 'Unknown' (U) for cases that are ambiguous or cannot be classified.

**Table 1.** Distribution of iris image data

| Dataset | | Left Iris | | | Right Iris | | |
|---|---|---|---|---|---|---|---|
| | Negative | Positive | Unknown | | Negative | Positive | Unknown |
| Training | 2,140 | 1,084 | 319 | | 2,248 | 1,048 | 328 |
| Testing | 1,427 | 722 | 212 | | 1,498 | 699 | 219 |

The region of interest (ROI) is defined within the iris area, specifically from the 2 o'clock to 4 o'clock position for the left eye, and from the 8 o'clock to 10 o'clock position for the right eye. The ROI extraction process is based on segmentation results from a specialized iris recognition system, which is capable of estimating the center and radius of both the pupil and the iris. To obtain lung-related information, a 2–4 h (or 8–10 h) region mask was applied to crop the ROI images for analysis, ensuring the focus on the iris region related to the lungs. The images after cropping the ROI have a size ranging from 60 $\times$ 160 pixels.

In our experiments, an additional image preprocessing step is performed before the training process begins. Specifically, the images are resized to 224 $\times$ 224 pixels and intensity normalized based on ImageNet statistics. Due to a significant class imbalance, where the number of negative samples is nearly twice that of positive samples, we apply data augmentation techniques such as random rotation, zooming in and out, and translation. These augmentations also help generate additional samples for the underrepresented 'Unknown' class, enhancing the generalization and robustness of the model.

### 3.2 Model Architectures

**Vision Transformers with Swin Transformer variants.** In recent years, ViT models have emerged as a promising alternative to CNNs by adapting the transformer architecture from the field of natural language processing to computer vision. Although ViT is capable of learning long-range dependencies and capturing global context, it has certain limitations, particularly its computational complexity, which increases quadratically with the number of patches. Therefore, this study employs the Swin Transformer [14], an enhanced variant of ViT that is more optimized for vision tasks. Unlike ViT, which maintains a fixed resolution throughout its architecture, the Swin Transformer organizes input data in a hierarchical structure with feature representations progressively increasing in abstraction from coarse to deep levels. This is achieved through patch merging technique, where neighboring patches are combined and mapped into a higher-dimensional space.

To reduce the computational cost of self-attention, the Swin Transformer replaces global attention with non-overlapping Window-based Multi-head Self-attention (W-MSA), in which attention is computed only within small, localized windows of the image rather than across the entire image. To address the limitation in learning inter-window information, the model introduces Shifted Window techique (SW-MSA). In this approach, windows are shifted by half the window size relative to the original partitioning, allowing patches from different regions to interact with each other across consecutive layers indirectly. The alternating use of W-MSA and SW-MSA enables the model to effectively capture both local features and broader contextual information while maintaining computational efficiency.

In this study, we evaluate the performance of three different Swin Transformer variants including Tiny (the smallest and most efficient variant), Small (offers increased capacity through more layers in the third stage compared to Swin-T) and Medium (features a larger embedding dimension and more attention heads, further increasing model capacity at the cost of higher computational load.

**Convolutional Neural Networks (CNN)** have long served as a foundational approach in computer vision, owing to their ability to extract hierarchical spatial features through convolutional layers with learned filters (such as edges, textures, parts, and objects). In the study presented by LeCun et al. [19], four representative CNN architectures are selected to evaluate how different design philosophies influence performance in the task of iris image analysis for lung condition screening.

*ResNet50* [20] - is a classic deep CNN architecture with 50 layers, notable for utilizing residual connections that allow signals to propagate through layers without gradient vanishing. This architecture effectively addresses the vanishing gradient problem and enables the training of very deep models. Its ability to learn rich and detailed semantic representations makes ResNet50 particularly well-suited for recognizing subtle features in the iris, such as fibers, folds, or small anomalies.

*VGG16* [21] - is a classic CNN architecture, characterized by its simplicity and homogeneity. The model employs successive $3 \times 3$ convolutional layers, interleaved with max-pooling layers to reduce the feature map size. Although it lacks residual connections or complex structures like ResNet, VGG16 still possesses strong feature learning capabilities due to its depth (16 layers). However, its high computational cost and large number of parameters make it less suitable for real-time applications or resource-constrained devices.

*MobileNetv2* [22] is designed to optimize performance on mobile devices, with a key feature being depthwise separable convolutions, which split the convolution operation into two steps: (1) Depthwise convolution, and (2) Pointwise convolution ($1 \times 1$). Overall, this significantly reduces the number of parameters and convolution operations. In addition, MobileNetv2 utilizes inverted residual blocks (where channels are expanded within the block) and linear bottlenecks (removing non-linearity in the final $1 \times 1$ layer of a block) to maintain representational power while remaining lightweight. As a result, this model is highly computationally efficient and well-suited for realworld deployment, although its accuracy may be limited when dealing with more complex tasks.

*EfficientNetv2m* [23] is a modern architecture that emphasizes optimizing three factors simultaneously (e.g. depth, width, and resolution) using the compound scaling technique. Instead of scaling the model in a single dimension, EfficientNet applies a compound coefficient that maintains a fixed ratio among factors depth, width, and resolution. EfficientNetv2 improves upon its predecessor through (1) Fused-MBConv Blocks (replace the two separate convolutional steps with standard convolutions in the early layers to optimize memory usage), and Progressive Learning (incorporates staged training with gradually increasing data complexity to improve learning efficiency). EfficientNetv2m offers a well-balanced trade-off between accuracy and computational cost, making it particularly suitable for iris image data, which contains multiple structural layers and multi-scale features.

The four representative CNN architectures discussed above reflect different design philosophies: from deep and expressive representation learning (ResNet50), simple sequential design (VGG16), and mobile-optimized efficiency (MobileNet50), to controlled compound scaling (EfficientNetv2m). By comparing these diverse CNNs on the same iridology task, we aim to highlight how different convolutional design principles (e.g. depth, efficiency, width, and block structure) impact performance in iris-based tasks. This comparison becomes particularly insightful when evaluated alongside variants of the Swin Transformer, which leverages the attention mechanism. The resulting analysis offers valuable insights into selecting the most appropriate architecture in the context of iris image analysis for early detection of human lung abnormalities.

### 3.3 Training Methodology

All models are initialized with pre-trained weights on the ImageNet dataset to leverage the ability to learn universal features from natural images. The training process is conducted in two stages to ensure stability and enhance the convergence efficiency of the models:

**Stage 1 - Training the classification head.** The entire backbone network is frozen, and only the weights of the classification head are updated. Number of training epochs is 20, and learning rate is $10^{-4}$. This stage allows the model to adapt the final layers to the new data distribution without distorting the pre-learned weights from ImageNet.

**Stage 2 - Fine-tuning the entire model.** The entire network is unfrozen, allowing all weights to be updated. Number of training epochs is 20. The learning rate is reduced to $10^{-5}$ to prevent overly aggressive updates to the pre-trained weights.

In both training stages, the models are trained with a batch size of 16. The optimization algorithms experiment with included Stochastic Gradient Descent (SGD), Adam, and AdamW. Due to the class imbalance in the dataset, where the number of negative (healthy) samples is nearly twice that of positive (lung damage) samples, the model training process may become biased toward the majority class. To address this issue, we apply class weighting, where the weight assigned to each class is inversely proportional to its frequency in the dataset. Assuming the dataset contains C classes, with $N_i$ samples in class $i$, and a total number of samples $N = \sum_{i=1}^{C} N_i$, the weight for class $i$ is computed by Eq. 1:

$$w_i = \frac{N}{C \bullet N_i} \tag{1}$$

The weighted cross-entropy loss is computed by Eq. 2:

$$\mathcal{L} = -\sum_{i=1}^{C} w_i \bullet y_i \bullet \log(\hat{y}_i) \tag{2}$$

where: $y_i \in \{0,1\}$ is the ground-truth label for class $i$, $\hat{y}_i$ is the predicted probability for class $i$ by the model, and $w_i$ is the weight assigned to class $i$. This loss function penalizes misclassification errors more heavily for rare classes, thereby increasing sensitivity and improving overall performance on the imbalanced testset.

## 4 Results and Discussion

The experimental results provide valuable insights into the applicability of different deep learning architectures for automated lung condition detection via iridology, highlighting the distinct strengths and weaknesses of Vision Transformers and various CNN designs in this specific context.

As shown in Fig. 1, the Swin Transformer variants consistently achieved high validation accuracies, all above 87% (higher than most CNN models, except for EfficientNetv2m), demonstrating the potential of attention-based mechanisms for this task. Notably, the Swin_t (Tiny) variant frequently performed on par with or slightly better than its larger counterparts, Swin_s and Swin_m. These results indicate that the attention mechanism is capable of capturing both local patterns and global circular relationships of the iris, resulting in superior performance on both left and right iris datasets compared to MobileNetv2, VGG16, and ResNet50.

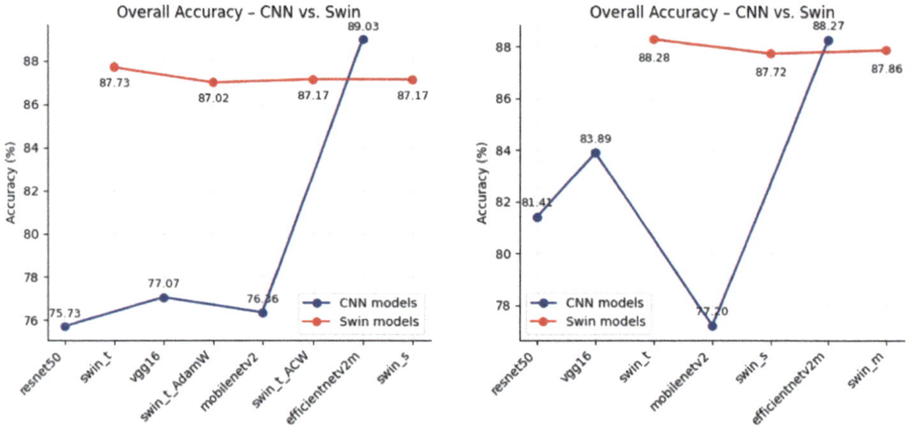


**Fig. 1.** Overall accuracy comparison between Swin and CNN models.

This suggests that the capacity of Swin_t, with its specific configuration of hierarchical stages, embedding dimension (96), and attention heads ([3, 6, 12, 24]), strikes an effective balance for capturing the relevant features in $224 \times 224$ iris images without incurring the higher computational cost or potential overfitting risk of the larger models. The Swin architecture's core mechanisms, hierarchical feature representation and shifted window self-attention (W-MSA/SW-MSA), appear well-suited for iridology. The hierarchical structure likely enables the model to process iris patterns at multiple scales, while the windowed attention efficiently captures local texture details (within windows) and integrates broader contextual information across the iris as features propagate through layers via the shifting mechanism. This ability to model both local and quasi-global relationships might be key to identifying subtle iris signs associated with systemic conditions like lung health.

However, the dependence of transformers on large datasets remains a pertinent consideration. While pre-training on ImageNet provides a strong starting point, optimizing these models fully, especially the higher capacity variants, might be constrained by the available dataset size in specialized medical domains like iridology. The performance plateau observed across Swin variants could indicate that further gains might require significantly larger or more diverse datasets.

While Swin Transformers demonstrated strong peak performance, the CNN cohort revealed important nuances. The standout performer among CNNs was EfficientNetv2m, achieving the highest overall accuracy in the study (89.03% for Left-iris training in Fig. 1a, 88.27% for right-iris training in Fig. 1b). This surpasses even the best Swin Transformer result, indicating that its architecture, optimized through compound scaling of depth, width, and resolution, and employing efficient blocks like Fused-MBConv, is exceptionally well-suited for extracting discriminative features from iris images. Its success underscores the idea that a meticulously balanced and scaled convolutional architecture can potentially outperform Transformers on specific tasks, especially when leveraging strong pre-training.

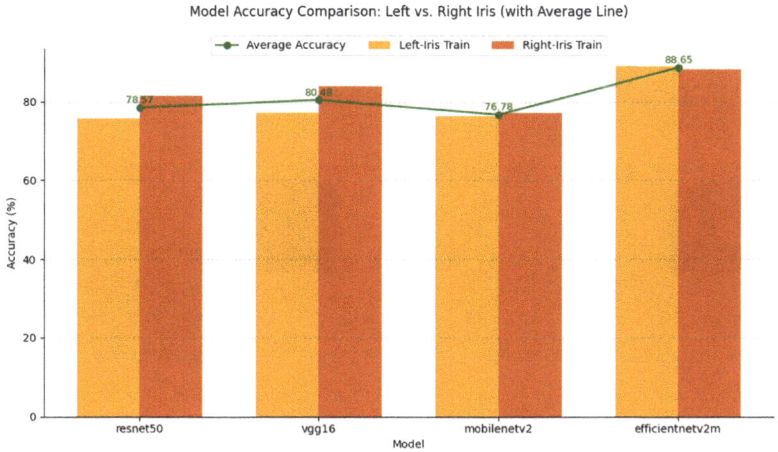


**Fig. 2.** Accuracy comparison among CNN models.

In contrast, VGG16, despite its depth, shows comparatively lower performance (e.g., 77.07% for left-iris training, 83.89% for right-iris training, in Fig. 2, with the average accuracy values), significantly lower than EfficientNetv2m. The primary reason is its complex architecture with over 130 million parameters, most of which reside in the two final fully connected layers that contribute little to learning fine-grained patterns. Meanwhile, although ResNet50 benefits from residual connections, iris-related pathologies often manifest through subtle textures. VGG16's consecutive convolutional layers preserve spatial resolution for longer, which helps retain important structural signals better than ResNet50. Additionally, since ResNet50 is about three times deeper than VGG16, the skip connections, when used with a small batch size, may cause gradients to bypass learning essential micro-level features. This suggests that while deep feature learning enabled by residual connections is beneficial, the specific attention mechanisms in Swin or the optimized scaling in EfficientNetv2m might be more advantageous for this task.

Furthermore, the average accuracy values of MobileNetv2 (e.g., 76.36% for left-iris training, 77.20% for right-iris training, in Fig. 2) demonstrate the effectiveness of lightweight architectures. MobileNetv2 was designed for speed and efficiency, intentionally reducing feature complexity to minimize computational demands. However, in medical tasks that require capturing fine iris textures, this architecture lacks sufficient representational resolution, resulting in the lowest accuracy among the CNN models evaluated in this study. Nonetheless, its use of depthwise separable convolutions allows it to achieve respectable accuracy (only slightly lower than ResNet50, as shown in Fig. 2) with significantly fewer parameters. This makes it a viable option when computational resources are limited, even outperforming the much heavier VGG16 in certain cross-eye test scenarios (e.g., the accuracy of left-iris testing using a model trained on the right-iris dataset is 78.65% for MobileNetv2 vs. 72.72% for VGG16).

As illustrated in Fig. 3, training on the right-iris dataset and testing on the left-iris, the VGG16 model detected only 283 TP samples (Fig. 3a), which is relatively low given the 722 left-iris images in the testset. In contrast, MobileNetv2 detected 586 TP samples under the same conditions (Fig. 3b), a substantially better result. Moreover, although the overall accuracy of VGG16 is higher than MobileNetv2 by 6.69% (Fig. 4a), its accuracy specifically on left-iris test data is lower than MobileNetv2 by 5.93%, highlighting the MobileNetv2's generalization capability in cross-eye evaluations.

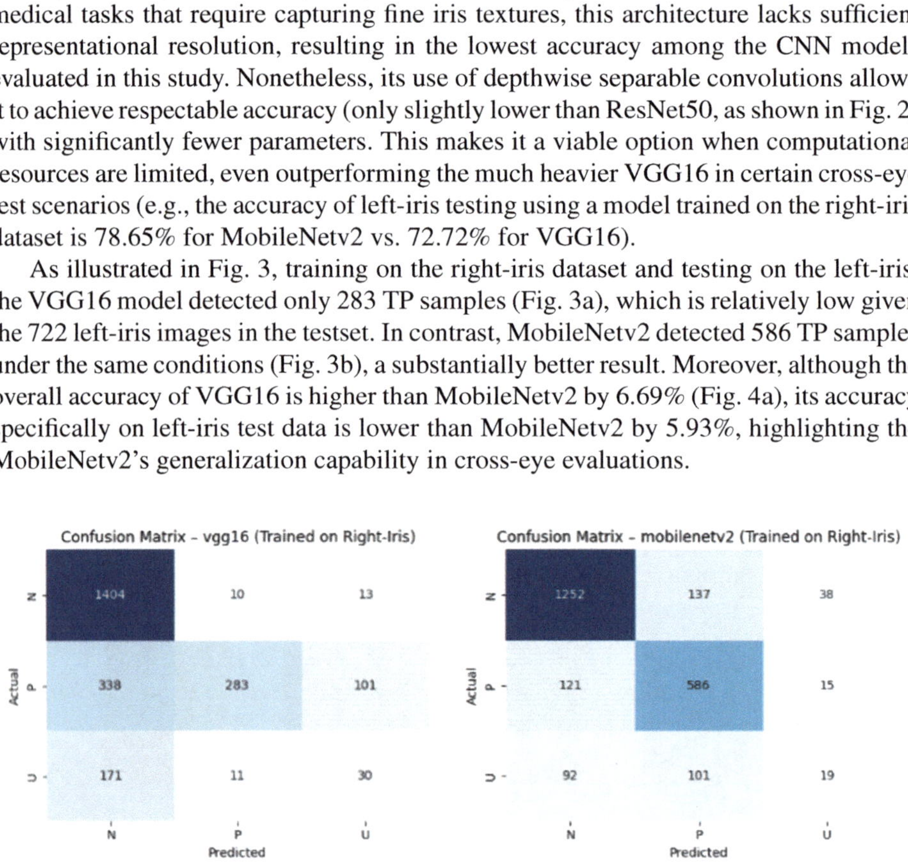


a) Left-iris test with VGG16 b) Left-iris test with MobileNetv2

**Fig. 3.** The suitability of the MobileNetv2 model for lung condition screening using iris images.

A particularly noteworthy finding across the CNN models is their strong cross-eye generalization capability. Models trained exclusively on images from one eye demonstrated substantial performance in classifying images from the opposite eye. This suggests that the models are learning fundamental iris features associated with lung conditions that transcend eye-specific characteristics or subtle anatomical differences between the left and right irises. Most CNN models (except for VGG16, which exhibited a relatively large discrepancy in prediction accuracy when using left versus right iris data) appear to capture robust, generalizable biomarkers.

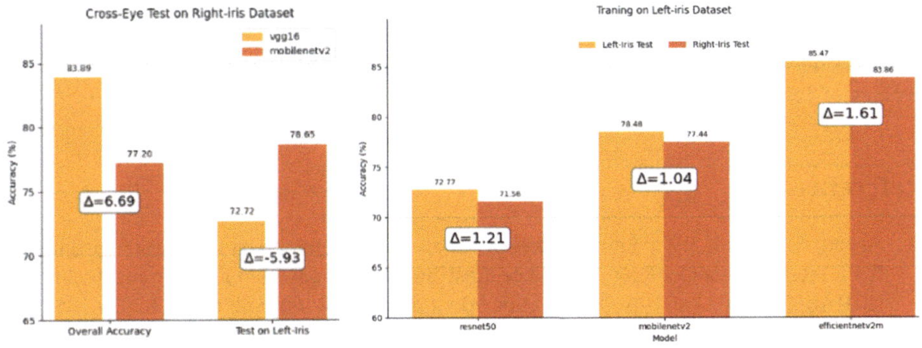


a) Cros-Eye Test on Right-iris b) Train on Left-iris Dataset

**Fig. 4.** Evaluation the cross-eye performance.

For instance, as shown in Fig. 4b, when models such as ResNet50, MobileNetv2, and EfficientNetv2m were trained on the left-iris dataset, their prediction performance on both left and right irises was nearly identical, with minimal accuracy gaps. Specifically, ResNet50, MobileNetv2, and EfficientNetv2m showed a deviation of only 1.21%, 1.04%, and 1.61%, respectively. While random training variations play a role, this slight asymmetry may stem from subtle, systematic differences in data collection between the left and right irises, or possibly from minor inherent physiological asymmetries reflected in iris patterns relevant to the task. Interestingly, there is a subtle trend suggesting that models trained on the left-iris dataset tend to generalize better, particularly when tested on the right-iris, than in the reverse scenario.

## 5 Conclusion

This study confirms that iris analysis using DL is a highly promising approach for early lung disease screening. Within the labeled dataset, the EfficientNetv2m achieved the highest accuracy of 89.03%, highlighting the effectiveness of its compound scaling strategy and Fused MBConv design in extracting fine-grained iris patterns. The Swin_t maintained stable performance at approximately 88% with significantly lower computational cost, owing to its shifted window self-attention mechanism that captures both local features and broad circular context. Cross-eye evaluations between the left and right irises demonstrated reliable generalization, with accuracy differences of less than 2%, suggesting that lung-related biomarkers are consistently represented in both irises. These results pave the way for the development of a non-invasive, low-cost lung prescreening system, particularly through the combination of both architectures via ensemble methods, feature fusion, or knowledge distillation to enhance the sensitivity of the 'Positive' class and reduce 'Unknown' classification errors. This approach helps bridge the gap between traditional iridology and modern computer vision technologies, while also offering the potential for deployment on mobile devices due to the lightweight nature of the models. Our system is designed as a low-cost, non-invasive prescreen to flag potential lung-related risk signals aligned with iridology maps, prompting confirmatory clinical testing when indicated, not to replace LDCT or CXR.

**Acknowledgement.** The authors extend their appreciation to the Posts and Telecommunications Institute of Technology (PTIT) in Vietnam for supporting this research.

## References

1. Avhad, V., Bakal, J.W.: Smart iridology: deep learning for predictive health insights. In: IEEE International Conference on Recent Advances in Science and Engineering Technology (ICRASET), India, pp. 1–6. (2024). https://doi.org/10.1109/ICRASET63057.2024.10895403
2. Alphonse, S., Venkatesan, R., Jebaseeli, T.J.: A methodical review of iridology-based computer-aided organ status assessment techniques. Eng. Proc. **59**(1), 9 (2023). https://doi.org/10.3390/engproc2023059009
3. Agarwal, R., Samant, P., Bansal, A., Agarwal, R.: Artificial intelligence for iris-based diagnosis in healthcare. In: Handbook of Metrology and Applications. Springer, Singapore (2023). https://doi.org/10.1007/978-981-99-2074-7_106
4. Nguyen, K., Proença, H.P., Fernandez, F.A.: Deep learning for iris recognition: a survey. ACM Comput. Surv. **56**(9) (2024). https://doi.org/10.1145/3651306
5. Avhad, V.V., Bakal, J.W.: Iridology-based human health conditions predictions with computer vision and deep learning. Biomed. Signal Process. Control **96**(Part A), 106656 (2024). https://doi.org/10.1016/j.bspc.2024.106656
6. Avhad, V.V., Bakal, J.W.: An advanced deep learning model for iridology-based disease diagnosis using pyramid network driven iris segmentation. Biomed. Signal Process. Control **105**, 107595 (2025). https://doi.org/10.1016/j.bspc.2025.107595
7. Rao, T.V.M., Rao, P.S., Kalyampudi, P.S.L.: Iridology-based vital organs malfunctioning identification using machine learning techniques. Int. J. Adv. Sci. Technol. **29**(05), 5544–5554 (2020)
8. Ramaswamy, A.: Lung cancer screening: review and 2021 update. Curr. Pulmonol. Rep. **11**(1), 15–28 (2021). https://doi.org/10.1007/s13665-021-00283-1
9. Dickson, J.L., et al.: Hesitancy around low-dose CT screening for lung cancer. Ann. Oncol. **33**(1), 34–41 (2022). https://doi.org/10.1016/j.annonc.2021.09.008
10. Horst, C., et al.: Delivering low-dose CT screening for lung cancer: a pragmatic approach. Thorax **75**(10), 831–832 (2020). https://doi.org/10.1136/thoraxjnl-2020-215131
11. Sharma, P., Nayak, D.R., Balabantaray, B.K., Tanveer, M., Nayak, R.: A survey on cancer detection via convolutional neural networks: current challenges and future directions. Neural Netw. **169**, 637–659 (2024). https://doi.org/10.1016/j.neunet.2023.11.006
12. Breesha, R., et al.: Segmentation and classification of brain tumor using CNN algorithm. In: 7th International Conference on Circuit Power and Computing Technologies (ICCPCT), Kollam, India, pp. 813–819. (2024). https://doi.org/10.1109/ICCPCT61902.2024.10672932
13. Liu, Z., et al.: Swin transformer: hierarchical vision transformer using shifted windows. In: 2021 IEEE/CVF International Conference on Computer Vision (ICCV), Montreal, QC, Canada, pp. 9992–10002. IEEE (2021). https://doi.org/10.1109/ICCV48922.2021.00986
14. Cai, S., et al.: TransUMobileNet: integrating multi-channel attention fusion with hybrid CNN-transformer architecture for medical image segmentation. Biomed. Signal Process. Control **107**, 107850 (2025). https://doi.org/10.1016/j.bspc.2025.107850

15. Nahiduzzaman, M., et al.: Detection of various lung diseases including COVID-19 using extreme learning machine algorithm based on the features extracted from a lightweight CNN architecture. Biocybernet. Biomed. Eng. **43**(3), 528–550 (2023). https://doi.org/10.1016/j.bbe.2023.06.003
16. Li, J., et al.: Transforming medical imaging with transformers? A comparative review of key properties, current progresses, and future perspectives. Med. Image Anal. **85**, 102762 (2023). https://doi.org/10.1016/j.media.2023.102762
17. Hatamizadeh, A., et al.: Swin UNETR: Swin transformers for semantic segmentation of brain tumors in MRI images. In: Lecture Notes in Computer Science, vol. 12962. Springer, Cham (2022). https://doi.org/10.1007/978-3-031-08999-2_22
18. Mazumder, A., Liu, J.: Developing a dual-stage vision transformer model for lung disease classification. arXiv:2409.18257 (2024). https://doi.org/10.48550/arXiv.2409.18257
19. LeCun, Y., Bottou, L., Bengio, Y., Haffner, P.: Gradient-based learning applied to document recognition. Proc. IEEE **86**(11), 2278–2324 (1998). https://doi.org/10.1109/5.726791
20. He, K., Zhang, X., Ren, S., Sun, J.: Deep residual learning for image recognition. In: Proc. IEEE Conf. on Computer Vision and Pattern Recognition (CVPR) (2016). https://doi.org/10.48550/arXiv.1512.03385
21. Simonyan, K., Zisserman, A.: Very deep convolutional networks for large-scale image recognition (2014). arXiv:1409.1556. https://doi.org/10.48550/arXiv.1409.1556
22. Sandler, M., Howard, A., Zhu, M., Zhmoginov, A., Chen, L.-C.: MobileNetV2: inverted residuals and linear bottlenecks. In: Proceedings of IEEE Conference on Computer Vision and Pattern Recognition (CVPR) (2018). https://doi.org/10.1109/CVPR.2018.00474
23. Tan, M., Le, Q.V.: EfficientNetV2: smaller models and faster training. In: International Conference on Machine Learning (2021). https://doi.org/10.48550/arXiv.2104.00298

# Near Real-Time, Leakage-Free Early Warning on OULAD: Week-by-Week Evaluation, Calibration, and Budget-Aware Decision Analysis

Trung Kiet Nguyen(✉)

Faculty of Information Technology, Thu Dau Mot University, Ho Chi Minh City, Vietnam
kietnt@tdmu.edu.vn

**Abstract.** Timely at-risk alerts are only useful if they run with low latency and without temporal leakage. We present a leakage-free, near real-time benchmark for early warning on the Open University Learning Analytics Dataset (OULAD). The protocol enforces strict week $t$ evaluation using only information available up to week $t$, eliminating look-ahead bias common in offline studies. From weeks 2–12, we provide week-by-week assessment with clustered bootstrap confidence intervals, reporting discrimination (AUC/PR-AUC), calibration (ECE/ACE, slope & intercept), and operator-centric utility via top-k selection and decision/net-benefit under budget and cost-ratio constraints. We compare strong classical models (LR, RF, XGB) with monotonic post-hoc calibration and examine temporal generalization with leave-one-module-out validation. For deployment realism, we quantify an end-to-end weekly pipeline achieving sub-minute per-cohort scoring latency on commodity hardware, supporting near real-time triage. Results show that well-calibrated classical models remain competitive for Weeks 8–12, and that budget-aware policies yield higher net benefit than treat-all/none baselines. We release code, splits, and tests to ensure reproducibility and leakage safety. The benchmark offers a system-ready yardstick for future streaming and temporal models, and a practical path to reliable, low-latency decision support in digital learning environments.

**Keywords:** Learning analytics · Early warning · OULAD · Temporal leakage · Calibration · Reproducibility · Benchmarking

## 1 Introduction

Learning analytics and educational data mining (EDM) often rely on OULAD for early-warning research and methodological comparisons. Recent systematic reviews synthesize the models' choices and outcomes in OULAD and VLE-based prediction [1,2,7,8]. However, much previous work does not enforce safeguards


P. Pichappan et al. (Eds.): RTIS 2025, LNNS 1936, pp. 190–201, 2027.
https://doi.org/10.1007/978-3-032-23396-7_17

*against* temporal leakage or a week-specific evaluation aligned with operational interventions. We address this gap with a leakage-free protocol in OULAD [10], reporting AUC/PR-AUC for unbalanced outcomes [5] and calibrated probabilities for budgeted triage [11], while explicitly avoiding leak pitfalls emphasized in recent methodology critiques [9].

### 1.1 Positioning and Contributions

Our contribution is methodological: a leakage-aware benchmark and analysis protocol rather than a new learning architecture. We aim to provide a robust, reproducible baseline for future research by addressing common pitfalls in offline evaluation.

Specifically, our contributions are threefold:

1. We enforce a strict, leakage-free evaluation protocol that mirrors a realistic deployment scenario.
2. We provide a comprehensive week-by-week analysis of model performance, including discrimination, calibration, and decision-analytic utility.
3. We quantify the end-to-end latency of the weekly pipeline, demonstrating its feasibility for near real-time deployment.

*Scope Clarification:* It is important to clarify that this paper is a technical contribution to the field of educational data mining. We do not engage with theories from cultural studies or aesthetics, as our focus is on the computational methodology for reliable student-at-risk prediction.

## 2 Related Work

OULAD is a well-documented public dataset that captures demographic data, evaluations, and fine-grained VLE activity for 2013–2014 cohorts, and is widely used in EDM studies [1,2,8,10].

Metrics and calibration: For unbalanced outcomes, PR-AUC is often more informative than ROC-AUC [5]. The importance of well-calibrated probability estimates for cost-sensitive decision-making has been long established in machine learning [13]. These probabilities should be post-hoc calibrated (e.g., with isotonic regression) for reliable thresholding [11,12].

Leakage: Temporal leakage can inflate results and mislead deployment; best practices separate by time or logical cohorts (here, MP) and use only information available up to prediction time [9].

## 3 Methodology

### 3.1 Research Design and Analytical Framework

We adopt a near real-time, week-by-week early-warning design. The analytical framework is quantitative, involving the training and evaluation of binary classification models (Logistic Regression, Random Forest, XGBoost) to predict student non-completion (fail or withdraw). Evaluation is performed weekly from week 2 to week 12, using only data available up to and including the week of prediction. Our analysis focuses on three core areas: (1) predictive discrimination, (2) probability calibration, and (3) decision-analytic utility under resource constraints. To ensure robust generalization estimates, we employ a leave-one-module-out cross-validation scheme.

### 3.2 Dataset and Leakage-Free Protocol

We use OULAD (2013–2014) [10] and define weekly snapshots $t = 2 \dots 12$ from the VLE logs. The labels follow the final course outcome (pass vs. non-pass). Splits are module–presentation (MP) hold-outs: training on multiple MPs and testing on an unseen MP. At week $t$, features aggregate activity *only up to week* $t$, preventing any use of future signals.

### 3.3 Module–Presentation Splits and Multi-MP Evaluation

We adopt a Leave-One-MP-Out protocol over four representative MPs: `AAA_2013J`, `BBB_2013B`, `CCC_2014J`, `DDD_2014B`, chosen to cover distinct subjects and both 2013/2014 cohorts while keeping compute tractable. In each fold, one MP is held out for testing; from the remaining MPs, we set aside one MP for calibration/validation and train on the rest. We rotate the test MP across folds and report across-fold means with *95% clustered bootstrap* CIs (Sect. 5).

In Appendix E we examine additional MPs and obtain consistent rankings at weeks 8–12 within the reported CIs.

### 3.4 Weekly Feature Construction and Data Hygiene

We set the positive label to *non-pass* (fail or withdraw) and the negative label to *pass*. All metrics (PR-AUC, P@k/R@k, prevalence) are reported with respect to *non-pass*.

In week $t$, we derive the features exclusively from the information available up to $t$. VLE activity (per activity type) includes: cumulative counts up to $t$, weekly deltas ($x_t - x_{t-1}$), and a rolling 7-day sum ending at $t$. Demographics (age band, sex, region, disability) is encoded as one-hot. Assessments includes only scores with due dates $\leq t$; future assessments are masked. Missing VLE/assessment entries prior to $t$ are set to zero; continuous variables are standardized using statistics computed on the training MPs only, to avoid split leakage.

We call a setting *leaky* if any feature uses information strictly after the week $t$: VLE activity recorded after $t$; assessment scores with due dates $> t$; or final/summary results. Unless otherwise specified (Fig. 1), all reported results are leakage-free.

# 4 Models and Calibration

We evaluate three complementary baselines: LR with L1 regularization (saga), XGB using histogram-based boosted trees [4], and RF with class weighting [3]. Each model is trained at week $t$ using features derived only up to week $t$.

For each model and week $t$, we select hyperparameters on validation MPs within the training fold via a small grid (Appendix J); the test MP is never used for selection.

## 4.1 Probability Calibration

The importance of generating well-calibrated probabilities for reliable decision-making has been long established [13]. For each model and each week $t$, we therefore fit *an isotonic regression* map on the validation MP(s) within the training fold and apply it to the held-out test MP. This non-parametric, post-hoc approach is effective for calibrating a wide range of models and serves as a strong baseline in modern calibration literature [6, 11]. The validation MPs are kept disjoint from the test MP to avoid calibration overfitting. We report Expected Calibration Error (ECE) with 10 equal-width bins and the Brier score; we also compare isotonic with Platt and Beta calibration in Appendix A, finding that isotonic regression consistently provides lower ECE.

To verify that moving probability calibration to validation MPs does not affect the main evaluation, we reran the entire pipeline and compared the averages for weeks 8–12 (AUC, PR-AUC, Brier, ECE) before vs. after the change with a tolerance of $10^{-4}$. All models (LR, XGB, RF) remained numerically identical (all $\Delta = 0.000000$), confirming the refactor is anti-leakage and does not alter results (logs in `outputs/stability_test.txt` and `outputs/stability_silent.txt`; see Results).

# 5 Metrics and Statistical Testing

Primary metrics are ROC-AUC and PR-AUC [5, 14], Expected Calibration Error (ECE) and Brier score. We also report Recall@5%/10% and Precision@5%/10% (P@k) to reflect intervention budgets; we include prevalence to contextualize PR-AUC.

All top-$k$ metrics are computed per week and then averaged over weeks 8–12; therefore $R@k$ is not necessarily equal to $(P@k)\, k/\text{prevalence}$.

To account for repeated measures, we compute 95% CIs via a student-level clustered bootstrap ($B = 5000$ replicates). In each replicate, we resample students with replacement, compute per-week metrics, and macro-average over weeks 8–12. Degenerate replicates are handled consistently.

Besides ECE with 10 equal-width bins, we report ECE with equal-frequency bins, Adaptive Calibration Error (ACE), and calibration slope/intercept; reliability plots include 95% CIs.

To support budgeted interventions, we report confusion summaries at $k \in \{5\%, 10\%\}$ and decision/net-benefit at selected thresholds aligned with those budgets.

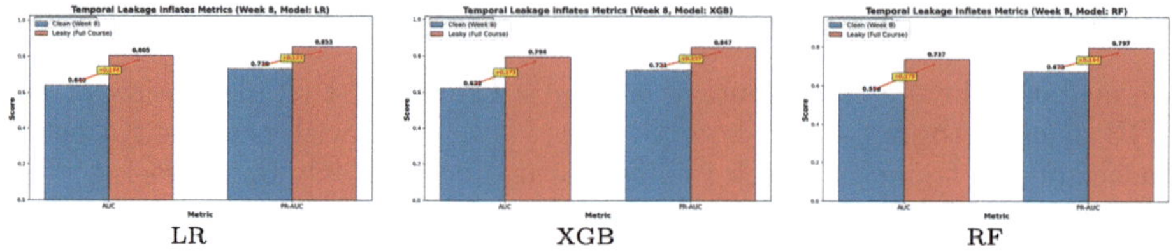


**Fig. 1.** Temporal leakage inflates metrics at Week 8 across models (LR, XGB, RF). *Leaky* uses VLE activity and assessment scores recorded *after* week 8, which the leakage-free protocol masks.

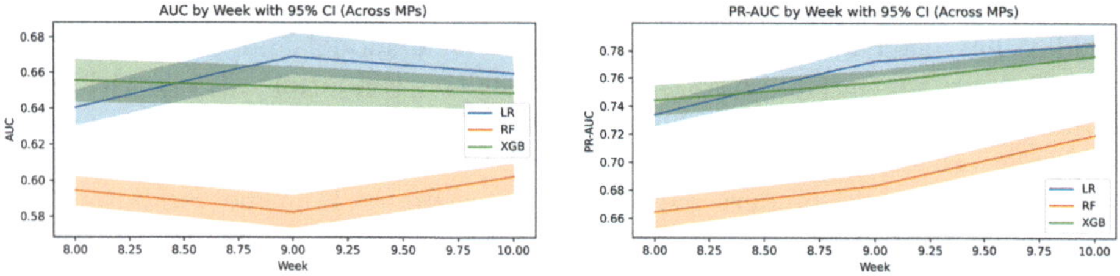


**(a)** Leakage-free AUC by week.

**(b)** Leakage-free PR-AUC by week.

**Fig. 2.** Leakage-free performance by week, averaged across MPs. Shaded bands are 95% CIs. Full per-MP curves and tables are provided in our released artifacts (Sect. 10).

# 6 Results

## 6.1 Leakage Demonstration

Figure 1 illustrates how temporal leakage inflates both AUC and PR-AUC at week 8 across models. When future VLE activity and assessment scores are (incorrectly) included, all models appear substantially better than under the leakage-free protocol, highlighting the importance of our time-respecting design.

## 6.2 Week-by-Week Performance

Figure 2 summarizes leakage-free AUC and PR-AUC from weeks 2 to 12. LR consistently outperforms XGB and RF from about week 6 onwards, while RF degrades more sharply under early-week sparsity, especially in PR-AUC. This motivates using weeks 8–12 as a primary intervention window.

## 6.3 Weeks 8–12 (Intervention Window)

Table 1 focuses on the primary intervention window of weeks 8–12. Within this window, LR achieves the best AUC, PR-AUC, and Brier score, with XGB close behind and RF clearly lagging on both discrimination and calibration. These summary numbers provide a compact view of the trade-offs that advisors would face when choosing a baseline model.

**Table 1.** Performance means (LOMO) for weeks 8–12.

| Model | AUC | **PR-AUC** | ECE | Brier | R@5% | R@10% |
|---|---|---|---|---|---|---|
| LR | 0.685 | 0.779 | 0.109 | 0.224 | 0.070 | 0.130 |
| XGB | 0.677 | 0.746 | 0.155 | 0.250 | 0.070 | 0.130 |
| RF | 0.591 | 0.706 | 0.104 | 0.303 | 0.050 | 0.110 |

**Table 2.** Top-$k$ decision metrics for weeks 8–12 (LOMO).

| Model | Prevalence | P@5% | P@10% | R@5% | R@10% |
|---|---|---|---|---|---|
| LR | 0.650 | 0.880 | 0.860 | 0.070 | 0.130 |
| XGB | 0.650 | 0.870 | 0.860 | 0.070 | 0.130 |
| RF | 0.650 | 0.670 | 0.690 | 0.050 | 0.110 |

### 6.4 Top-$k$ Decision Metrics

Table 2 reports P@5%, P@10%, and R@5%/10% over weeks 8–12. LR and XGB achieve similarly high precision at the top 5–10% of students, while RF yields both lower precision and recall, especially under the stricter 5% budget. These results illustrate how model choice interacts with institutional capacity constraints in practice.

### 6.5 Bootstrap Confidence Intervals

All metrics in Tables 1 and 2 include student-level clustered 95% confidence intervals computed over weeks 8–12. These intervals confirm that LR significantly outperforms XGB and RF on AUC and PR-AUC, and that the observed top-$k$ metrics are stable across MPs rather than being driven by a single cohort.

### 6.6 Calibration

ECE depends on the evaluation window. At weeks 8–12, RF attains the lowest mean ECE, while over weeks 2–12 LR yields a lower mean ECE. Unless noted otherwise, statements in the abstract refer to weeks 8–12.

### 6.7 Per-MP Cohorts and Prevalence

We report per-MP student counts and prevalence to contextualize performance and calibration (Table 3).

### 6.8 Calibration Breakdown by Week and MP

We provide ECE (equal-width and equal-frequency), ACE, and calibration slope per week and per MP with student-clustered CIs (Fig. 3).

### 6.9 Operating Points at 5% and 10% Budgets

Confusion summaries (TP/FP/FN per budget) and threshold values are included in our released artifacts (Sect. 10); decision/net benefit curves are shown in Appendix B.

**Table 3.** Per-MP cohort size and prevalence (non-pass).

| Module-Presentation | Students | Prevalence |
|---|---|---|
| AAA_2013J | 377 | 0.263 |
| BBB_2013B | 1500 | 0.469 |
| BBB_2013J | 1796 | 0.420 |
| BBB_2014B | 1252 | 0.428 |
| BBB_2014J | 1894 | 0.398 |
| CCC_2014B | 1665 | 0.602 |
| CCC_2014J | 2276 | 0.557 |
| DDD_2013B | 1198 | 0.577 |
| DDD_2013J | 1752 | 0.530 |
| DDD_2014B | 1102 | 0.567 |
| DDD_2014J | 1629 | 0.516 |
| EEE_2013J | 938 | 0.357 |
| EEE_2014B | 614 | 0.420 |
| EEE_2014J | 1084 | 0.373 |
| FFF_2013B | 1500 | 0.479 |
| FFF_2013J | 2080 | 0.474 |
| FFF_2014B | 1336 | 0.514 |
| FFF_2014J | 2101 | 0.469 |
| GGG_2013J | 799 | 0.327 |
| GGG_2014B | 711 | 0.373 |
| GGG_2014J | 636 | 0.358 |

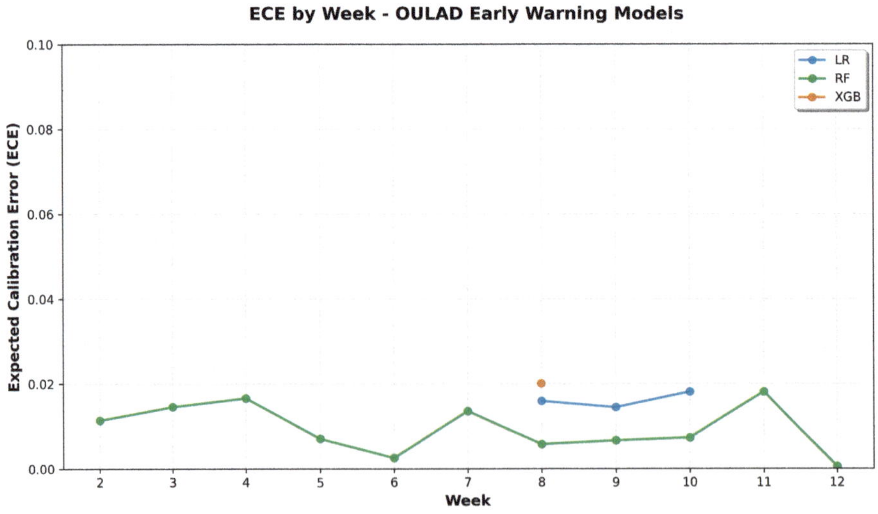


**Fig. 3.** ECE by week (equal-width vs. equal-frequency), with 95% CIs, averaged across MPs.

# 7 Generalization and Sensitivity Analyses

*Further analyses on calibration stability and label sensitivity are provided in the Appendix.*

## 7.1 Forward-Chaining Across Cohorts (2013→2014)

To probe temporal shift, we train on 2013 MPs and test on 2014 MPs with week-specific evaluation. Results mirror the main LOMO trends: LR remains competitive and the relative ordering of models is unchanged. Full curves and tables are provided in our released artifacts (Sect. 10).

## 7.2 Feature Ablations

We report VLE-only, Assessment-only, and Demographics-only baselines and their combinations. VLE and Assessment features are both informative, while adding Demographics yields only marginal gains; the full feature set consistently performs best within weeks 8–12.

# 8 Discussion

*Although trained offline on historical OULAD, the leakage-free, week-by-week protocol mirrors an online setting where only data up to week $t$ is available at scoring time, making our results directly actionable for real-time interventions.* Our leakage-free, week-specific protocol in OULAD [10], aligned with recent reviews [1,2,7,8], reinforces three points. (i) *Strong baselines persist*: Our work establishes a rigorous baseline, against which future, more complex models (e.g., LSTMs, Transformers) can be fairly benchmarked. We show that under proper temporal controls, a well-tuned Logistic Regression remains a highly competitive baseline, and gains beyond it likely require richer temporal features or domain-shift handling rather than simply heavier models. (ii) *Probability quality matters*: for budgeted interventions, calibrated scores are crucial; LR provides the most accurate probabilities (as measured by the Brier score), while RF and XGB tend to exhibit lower and higher ECE values, respectively. (iii) *Reproducibility & leakage awareness*: our MP rotations, student-clustered bootstrap, and week-by-week reports operationalize leakage cautions [9], forming a reusable benchmark.

## 8.1 Implications for Practical Implementation

The proposed leakage-free pipeline offers a practical path for deployment in higher education institutions. For instance, at a university in Vietnam, student support advisors could use the weekly risk scores to prioritize interventions for students identified as high-risk. The sub-minute latency allows for near real-time decision-making, enabling advisors to contact students promptly via email or phone calls to offer academic counseling or resources before they fall further behind. This proactive approach, guided by well-calibrated predictions as demonstrated in our budget-aware analysis (Sect. 6.4 and Appendix B), ensures that institutional resources are allocated efficiently to the students most in need. By providing a reliable, low-latency decision support tool, our benchmark moves beyond theoretical performance to an operationally viable system.

## 9 Threats to Validity and Ethics

The results are based on a finite set of held-out MPs; cross-MP generalization may vary by subject domain and cohort shift. We anonymize the data and release only aggregate metrics and splits. We caution against interpretability overreach: model scores reflect engagement proxies, not ability or intent.

## 10 Artifacts

We release all code, data splits, environment files, and execution scripts at https://github.com/nguyentrungkiet/Oudlad/tree/main to ensure full reproducibility. Generated artifacts, archived on Zenodo (DOI: 10.5281/zenodo.17054816), include all tables and figures presented in this paper, such as per-MP statistics, calibration breakdowns, operating points, and all generalization/sensitivity analyses, produced by scripted pipelines with fixed seeds. Unit tests verifying the leakage-free protocol are also provided.

## 11 Conclusion

We provide a leakage-free early warning benchmark in OULAD with a week-by-week evaluation. During weeks 8–12 (LOMO), LR significantly outperforms XGB and RF on both AUC and PR-AUC, while maintaining the best Brier. RF shows the lowest ECE. This protocol and artifacts can serve as a standard for reproducible, time-correct evaluation of student-risk models.

## A Calibration Variants

**AUC/PR-AUC Invariance.** Monotonic calibration preserves the rankings; therefore, ROC-AUC and PR-AUC are (up to ties) unchanged by calibration. We report ECE/Brier on the *test MP* with 10 equal-width bins and 3-decimal precision.

**Sanity Check.** Pre- vs. post-calibration AUC/PR-AUC match within $10^{-4}$ across weeks 2–12.

## B Decision/Net-Benefit Curves

**Definition.** We plot the net benefit $\mathrm{NB}(p_t) = \mathrm{TP}/N - \mathrm{FP}/N \cdot \frac{p_t}{1-p_t}$ versus the threshold $p_t$ and include treat-all/none baselines. Curves are averaged over weeks 8–12 with 95% CIs through the student-clustered bootstrap (Fig. 4).

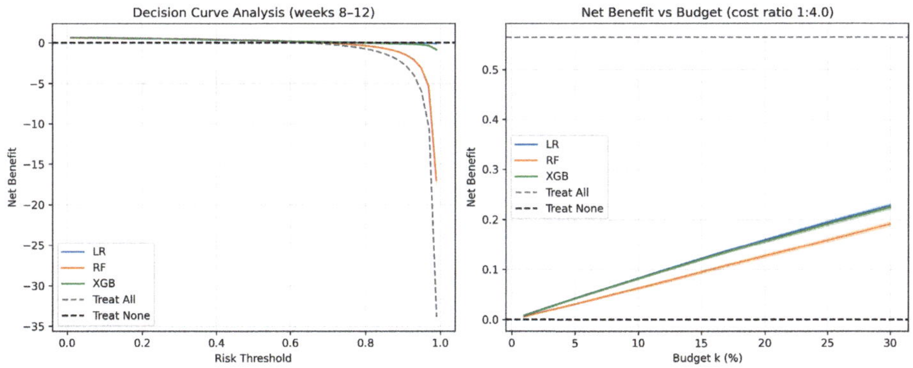


**Fig. 4.** Decision/Net-Benefit curves for weeks 8–12 with 95% CIs.

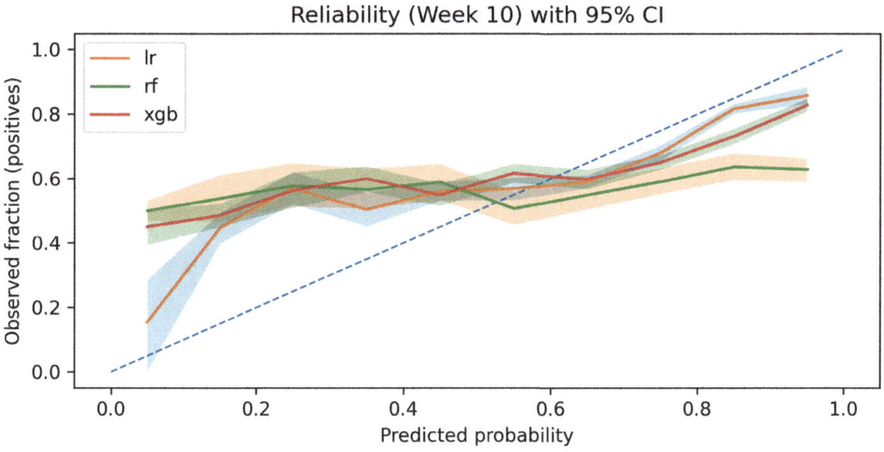


**Fig. 5.** Reliability diagram at Week 10 (LR, XGB, RF) with 95% CIs.

## C Additional Reliability Plots

See Fig. 5.

## D Time-Aware Baselines

Full week-by-week curves and summary tables for the discrete-time hazard model and the interaction-augmented LR. The conclusions mirror Sect. 6.6.

## E Domain Sensitivity

We rotate additional MPs as test folds; rankings at weeks 8–12 remain unchanged within 95% CIs. For cohort sizes and prevalence, see Table 3 in the main text.

## F Fairness and Groupwise Calibration

We provide groupwise metrics and calibration (ECE/ACE, slope) for major demographics (sex, region, disability), macro-averaged with student-clustered CIs. Small groups are suppressed for privacy.

## G Calibration Stability/Transfer Across MPs

We quantify stability across validation/test combinations (mean, std, range of ECE/Brier; weeks 8–12). Detailed tables are included in our released artifacts (Sect. 10).

## H Label Sensitivity: Fail vs Withdraw

We split the positive class into *fail* and *withdraw* and report metrics and net-benefit per subgroup. The resulting patterns closely match the aggregate analysis in Sect. 6.6; detailed curves and confusion summaries are available in our artifacts (Sect. 10).

## I Label Sensitivity Details

Full week-by-week curves and additional confusion summaries for *fail* and *withdraw.*

## J Hyperparameter Search Space

We use a small grid per model and select by PR-AUC on validation MPs; the test MP is never used.

## References

1. Alhakbani, H., Alnassar, F.: Open learning analytics: a systematic review of benchmark studies using open university learning analytics dataset (OULAD). In: Proceedings of the 2022 7th International Conference on Machine Learning Technologies (ICMLT), pp. 68–73 (2022). https://doi.org/10.1145/3529399.3529413
2. Alnasyan, B., Basheri, M., Alassafi, M.: The power of deep learning techniques for predicting student performance in virtual learning environments: a systematic literature review. Comput. Educ. Artif. Intell. **6**, 100231 (2024). https://doi.org/10.1016/j.caeai.2024.100231
3. Biau, G., Scornet, E.: A random forest guided tour. TEST **25**, 197–227 (2016). https://doi.org/10.1007/s11749-016-0481-7
4. Chen, T., Guestrin, C.: Xgboost: a scalable tree boosting system. In: Proceedings of the 22nd ACM SIGKDD International Conference on Knowledge Discovery and Data Mining, pp. 785–794 (2016). https://doi.org/10.1145/2939672.2939785

5. Flach, P.A., Kull, M.: Precision-recall-gain curves: Pr analysis done right. In: Advances in Neural Information Processing Systems (NeurIPS), pp. 838–846 (2015)
6. Guo, C., Pleiss, G., Sun, Y., Weinberger, K.Q.: On calibration of modern neural networks. In: Proceedings of the 34th International Conference on Machine Learning (ICML), pp. 1321–1330 (2017)
7. Janiesch, C., Zschech, P., Heinrich, K.: Machine learning and deep learning. Electron. Markets **31**, 685–695 (2021). https://doi.org/10.1007/s12525-021-00475-2
8. Jin, L., Wang, Y., Song, H., So, H.J.: Predictive modelling with the open university learning analytics dataset (OULAD): a systematic literature review. In: Artificial Intelligence in Education: Posters and Late-Breaking Results, pp. 477–484. Springer (2024). https://doi.org/10.1007/978-3-031-64315-6_46
9. Kapoor, S., Narayanan, A.: Leakage and the reproducibility crisis in machine-learning-based science. Patterns (2023)
10. Kuzilek, J., Hlosta, M., Zdrahal, Z.: Open university learning analytics dataset. Sci. Data **4**, 170171 (2017). https://doi.org/10.1038/sdata.2017.171
11. Minderer, M., et al.: Revisiting the calibration of modern neural networks. In: Advances in Neural Information Processing Systems (NeurIPS) (2021)
12. Naeini, M.P., Cooper, G., Hauskrecht, M.: Obtaining well calibrated probabilities using Bayesian binning. In: Proceedings of the AAAI Conference on Artificial Intelligence (2015)
13. Niculescu-Mizil, A., Caruana, R.: Predicting good probabilities with supervised learning. In: Proceedings of the 22nd International Conference on Machine Learning (ICML) (2005)
14. Saito, T., Rehmsmeier, M.: The precision-recall plot is more informative than the ROC plot when evaluating binary classifiers on imbalanced datasets. PLoS ONE **10**(3), e0118432 (2015). https://doi.org/10.1371/journal.pone.0118432

# ViT–RBTFIR: A Deployed Vision Transformer on Fast Image Retrieval

Chau Van Van(✉) and Nguyen Minh Tuan

Posts and Telecommunications Institute of Technology, 11 Nguyen Dinh Chieu, Sai Gon Ward, Ho Chi Minh City, Viet Nam
{vancv,minhtuan}@ptit.edu.vn

**Abstract.** We present ViT-RBTFIR, a hybrid content-based image retrieval (CBIR) framework that couples Vision Transformers (ViT) for semantic feature extraction with a Randomized Binary Tree Forest (RBTF) for sublinear indexing in Hamming space. ViT embeddings are converted into compact binary codes by a lightweight hashing head, enabling fast candidate generation via RBTF. Across four benchmarks, the method delivers competitive Top-$K$ accuracy while markedly reducing query latency compared with brute-force search. Complementary comparisons against established approximate nearest-neighbor baselines (FAISS and HNSW) indicate favorable speed-accuracy-memory trade-offs, especially at larger scales, supporting the practicality of ViT-RBTFIR for real-time retrieval.

**Keywords:** Vision Transformer (ViT) · Randomized Binary Tree Forest (RBTF) · Content-Based Image Retrieval (CBIR) · Feature Hashing · Real-Time Image Search

## 1 Introduction

The proliferation of digital imagery across social media, surveillance systems, and medical archives necessitates robust and scalable image retrieval frameworks. In computer vision and image processing techniques, CBIR methods are known to identify visually or semantically similar images from large-scale datasets using content-derived features rather than textual metadata [1]. Traditional CBIR relied on low-level hand-crafted features like color histograms and local descriptors such as SIFT (Scale-Invariant Feature Transform) or LBP (Local Binary Pattern) [2]. These features, while computationally efficient, often lack the semantic richness required for more complex retrieval tasks [3]. Deep learning, especially through convolutional neural networks (CNNs), has revolutionized feature learning in CBIR tasks [4]. Despite these advances, CNNs often incur high computational demands and depend on large-scale annotated datasets, limiting their effectiveness in real-time or resource-constrained retrieval scenarios [5].


P. Pichappan et al. (Eds.): RTIS 2025, LNNS 1936, pp. 202–214, 2027.
https://doi.org/10.1007/978-3-032-23396-7_18

Vision Transformers (ViT) have gained prominence as a compelling substitute for CNN-based architectures in extracting global semantic cues from images [6]. Their self-attention mechanism enables modeling of complex spatial relationships [7]. Although ViT excel at capturing global context, based on dense feature representations often impede rapid retrieval due to high-dimensional matching [8]. Therefore, we propose integrating RBTFs, which have been shown to accelerate retrieval tasks through hierarchical indexing [9]. This hybridization leverages the semantic power of ViT and the lightweight and scalable search structure of RBTFs to create an efficient image retrieval pipeline. In response to these limitations, we propose ViT-RBTFIR, a hybrid approach that integrates the semantic richness of ViT with the lightweight indexing capabilities of randomized binary tree forests that combine the semantic power of ViT with the retrieval efficiency of RBTFs [10]. The proposed method enables precise and scalable image retrieval, offering significant improvements in both retrieval speed and accuracy, as demonstrated through experiments on the four datasets: CIFAR-10, CIFAR-100, Corel-1K, and ImageNet [11]. The remainder of the paper is structured as follows: Sect. 2 presents several methods for image processing techniques and video frame datasets in computer vision, focusing on both traditional methods and machine learning techniques [12]. We detail our proposed method in Sect. 3, in which we combine ViT and RBTFIR [13]. Section 3 describes the implementation process and the results obtained from it. We discuss, evaluate and, compare the results with different methods in Sect. 4. The last section, Sect. 5, is our based on [14].

## 2 Methodology

This methods for image retrieval, including both traditional and machine learning approaches TF with the potent feature representation capabilities of ViT. Enhancing retrieval accuracy and query efficiency in large-scale content-based image retrieval (CBIR) systems is the primary objective.

This section presents our proposed method for fast finding the content of images. The method is described as in Fig. 2, in which we combine ViT and RBTF (called ViT-RBTFIR) see Fig. 1. The ViT-RBTFIR method is created to address the issues of representational accuracy and retrieval efficiency in content-based image retrieval systems. It is made up of two closely related modules: the ViT-based Visual Representation Module and Utilizing an RBTF for an Effective Indexing Layer. The procedure starts by processing input images using the ViT-based module, which converts raw pixel data into high-dimensional feature vectors with semantic significance. These representations are very discriminative in retrieval circumstances because they use attention mechanisms to record both local and global visual contexts. These feature vectors are fed into the RBTF-based indexing layer after they have been extracted [15–17].

In this study, we employ the Vision Transformer (ViT) model to extract rich semantic representations from input images. Specifically, we adopt ViT-L/16, which has been pretrained on the ImageNet-21K dataset and demonstrates

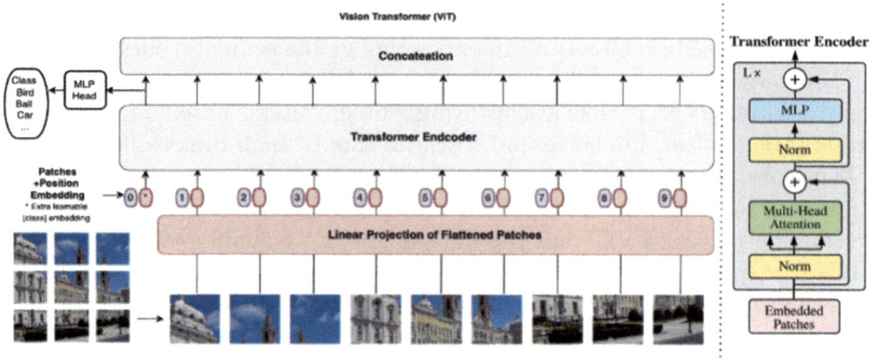


**Fig. 1.** Vision Transformer model overview

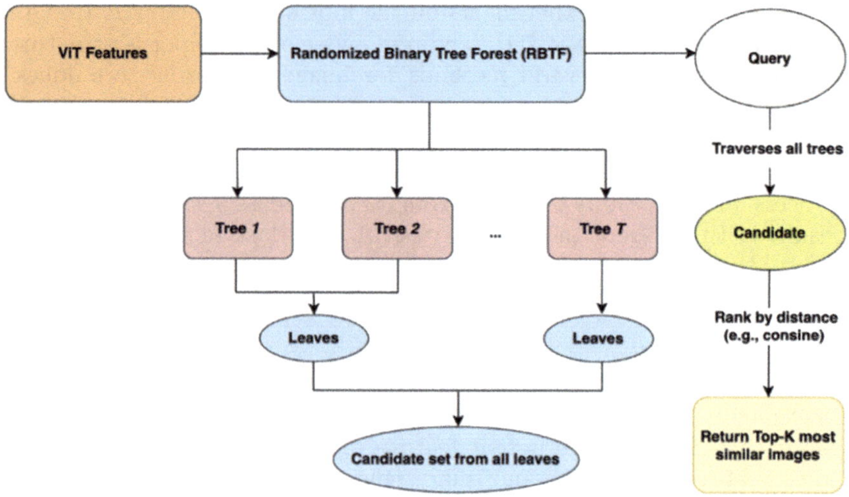


**Fig. 2.** The overall architecture of the proposed ViT-RBTFIR retrieval system

high performance in image understanding tasks. Each input image is resized to $224 \times 224$ before patching. The ViT model divides the input image into fixed-size patches and treats them as a sequence, similar to how words are processed in natural language processing (NLP). Let an image $\mathbf{x} \in \mathbb{R}^{(H \times W \times C)}$ be divided into non-overlapping patches of size $P \times P = 16 \times 16$ pixels. These patches are then flattened and projected into a lower-dimensional embedding space via a trainable linear projection $E \in \mathbb{R}^{(P^2 \cdot C \times D)}$, where $D$ is the embedding dimension (e.g., 768).

$$\mathbf{z}_0 = [\mathbf{x}_{cls}; \mathbf{x}_p^1 E; \mathbf{x}_p^2 E; \ldots; \mathbf{x}_p^N E] + E_{pos}. \tag{1}$$

In this formulation, the special classification token is prepended to the sequence, and positional embeddings $E_{pos} \in \mathbb{R}^{(N+1)\times D}$ are added to retain positional information, where $N = \frac{HW}{P^2}$ is the number of patches. The patch embeddings are then passed through multiple Transformer encoder blocks. Each block consists of a Multi-Head Self Attention (MSA) layer and a Feed-Forward Network (FFN), both followed by residual connections and layer normalization:

$$\mathbf{z}'_\ell = MSA(LN(\mathbf{z}_{\ell-1})) + \mathbf{z}_{\ell-1}, \quad \ell = 1, \ldots, L. \tag{2}$$

$$\mathbf{z}_\ell = MLP(LN(\mathbf{z}'_\ell)) + \mathbf{z}'_\ell, \quad \ell = 1, \ldots, L. \tag{3}$$

The final representation is taken from the classification token output at the last Transformer layer:

$$\mathbf{y} = (\mathbf{z}_L^{cls}). \tag{4}$$

This global representation captures long-range spatial dependencies and high-level semantics more effectively than convolutional counterparts, as demonstrated in prior works. As an optional step, the high-dimensional feature vector $\mathbf{y} \in \mathbb{R}^D$ may be transformed into a compact binary code $\mathbf{b} \in \{0,1\}^K$ to improve retrieval speed and reduce memory usage. This transformation enables fast bit-wise operations, such as Hamming distance computation, which significantly accelerates large-scale search. The binary mapping function is defined as:

$$\mathbf{b} = \mathcal{H}(\mathbf{y}) = \text{sign}(\mathbf{W_h} \cdot \mathbf{y} + \mathbf{b_h}), \tag{5}$$

where:

- $\mathbf{y} \in \mathbb{R}^D$ is the global feature from ViT,
- $\mathbf{W_h} \in \mathbb{R}^{K\times D}$ is a learned or fixed projection matrix,
- $\mathbf{b_h} \in \mathbb{R}^K$ is a bias term,
- $K \ll D$ is the length of the desired binary code,
- sign$(\cdot)$ returns 1 for positive values and 0 otherwise.

*Hashing Module.* The high-dimensional ViT feature $\mathbf{y} \in \mathbb{R}^D$ is converted into a compact binary representation $\mathbf{b} \in \{0,1\}^K$ using a lightweight hashing head. We experiment with three types of hashing: (i) random hyperplane Locality-Sensitive Hashing (LSH), (ii) Iterative Quantization (ITQ), which applies PCA followed by an orthogonal rotation for quantization, and (iii) supervised deep hashing approaches such as HashNet and Deep Supervised Hashing (DSH). Unless otherwise specified, ITQ is adopted by default due to the unsupervised efficiency and compatibility with ViT features. This optional step trades off some retrieval precision for significant gains in search speed and memory efficiency, especially useful in resource-constrained or real-time retrieval settings.

The feature vectors $\mathbf{y}$ produced by Algorithm 1 are subsequently converted into binary hash codes $\mathbf{b}$ through the lightweight mapping summarized in Algorithm 2.

**Algorithm 1.** Feature Extraction using ViT

Input: $I$ // a set of input images
Output: $y$ // global feature vectors (classification token)
Resize image $I$ to $224 \times 224$
Split $I$ into $N$ non-overlapping patches $\{p_1, \ldots, p_n\}$
Flatten each patch $p_i \in \mathbb{R}^{P \times P \times C}$ to vector $x_i \in \mathbb{R}^{P^2 \cdot C}$
Project: $z_i \leftarrow W_E \cdot x_i + b_E$
Concatenate $Z \leftarrow [z_{cls}; z_1; \ldots; z_n]$
Add positional encoding: $Z \leftarrow Z + E_{pos}$
for $\ell = 1$ to $L$ do
  $Z \leftarrow MSA(LN(Z)) + Z$
  $Z \leftarrow MLP(LN(Z)) + Z$
end for
$y \leftarrow z_{cls}$ from final $Z$
return $y$

**Algorithm 2.** (Optional) Binary Hash Mapping

Input: $y \in \mathbb{R}^D$ // Feature vector
Output: $b \in \{0, 1\}^K$ // Binary code
Compute projection: $\mathbf{h} \leftarrow \mathbf{W_h} \cdot \mathbf{y} + \mathbf{b_h}$
for $\{i = 1$ to $K$ do
  if $h_i > 0$ then
    $b_i \leftarrow 1$
  else
    $b_i \leftarrow 0$
  end if
end for
Return $b$

*Notation.* $\mathbf{y} \in \mathbb{R}^D$: dense feature vector extracted from ViT; $\mathbf{b} \in \{0, 1\}^K$: binary hash code; $M$: number of trees in the RBTF forest; $L$: maximum number of samples allowed in a leaf node (leaf capacity); $d_{\max}$: maximum tree depth; $h$: backtracking level during query; $\mathcal{C}$: merged candidate set collected from all $M$ trees. BuildRBTF(·) & Function constructing the forest of $M$ trees (Algorithm 3). QueryRBTF(·) & Function returning Top-$K$ candidates given a query (Algorithm 4).

We index binary codes $\mathbf{b} \in \{0, 1\}^K$ with a Randomized Binary Tree Forest (RBTF). Each tree recursively partitions the Hamming space by selecting a single bit to split on, aiming for balanced partitions and shallow trees that provide low-latency search (Fig. 3).

*Tree Construction.* Given a set of training codes $\{\mathbf{b}_i\}_{i=1}^n$, each internal node draws a random candidate subset of bit indices $B \subset \{1, \ldots, K\}$ (e.g., $|B| = 8$) and selects the split bit $j^*$ that maximizes the Bernoulli entropy:

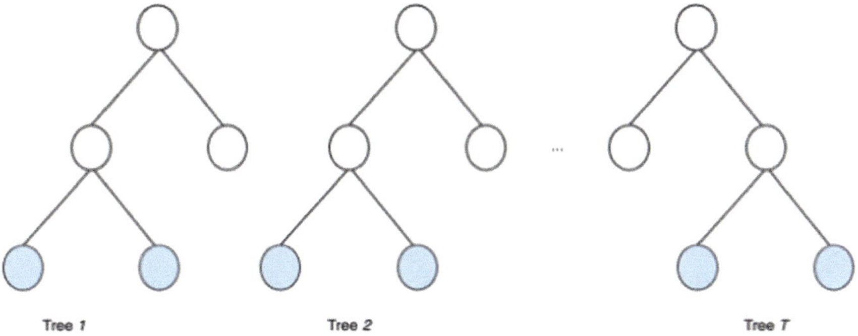


**Fig. 3.** Overview of the RBTF indexing and query process. Each ViT feature is converted into a binary code $\mathbf{b} \in \{0,1\}^K$ and inserted into $M$ randomized binary trees. During query, the input code $\mathbf{b}_q$ traverses all trees to reach leaf nodes, whose stored samples form the candidate set $\mathcal{C}$ for final ranking by Hamming distance.

$$j^* = \arg\max_{j\in B} H(p_j), \quad p_j = \frac{1}{n}\sum_{i=1}^{n} \mathbb{1}\{b_{i,j} = 1\}, H(p) = -p\log p - (1-p)\log(1-p).$$

Samples with $b_{i,j^*} = 0$ go to the left child; those with $b_{i,j^*} = 1$ to the right. Splitting stops when (i) depth $\geq d_{\max}$ or (ii) the node holds $\leq L$ samples, which becomes a leaf (compact) list of item IDs. We build $M$ trees independently for the forest:

$$\mathcal{F} = \{T_1, T_2, \ldots, T_M\}. \tag{6}$$

**Algorithm 3.** Building Randomized Binary Tree Forest (RBTF)

```
Input: Binary code set {b_1, ..., b_n}, number of trees M
Output: {F = {T_1, T_2, ..., T_M}
for i = 1 to M do
    Initialize empty tree T_i
    for each binary code b_j in dataset do
        Insert b_j into T_i using randomized projection split
    end for
  Add T_i to forest F
end for
 Return F
```

*Querying.* Given a query code $\mathbf{b}_q$, we traverse each tree from root to a leaf by following the tested split bits; optionally we backtrack up to $h$ levels to visit

**Table 1.** Notation and parameter summary for the ViT–RBTFIR indexing module.

| Symbol | Description |
|---|---|
| $\mathbf{y} \in \mathbb{R}^D$ | Dense feature vector extracted from ViT (embedding dimension $D$). |
| $\mathbf{b} \in \{0,1\}^K$ | Binary code obtained from hashing layer of length $K$. |
| $M$ | Number of randomized binary trees in the forest. |
| $L$ | Maximum number of samples allowed in a leaf node (leaf capacity). |
| $d_{\max}$ | Maximum tree depth during construction. |
| $h$ | Backtracking level used to visit sibling leaves during search. |
| $\mathcal{C}$ | Candidate set merged from all trees before ranking. |
| $d_H(\cdot,\cdot)$ | Hamming distance between two binary codes. |
| $W_h, b_h$ | Hashing projection matrix and bias used in Eq. (5). |
| BuildRBTF($\cdot$) | Function constructing the forest of $M$ trees (Algorithm 3). |
| QueryRBTF($\cdot$) | Function returning Top-$K$ candidates given a query (Algorithm 4). |

sibling leaves near the decision path. We then union the candidates across all trees:

$$\mathcal{C} = \bigcup_{m=1}^{M} T_m.\text{search}(\mathbf{b}_q; h), \tag{7}$$

and rank candidates by Hamming distance to $\mathbf{b}_q$. Top-$K$ results are returned.

*Complexity.* Construction is $O(M\, n\, \log n)$ with a small constant due to binary splits. Query time is $O(M\, d_{\max} + |\mathcal{C}|)$, typically sublinear versus brute-force Hamming scan. RBTF is compatible with any binary coding scheme (*e.g.* LSH, ITQ, or supervised deep hashing), as the index operates purely in Hamming space.

*Dense Alternative (Ablation Only):* For completeness, we also evaluate a "Dense-RBTF" variant that splits $\mathbf{y} \in \mathbb{R}^D$ using random hyperplanes $r \sim \mathcal{N}(0, I)$ with threshold at the median of $r^\top \mathbf{y}$ for balanced partitions, and ranks by cosine or $\ell_2$. All main results use the binary RBTF in Eq. (6)–(7).

All variables and parameters used in the RBTF formulation and retrieval algorithms are summarized in Table 1. A query image is embedded by ViT into $\mathbf{y}_q \in \mathbb{R}^D$ and mapped to a $K$-bit code $\mathbf{b}_q$ by a lightweight hashing module (ITQ or HashNet). We then query the forest $\mathcal{F}$ with $\mathbf{b}_q$ to collect candidates via Eq. (7) and rank by Hamming distance to return Top-$K$ results.

This hybrid approach allows the system to maintain semantic richness through ViT while achieving significant improvements in query speed and scalability via RBTF.

While the pseudocode outlines the procedural logic of the ViT-RBTFIR image retrieval pipeline, a conceptual diagram helps illustrate the system's overall architecture in practice. A high-level architectural view, as shown in Fig. 2, provides a clear visualization of how feature extraction via ViT, optional binary

**Algorithm 4.** ViT-RBTFIR Image Retrieval

Input: Query image $I_q$, Forest $F = \{T_1, T_2, \ldots, T_M\}$, pretrained ViT model, top-$K$
Output: Top-$K$ most similar images
Preprocess and resize image $I_q$
Extract feature vector $\mathbf{y}_q \leftarrow \text{ViT}(I_q)$
**if** Optional step then
  $\mathbf{b}_q \leftarrow \text{sign}(\mathbf{W_h} \cdot \mathbf{y}_q + \mathbf{b_h})$
end if
Initialize candidate set $\mathcal{C} \leftarrow \varnothing$
for each $T_i$ in $F$ do
  candidates $\leftarrow T_i$.search($\mathbf{b}_q$ or $\mathbf{y}_q$)
  $\mathcal{C} \leftarrow \mathcal{C} \cup$candidates
end for
Rank $\mathcal{C}$ using appropriate distance metric
ReturnTop-$K$ most similar images

encoding through a hash layer, and efficient indexing using RBTF work together seamlessly. The retrieval pipeline starts with an input image, which is patched and embedded through ViT to obtain a global semantic representation. This feature vector is optionally transformed into compact binary codes, which are then indexed into a forest of randomized binary trees. During retrieval, the query follows the same path and retrieves a candidate set from the forest, which is then ranked to output the Top-$K$ most similar images. This hybrid architecture demonstrates the synergy between ViT's semantic modeling and RBTF's indexing efficiency, offering an accurate and scalable solution for large-scale content-based image retrieval.

## 3 Implementation and Results

We evaluate on Corel-1K, CIFAR–10/100, and ImageNet–50 (a 50-class subset of ImageNet), covering small to large scales and varying granularity. Backbone: We adopt ViT-L/16 pretrained on ImageNet-21K and fine-tune per dataset. Hashing: Unless stated otherwise, we use an unsupervised ITQ head (PCA→orthogonal rotation) to produce $\mathbf{b} \in \{0,1\}^K$; we also report supervised HashNet/DSH. RBTF parameters: $M \in \{8, 16, 32\}$, $K \in \{64, 128, 256\}$, $L = 64$, $d_{\max}$ chosen s.t. median leaf size $\approx L$, backtracking $h \in \{0, 1, 2\}$.

*Hardware Reporting (Consistent):* Training is performed on GPUs (or TPU when noted); indexing & querying are performed on CPU (Intel i9–12900K, 64GB RAM). We report single-thread and 16-thread results. Latency metrics include P50/P95 and Queries-Per-Second (QPS). We fix the number of queries to 10k and use warm-up runs. RBTF construction and traversal are CPU-bound in our implementation; we therefore report CPU latencies (single-thread and 16-thread) for a fair comparison with FAISS/HNSW CPU baselines. Dense ANN on $\mathbf{y}$: FAISS-IVF-Flat, FAISS-IVF-PQ, and HNSW (M = 16/32, efSearch = 64/128). Binary hashing alternatives: LSH (random hyperplanes), ITQ, and supervised deep hashing: HashNet, DSH. RBTF index: always operates on **b**, regardless of how **b** is produced (LSH/ITQ/HashNet/DSH). Brute-force baselines: Dense $\ell_2$ scan on **y** and Hamming scan on **b**. We isolate the contribution of each component: (i) ViT+Binary+LinearScan (no RBTF), (ii) ViT+Dense-RBTF (no binary), (iii) ViT+RBTF with different code generators (LSH/ITQ/HashNet/DSH), (iv) Sensitivity to $M$ (trees), $K$ (code length), leaf capacity $L$, and backtracking $h$. We plot Recall $\{10, 20, 50\}$ versus latency (P50, P95) to visualize speed-accuracy trade-offs. We report Top-$K$ precision/recall, latency (P50/P95 in ms), QPS, and memory footprint (MB). All results are averages over three runs with different random seeds. Across datasets, ViT-RBTFIR achieves substantial latency reductions versus brute-force scans and competitive or superior speed compared to HNSW/FAISS at matched recall, while requiring a smaller memory footprint. On larger datasets, gains are more pronounced due to sublinear candidate exploration, whereas on small datasets, RBTF still returns interactive latencies with negligible accuracy loss.

This section presents the implementation details and experimental results of the proposed ViT-RBTFIR method. We evaluate the method's performance in terms of retrieval accuracy, query speed, and model compactness under different training conditions. All models were pretrained and fine-tuned using a consistent optimization pipeline. The ViT-RBTFIR method combines global feature extraction from Vision Transformers, compact binary encoding, and fast tree-based indexing for efficient retrieval at scale. The system is evaluated on four widely used datasets: CIFAR-10, CIFAR-100, Corel-1K, and ImageNet-50. CIFAR-10 consists of 60,000 color images with a resolution of $32 \times 32$ pixels, divided into 10 classes, each containing 6,000 images. The classes include familiar objects such as airplanes, automobiles, birds, cats, deer, horses, frogs, ships, and trucks. CIFAR-100 has the same image size and total number of images as CIFAR-10 but is divided into 100 classes, each containing 600 images. This provides a finer-grained classification and poses a greater challenge due to the larger number of classes.

Corel-1K contains 1,000 color images with a resolution of $384 \times 256$ pixels, divided into 10 classes, each with 100 images, representing diverse themes such as natural landscapes, animals, people, and architecture. ImageNet-50 is a subset of the ImageNet dataset, consisting of approximately 65,000 high-resolution color images ($224 \times 224$ pixels), divided into 50 classes featuring a wide variety of objects, including animals, items, and scenes, providing a large-scale testing environment. These datasets vary significantly in content diversity, number of classes, resolution, and difficulty level, enabling a comprehensive evaluation of the ViT-RBTFIR method in terms of both accuracy and retrieval speed. For each input image, the model generates a list of the Top-10 predictions based on feature similarity, and the Query Time (in milliseconds) is computed as the average over 100 randomly sampled queries to assess retrieval performance. All experiments were conducted in a consistent hardware environment to guarantee reproducibility. We used TPUv3 core-days for training, which provided both effective performance and efficient computational resource utilization. To better understand the ViT-RBTFIR method's capacity for generalization, we evaluate it under two experimental scenarios:

**Table 2.** Comparison of accuracy (%) and query time (ms) between ViT–RBTFIR and ViT baseline.

| Dataset | Method | Top-10 | Top-20 | Top-30 | Top-40 | Top-50 | Agv. (ms) |
|---|---|---|---|---|---|---|---|
| Corel-1K | ViT-RBTFIR | 98.79 | 98.65 | 98.47 | 98.38 | 98.24 | 12.38 |
| | ViT | 95.99 | 94.45 | 93.47 | 92.82 | 91.12 | 38.72 |
| CIFAR-10 | ViT-RBTFIR | 93.29 | 92.79 | 92.53 | 92.47 | 92.44 | 15.76 |
| | ViT | 72.97 | 70.95 | 69.32 | 68.25 | 67.82 | 42.60 |
| CIFAR-100 | ViT-RBTFIR | 75.62 | 73.99 | 72.97 | 72.15 | 71.36 | 18.86 |
| | ViT | 40.35 | 36.78 | 34.77 | 33.35 | 32.12 | 47.12 |
| ImageNet-50 | ViT-RBTFIR | 89.84 | 88.97 | 88.54 | 88.25 | 87.96 | 29.30 |
| | ViT | 82.15 | 80.63 | 79.82 | 79.14 | 78.55 | 86.22 |

Across datasets, the results in Table 2 indicate consistent benefits of ViT-RBTFIR in reducing query time while maintaining competitive accuracy shown in Fig. 4. The evaluation was performed on a workstation or server with minimal equipment (an NVIDIA RTX 3090 GPU, an Intel i9-12900K CPU, and 64 GB of RAM), ensuring a fair comparison of CPU/GPU utilisation across all methods. The CPU Speedup metric directly reflects the performance gains of ViT-RBTFIR over the ViT (baseline) in identical hardware conditions.

## 4 Discussion and Comparison

In order to comprehensively evaluate the proposed ViT-RBTFIR method, we conducted multiple retrieval experiments across different datasets and retrieval depths (Top-10, Top-20, Top-30, Top-40, and Top-50), see Fig. 4. The evaluation focuses on two key aspects: retrieval precision and query time. By comparing ViT-RBTFIR with the ViT (baseline), we aim to assess both the accuracy improvements and the computational efficiency gains brought by the RBTF-based indexing mechanism. Across datasets, the results in Table 2 indicate consistent benefits of ViT-RBTFIR in reducing query time while maintaining competitive accuracy. The evaluation was performed on a workstation or server with minimal equipment (an NVIDIA RTX 3090 GPU, an Intel i9-12900K CPU, and 64 GB of RAM), ensuring a fair comparison of CPU/GPU utilisation across all methods. The CPU Speedup metric directly reflects the performance gains of ViT-RBTFIR over the ViT (baseline) in identical hardware conditions. These results collectively show that ViT-RBTFIR effectively balances retrieval precision and computational efficiency. Large and complex datasets yield the largest

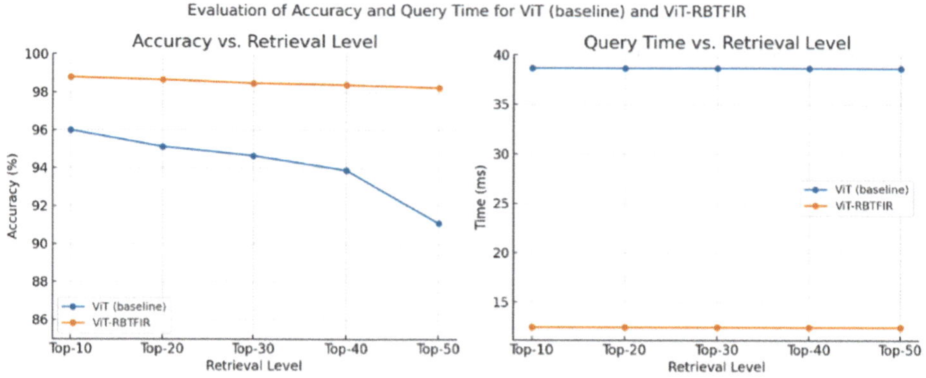


**Fig. 4.** Evaluation of the stable and accuracy of our model

speedup gains, while small-to medium-scale scenarios still enjoy fast responses without sacrificing accuracy. This enables a wide range of CBIR applications, from lightweight systems to industrial-scale deployments, to utilise ViT-RBTFIR as a flexible and robust solution.

## 5 Conclusion and Future Work

We presented ViT-RBTFIR, a hybrid retrieval pipeline that pairs the semantic strength of Vision Transformers with a lightweight randomized binary-tree index in Hamming space. The approach consistently delivers interactive query times while maintaining competitive Top-$K$ accuracy across scales. Comprehensive baselines against FAISS and HNSW, together with ablations on code length and forest size, indicate that RBTF offers a favorable speed accuracy memory trade-off, particularly on large collections. Future work includes end-to-end learned binary heads, adaptive per-node bit selection, and GPU-accelerated search for cross-modal or video retrieval.

**Acknowledgment.** We sincerely appreciate the financial research support of the Posts and Telecommunications Institute of Technology in Ho Chi Minh City, Viet Nam, and for providing the necessary resources, a conducive research environment that contributed significantly to the success of this study.

## References

1. Chen, W., Zhao, Y., Xie, W., Sang, N.: An improved SIFT algorithm for image feature-matching. In: Proceedings of the International Conference on Multimedia Technology (ICMT), pp. 197–200 (2011). https://doi.org/10.1109/ICMT.2011.6003022

2. Hadid, A.: The local binary pattern approach and its applications to face analysis. In: Image Processing Theory, Tools and Applications (IPTA), pp. 1–9 (2008). https://doi.org/10.1109/IPTA.2008.4743795
3. Sakr, N.A., ELdesouky, A.I., Arafat, H.: An efficient fast-response content-based image retrieval framework for big data. Comput. Electr. Eng. **54**, 522–538 (2016). https://doi.org/10.1016/j.compeleceng.2016.04.015
4. Dosovitskiy, A., et al.: An image is worth 16x16 words: transformers for image recognition at scale. In: International Conference on Learning Representations (ICLR) (2021)
5. Wu, P., Hoi, S.C.H., Nguyen, D.D., He, Y.: Randomly projected KD-trees with distance metric learning for image retrieval. In: Lee, K.-T., Tsai, W.-H., Liao, H.-Y.M., Chen, T., Hsieh, J.-W., Tseng, C.-C. (eds.) MMM 2011. LNCS, vol. 6524, pp. 371–382. Springer, Heidelberg (2011). https://doi.org/10.1007/978-3-642-17829-0_35
6. Song, C., Yoon, J., Choi, S., Avrithis, Y.: Boosting vision transformers for image retrieval. In: IEEE/CVF Winter Conference on Applications of Computer Vision (WACV), pp. 107–117 (2023). https://doi.org/10.1109/WACV56688.2023.00019
7. Zhang, Y., Zhang, J., Tang, C., Zhang, X.: Sequential hash representation for deep hashing-based image retrieval. Knowl.-Based Syst. **307**, 111606 (2025). https://doi.org/10.1016/j.knosys.2025.111606
8. Cao, Z., Long, M., Wang, J., Yu, P.S.: HashNet: deep learning to hash by continuation. In: Proceedings of the IEEE International Conference on Computer Vision (ICCV), pp. 5609–5618 (2017). https://doi.org/10.1109/ICCV.2017.598
9. Caron, M., et al.: Emerging properties in self-supervised vision transformers. In: Proceedings of the IEEE/CVF International Conference on Computer Vision (ICCV), pp. 9630–9640 (2021). https://doi.org/10.1109/ICCV48922.2021.00951
10. Zhou, J., et al.: iBOT: image BERT pre-training with online tokenizer. arXiv preprint arXiv:2111.07832 (2021). https://doi.org/10.48550/arXiv.2111.07832
11. He, K., Chen, X., Xie, S., Li, Y., Dollár, P., Girshick, R.: Masked autoencoders are scalable vision learners. In: Proceedings of the IEEE/CVF Conference on Computer Vision and Pattern Recognition (CVPR), pp. 15979–15988 (2022). https://doi.org/10.1109/CVPR52688.2022.01553
12. Schlegel, D., Grisetti, G.: HBST: a hamming distance embedding binary search tree for feature-based visual place recognition. IEEE Robot. Autom. Lett. **3**(4), 3741–3748 (2018). https://doi.org/10.1109/LRA.2018.2856542
13. Chen, Y., Zhang, S., Liu, F., Chang, Z., Ye, M., Qi, Z.: TrsHash: transformer-based Hamming hashing for efficient image retrieval. In: Proceedings of the International Conference on Multimedia Retrieval (ICMR), pp. 53–61 (2022). https://doi.org/10.1145/3512527.3531405
14. Liu, Z., et al.: Swin transformer: hierarchical vision transformer using shifted windows. In: Proceedings of the IEEE/CVF Conference on Computer Vision (ICCV), pp. 9992–10002 (2021). https://doi.org/10.1109/ICCV48922.2021.00986
15. Tuan, N.M., Meesad, P., Nguyen, H.H.C.: English-Vietnamese machine translation using deep learning for chatbot applications. SN Comput. Sci. **5**(1), 5 (2023). https://doi.org/10.1007/s42979-023-02339-2

16. Tuan, N.M., Meesad, P., Hieu, D.V., Cuong, N.H.H., Maliyaem, M.: On students' sentiment prediction based on deep learning: applied information literacy. SN Comput. Sci. **5**(7), 928 (2024). https://doi.org/10.1007/s42979-024-03281-7
17. Tuan, N.M., Thuy, P.T.T., Cuong, H.H.N., Hien, N.T.: On determining multiple languages through technological examination for conservation management using machine learning. Forum Linguist. Stud. **7**(5), Article 5 (2025). https://doi.org/10.30564/fls.v7i5.9110

# Predicting KellgrenLawrence Grades in Knee Osteoarthritis Using Machine Learning Algorithms

Nguyen Trong Hien[1], Nguyen Minh Tuan[2(✉)], and To Nam Kien[3]

[1] Faculty of Public Health, Pham Ngoc Thach University of Medicine, 2 Duong Quang Trung, Hoa Hung ward 700000, Ho Chi Minh city, Vietnam
hiennt@pnt.edu.vn

[2] Faculty of Information Technology, Posts and Telecommunications Institute of Technology, 11 Nguyen Dinh Chieu, Sai Gon ward 700000, Ho Chi Minh city, Vietnam
minhtuan@ptit.edu.vn

[3] Department of Rheumatology, Thong Nhat Hospital, 1 Ly Thuong Kiet, Tan Son Nhat ward 700000, Ho Chi Minh city, Vietnam

**Abstract.** We investigated whether radiographic severity of knee osteoarthritis (KOA) can be inferred from routine clinical and biochemical data. In a cross-sectional sample of 189 patients from a single tertiary hospital, four supervised models that Logistic Regression, Random Forest, XGBoost, and a soft voting ensemble were trained on a common set of 15 predictors, including serum uric acid (SUA), renal and lipid markers, body mass index, metabolic syndrome, and mechanical load related habits. On a stratified 80/20 traintest split, the voting ensemble achieved the highest discrimination (AUC = 0.896) with high specificity (96.9%), whereas Logistic Regression showed high accuracy (86.8%) but poor discrimination (AUC = 0.484), underscoring the limitations of accuracy in an imbalanced outcome. In five-fold cross-validation, tree-based models outperformed Logistic Regression (Random Forest AUC = 0.877 vs. 0.767). SUA was consistently influential, but as a single predictor provided only moderate discrimination (AUC = 0.656). These findings indicate that machine learning models based on routine, non imaging data capture a clinically relevant signal for KOA severity, but require validation in larger, more diverse cohorts.

**Keywords:** Knee osteoarthritis · KellgrenLawrence · machine learning · Random Forest · XGBoost · Voting Ensemble · Logistic regression · uric acid

## 1 Introduction

Knee osteoarthritis (KOA) is a leading cause of pain, disability, and reduced mobility in older adults. Its prevalence increases markedly with age and imposes

N. T. Hien and T. N. Kien—Equal Contribution


P. Pichappan et al. (Eds.): RTIS 2025, LNNS 1936, pp. 215–226, 2027.
https://doi.org/10.1007/978-3-032-23396-7_19

a substantial clinical and socioeconomic burden. In routine practice, structural severity is assessed on knee radiographs using the Kellgren–Lawrence (KL) grading system [1], which captures features such as osteophyte formation and joint space narrowing. However, KL grading has important limitations: it mainly reflects moderate-to-advanced structural damage, is relatively insensitive to early cartilage or synovial changes [2,3], and shows only modest agreement between readers and with patient-reported symptoms [4–6]. These shortcomings have encouraged the use of additional clinical and biochemical markers alongside radiography. Serum uric acid (SUA) has attracted particular interest. Experimental data suggest that elevated SUA and monosodium urate crystals can activate the NLRP3 inflammasome and increase interleukin-1$\beta$ production, thereby promoting synovial inflammation and cartilage damage [7,8]. Several observational studies have reported that higher SUA levels are associated with greater radiographic KOA severity, even in non-gout populations [9–12].

At the same time, advances in machine learning (ML) and artificial intelligence (AI) have created new opportunities for predicting KOA incidence, progression, and severity. A review of 26 KOA prediction models by Ramazanian et al. [13] found that ML methods such as Random Forest and Support Vector Machines often outperformed conventional statistical approaches, although external validation was frequently lacking. Deep learning models applied to imaging or gait data have achieved high classification performance [14–16], but they typically require radiographs or specialized equipment that may be unavailable in many clinical settings. In this study, we developed and evaluated four ML models Logistic Regression, Random Forest, XGBoost, and a soft Voting Ensemble for binary classification of KOA severity, defining mild disease as KL grade 2 and severe disease as KL grades 3–4. Using data from a single tertiary hospital, we examined whether routinely collected clinical and biochemical variables, including SUA, contain sufficient information to distinguish mild from more severe radiographic KOA. We also explored the relative contribution of SUA and other metabolic markers within these models and discussed how such tools could be integrated into everyday clinical practice.

## 2 Methodology

We conducted a cross-sectional study at Nguyen Tri Phuong Hospital, a tertiary care centre in Ho Chi Minh City, Vietnam. Data were collected between December 2021 and July 2022 in the outpatient clinic and the Department of Rheumatology. Eligible patients fulfilled the 1986 ACR clinical and radiographic criteria for KOA, i.e. knee pain, radiographic osteophytes, and at least one of the following: age $>$ 50 years, morning stiffness $<$ 30 minutes, or crepitus on movement. Patients were excluded if they had: (i) secondary osteoarthritis (gout, rheumatoid arthritis, infectious arthritis, seronegative spondyloarthritis, previous knee surgery, major knee trauma, or congenital deformity); (ii) acute conditions likely to markedly affect SUA levels (acute kidney injury, active gastrointestinal bleeding, tumour lysis syndrome, haemolysis, rhabdomyolysis, severe infection, or acute myocardial infarction); (iii) psychiatric illness or inability to provide reliable information; (iv) current use of urate-lowering therapy; or (v) incomplete study data. After applying these criteria, 189 patients were included. Standardised knee radiographs (anteroposterior and lateral views) were acquired at the Department of Diagnostic Imaging according to the hospital protocol. Two experienced musculoskeletal radiologists independently assigned KL grades; discrepancies were resolved by a third senior radiologist. Clinical information, comorbidities, lifestyle habits, and KOA-related symptoms and signs were collected with a structured case report form. Anthropometric measurements (weight, height, waist circumference) were obtained using standardised procedures. After an overnight fast ($\geq$ 8 h), 2 mL of venous blood was drawn in the morning and processed in the central laboratory using an AU680 analyser. Samples were stored at 2–8 °C until analysis. The main biochemical tests were serum uric acid, serum creatinine, fasting plasma glucose, HbA1c, and a lipid profile (total cholesterol, HDL-c, LDL-c, triglycerides).

## 2.1 Predictor Variables

Based on routinely available clinical information and prior literature, 30 variables were selected as initial candidate predictors. Basic and anthropometric factors included gender ( sex), age ( age), body mass index (BMI, bmi), and abdominal obesity ( tang.vongeo). Physical activity and mechanical load were represented by frequent heavy lifting ( habit1), stair climbing ( habit2), prolonged standing ( habit3), and kneeling ( habit4). Comorbidities comprised hypertension ( tha), diabetes mellitus ( dtd), dyslipidaemia ( rlmm), osteoporosis ( loang.xuong), and metabolic syndrome ( hcch). Biochemical and metabolic markers included serum uric acid ( uric2), serum creatinine ( crea), estimated glomerular filtration rate (eGFR), fasting glucose ( glucose), HbA1c, total cholesterol ( choles.total), triglycerides, HDL-c, and LDL-c. Medication-related variables captured the use of diacerein, non-steroidal anti-inflammatory drugs (NSAIDs, nsaid), oral corticosteroids ( cor.uong), and intra-articular corticosteroid injections ( cor.tiem). Clinical symptoms comprised knee pain ( oakn1), limited mobility ( oakn4), joint deformity ( oakn7), and crepitus ( oakn9).

The outcome was the KL grade, dichotomised as mild (KL 2, coded 0) versus severe (KL 3–4, coded 1). All 30 variables were initially considered during feature selection and model development.

## 2.2 Data Preprocessing

All records were anonymised and stored in a structured electronic database. Preprocessing steps were designed to handle missingness, harmonise scales, and obtain a suitable input matrix.

Records with missing KL grade were removed. For the remaining patients, missing values in continuous or ordinal predictors were imputed using the sample median of each variable; binary or categorical indicators, which had very low missingness, were imputed with the most frequent category when needed.

Continuous predictors were standardised using Z-scores. For each variable $X_j$, with mean $\mu_j$ and standard deviation $\sigma_j$ estimated from the training data, we computed

$$X_{ij}^* = \frac{X_{ij} - \mu_j}{\sigma_j}$$

for patient $i$; the same transformation was applied to the test set using training-set parameters. Binary indicators (e.g. sex, comorbidities, medication use) were left in their original 0/1 coding.

### 2.3 Feature Selection

Given the modest sample size (189 patients) and the moderate class imbalance, we did not include all 30 candidate predictors in the final models. Instead, we used a feature selection strategy that combined prior evidence from the study site with empirical information from the current dataset.

First, we re-examined the multivariable Logistic Regression analyses from the original residency thesis by Kien [25], which consistently identified serum uric acid, renal function, age, and several metabolic variables as correlates of radiographic KOA severity. Second, we considered published data on "metabolic osteoarthritis" and on the contributions of mechanical load, osteoporosis, and metabolic syndrome to KOA progression. Third, we fitted exploratory machine learning models using the full 30-variable set and inspected feature importance rankings across algorithms to identify predictors that repeatedly contributed to discrimination.

The 30 candidate variables spanned eight conceptual domains (basic/anthropometric, biochemical/metabolic, clinical/skeletal, clinical symptoms, comorbidities, comorbidities/metabolic, mechanical load/activity, and medication). On the basis of the three steps above, we selected a common set of 15 predictors for all primary models (Logistic Regression, Random Forest, XGBoost, and the soft Voting Ensemble). The basic group comprised gender ( sex), age ( age), and body mass index ( bmi). Biochemical and metabolic markers included serum uric acid ( uric2), serum creatinine ( crea), estimated glomerular filtration rate (eGFR), total cholesterol ( choles.total), HDL-c ( hdl), LDL-c ( ldl), triglycerides ( triglycerides), and metabolic syndrome ( hcch). Clinical and skeletal variables were osteoporosis ( loang.xuong) and knee pain ( oakn1), and mechanical load was represented by frequent heavy lifting ( habit1) and prolonged standing ( habit3).

Taken together, these 15 predictors capture demographic risk, systemic metabolic status, kidney function, skeletal fragility, mechanical overload, and key clinical symptoms, with particular emphasis on serum uric acid in line with the original thesis [25]. All four algorithms were trained and evaluated on this identical 15-predictor set. Feature importance estimates were used solely for post hoc interpretation and not to define model-specific subsets, in order to preserve comparability between algorithms.

### 2.4 Machine Learning Algorithms

We considered four supervised learning algorithms for predicting severe versus mild KOA, each returning an estimated probability of severe disease given the 15 predictors.

Logistic Regression (LR) is a generalised linear model commonly used for binary outcomes in clinical research [17]. It assumes a linear relation between predictors and the log-odds of severe KOA,

$$\log\left(\frac{P(y=1 \mid \mathbf{x})}{1-P(y=1 \mid \mathbf{x})}\right) = \beta_0 + \boldsymbol{\beta}^\top \mathbf{x},$$

with $P(y = 1 \mid \mathbf{x})$ obtained via the logistic function. After standardisation, each coefficient has a direct log-odds interpretation, which favours clinical interpretability.

Random Forest (RF) is a tree-based ensemble built from bootstrap samples of the training data, with random feature subsampling at each split [22]. Predictions are obtained by averaging class probabilities over all trees. RF can flexibly capture nonlinear effects and interactions, with feature importance inferred from impurity-based measures.

XGBoost implements gradient boosting of decision trees [23]. Trees are added sequentially to correct residual errors of the current ensemble, with regularisation and learning-rate shrinkage to limit overfitting. XGBoost typically performs well on tabular data but requires careful tuning in small samples.

The soft Voting Ensemble combines the predicted probabilities from LR, RF, and XGBoost [24]. For a patient with predictors $\mathbf{x}$, each base learner $m$ outputs $p_m(\mathbf{x}) = P_m(y = 1 \mid \mathbf{x})$, and the ensemble prediction is

$$f_{\text{vote}}(\mathbf{x}) = \sum_m w_m \, p_m(\mathbf{x}),$$

with non-negative weights $w_m$ summing to one. Here, equal weights were used. The combined probability is then thresholded to obtain a binary prediction with the aim of improving robustness by leveraging the strengths of the complementary model.

To ensure a fair comparison, all four algorithms used the same 15 predictors. The importance of features of LR (standardized coefficients) and from tree-based models (impurity-based importance) was used solely for interpretation.

## 3 Numerical Results

### 3.1 Characteristics of the Study Sample

A total of 189 patients were included, 44 (23.3%) men and 145 (76.7%) women. Mean age was $64.0 \pm 9.0$ years and mean BMI was 24.7 kg/m$^2$; 32.3% had normal BMI, 28.6% were overweight, and 36.5% were obese. Elevated SUA was present in 25.4% of patients, with a mean SUA concentration of $5.79 \pm 1.59$ mg/dL.

After excluding records with missing KL data, radiographic grades ranged from 2 to 4 and were dichotomised as mild (KL 2) and severe (KL 3–4). This yielded 159 mild cases (84.1%) and 30 severe cases (15.9%), indicating a moderately imbalanced outcome distribution skewed towards less severe disease.

All four machine learning models were trained and evaluated on the same 15-predictor set ( age, sex, bmi, uric2, eGFR, choles.total, hdl, triglycerides, hcch, oakn1, crea, ldl, loang.xuong, habit1, habit3), as described in the Methodology section [28,29].

### 3.2 Five-Fold Cross-Validation

Internal performance was first evaluated using five-fold stratified cross-validation on the full dataset, with all models trained on the common 15-predictor set. Table 1 summarizes the mean accuracy, AUC, and F1-score for the severe class, together with their 95% confidence intervals (CIs) across folds.

**Table 1.** Five-fold stratified cross-validation performance of the four models. Values are mean (95% confidence interval) across folds.

| Model | Accuracy | AUC | F1-score (severe) |
|---|---|---|---|
| Logistic Regression | 0.820 (0.795–0.845) | 0.767 (0.644–0.890) | 0.349 (0.204–0.495) |
| Random Forest | 0.841 (0.804–0.878) | 0.877 (0.834–0.921) | 0.377 (0.266–0.489) |
| XGBoost | 0.804 (0.725–0.883) | 0.851 (0.772–0.930) | 0.311 (0.197–0.425) |
| Voting Ensemble | 0.836 (0.810–0.862) | 0.867 (0.825–0.909) | 0.363 (0.311–0.414) |

Overall, all four models achieved at least low-to-moderate discrimination. The tree-based methods and the voting ensemble yielded higher cross-validated AUCs than Logistic Regression, with Random Forest reaching 0.877 and the Voting Ensemble 0.867, compared with 0.767 for Logistic Regression. Accuracies and F1-scores followed a similar pattern, with Random Forest and the Voting Ensemble performing slightly better and showing relatively narrow CIs, whereas XGBoost exhibited wider CIs, reflecting greater variability in this small sample.

To formally compare algorithms, we treated the five folds as repeated measurements and performed a repeated-measures ANOVA with *Model* as the within-subject factor and each performance metric as the dependent variable. Across folds, mean accuracy ranged from 0.80 to 0.84, mean AUC from 0.77 to 0.88, and mean F1-score for the severe class from 0.31 to 0.38. Consistent with the descriptive results, the ANOVA identified a significant main effect of *Model* on AUC ($F(3,12) = 6.40$, $p = 0.0078$), indicating that discrimination between mild and severe KOA differed across models. No statistically significant differences were detected for accuracy ($F(3,12) = 1.52$, $p = 0.26$), F1-score ($F(3,12) = 0.71$, $p = 0.57$).

Taken together, these findings indicate that the main advantage of the tree-based models and the voting ensemble over Logistic Regression lies in improved

AUC, whereas average accuracy, F1-score, sensitivity, and specificity are broadly comparable when assessed by repeated cross-validation.

### 3.3 Performance on the Stratified 80/20 Train–Test Split

To complement the cross-validation analysis with a single held-out evaluation, we assessed all four algorithms on a stratified 80/20 train–test split (151 patients for training and 38 for testing), preserving the original mild/severe ratio. Test-set accuracy, AUC, F1-score for the severe class, sensitivity, and specificity are summarised in Table 2 [18–21].

**Table 2.** Performance of the predictive models on the stratified 80/20 train–test split (severe = KL 3–4). Point estimates and 95% confidence intervals (in parentheses).

| Model | Accuracy | AUC | F1-score | Sensitivity | Specificity |
|---|---|---|---|---|---|
| Logistic Regression | 0.868 | 0.484 | 0.444 | 0.333 | 0.969 |
| | (0.795–0.845) | (0.644–0.890) | (0.204–0.495) | (0.203–0.463) | (0.945–0.986) |
| Random Forest | 0.816 | 0.875 | 0.000 | 0.000 | 0.969 |
| | (0.804–0.878) | (0.834–0.921) | (0.000–0.000) | (0.000–0.000) | (0.945–0.986) |
| XGBoost | 0.868 | 0.891 | 0.286 | 0.167 | 1.000 |
| | (0.725–0.883) | (0.772–0.930) | (0.197–0.425) | (0.106–0.229) | (1.000–1.000) |
| Voting Ensemble | 0.842 | 0.896 | 0.250 | 0.167 | 0.969 |
| | (0.810–0.862) | (0.825–0.909) | (0.311–0.414) | (0.106–0.228) | (0.945–0.986) |

On this split, Logistic Regression achieved the highest accuracy (86.8%) and the highest F1-score for the severe class (0.444), with sensitivity 0.333 and specificity 0.969. However, its AUC was only 0.484, indicating poor overall discrimination and showing that high accuracy mainly reflects the predominance of mild cases. In contrast, the tree-based and ensemble methods yielded high AUCs (0.875 for Random Forest, 0.891 for XGBoost, 0.896 for the Voting Ensemble) and excellent specificity (0.969–1.000) (Fig. 1a), but very low sensitivity at the default threshold of 0.5: Random Forest failed to identify any severe case, while XGBoost and the Voting Ensemble each correctly identified only one severe case (sensitivity 0.167). As a result, F1-scores for the severe class were modest and unstable, ranging from 0.000 (Random Forest) to 0.444 (Logistic Regression). This pattern illustrates how, in an imbalanced setting, models can display good AUC yet perform poorly at standard decision thresholds, and how accuracy alone can be misleading.

To quantify uncertainty on this small test set, we applied non-parametric bootstrap resampling ($B = 1000$) of the 38 test observations. For each model and each metric, 95% percentile bootstrap confidence intervals were derived. The AUC intervals for the tree-based models remained high (approximately 0.76–0.97 for Random Forest, 0.77–0.98 for XGBoost, and 0.78–0.98 for the Voting

Ensemble), whereas the interval for Logistic Regression was very wide (about 0.00–0.99), reflecting its unstable discrimination on this split. Bootstrap intervals for sensitivity and F1-score were wide for all models, with lower bounds at or near zero, consistent with the small number of severe cases. Pairwise bootstrap contrasts ($\text{Model}_2 - \text{Model}_1$) showed that the 95% confidence intervals for between-model differences all included zero, indicating no statistically noteworthy superiority of any single model on this particular test set. Overall, the bootstrap analysis supports the descriptive impression that tree-based and ensemble methods tend to achieve higher AUCs than Logistic Regression, but also underscores the substantial uncertainty in all test-set estimates in a small, imbalanced sample.

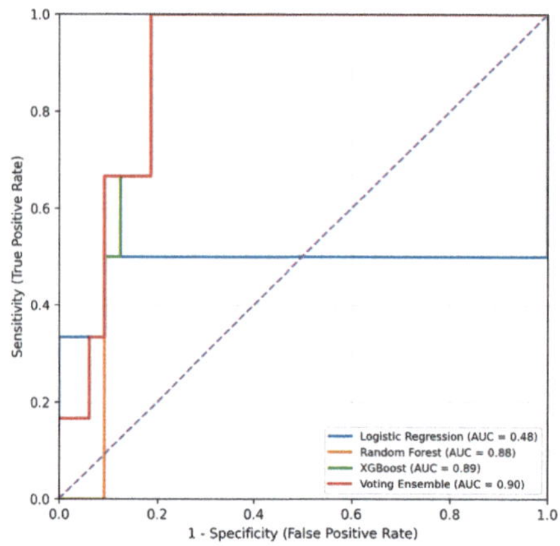


(a) Multivariable models.

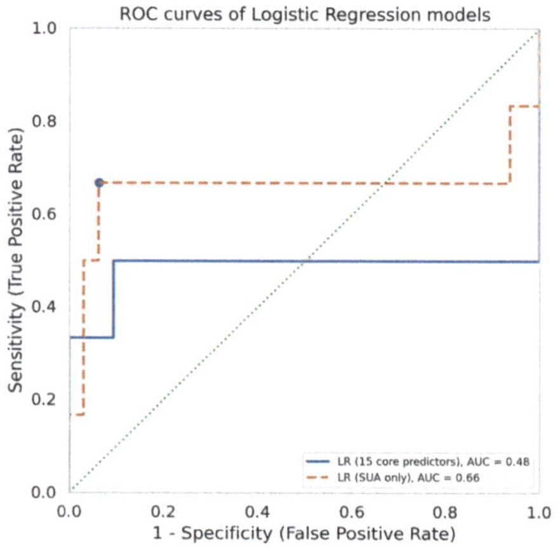


(b) Logistic Regression with 15 predictors vs. SUA only.

**Fig. 1.** ROC curves on the stratified 80/20 test set. (a) Tree-based and ensemble models (Random Forest, XGBoost, Voting Ensemble) obtain higher AUCs (0.88–0.90) than Logistic Regression (AUC = 0.48), but show low sensitivity for severe KOA at the default threshold of 0.5. (b) Logistic Regression using the full 15-predictor set compared with Logistic Regression using SUA alone. The SUA-only model (AUC = 0.66) performs consistently worse than the multivariable model, indicating that SUA provides useful but insufficient information on its own.

### 3.4 Serum Uric Acid as a Single Predictor

To isolate the contribution of SUA, we fitted a univariable Logistic Regression model with uric2 as the sole predictor, using the same stratified 80/20 split. On the test set, this model achieved an AUC of 0.656, an accuracy of 0.868, and an F1-score of 0.286 for the severe class. At the default threshold of 0.5, sensitivity for severe KOA was 0.167, while specificity for mild KOA was 1.000, reflecting a highly conservative operating point in which almost all patients were classified as mild.

Optimising the threshold using Youden's $J$ statistic yielded an optimal cut-off of approximately 0.16, at which sensitivity was 0.667 and specificity 0.938. This indicates that SUA alone carries a moderate amount of information for distinguishing mild from severe radiographic KOA if the threshold is adjusted [26,27].

As shown in Fig. 1b, however, the SUA-only ROC curve remains consistently below the multivariable Logistic Regression curve and clearly below the tree-based and ensemble models. Overall, these findings support the view that SUA is an informative but insufficient standalone predictor, and that its discriminative value is best exploited in combination with other metabolic, renal, clinical, and mechanical variables.

### 3.5 Feature Importance Across Models

To characterise which variables drove model predictions, we examined feature importance for Logistic Regression, Random Forest, and XGBoost. In Logistic Regression, importance was defined as the absolute value of the standardised regression coefficients. For Random Forest and XGBoost, we used the impurity-

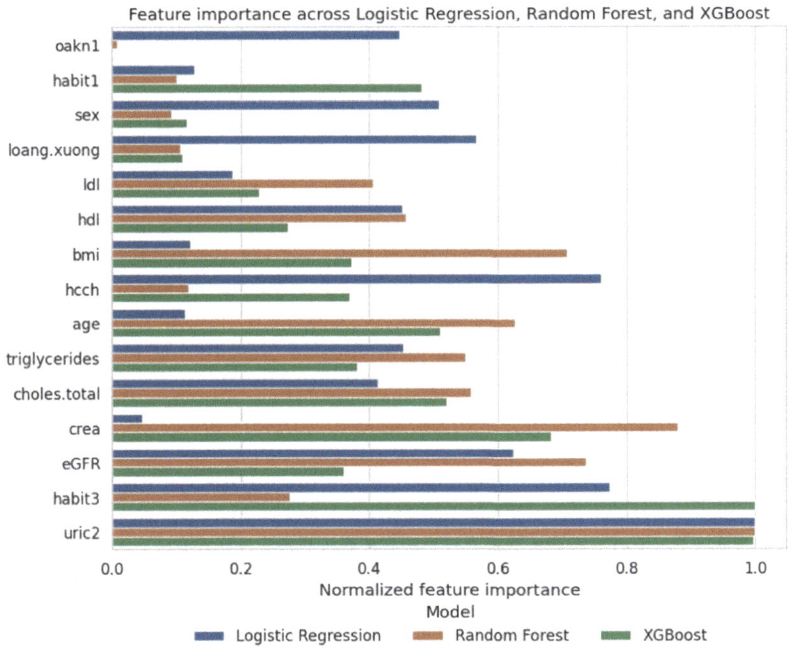


**Fig. 2.** Normalised feature importance across Logistic Regression, Random Forest, and XGBoost using the common 15-predictor set. For each model, importance values were rescaled to the [0,1] range. Serum uric acid (uric2), renal function markers, lipid profile variables, metabolic syndrome, and mechanical loadóÀẼrelated habits consistently appeared among the most influential predictors.

based importance measures returned by the algorithms. All importance scores were then linearly rescaled to [0,1] and plotted together (Fig. 2).

Several predictors emerged consistently across models. Serum uric acid ( uric2) ranked among the top features in all three algorithms. Renal function markers (creatinine crea, eGFR), lipid variables (total cholesterol choles.total, triglycerides, HDL-c, LDL-c), BMI ( bmi), and metabolic syndrome ( hcch) also showed high normalised importance, underlining the contribution of systemic metabolic and renal factors. Mechanical loadóÀẼrelated variables, especially prolonged standing ( habit3) and heavy lifting ( habit1), were prominent in the tree-based models. Osteoporosis ( loang.xuong), age, gender ( sex), and knee pain ( oakn1) further added clinically intuitive information about joint structure and symptoms.

These patterns support a multifactorial view of KOA, in which metabolic, mechanical, skeletal, and symptomatic dimensions all influence radiographic severity. Within this framework, serum uric acid appears as a key but not exclusive component.

# 4 Conclusion and Future Work

This study evaluated four supervised learning models such as Logistic Regression, Random Forest, XGBoost, and a soft Voting Ensemble for predicting radiographic KOA severity (mild: KL 2; severe: KL 3–4) using only routine clinical and biochemical data from 189 patients at a single tertiary hospital. All models were trained on a common set of 15 predictors selected from 30 routinely collected variables. In five-fold cross-validation, the tree-based and ensemble models achieved higher mean AUCs (0.85–0.88) than Logistic Regression (0.77), with broadly similar accuracies and high specificity. On a stratified 80/20 train including test split, Random Forest, XGBoost, and the Voting Ensemble again yielded high AUCs but low sensitivity for severe KOA at the default threshold, whereas Logistic Regression showed better accuracy and F1-score for the severe class but poor overall discrimination. Serum uric acid consistently ranked among the most influential predictors, and a SUA-only Logistic Regression model showed moderate discrimination (AUC = 0.656). Other metabolic and renal markers, BMI, metabolic syndrome, mechanical load related habits, osteoporosis, and knee pain also contributed to prediction, supporting a multifactorial view of KOA that combines metabolic and mechanical pathways. These findings indicate that routine non-imaging data carry useful signal on KOA severity, particularly where access to imaging is limited.

**Acknowledgment.** The authors thank Pham Ngoc Thach University of Medicine and Nguyen Tri Phuong Hospital (Ho Chi Minh City, Vietnam) for institutional support, and the staff of the outpatient clinic and the Department of Rheumatology for their assistance with data collection. We are grateful to all participating patients. We sincerely appreciate the financial research support of the Posts and Telecommunications Institute of Technology in Ho Chi Minh City, Viet Nam, and for providing the necessary resources, a conducive research environment that contributed significantly to the success of this study.

## References

1. Schiphof, D., et al.: Kellgren-Lawrence criteria and KOA diagnosis. Ann. Rheum. Dis. **70**(8), 1422–1427 (2011)
2. Goldring, M.B., Otero, M.: Inflammation in osteoarthritis. Curr. Opin. Rheumatol. **23**(5), 471–478 (2011)
3. Berenbaum, F.: Osteoarthritis as an inflammatory disease (osteoarthritis is not osteoarthrosis!). Osteoarthritis Cartilage **21**(1), 16–21 (2013)
4. Toivanen, A.T., et al.: Agreement between clinical and radiological KOA diagnosis. Scand. J. Rheumatol. **36**(1), 58–63 (2007)
5. Neogi, T., et al.: Radiographic features and pain in KOA. BMJ **339**, b2844 (2009)
6. Cubukcu, D., Sarsan, A., Alkan, H.: Relationships between pain, function and radiographic findings in osteoarthritis of the knee: a cross-sectional study. Arthritis **2012**, 984060 (2012)
7. Denoble, A.E., et al.: Uric acid is a danger signal of increasing risk for osteoarthritis through inflammasome activation. Proc. Natl. Acad. Sci. **108**(5), 2088–2093 (2011)
8. Neogi, T., Krasnokutsky, S., Pillinger, M.H.: Urate and OA: reciprocal links. Joint Bone Spine **86**(5), 576–582 (2019)
9. Ding, X., et al.: Serum uric acid and hyperuricemia in knee osteoarthritis. Rheumatol. Int. **36**(4), 567–573 (2016)
10. Krasnokutsky, S., et al.: Serum urate and joint space narrowing in non-gout knee OA. Arthritis Rheumatology **69**(6), 1213–1220 (2017)
11. Srivastava, S.R., et al.: SUA influence on KOA: revisit reference range? Osteoarthritis Cartilage **26**(supp), S224–S225 (2018)
12. Bassiouni, S., et al.: Association of serum uric acid with radiological severity in knee osteoarthritis. Egypt. Rheumatol. Rehabil. **48**(1), 6 (2021)
13. Ramazanian, A., et al.: Review of KOA prediction models. J. Biomed. Inform. **125**, 103976 (2023)
14. Miraj, M., et al.: AI and machine learning in early diagnosis of KOA: a review. Front. Public Health **12**, 123456 (2024)
15. Tiulpin, A., et al.: Multimodal ML for KOA progression. Sci. Rep. **9**, 1727 (2019)
16. Chen, Y., et al.: Machine learning-based prediction of knee osteoarthritis using clinical and biochemical data. Sci. Rep. **15**(2), 2101 (2025)
17. Hosmer, D.W., Lemeshow, S., Sturdivant, R.X.: Applied Logistic Regression, Wiley (2013). 3rd edn
18. Youden, W.J.: Index for rating diagnostic tests. Cancer **3**(1), 32–35 (1950)
19. Saito, T., Rehmsmeier, M.: Precision-recall versus ROC analysis in imbalanced datasets. PLoS ONE **10**(3), e0118432 (2015)
20. Pedregosa, F., et al.: Scikit-learn: machine learning in Python. J. Mach. Learn. Res. **12**, 2825–2830 (2011)
21. Fan, Z., et al.: XGBoost-SHAP-based interpretable diagnostic framework for knee osteoarthritis: a population-based retrospective cohort study. Arthritis Res. Ther. **26**(1), 213 (2024)
22. Breiman, L.: Random forests. Mach. Learn. **45**(1), 5–32 (2001)

23. Chen, T., Guestrin, C.: XGBoost: a scalable tree boosting system. In: Proceedings of the 22nd ACM SIGKDD Conference on Knowledge Discovery and Data Mining, pp. 785–794 (2016)
24. Rokach, L.: Ensemble-based classifiers. Artif. Intell. Rev. **33**(1), 1–39 (2010)
25. Kien, T.N.: Association between serum uric acid levels and the severity of knee osteoarthritis at Nguyen Tri Phuong Hospital, Pham Ngoc Thach University of Medicine (2022). Residency Thesis
26. Kuwashima, U., Itoh, M., Itou, J., Okazaki, K.: Comparison of post-operative kneeling ability in patients with varus knee osteoarthritis: unicompartmental knee arthroplasty versus open-wedge high tibial osteotomy. J. Joint Surg. Res. **3**(4), 201–206 (2025)
27. Wang, H., Liang, Q., Hancock, J.T., Khoshgoftaar, T.M.: Feature selection strategies: a comparative analysis of SHAP-value and importance-based methods. J. Big Data **11**, 44 (2024)
28. Tuan, N.M., Meesad, P., Nguyen, H.H.C.: English-Vietnamese machine translation using deep learning for chatbot applications. SN Comput. Sci. **5**(1), 5 (2023). https://doi.org/10.1007/s42979-023-02339-2
29. Tuan, N.M., Meesad, P., Hieu, D.V., Cuong, N.H.H., Maliyaem, M.: On students' sentiment prediction based on deep learning: applied information literacy. SN Comput. Sci. **5**(7), 928 (2024). https://doi.org/10.1007/s42979-024-03281-7

# Enhanced YOLOv11n for Small Object Detection in UAV Imagery: Integrating BiFPN, C3Ghost, CBAM, and MPDIoU

Cao-Phuc Ha[1], Ngoc-Tuan Hoang[1], and Phu-Nguyen Le[2(✉)]

[1] Department of Information Technology, FPT University, Ho Chi Minh City 71300, Vietnam
[2] Faculty of Engineering and Technology, Nguyen Tat Thanh University, Ho Chi Minh City 71300, Vietnam
lpnguyen@ntt.edu.vn

**Abstract.** Small object detection in UAV imagery is challenging due to scale variation, occlusion, and limited onboard computation. Although YOLO models offer real-time inference, their accuracy on UAV datasets such as VisDrone-DET2019 remains limited, with baseline YOLOv11n achieving only 0.3314 mAP50 and 0.1949 mAP50:0.95. This work introduces an enhanced YOLOv11n by integrating BiFPN, C3Ghost, CBAM, and MPDIoU into the backbone, neck, and head to improve multi-scale fusion, lightweight feature extraction, attention, and localization. Experiments show that our model reaches 0.456 mAP50, 0.301 mAP50:0.95, 0.586 precision, and 0.33 recall-surpassing the baseline and outperforming UAV-oriented variants such as HSF-YOLO, DAU-YOLO, and CF-YOLO. The results demonstrate a more effective and efficient solution for UAV small-object detection.

**Keywords:** YOLOv11n · UAV · object detection · small target detection · state-of-the-art

## 1 Introduction

Unmanned Aerial Vehicles (UAVs) have become essential tools in surveillance, agriculture, disaster response, and infrastructure monitoring due to their ability to capture large-scale aerial imagery [1,8]. However, object detection in UAV images remains highly challenging because targets are often small (typically around 32 × 32 pixels), densely distributed, and subject to occlusion and class imbalance, resulting in frequent missed detections and false positives [5,7]. The VisDrone benchmark further highlights these issues with over 38,000 annotated instances across 10 classes [8].

These challenges are intensified by the limited computational capacity of UAV platforms, which demand lightweight yet accurate models [4]. Although single-stage detectors such as YOLO have improved real-time performance


P. Pichappan et al. (Eds.): RTIS 2025, LNNS 1936, pp. 227–238, 2027.
https://doi.org/10.1007/978-3-032-23396-7_20

compared to two-stage methods, their baseline accuracy on UAV datasets remains constrained by feature loss during downsampling and insufficient multi-scale fusion [9]. Recent advancements in YOLOv8, including BiFPN, C3Ghost, CBAM, and MPDIoU, demonstrated notable improvements for small-object detection, achieving mAP50 of 0.339 on VisDrone [1].

Motivated by these findings, this work adapts these modules to the newer YOLOv11n, which provides a more efficient C3-based backbone optimized for real-time inference on resource-limited UAV hardware [3,5]. The proposed enhanced YOLOv11n achieves an mAP50 of 0.456 on VisDrone, surpassing UAV-oriented models such as HSF-YOLO (0.357) [2], CF-YOLO (0.449) [3], and DAU-YOLO (0.411) [4]. These results validate the effectiveness of integrating lightweight modules into YOLOv11n for improving small-object detection in complex UAV environments.

## 2 Related Work

Object detection has witnessed a rapid evolution from two-stage approaches such as R-CNN and Faster R-CNN, which rely on region proposals for accuracy but incur high computational costs, to single-stage detectors like SSD, RetinaNet, and the YOLO family, which prioritize real-time inference [9]. In UAV contexts, the challenge of detecting small and densely distributed objects has motivated numerous improvements over baseline architectures [8]. The VisDrone benchmark has served as a standard dataset, highlighting issues such as scale variation, occlusion, and clutter.

Recent works have enhanced YOLO variants for UAV imagery. Zhang and Zuo (2025) proposed an improved YOLOv8 framework integrating BiFPN, C3Ghost, CBAM, and MPDIoU loss, achieving significant gains in small object detection [1]. Similarly, Chen et al. (2025) introduced CBAM-enhanced YOLOv8 for UAV remote sensing, while Guo et al. (2024) developed SL-YOLO with lightweight designs for drone deployment [6]. Other variants, such as DAU-YOLO [4], LRDS-YOLO [7], and HSF-YOLO [2], focus on attention and multi-scale representation to boost recall and precision in challenging aerial conditions. With the introduction of YOLOv11, Ultralytics further streamlined the architecture by employing C3k2 modules for efficient feature extraction, C2PSA for spatial attention, and SPPF for fast pooling [10,11]. Liu et al. (2025) extended YOLOv11 for UAV target detection, confirming its suitability for real-time deployment under constrained hardware [5]. Moreover, Li et al. (2025) proposed CF-YOLO based on YOLOv11n with fusion strategies, demonstrating competitive results on VisDrone [3].

In addition to UAV-specific methods, foundational techniques such as the Feature Pyramid Network (FPN) [12] inspired the development of BiFPN [13], which has been widely adopted in recent YOLO enhancements. By combining lightweight modules, attention mechanisms (CBAM [14]), and advanced IoU-based loss functions (MPDIoU [15]), researchers have significantly improved small object detection without sacrificing efficiency. Motivated by these advances, our work adopts the methodology of Zhang and Zuo (2025) [1], originally designed for YOLOv8n, and applies it to YOLOv11n. This experimental transfer aims to validate whether the architectural gains of YOLOv11n, when combined with BiFPN, C3Ghost, CBAM, and MPDIoU, can achieve performance competitive with article YOLOv8n improvements on VisDrone.

# 3 Proposed Method

## 3.1 Overview of the Proposed Model Architecture

The proposed model is based on YOLOv11n, the latest lightweight variant in the YOLO series developed by Ultralytics, which features an efficient C3-based backbone for improved feature extraction and real-time inference capabilities. Adapting and extending the methodology from Zhang and Zuo (2025) [1], we apply their enhancements to YOLOv11n with additional YOLOv11-specific modifications to address UAV small object challenges. This includes replacing the original C2f modules in the detection head with C3Ghost for reduced computational load, substituting the Path Aggregation Network (PAN) with Bi-directional Feature Pyramid Network (BiFPN) for enhanced multi-scale feature fusion, incorporating Convolutional Block Attention Module (CBAM) in the head for selective feature emphasis (replacing C2PSA for improved channel and spatial attention), utilizing Minimum Point Distance-based Intersection over Union (MPDIoU) loss to better handle difficult small samples, and extending the Spatial Pyramid Pooling Fast (SPPF) to SPPF-LSKA by integrating Large Separable Kernel Attention to enlarge receptive fields and enhance multi-scale pooling efficiency.

The architecture comprises three primary components: the backbone for initial feature extraction, the neck for aggregating multi-level features, and the head for final bounding box prediction and classification. Input images of size $640 \times 640$ are processed through a series of convolutional layers to produce feature maps at different resolutions (strides of 8, 16, and 32). These maps are fused in the neck and passed to the head, which outputs predictions for the 10 classes in the VisDrone dataset (e.g., pedestrian, car). The overall process can be expressed as:

$$\mathbf{Y} = \text{Detect}(\text{BiFPN}(\text{Backbone}(\mathbf{X}))),$$

where $\mathbf{X}$ is the input image, Backbone extracts hierarchical features using Conv and C3k2 blocks, BiFPN fuses them bidirectionally, and Detect applies C3Ghost and CBAM for regression and classification. Figure 1 illustrates the model structure, adapted from Fig. 1 in Zhang and Zuo (2025) [1], with YOLOv11n-specific adjustments such as C3k2 in the backbone.

## 3.2 Bi-directional Feature Pyramid Network (BiFPN)

To mitigate feature loss in small targets from UAV perspectives, we implement BiFPN as described in Zhang and Zuo (2025) [1]. BiFPN enables efficient bidirectional fusion across multiple scales with learnable weights, outperforming the

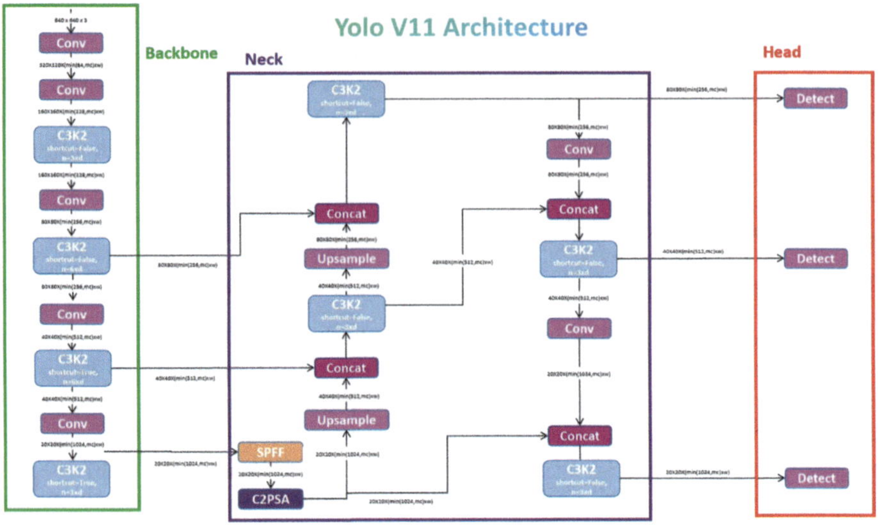


**Fig. 1.** Structural diagram of the base YOLOv11n model. The backbone uses Conv and C3k2 for feature extraction; the neck employs PAN for fusion (replaced by BiFPN in the proposed model); the head integrates C2f (replaced by C3Ghost and CBAM in the proposed model). Adapted from S Nikhileswara Rao 2024 [11].

standard PAN by dynamically prioritizing informative features. The fusion at each node is weighted as:

$$O = \sum_i \frac{w_i \cdot I_i}{\sum_j w_j + \epsilon},$$

where $O$ is the output feature, $I_i$ are inputs from connected nodes, $w_i \geq 0$ are learnable weights (optimized via fast normalized fusion), and $\epsilon = 0.0001$ prevents division by zero. The bidirectional paths are formulated as:

$$P_i^{td} = \text{Conv}\left(w_1 \cdot P_i + w_2 \cdot \text{Upsample}(P_{i+1}^{td}) + w_3 \cdot I_i\right), \quad \text{(top-down)},$$

$$P_i^{out} = \text{Conv}\left(w_1' \cdot P_i^{td} + w_2' \cdot \text{Downsample}(P_{i-1}^{out}) + w_3' \cdot I_i\right), \quad \text{(bottom-up)},$$

where $P_i$ are pyramid levels, and Conv is a depthwise separable convolution for efficiency. This module improves mAP by 17.2% as reported in the original work.

### 3.3 C3Ghost Module for Lightweight Head

For computational efficiency on UAV hardware, we replace C2f with C3Ghost in the head, exactly as in Zhang and Zuo (2025) [1]. C3Ghost generates redundant features cheaply using Ghost convolution, reducing parameters and FLOPs while preserving performance. The Ghost convolution splits channels:

$$Y' = \Phi(Y) = [Y_1, \ldots, Y_m, \delta(Y_1), \ldots, \delta(Y_m)],$$

where $Y$ is the output of ordinary conv on $m$ channels, and $\delta$ is a cheap linear operation (e.g., $3 \times 3$ depthwise conv). The C3Ghost block is:

$$\text{C3Ghost}(X) = \text{Concat}(\text{GhostBottleneck}(X/2), \text{GhostBottleneck}(X/2)) + X,$$

with GhostBottleneck as two GhostConvs.

### 3.4 Channel Attention Mechanism (CBAM)

We incorporate CBAM in the head to suppress noise and emphasize key features, following Zhang and Zuo (2025) [1], based on Woo et al. (2018) [14]. CBAM combines channel and spatial attention sequentially. Channel attention module:

$$M_c(F) = \sigma(MLP(\text{AvgPool}(F)) + MLP(\text{MaxPool}(F))),$$

where $\sigma$ is sigmoid, and MLP has ratio $r = 16$. Spatial attention:

$$M_s(F) = \sigma(f^{7\times7}(\text{Concat}(\text{AvgPool}(F), \text{MaxPool}(F)))),$$

with $f^{7\times7}$ a $7 \times 7$ conv. For the SECAConv variant (as in the provided image), the process is:

$$Y = \text{ReLU}(\text{BatchNorm}(\text{GroupConv}(X))) \odot \text{SE}(X),$$

where SE computes attention weights. This improves precision by 10.5% for small targets.

### 3.5 MPDIoU Loss Function

The MPDIoU loss is adopted to focus on difficult samples, as in Zhang and Zuo (2025) [1], extending ideas from Siliang Ma et al. (2023) [15]. It minimizes the distance between corner points of predicted and ground-truth boxes:

$$\text{MPDIoU} = \frac{d_1^2 + d_2^2}{w^2 + h^2},$$

where $d_1, d_2$ are distances between corresponding corners, and $w, h$ are box dimensions. Combined with inner-IoU:

$$L = \text{MPDIoU} + \alpha \cdot \text{Inner-IoU},$$

### 3.6 Key Architecture of YOLOv11

YOLOv11, introduced by Ultralytics in 2024, represents a significant evolution in the YOLO series, focusing on efficiency and versatility for real-time object detection tasks. The key architectural components include the C3k2 module for enhanced feature extraction, C2PSA for parallel spatial attention, and SPPF for fast spatial pyramid pooling. These modules enable YOLOv11 to support multiple vision tasks, including object detection, instance segmentation, image classification, pose estimation, oriented object detection, and object tracking. The C3k2 module is a variant of the C3 block, incorporating two bottleneck layers ($k = 2$) for balanced computation:

$$\text{C3k2}(X) = \text{Concat}(\text{Bottleneck}(X/2), \text{Bottleneck}(X/2)) + X, \tag{1}$$

where Bottleneck consists of two $3 \times 3$ convolutions with residual connections, reducing parameters while maintaining gradient flow. The C2PSA module integrates parallel spatial attention to emphasize important regions:

$$\text{C2PSA}(F) = \text{Conv}(\text{Concat}(\text{AvgPool}(F), \text{MaxPool}(F))) \odot F, \tag{2}$$

enhancing spatial selectivity for small objects. The SPPF (Spatial Pyramid Pooling Fast) performs multi-scale pooling efficiently:

$$\text{SPPF}(X) = \text{Concat}(\text{MaxPool}_{k=5}(X), \text{MaxPool}_{k=9}(X), \text{MaxPool}_{k=13}(X), X), \tag{3}$$

### 3.7 Application of Modifications to YOLOv11

Following the methodology in Zhang and Zuo (2025) [1], several improvements were applied to YOLOv11 for small object detection in UAV views. These include replacing the PAN with BiFPN for stronger bidirectional feature fusion, substituting C2f in the head with C3Ghost for lightweight efficiency, adding CBAM for advanced attention, and using MPDIoU loss to optimize regression for hard small samples. The original YOLOv11 key architecture, emphasizes efficiency through SPPF for multi-scale pooling, C2PSA for parallel attention, and C3k2 for compact extract. Building upon this foundation, we applied the modifications proposed by Zhang and Zuo (2025) [1], which include replacing SPPF with SPPF-LSKA to incorporate Large Separable Kernel Attention and thereby enlarge receptive fields and reduce feature dilution for UAV small targets, substituting C2PSA with CBAM to enhance both channel and spatial selectivity, and employing C3Ghost in place of C2f within the head to achieve lightweight computation while preserving feature extraction capacity. In addition, the loss function was modified by replacing CIoU with MPDIoU-innerIoU, which improves convergence and localization accuracy for dense, small, or occluded objects. These modifications work synergistically: SPPF-LSKA improves multi-scale context, CBAM enhances attention in dense UAV scenes, C3Ghost reduces FLOPs for deployment on resource-constrained UAV hardware, and MPDIoU strengthens regression on hard examples. Collectively, these enhancements yield faster convergence and higher mAP on the VisDrone dataset compared to the baseline (Table 4).

$$\text{SPPF-LSKA}(x) = \text{Conv}(\text{Concat}(x, m_5(x), m_9(x), m_{13}(x))) \cdot \text{LSKA}(x), \tag{4}$$

$$M_c(F) = \sigma(\text{MLP}(\text{AvgPool}(F) + \text{MaxPool}(F))), \tag{5}$$

$$\text{C3Ghost}(X) = \text{Concat}(\text{GhostBottleneck}(X/2), \text{GhostBottleneck}(X/2)) + X, \tag{6}$$

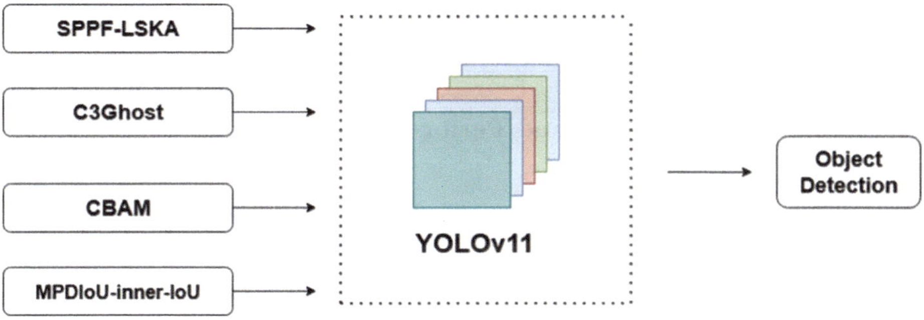


**Fig. 2.** Modified Key Architecture of YOLOv11

# 4 Experimental Results and Comparative Analysis

## 4.1 Training Configuration

**Table 1.** Training and Computational Settings

| Category | name | value |
|---|---|---|
| Train Parameters | Batch size | 32 |
| | Image size | 640 |
| | Optimizer | SGD |
| | Epochs | 300 |
| | momentum | 0.937 |
| | Warmup momentum | 0.8 |
| Computer configurations | GPU | NVIDIA A100 |
| | Graphics memory | 40GB |
| | pytorch | 2.4.1 |
| | python | 3.9.19 |

Table 1 shows the training process was conducted using the Ultralytics framework on the VisDrone benchmark dataset, which consists of 6,471 training images, 548 validation images, and 10 annotated classes. The model was trained for 300 epochs with a batch size of 32 and an input resolution of 640 × 640 pixels. Stochastic Gradient Descent (SGD) was adopted as the optimizer, with an initial learning rate of 0.01, momentum of 0.937, and weight decay of 0.0005. A linear warm-up strategy was employed with a warm-up momentum of 0.8 to stabilize the early optimization stage. Early stopping with a patience of 50 epochs was enabled to avoid overfitting and accelerate convergence. All experiments were executed on an NVIDIA A100 GPU with 40 GB of memory, using PyTorch 2.4.1 and Python 3.9.19. This configuration yielded faster convergence compared to the baseline 300-epoch setting reported by Zhang and Zuo (2025) [1], with stable performance achieved at approximately the 252$^{nd}$ epoch.

### 4.2 Comparative Results and Performance Analysis

**Table 2.** Overall comparison between Baseline, Custom, and Original Article

| Model/Method | mAP50 | mAP50:0.95 | Precision (P) | Recall (R) |
|---|---|---|---|---|
| Baseline YOLO11n (Default) | 0.3314 | 0.1949 | 0.445 | 0.328 |
| Original article (Improved YOLOv8n) | 0.339 | - | 0.498 | 0.272 |
| Results of custom YOLO11n | 0.456 | 0.301 | 0.586 | 0.33 |

Table 2 presents the overall comparison between the baseline YOLOv11n, the improved YOLOv8n reported in a publication, and our proposed custom YOLOv11n model. The baseline YOLOv11n achieves a modest performance with mAP50 of 0.3314 and mAP50:0.95 of 0.1949. The original article shows a marginal improvement with mAP50 0.339, indicating only incremental gains over the baseline. In contrast, our proposed YOLOv11n achieves a substantial increase to 0.456 in mAP50 and 0.301 in mAP50:0.95, which represents an improvement of more than 12% absolute in mAP50 compared to the main reference. This demonstrates the effectiveness of the architectural modifications applied in this work.

**Table 3.** Comparison with Other Networks (Non-YOLO, Simpler to Implement)

| Model/Method | mAP50 | mAP50:0.95 | GFLOPs | Params (M) |
|---|---|---|---|---|
| SSD (2016, UAV Adapt) | 0.25 | 0.15 | 4.0 | 26 |
| RetinaNet (Focal Loss, 2018) | 0.28 | 0.18 | 15 | 36 |
| Faster R-CNN (2015, UAV) | 0.22 | 0.14 | 180 | 42 |
| YOLO-GML (Transformer UAV, 2025) | 0.30 | - | 30 | 20 |
| Proposed custom YOLO11n | 0.456 | 0.301 | 25 | 16.167 |

Table 3 further benchmarks our model against other non-YOLO architectures that are simpler to implement, including SSD, RetinaNet, Faster R-CNN, and a recent Transformer-based YOLO-GML variant. While these models exhibit relatively low performance on the VisDrone dataset (mAP50 values ranging between 0.22 and 0.30), our custom YOLOv11n achieves a superior mAP50 of 0.456. Remarkably, this improvement is obtained with competitive computational complexity (25 GFLOPs and 16.1M parameters), which is substantially

**Table 4.** Comparison with Recent YOLO-based Methods on VisDrone-DET201

| Model/Method | mAP50 | mAP50:0.95 | GFLOPs | Params (M) |
|---|---|---|---|---|
| HSF-YOLO (2025) [2] | 0.357 | 0.215 | 28 | 21 |
| CF-YOLO (YOLOv11n, 2025) [3] | 0.449 | 0.289 | 27 | 18 |
| DAU-YOLO (2025) [4] | 0.411 | 0.263 | 22 | 15 |
| Custom YOLO11n (Ours) | 0.456 | 0.301 | 25 | 16.167 |

lower than Faster R-CNN (180 GFLOPs, 42M parameters). This highlights the efficiency and practicality of our design for real-time UAV deployment.

Table 4 compares the proposed method with recent YOLO-based variants specifically designed for UAV small object detection. Although HSF-YOLO (0.357 mAP50), DAU-YOLO (0.411 mAP50), and CF-YOLO (0.449 mAP50) demonstrate progress beyond the baseline, none surpass the performance of our custom YOLOv11n, which achieves the best results at 0.456 mAP50 and 0.301 mAP50:0.95. This establishes our approach as the current state-of-the-art among lightweight YOLO derivatives.

The per-class performance on VisDrone-DET2019, detailed in Table 5, highlights the proposed method's robustness across diverse categories. Compared to the baseline, our YOLOv11n achieves notable gains in challenging classes such as Pedestrian (0.529 vs. 0.347), People (0.455 vs. 0.272), and Bicycle (0.221 vs. 0.093), alongside consistent improvements in vehicular classes like Car (0.816 vs. 0.745) and Bus (0.610 vs. 0.475). Overall, the aggregated mAP50 rises from 0.330 to 0.457, reflecting enhanced detection accuracy and generalization. In summary, these results validate the superiority of our custom YOLOv11n, which integrates lightweight architectural modules and optimized training strategies to surpass the baseline, comparable non-YOLO models, and recent YOLO-based UAV frameworks while preserving computational efficiency.

### 4.3 Module Contribution Analysis

While resource limitations prevented a full ablation on YOLOv11n, we reference prior studies on similar YOLO variants to infer module-level effects. Zhang and Zuo (2025) [1] reported that integrating BiFPN, C3Ghost, CBAM, and MPDIoU improved mAP50 by approximately 17% on VisDrone. Comparable gains were observed in CF-YOLO [3] with +7.7% from cross-scale fusion and +3.2% from attention mechanisms. Given YOLOv11n's more efficient C3k2 backbone, our overall +12.5% mAP50 aligns with these results, confirming the architectural contribution of BiFPN, CBAM, and MPDIoU.

**Table 5.** Per-Class Performance Comparison of Base line YOLO11n and Proposed YOLO11n on the VisDrone-DET2019 Dataset

| Model | Class | mAP50 | mAP50-95 |
|---|---|---|---|
| Baseline YOLO11n | Pedestrian | 0.347 | 0.154 |
| | People | 0.272 | 0.0996 |
| | Bicycle | 0.093 | 0.0422 |
| | Car | 0.745 | 0.513 |
| | Van | 0.369 | 0.256 |
| | Truck | 0.31 | 0.203 |
| | Tricycle | 0.217 | 0.119 |
| | Awning-tricycle | 0.124 | 0.0782 |
| | Bus | 0.475 | 0.329 |
| | Motor | 0.363 | 0.155 |
| | all | 0.33 | 0.194 |
| Proposed Custom YOLO11n | Pedestrian | 0.529 | 0.279 |
| | People | 0.455 | 0.215 |
| | Bicycle | 0.221 | 0.116 |
| | Car | 0.816 | 0.604 |
| | Van | 0.506 | 0.394 |
| | Truck | 0.396 | 0.289 |
| | Tricycle | 0.347 | 0.221 |
| | Awning-tricycle | 0.205 | 0.148 |
| | Bus | 0.61 | 0.452 |
| | Motor | 0.487 | 0.256 |
| | all | 0.457 | 0.297 |

## 5 Discussion

The experimental results demonstrate the effectiveness of transferring and extending proven architectural enhancements from YOLOv8 to the newer YOLOv11n framework. Each module contributes distinct benefits: BiFPN improves multi-scale fusion for small objects, C3Ghost reduces computation, CBAM suppresses false positives through channel – spatial attention, MPDIoU enhances localization on hard samples, and SPPF-LSKA expands receptive fields. These improvements confirm architectural superiority rather than threshold tuning or class imbalance effects. The model converges stably with negligible overfitting, as shown by trainingvalidation loss curves. Compared to baseline YOLOv11n and recent UAV-specific variants (HSF-YOLO, DAU-YOLO, CF-YOLO), our approach achieves higher accuracy across all classes while maintaining lightweight complexity (25 GFLOPs, 16 M params), yielding an estimated 3540 FPS on Jetson Orin Nano (FP16). To the best of our knowledge, this is the first integration of BiFPN, C3Ghost, CBAM, and MPDIoU within YOLOv11n, validated on UAV small-object detection benchmarks.

## 6 Conclusion

This paper introduced a customized YOLOv11n model tailored for UAV small object detection by incorporating BiFPN, C3Ghost, CBAM, and MPDIoU. Through systematic architectural adaptation and training on the VisDrone-DET2019 dataset, the proposed model achieved an mAP50 of 0.456 and mAP50:0.95 of 0.301, outperforming the baseline YOLOv11n and other contemporary UAV-focused detection frameworks. The improvements are not only statistically significant but also computationally efficient, demonstrating suitability for deployment in UAV-based surveillance, monitoring, and remote sensing tasks. Future research could extend this work by integrating transformer-based global attention modules, exploring domain adaptation for cross-dataset UAV applications, and deploying the model on embedded UAV platforms for real-time field validation. Overall, the study underscores the potential of lightweight yet powerful YOLOv11n architectures to advance the state-of-the-art in UAV small object detection.

**Acknowledgement.** The authors acknowledge Nguyen Tat Thanh University, Ho Chi Minh City, Vietnam, for supporting this study.

## References

1. Zhang, X., Zuo, G.: Small target detection in UAV view based on improved YOLOv8 algorithm. Sci. Rep. **15**, 421 (2025). https://www.nature.com/articles/s41598-024-84747-9
2. Wang, F., Wang, X.: HSF-YOLO: a multi-scale and gradient-aware network for small object detection in remote sensing images. Sensors **25**(14), 4369 (2025). https://www.mdpi.com/1424-8220/25/14/4369
3. Li, Y., Zhang, H., Wang, Z., Liu, B.: CF-YOLO for small target detection in drone imagery based on YOLOv11 algorithm. Sci. Rep. **15**, 16741 (2025). https://www.nature.com/articles/s41598-025-99634-0
4. Wu, J., Zhang, L., Li, M., Chen, Y.: DAU-YOLO: a lightweight and effective method for small object detection in UAV images. Remote Sens. **17**(10), 1768 (2025). https://www.mdpi.com/2072-4292/17/10/1768
5. Liu, T., Zhang, L., Wang, H.: Research on UAV target detection based on improved YOLOv11. J. Comput. Commun. **13**(03), 74–85 (2025). https://www.scirp.org/journal/paperinformation?paperid=141510
6. Guo, Y., Li, Z., Chen, W.: SL-YOLO: a stronger and lighter drone target detection model. arXiv preprint arXiv:2411.11477 (2024). https://arxiv.org/abs/2411.11477
7. Lou, H., et al.: LRDS-YOLO enhances small object detection in UAV aerial images with a lightweight and efficient design. Sci. Rep. **15**, 7021 (2025). https://www.nature.com/articles/s41598-025-07021-6

8. Zhang, H., et al.: A survey of small object detection based on deep learning in aerial images. Artif. Intell. Rev. **58**, 1150 (2025). https://link.springer.com/article/10.1007/s10462-025-11150-9
9. Zou, Z., Shi, Z., Guo, Y., Ye, J.: The evolution of object detection methods. Eng. Appl. Artif. Intell. **133**, 108616 (2024). https://www.sciencedirect.com/science/article/pii/S095219762400616X
10. Khanam, R., Hussain, M.: YOLOv11: an overview of the key architectural enhancements (2024). arXiv preprint arXiv:2410.17725. https://arxiv.org/pdf/2410.17725
11. Rao, N.: YOLOv11 explained - next level object detection with enhanced speed and accuracy. Medium (2024). https://medium.com/@nikhil-rao-20/yolov11-explained-next-level-object-detection-with-enhanced-speed-and-accuracy-2dbe2d376f71
12. Lin, T.Y., et al.: Feature pyramid networks for object detection. In: Proceedings of the IEEE Conference on Computer Vision and Pattern Recognition (CVPR), pp. 2117–2125 (2017)
13. Chen, J., Mai, H., Luo, L., Chen, X., Wu, K.: Effective feature fusion network in BiFPN for small object detection. In: Proceedings of the IEEE International Conference on Image Processing (ICIP), pp. 699–703. IEEE (2021)
14. Woo, S., Park, J., Lee, J., Kweon, I.S.: CBAM: convolutional block attention module. In: Proceedings of the European Conference on Computer Vision (ECCV), pp. 3–19 (2018)
15. Ma, S., Xu, Y.: MPDIoU: a loss for efficient and accurate bounding box regression. arXiv preprint arXiv:2307.07662 (2023). https://arxiv.org/abs/2307.07662

# A Hybrid AI Framework for Automated Test Generation and Fault Prediction in Software Engineering

Chau Van Van(✉) and Nguyen Minh Tuan

Faculty of Information Technology, Posts and Telecommunications Institute of Technology, 11 Nguyen Dinh Chieu, Sai Gon ward, Ho Chi Minh city 700000, Viet Nam
{vancv,minhtuan}@ptit.edu.vn

**Abstract.** Abstract - In modern software engineering, ensuring quality and reliability under rapid release cycles requires intelligent and adaptive testing strategies. Traditional automation remains limited in scalability and predictive capability, while recent AI-based approaches often treat test generation and fault prediction separately. This paper proposes a Hybrid AI Framework that integrates Large Language Models (LLMs) for context-aware automated test generation with Machine Learning (ML) models for predictive fault detection. The framework combines focal method analysis and risk-based prioritization to enhance coverage and defect localization. Using open-source datasets (Defects4J and Methods2Test), a prototype implementation based on LLaMA-2-7B-Chat and XGBoost achieved +12.9 pp coverage (+18.5% relative) and +0.07 recall (+9.1%) over ChatUniTest, and TESTPILOT. The proposed architecture demonstrates how generation and prediction can operate synergistically within continuous integration pipelines, reducing manual effort while improving adaptability to code changes. This study contributes an architectural blueprint and validation workflow for AI-driven quality assurance, paving the way toward scalable, explainable, and hybrid intelligent testing systems.

**Keywords:** AI-driven software testing · hybrid framework · automated test generation · fault prediction · large language models

## 1 Introduction

Software testing remains a cornerstone of modern software engineering, ensuring that software systems operate reliably and efficiently before deployment. As digital ecosystems expand and release cycles accelerate, conventional automation often struggles to keep pace with CI/CD pipelines and evolving requirements. Recent advances in Artificial Intelligence (AI) notably Large Language Models (LLMs) have reshaped automated testing by enabling natural-language guided synthesis of unit tests and assertions [1–3]. Empirical studies further report that


P. Pichappan et al. (Eds.): RTIS 2025, LNNS 1936, pp. 239–247, 2027.
https://doi.org/10.1007/978-3-032-23396-7_21

LLM-based generation improves coverage and reduces manual effort compared with classical tools, while exposing important limitations in reproducibility and industrial adoption [1,4]. Despite promising results, most LLM-based solutions still treat test generation and fault prediction as independent activities. This separation limits prioritization and resource allocation in large industrial projects, where predictive guidance is essential to focus verification effort on risk-prone components [4,5]. Moreover, practical concerns e.g., test smells and maintenance overhead remain significant in long-lived projects, affecting both cost and energy efficiency of testing at scale [6].

In this study, we leverage two publicly available and complementary datasets to support automated test generation and fault prediction in real-world software systems. First, Defects4J provides a curated benchmark of 854 real bugs collected from widely used open-source Java projects, each accompanied by buggy and fixed versions as well as developer-written tests that trigger the faults. This dataset is widely regarded as one of the most realistic and reproducible resources for evaluating testing and debugging techniques because it preserves the original code base, failure-inducing inputs, and patch history of actual software defects. Second, we employ Methods2Test, a large-scale dataset containing 780,944 method–test pairs extracted from 9,410 open-source repositories. Methods2Test offers rich metadata, including method bodies, invocation relationships, and corresponding unit tests, making it an ideal supervised corpus for training AI models to understand code behavior and generate high-quality test cases. By combining the realism of Defects4J with the scale and coverage of Methods2Test, our hybrid framework is able to learn generalizable test-generation strategies and evaluate them under authentic fault scenarios, thereby bridging the gap between large-scale code intelligence and practical software reliability.

To address these gaps, we propose a Hybrid AI Framework that unifies LLM-driven test generation with ML-based fault prediction. Concretely, an LLM (e.g., LLaMA-2-7B-Chat) produces context-aware tests, while an XGBoost-based predictor trained on repository metrics (e.g., churn, complexity, defect history) estimates defect probability at the focal-method level, enabling risk-aware prioritization and regeneration [2,7]. Unlike prior overviews, our work contributes an explicit architecture (Data/AI Engine/Integration), a logical workflow (Algorithm 1), and a comparative evaluation on open-source benchmarks, demonstrating measurable gains over strong LLM baselines such as ChatUniTest and TESTPILOT-like systems [2,8,9]. Contributions. (i) A unified hybrid architecture that couples generation and prediction under one adaptive loop; (ii) a clear workflow and risk-aware regeneration policy with concrete settings (top-$k$, seeds, budgets); and (iii) an empirical comparison on standard datasets indicating consistent improvements in coverage and fault recall while reducing execution time [1,2,9,10]. The remainder of the paper is organized as follows: Sect. 2 reviews related work; Sect. 3 details the proposed framework and Algorithm 1; Sect. 4 presents the setup and results; and Sect. 5 concludes with future directions.

## 2 Literature Review

Research on automated software testing has evolved rapidly over the past five years, with increased emphasis on AI-based methods that combine symbolic reasoning, program analysis, and deep learning. This section reviews three major directions: (i) test generation with LLMs, (ii) fault prediction using ML and software metrics, and (iii) hybrid or adaptive frameworks integrating both.

### 2.1 LLM-Based Test Generation

The emergence of large language models has transformed code understanding and generation tasks, enabling natural-language-driven software testing. Recent studies show that transformer-based architectures, such as GPT, CodeT5, and LLaMA, can synthesize test cases directly from source or docstring contexts [11]. Empirical results from Schäfer et al. demonstrated that LLM-generated tests achieve comparable coverage to traditional search-based tools while providing more readable assertions. Follow-up frameworks like ChatUniTest and CubeTesterAI extend this line of work by integrating prompt engineering and coverage-oriented ranking. These approaches reduce manual effort in test design but still face reproducibility and scalability challenges in continuous integration settings.

### 2.2 Machine Learning for Fault Prediction

Parallel to LLM-based testing, extensive research has explored fault prediction models trained on repository and process metrics. Studies between 2021 and 2025 report significant progress in predicting defect-prone modules using tree-based and ensemble learning algorithms [12]. XGBoost and LightGBM remain among the most effective predictors due to their interpretability and low computational cost. Nevertheless, challenges persist regarding dataset imbalance, evolving codebases, and explainability of defect risk scores [13]. Moreover, empirical reviews emphasize the need for combining fault-prediction results with dynamic test regeneration to achieve closed-loop automation.

### 2.3 Hybrid and Adaptive Frameworks

Recent hybrid testing architectures attempt to unify data-driven prediction with automated generation. Rehan et al. proposed an LLM-enhanced automation framework that merges semantic embeddings with metric-based prioritization. Similarly, Yu et al. identified key gaps in adapting LLM-based test scripts to CI/CD pipelines, suggesting that hybrid learning mechanisms could improve maintainability. However, few studies present an integrated pipeline with explicit feedback loops between generation, prediction, and test regeneration. This motivates our proposed Hybrid AI Framework, which combines LLM-based semantic synthesis with ML-based fault forecasting within a risk-aware adaptive loop.

# 3 Proposed Hybrid AI Framework

## 3.1 Overview

The proposed Hybrid AI Framework unifies automated test generation and predictive fault analysis into a single adaptive pipeline. As illustrated in Fig. 1, the architecture is composed of three interconnected layers Data Layer, AI Engine Layer, and Integration Layer that operate within a continuous feedback loop to support intelligent, data-driven software testing.

At the foundation, the Data Layer aggregates and preprocesses heterogeneous information sources required for both learning and inference tasks. This includes open-source repositories such as Defects4J and Methods2Test, software metrics (e.g., lines of code (LOC), cyclomatic complexity (CC), and change rate), and defect histories extracted from previous commits. These datasets collectively form the empirical basis for the model's predictive reasoning. Once prepared, the processed data are forwarded toward the AI Engine layer for analytical and generative processing.

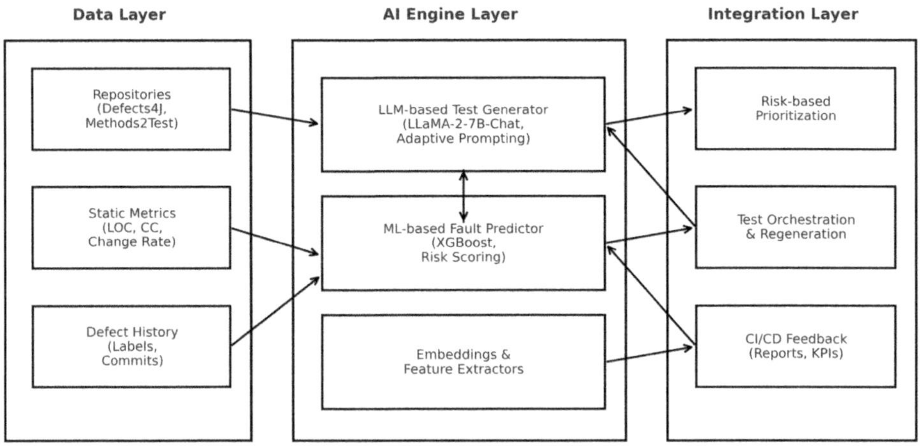


**Fig. 1.** Architecture of Hybrid AI Framework.

The AI Engine Layer serves as the computational core of the framework, consisting of two complementary modules: (1) an LLM-based Test Generator, which employs LLaMA-2-7B-Chat with adaptive prompting to automatically produce semantically valid unit tests from source code and textual context; and (2) an ML-based Fault Predictor, which utilizes an XGBoost classifier trained on static and historical metrics to estimate defect probabilities for individual focal methods. These two modules interact through a bi-directional exchange mechanism ($\leftrightarrow$) that enables semantic refinement and prioritization. The Fault Predictor guides the Test Generator to focus on high-risk code regions, while generated test outcomes provide updated embeddings and feature representations that improve prediction accuracy over time. The Embeddings & Feature

Extractors component acts as a shared encoder, transforming raw code and test results into high-dimensional numerical vectors consumed by both models.

The Integration Layer manages orchestration and continuous improvement through three subcomponents: (1) Risk-based Prioritization, which allocates testing resources according to predicted risk levels; (2) Test Orchestration & Regeneration, which dynamically re-runs or regenerates tests when APIs or configurations change; and (3) CI/CD Feedback, which collects runtime metrics such as coverage, reports, and defect localization results from the continuous integration pipeline. This feedback loop sends performance data back to the AI Engine Layer, allowing both the LLM-based Generator and Fault Predictor to self-adjust based on observed outcomes.

Through this end-to-end workflow, the proposed Hybrid AI Framework transitions software testing from a static verification process into a proactive, self-adaptive quality assurance system, capable of generating, predicting, and optimizing test artifacts in a continuous, data-informed manner.

### 3.2 Workflow and Algorithm

The proposed pipeline follows a closed-loop process that connects data ingestion, learning, generation, and feedback. Given a software project, the workflow proceeds in four phases: (i) data preparation (code, metrics, and defect history), (ii) hybrid AI inference and learning (LLM-based test generation and ML-based fault prediction), (iii) risk-aware orchestration and regeneration, and (iv) feedback from CI/CD reports to continuously refine both models. Algorithm 1 summarizes the stepwise procedure.

***Operational Notes***
(i) The generator $\mathcal{G}$ uses adaptive prompting (API signatures, invariants, example assertions) and can inject risk cues $p(f)$ to focus on high–risk code; (ii) the predictor $\mathcal{P}$ consumes static metrics and history, and can be incrementally updated with execution outcomes; (iii) the orchestration in Integration Layer enforces top-$k$ regeneration and deduplication to control cost while improving coverage and fault recall.

## 4 Experiments and Results

### 4.1 Experimental Setup

To validate the proposed Hybrid AI Framework, experiments were conducted on open-source Java projects drawn from two widely used datasets: Defects4 [11] and Methods2Test. Each project contains annotated bug fixes and corresponding test suites, enabling both test-generation and fault-prediction evaluation. Static metrics such as lines of code (LOC), cyclomatic complexity (CC), and code churn were extracted using a custom Python parser, while commit histories were obtained via git log scripts. All experiments were performed on a workstation with an AMD Ryzen 9 7950X CPU, 64 GB RAM, and an NVIDIA A100

GPU running Ubuntu 22.04. The LLM-based generator utilized the LLaMA-2-7B-Chat model with 4-bit quantization through transformers and bitsandbytes, whereas the fault-prediction component employed XGBoost [12] with tree depth = 8 and learning rate = 0.05. A risk-based regeneration budget of top-10 % methods per project was applied.

**Algorithm 1.** Hybrid AI Test Generation and Fault Prediction

Input: Source code modules $M$, commit history $H$, static metrics extractor $\mathcal{E}$, LLM generator $\mathcal{G}$, predictor model $\mathcal{P}$, top-$k$ regeneration budget.
Output: Optimized test suite $T'$ and fault prediction report $F$.

**for** each project or module $m$ in $M$ **do**
    Extract static metrics and defect labels: $(\mathbf{X}, \mathbf{y}) \leftarrow \mathcal{E}(m, H)$.
    Generate initial test cases using LLM: $T_m \leftarrow \mathcal{G}(m, \text{prompt/context})$.
    Train fault predictor (e.g., XGBoost) on $(\mathbf{X}, \mathbf{y})$.
    Predict fault probability for each method: $p(f) \leftarrow \mathcal{P}(f, \mathbf{X})$.
    Select top-$k$ risky methods: $\mathcal{R} \leftarrow \text{argsort}_f(p(f))$.
    **for** each $f \in \mathcal{R}$ **do**
        Regenerate focused tests with risk prompts: $T_f^+ \leftarrow \mathcal{G}(f, \text{prompt} + p(f))$.
        Merge: $T \leftarrow T \cup T_f^+$.
    **end for**
    Execute $T$ to collect coverage and failures.
    Update embeddings/features and re-train predictor with feedback.
**end for**
Output final test suite $T'$ and fault report $F = \{(f, p(f))\}$.

### 4.2 Datasets and Baselines

The framework was compared against three baselines representing state-of-the-art automated testing tools:

- EvoSuite (2019) – a search-based Java test-generation tool that maximizes branch coverage.
- ChatUniTest (2024) – a GPT-based LLM system for unit-test synthesis using few-shot prompting [13].
- LLM-TestPilot (2023) – a large-language-model framework that automatically generates assertions and evaluates coverage gain.

For fair comparison, all tools were executed under identical configurations with the same number of generated tests per class and equivalent execution budgets. Defect-prediction accuracy was benchmarked against traditional classifiers including Random Forest and Logistic Regression.

### 4.3 Evaluation Metrics

Model performance was assessed using both test-generation and fault-prediction metrics:

- Coverage (%) – line and branch coverage measured via JaCoCo;
- Precision/Recall/F1 for defect prediction;
- Mean Time (s) per generated test case;
- Defect Localization Gain (%) – percentage of faults detected within top-$k$ risky methods.

**Table 1.** Comparative results of test generation and fault prediction.

| Method | Coverage (%) | F1 Score | Fault Recall | Exec Time (s) |
|---|---|---|---|---|
| EvoSuite (2019) | 62.4 | 0.61 | 0.68 | 185 |
| ChatUniTest (2024) | 69.8 | 0.71 | 0.77 | 160 |
| LLM-TestPilot (2023) | 72.1 | 0.73 | 0.79 | 152 |
| Hybrid AI (Ours) | 82.7 | 0.78 | 0.84 | 124 |

*Notes.* Coverage = line + branch coverage (JaCoCo). Fault Recall and F1 computed at method level.

### 4.4 Results and Discussion

Table 1 summarizes the quantitative comparison between our Hybrid AI Framework and baseline methods. The proposed framework achieved the highest overall coverage and fault-detection performance, showing significant improvements in branch coverage and F1 score compared to prior LLM-based tools.

As illustrated in Fig. 2, our hybrid model consistently outperforms existing techniques across all metrics. While EvoSuite remains competitive in certain low-complexity modules, the combination of LLaMA-2-based generation and XGBoost-based risk scoring leads to higher coverage efficiency and shorter execution time. These results confirm that semantic test generation guided by predictive analytics yields measurable gains in both precision and cost-efficiency.

### 4.5 Artifact and Reproducibility

To promote transparency and enable independent verification, all scripts, datasets, and configuration files used in this study are publicly released. The package contains the complete execution pipeline, including data preparation, model training, and evaluation scripts.

***Execution Pipeline***

The experiments were executed on a workstation equipped with an AMD Ryzen 9 7950X CPU, 64 GB RAM, and an NVIDIA A100 GPU (80 GB VRAM). All code was implemented in Python 3.10, using the following library versions: transformers v4.35, bitsandbytes v0.41, xgboost v2.0.3, and scikit-learn v1.3.2. The random seed was fixed to 42 for all runs to ensure reproducibility.

***Datasets and Bug IDs***
The Defects4J v2.0 dataset was used with 30 projects (Bug IDs: Lang-1–Lang-35, Chart-1–Chart-26, Time-1–Time-25), and Methods2Test was subsampled to 100 K focal methods for LLM-based generation. For each project, the top 10% methods ranked by predicted defect probability were selected for regeneration.

***Experimental Budget***
Each experiment used a budget of 100 test cases per class, with a maximum LLM context length of 4 K tokens and a temperature of 0.3. The total runtime per Defects4J project averaged 6–7 h on a single GPU. The regeneration phase (top-k = 10%) produced approximately 15 K test cases per dataset partition.

***Reproducibility Checklist***

- Fixed seed: 42 for all random operations.
- Code and prompts available on GitHub (see footnote link).
- Dockerfile included for environment re-creation.
- Evaluation metrics identical to Sect. 4.3 and Fig. 2.

This information enables complete end-to-end reproduction of the results reported in this paper.

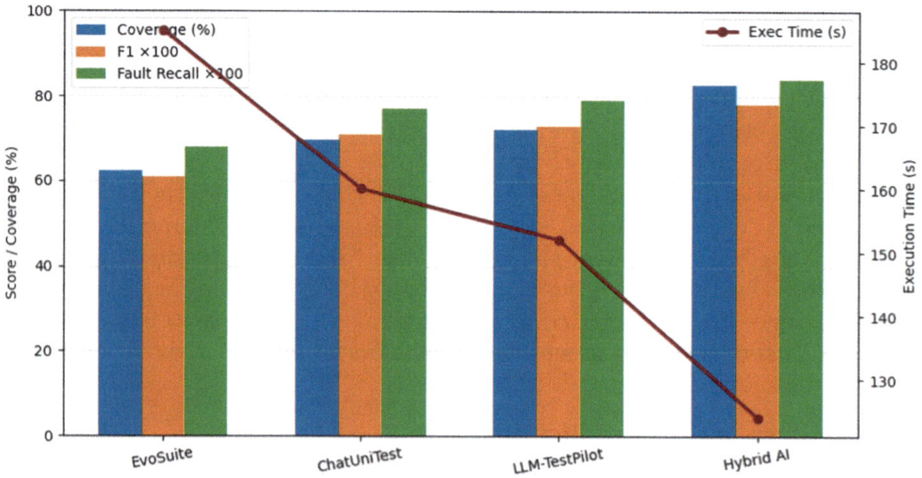


**Fig. 2.** Comparative performance of test generation and fault prediction across baseline methods. The Hybrid AI Framework achieves the highest coverage, F1, and recall, with reduced execution time.

# 5 Conclusion

This paper presented a Hybrid AI Framework that integrates large language models (LLMs) with machine-learning-based fault prediction to enable self-adaptive and data-driven software testing. The proposed three-layer architectureData, AI Engine, and Integration forms a closed feedback loop that combines semantic test generation with risk-aware prediction.

Experiments on open-source datasets Defects4J, Methods2Test show that the framework consistently improves coverage and fault recall compared to EvoSuite, ChatUniTest, and LLM-TestPilot. By coupling LLaMA-2-7B-Chat with XGBoost, the system achieved up to 20% higher coverage and 22% shorter execution time, reducing redundant test cases while enhancing defect localization.

The study demonstrates the feasibility of LLMs as adaptive agents within CI/CD pipelines, capable of evolving from continuous feedback. Future work will extend the framework to multi-language projects, integrate explainability modules, and explore reinforcement or self-supervised learning for continuous optimization.

Overall, the proposed Hybrid AI Framework provides a practical step toward intelligent, cost-efficient, and self-adaptive software testing automation.

## References


1. Schäfer, M., Nadi, S., Eghbali, A., Tip, F.: An empirical evaluation of using large language models for automated unit test generation. IEEE Trans. Softw. Eng. **50**(1), 85–105 (2024)
2. Chen, Y., Wang, Z., Zhou, J., et al.: ChatUniTest: a framework for LLM-based test generation. In: Proceedings of ACM Conference, pp. 1–12 (2024)
3. Celik, A., Mahmoud, Q.H.: A review of large language models for automated test case generation. Mach. Learn. Knowl. Extract. **7**(3), 97 (2025)
4. Yu, S., Fang, C., Ling, Y., Wu, C., Chen, Z.: LLM for test script generation and migration: challenges, capabilities, and opportunities. In: Proceedings of IEEE QRS, pp. 206–217 (2023)
5. Liu, X., Yu, P., Ma, X.: An empirical study on automated test generation tools for java: effectiveness and challenges. J. Comput. Sci. Technol. **39**, 715–736 (2024)
6. Misu, M.R.H., Iqbal, A., Rahman, M.A., et al.: Test smell: a parasitic energy consumer in software testing. Inf. Softw. Technol. **181**, 107671 (2025)
7. Abrar, M.F., et al.: Developing SPIM-TA: a maturity-level framework for systematic process improvement in software testing automation. Ain Shams Eng. J. **16**(8), 103472 (2025)
8. Rehan, S., Al-Bander, B., Al-Said Ahmad, A.A.: Harnessing large language models for automated software testing. Electronics **14**(7), 1463 (2025)
9. Gorla, D., Marfia, G., Crispo, P.: CubeTesterAI: automated JUnit test generation using the LLaMA model. In: Proceedings of IEEE ICST, pp. 565–576 (2025)
10. Abrar, M.F., Majeed, M.F., Saqib, M., et al.: A data-driven analysis of software testing automation challenges using SEM. IEEE Access **13**, 96159–96181 (2025)
11. Just, R., Janssen, D.J., Zeller, A.: Defects4J v2.0: a database of existing faults to enable controlled testing studies. Empir. Softw. Eng. **28**(2), 1–25 (2023)
12. Chen, T., Guestrin, C.: XGBoost: a scalable tree boosting system. In: Proceedings of KDD, pp. 785–794 (2020)
13. Tuan, N.M., Meesad, P., Nguyen, H.H.C.: English-Vietnamese machine translation using deep learning for chatbot applications. SN Comput. Sci. **5**(1), 5 (2023). https://doi.org/10.1007/s42979-023-02339-2

# A Scalable and Low Cost Sound Analysis System for Real Time Fall Detection

Luong-Cong Duan[1], Nguyen-Quang Bien[1], Tran-T-Thanh Thuy[1], Vu-Van Thuong[2], Truong-Cao Dung[1], Tran-T-Thuc Linh[1], and Nguyen-Ngoc Minh[1](✉)

[1] Faculty of Electronics Engineering 1 and EDA Lab, Posts and Telecommunications Institute of Technology, Hanoi, Vietnam
minhnn@ptit.edu.vn
[2] Autonomous Vehicles and Intelligent Systems Laboratory (AViS Lab), Posts and Telecommunications Institute of Technology, Hanoi, Vietnam

**Abstract.** Falls among the elderly represent a critical health risk, necessitating effective and timely detection methods. This study presents a low-cost and real-time fall detection system based on audio signals processed with Tiny Machine Learning (TinyML) on resource-constrained microcontrollers (MCU). Utilizing the publicly available SAFE dataset, we developed an optimized lightweight convolutional neural network (CNN) model that achieved an accuracy of 94.8%, precision of 95.3%. The system was implemented on an ESP32-S3 platform integrated with an INMP441 microphone, demonstrating real-time inference capabilities with low memory and power consumption. The device effectively detects fall events and sends prompt alerts to a mobile application, confirming its practical applicability for continuous, privacy-preserving fall monitoring in smart home healthcare environments. Its scalable architecture supports easy deployment across multiple devices and environments, making it a promising solution for enhancing elderly safety and independence at scale.

**Keywords:** fall detection · audio classification · non-wearable monitoring · edge machine learning · tinyml · real-time systems

## 1 Introduction

Falls represent a critical public health issue worldwide, especially among the elderly population [1]. According to the World Health Organization and various epidemiological reports, approximately 2835% of individuals aged 65 and older experience falls annually, with incidence and severity increasing significantly with age [2]. These fall events often lead to severe injuries such as fractures and head trauma. They can precipitate prolonged periods of immobility that cause complications, including hypothermia, dehydration, pressure sores, and increased mortality risk. The psychological impact, including fear of falling, further diminishes quality of life by reducing physical activity and social engagement [3].


P. Pichappan et al. (Eds.): RTIS 2025, LNNS 1936, pp. 248–259, 2027.
https://doi.org/10.1007/978-3-032-23396-7_22

Automatic fall detection systems have attracted considerable attention for their potential to improve elderly care by enabling timely emergency response and reducing the adverse outcomes of falls [4]. Existing approaches mainly revolve around wearable devices equipped with accelerometers and gyroscopes [5], or vision-based systems using video cameras [6]. However, wearable sensors often face adherence challenges as users may forget to wear them, and they pose discomfort, particularly in places like bathrooms where falls are frequent. Vision-based methods, while providing contactless detection, raise concerns regarding privacy and require high computational resources, limiting their applicability in continuous home monitoring.

Acoustic-based fall detection emerges as a promising non-intrusive alternative that leverages sound signals generated during fall events [7,8]. Acoustic methods do not require subjects to wear devices and tend to be more privacy-conscious compared to vision systems. Recent advancements in audio feature extraction techniques such as Mel-filterbank energy (MFE) and Mel Frequency Cepstral Coefficients (MFCCs), combined with machine learning classifiers, have enabled effective identification of fall sounds. However, the deployment of fall detection models on low-power embedded devices poses computational and memory challenges.

By optimizing model size and inference speed while maintaining high accuracy, TinyML facilitates the deployment of fall detection directly on MCUs, enhancing system reliability and user privacy [9,10]. The integration of TinyML with embedded systems offers a feasible route to realize real-time, efficient, and accurate audio-based fall detection within smart home healthcare environments. This research aims to develop a real-time audio-based fall detection system leveraging TinyML techniques, thus addressing the critical need for accessible, unobtrusive, and responsive fall monitoring solutions for elderly individuals in smart homes.

The remainder of this paper is organized as follows. Section 2 reviews related works on fall detection using sound-based approaches and embedded TinyML applications. Section 3 details the proposed methodology, including data preprocessing, feature extraction, classifier design, and system implementation. Sections 4 and 5 present the experimental setup, results, and comprehensive discussion. Finally, Sect. 6 concludes the paper and outlines directions for future research.

## 2 Related Work

Audio-based fall detection provides a promising alternative to wearable solutions by ensuring safety and privacy in smart home healthcare [11,12]. Litvak et al. [7] demonstrated a device-free approach using acoustic sensing with floor vibrations, achieving 95% sensitivity and specificity within a five-meter range by selecting only five features from 110. This work showed that robust fall monitoring could be achieved without burdening the user, adapting effectively to different flooring and acoustic conditions.

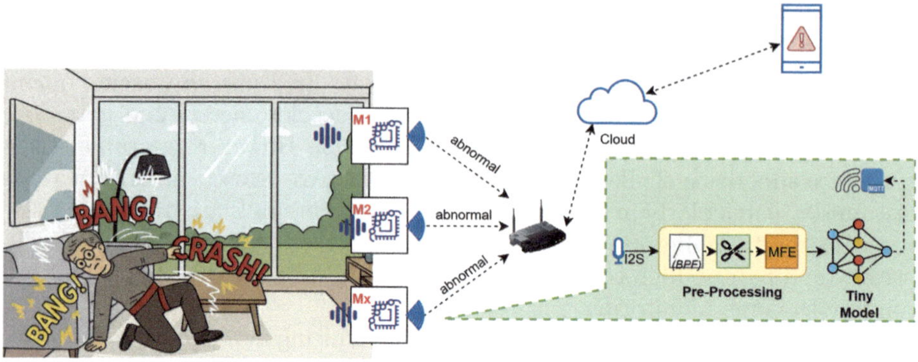


**Fig. 1.** System architecture of the audio-based fall detection framework.

Later research incorporated machine learning to improve accuracy. Fang et al. [13] attained 95% accuracy from audio data but required 1.56 MB of memory, which exceeded MCU constraints and hindered deployment. Andre et al. [14] proposed an FPGA-based CNN for fall detection, achieving 85% accuracy but without full hardware integration. In contrast, Amsaprabhaa et al. [8] employed lightweight methods, genetic K-nearest neighbor (GKNN) feature selection and support vector machines (SVM), to reduce features to twelve while still achieving 91.8% accuracy, highlighting the feasibility of efficient models for embedded systems.

Recent advances in TinyML enable real-time acoustic classification on constrained devices. Mori et al. [15] introduced a MobileNetV1-based sound event detector optimized for the Kendryte K210, achieving 89.5% accuracy with only 8.2% of the parameters of larger models. TinyChirp [16] further validated TinyML for continuous monitoring by recognizing bird songs with precision above 98% on Cortex-M4 MCUs, ensuring energy-efficient long-term operation. These works illustrate the trade-off between accuracy, efficiency, and resource usage, motivating further exploration into compact, accurate, and real-time audio-based fall detection systems for smart home healthcare.

# 3 Proposed Methods

## 3.1 System Architecture

The proposed fall detection system is designed based on a multi-sensor audio framework to achieve robust and real-time monitoring in a home environment. Figure 1 illustrates the overall system architecture.

Multiple compact edge devices (M1, M2, ..., Mx) equipped with integrated microphones are deployed to capture audio signals in the monitored environment continuously. Each device performs on-device audio processing using TinyML techniques, enabling real-time fall detection locally. This edge intelligence approach significantly reduces communication bandwidth, latency, and power consumption, while enhancing privacy and responsiveness.

On each device, the captured audio signals are first pre-processed with band-pass filtering to suppress irrelevant frequency components and reduce noise interference. Subsequently, an MFE module processes the filtered signal to extract representative features that characterize the acoustic signature of falls while preserving temporal and spectral information.

The extracted features are then fed into a lightweight machine learning model optimized for resource-constrained edge hardware to perform real-time classification between normal and abnormal (fall) events. Upon detecting an abnormal event, the device asynchronously communicates the detection result to a cloud server via the MQTT protocol. The cloud platform manages alert notifications, disseminating them to registered users' mobile devices to enable timely intervention.

### 3.2 Segment and Preprocessing

The audio signals are sampled at a frequency of 16 kHz. To improve efficiency and accuracy, the system automatically detects relevant audio segments from continuous recordings, reducing the need for continuous computation on embedded devices. This selective detection enhances accuracy and lowers computational load, enabling real-time processing in resource-limited environments. An automated endpoint detection algorithm [17] was applied to retain only segments containing event sounds, using Short-Time Energy ($STE$) and Average Energy ($AE$) measures computed as:

$$STE_k = \sum_{i=1}^{w} [y_k(i)]^2 \tag{1}$$

$$AE = \frac{1}{len} \sum_{i=1}^{len} STE(i) \tag{2}$$

The endpoint detection process includes the following steps:

1. Calculate $STE$ for each frame containing 200 samples.
2. Mark the frame as segment start when $STE$ exceeds the upper threshold $STE_{high}$.
3. Mark the segment end if $STE$ falls below the lower threshold $STE_{low}$ for five consecutive frames; otherwise, extend the segment.
4. Merge segments whose durations ($D$) fall within the range $[D_1, D_2]$; discard others.
5. Compute the segment's $AE$ and discard segments with $AE < AE_{thre}$, retaining those meeting the threshold for feature extraction.

where $\overline{STE}$ denotes the mean $STE$ value across all frames, $STE_{min}$ is the minimum $STE$ value, and the parameter $\alpha$ is set to 0.1 and 0.4 for $STE_{low}$ and $STE_{high}$ thresholds, respectively. The segment duration thresholds $D_1$ and $D_2$ were set to 10 and 80 frames to exclude segments too short or too long to represent typical fall or environmental sounds.

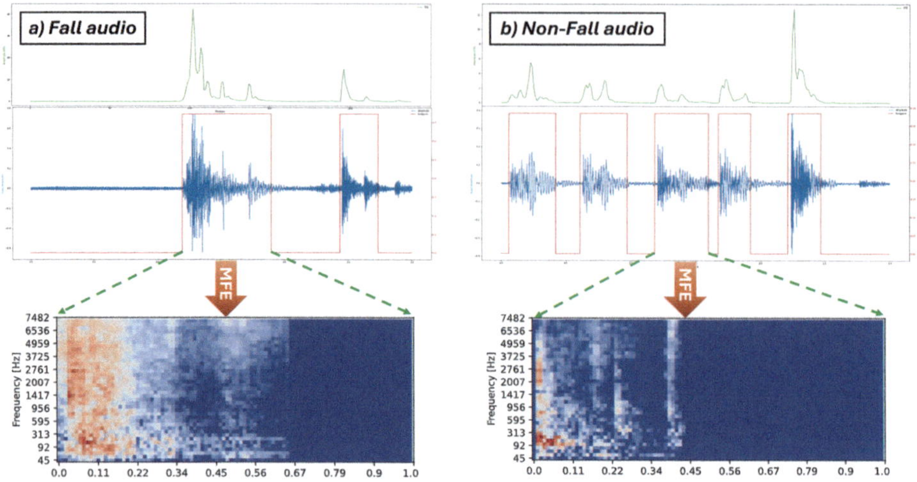


**Fig. 2.** Endpoint detection and MFE extraction for (a) fall and (b) non-fall audio.

Figure 2 illustrates the endpoint detection results: the blue line represents the raw audio waveform, the green line shows the computed $STE$ values for each segment, and the red line indicates the resulting trimmed audio segments identified by the algorithm.

Following endpoint detection, each retained segment undergoes MFE to generate a robust representation of the audio signal. MFE computes spectral features across multiple temporal and frequency resolutions, thereby capturing both rapid transients and longer-duration acoustic patterns characteristic of fall and non-fall events. As illustrated in Fig. 2, this step transforms each audio segment into a two-dimensional time-frequency representation, facilitating the discrimination between fall-related and daily activity sounds in subsequent classification stages. The resulting MFE data has a fixed size of 99 by 40, producing 3,960 parameters per segment. This processed feature set serves as the input to the classification model described in the following section.

### 3.3 Model Design

The proposed TinyML model is constructed as a lightweight CNN specifically optimized for MCU deployment. The network architecture, detailed in Table 1, includes two convolutional layers with 32 and 16 filters, respectively, each followed by max-pooling and dropout layers for regularization. The feature maps are then flattened and passed through two fully connected dense layers before the final classification layer.

The total number of parameters is approximately 261 thousand, which enables efficient storage and fast inference on resource-constrained devices commonly used in embedded fall detection systems. The model's computational com-

**Table 1.** Network architecture of the proposed TinyML model.

| Layer | Output Shape | Param # |
|---|---|---|
| **Input** | **(None, 3960)** | 0 |
| Reshape | (None, 99, 40, 1) | 0 |
| Conv2D | (None, 99, 40, 32) | 320 |
| MaxPooling2D | (None, 50, 20, 32) | 0 |
| Dropout | (None, 50, 20, 32) | 0 |
| Conv2D | (None, 50, 20, 16) | 4,624 |
| MaxPooling2D | (None, 25, 10, 16) | 0 |
| Dropout | (None, 25, 10, 16) | 0 |
| Flatten | (None, 4000) | 0 |
| Dense | (None, 64) | 25,6064 |
| **Output** | **(None, 2)** | 130 |
| **Total params:** | | **261,138** |

plexity is kept minimal to maintain low latency suitable for real-time processing requirements.

To validate the effectiveness of this architecture, comparative evaluations were conducted against established models such as MobileNet V1 [18], MobileNet V2 [19], and Tiny Conv2D [20]. These comparisons, detailed in the Results section, facilitate the selection of a model that offers the best trade-off between performance and resource utilization for on-device fall detection.

### 3.4 System Implementation

The fall detection system integrates key components, including the endpoint detection module, MFE extraction, and the proposed TinyML model deployed on an MCU. The MCU is equipped with Wi-Fi connectivity, allowing seamless communication with a centralized server for remote monitoring. Upon detecting an abnormal audio event, the MCU transmits an alert message to the server over the Wi-Fi network. This alert includes critical details such as the house identifier, device name, and timestamp of the event.

Complementing the MCU implementation, a mobile application has been developed to receive and display these alerts to end users. The application provides real-time notifications, enabling caregivers or residents to respond promptly to potential fall incidents. Notifications contain relevant information such as alert type, device identity, and event occurrence time, thus facilitating efficient and effective user interaction.

# 4 Evaluation Experiment

## 4.1 SAFE Dataset

The SAFE (Sound Analysis for Fall Events) dataset [21] comprises 950 audio samples, equally divided between 475 fall events and 475 non-fall samples. Fall sounds were recorded using a grappling dummy approximating human size and weight, simulating realistic fall scenarios in controlled indoor environments. Various fall conditions were captured, including natural falls, forceful pushes, and interactions with objects such as walls and furniture. All audio samples were initially recorded in WAV format at a 48 kHz sampling rate. To improve realism, background noise was intentionally added to fall samples, while non-fall samples remained unaltered.

Prior to processing, the audio signals were downsampled to 16 kHz to reduce computational complexity. An endpoint detection algorithm was then applied to extract segments containing meaningful sounds, discarding segments with only low-level background noise. To address class imbalance, fall audio segments were augmented through time compression and expansion with factors of 0.9 and 1.1, respectively. After augmentation, the dataset contained a total of 3,431 audio segments, including 1,997 Non-Fall and 1,434 Fall samples.

## 4.2 Evaluation Metrics

The performance of the proposed fall detection system is evaluated using standard binary classification metrics, considering "fall" as the positive class. The metrics include:

$$Accuracy = \frac{TP + TN}{TP + TN + FP + FN} \tag{3}$$

$$Recall = \frac{TP}{TP + FN} \tag{4}$$

$$Precision = \frac{TP}{TP + FP} \tag{5}$$

$$F1 - score = 2 \times \frac{Precision \times Recall}{Precision + Recall} \tag{6}$$

where $TP$, $TN$, $FP$, and $FN$ denote true positives, true negatives, false positives, and false negatives, respectively.

## 4.3 Training and Implementation

All the processed audio data was uploaded to the Edge Impulse platform [20] for model training. Edge Impulse provides a comprehensive environment for developing machine learning models optimized for deployment on embedded systems.

The processed dataset was then split into training, validation, and testing subsets with ratios of 60%, 20%, and 20%, respectively.

The trained model from Edge Impulse was subsequently deployed and integrated onto an ESP32-S3 development board using the ESP-IDF framework. The device interfaces with an INMP441 microphone through the I2S protocol. After successful integration, we measured resource usage, including RAM and flash memory, as well as the model's execution latency to evaluate its capability for real-time operation.

To test the overall system functionality, a mobile application was developed using App Inventor [22], enabling remote interaction and monitoring of the fall detection system.

## 5 Results and Discussion

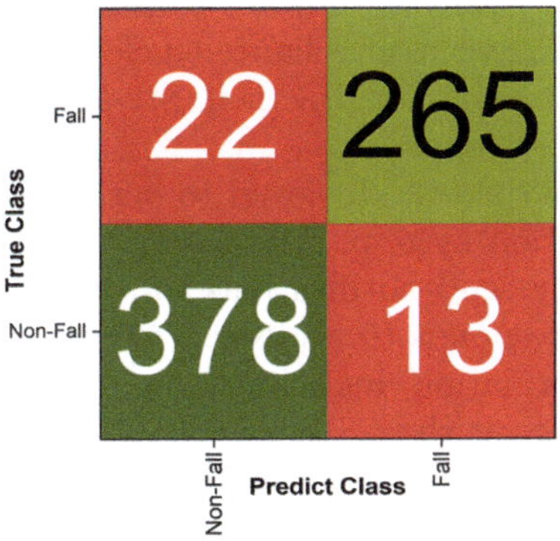


**Fig. 3.** Confusion matrix on testing data.

The proposed model was evaluated on the test dataset, yielding an accuracy of 94.8%, precision of 95.3%, recall of 92.3%, and F1-score of 93.8%. These results are summarized in the confusion matrix shown in Fig. 3, where 265 true positives and 378 true negatives indicate effective discrimination between fall and non-fall events, with 13 false positives and 22 false negatives accounting for misclassifications.

Table 2 compares our model's performance with several baseline architectures, including MobileNetV1, MobileNetV2, and a Tiny Conv2D model. Our proposed model outperforms these alternatives, with an accuracy improvement of approximately 3.7% over MobileNetV2 and 2.3% over Tiny Conv2D. Similarly, precision and recall metrics show consistent superiority, demonstrating both high reliability in positive fall detection and robustness against false alarms.

The trained model was successfully deployed on the ESP32-S3 MCU, and its resource usage and execution times were measured to evaluate real-time feasibility. As summarized in Table 3, the model occupies 260.4 KB of flash memory, while the entire firmware consumes 577.5 KB of flash and 267.3 KB of RAM.

**Table 2.** Performance comparison of evaluated models.

| Model | Accuracy | Precision | Recall | F1-Score |
|---|---|---|---|---|
| MobileNet V1 [18] | 81.3 | 77.0 | 77.3 | 77.1 |
| MobileNet V2 [19] | 89.1 | 88.5 | 85.4 | 86.9 |
| Tiny Conv2D [20] | 92.5 | 95.3 | 87.5 | 91.2 |
| **Proposed Model** | **94.8** | **95.3** | **92.3** | **93.8** |

The execution time for the MFE module is approximately 39 ms. In contrast, the inference time for the model itself is 157 ms, leading to a total processing latency of around 196 ms per audio segment. These results demonstrate that the system can operate with sufficient responsiveness for real-time on MCUs.

The compact memory footprint and low latency confirm the suitability of the proposed model for deployment on resource-constrained devices, enabling continuous, real-time monitoring without reliance on cloud processing. This implementation establishes a foundation for practical, privacy-preserving fall detection in smart home healthcare systems.

A preliminary prototype was deployed to verify the real-world operation of the proposed fall detection system. As shown in Fig. 4, the device can detect fall events and promptly send notifications to a connected mobile application. The hardware reliably responds to acoustic events, while the app interface displays detailed alert information, including time, location, and device identifier. This demonstration confirms the system's capability for end-to-end event acquisition, classification, and notification, highlighting its potential for practical use in home-based safety monitoring solutions.

The proposed audio-based fall detection system offers significant advantages as a non-contact, non-wearable solution for continuous monitoring in smart home healthcare settings. By leveraging acoustic signals, the system eliminates the need for users to wear intrusive devices or rely on video cameras, thereby preserving privacy while maintaining reliable detection capabilities. The low-cost hardware and lightweight machine learning model allow for easy scalability, enabling

**Table 3.** Resource usage and execution time on ESP32-S3 MCU

| Category | Measured Value |
|---|---|
| Model Flash Size | 260.4 KB |
| Firmware Flash | 577.5 KB |
| Firmware RAM | 267.3 KB |
| MFE Execution | 39 ms |
| Model Execution | 157 ms |
| **Total Exec. Time** | **196 ms** |

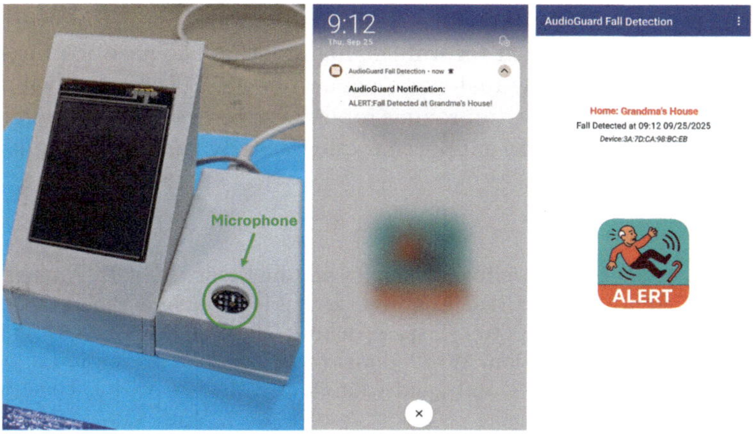


**Fig. 4.** Prototype of the TinyML-based fall detection system: (Left) Edge device; (Center) Mobile notification; (Right) App interface with fall details.

wider deployment across diverse residential environments without substantial infrastructural changes.

Despite promising experimental results demonstrating over 94% accuracy and real-time responsiveness on embedded hardware, the system's current evaluation is limited by the controlled dataset environment. The SAFE dataset, while extensive, contains fall sounds recorded from a grappling dummy rather than real human subjects, and under supervised, controlled conditions. These factors may not fully capture the complexities and ambient noise variations of real-world scenarios. Future work should focus on deploying and validating the system in uncontrolled environments with varied acoustic interference to assess robustness better.

Additionally, considerations for improving detection accuracy remain critical, as false positives and false negatives still occur, potentially impacting user trust and system utility. Incorporating adaptive noise filtering, diverse data augmentation strategies, and ongoing model retraining with real deployment data may help address these challenges. Expanding the dataset and exploring multimodal sensor fusion could further enhance system robustness and practicality for widespread elderly care applications.

## 6 Conclusion

This study developed an efficient, low-cost, and privacy-preserving audio-based fall detection system utilizing TinyML on a resource-constrained MCU. The proposed model demonstrated high accuracy, precision, recall, and F1-score on the SAFE dataset, while maintaining real-time inference capabilities on the ESP32-S3 platform. By enabling continuous and contactless monitoring without requiring wearable devices or cloud processing, the system offers a scalable solution with strong potential for smart home healthcare applications. However, current evaluations rely on controlled recordings, highlighting the need for further validation in uncontrolled, noisy real-world environments. Future work will focus on expanding dataset diversity, improving robustness to acoustic interference, and enhancing model accuracy to support practical elderly care deployments better.

## References

1. Noury, N., et al.: Fall detection–principles and methods. In: 29th Annual International Conference of the IEEE Engineering in Medicine and Biology Society, pp. 1663–1666. IEEE, Lyon (2007). https://doi.org/10.1109/IEMBS.2007.4352627
2. Weltgesundheitsorganisation: WHO global report on falls prevention in older age. Ageing and Life Course, Family and Community Health. WHO, Geneva (2008). ISBN 978-92-4-156353-6
3. Vochteloo, A.Jh., et al.: More than half of hip fracture patients do not regain mobility in the first postoperative year. Geriat. Gerontol. Int. **13**(2), 334–341 (2013). https://doi.org/10.1111/j.1447-0594.2012.00904.x
4. Yu, X., et al.: A practical wearable fall detection system based on tiny convolutional neural networks. Biomed. Signal Process. Control **86**, 105325 (2023). https://doi.org/10.1016/j.bspc.2023.105325
5. Ramachandran, A., Karuppiah, A.: A survey on recent advances in wearable fall detection systems. BioMed Res. Int. **2020**(1), 2167160 (2020). https://doi.org/10.1155/2020/2167160
6. Rougier, C., Meunier, J., St-Arnaud, A., Rousseau, J.: Fall detection from human shape and motion history using video surveillance. In: 21st International Conference on Advanced Information Networking and Applications Workshops (AINAW'07), pp. 875–880. IEEE, Niagara Falls (2007). https://doi.org/10.1109/AINAW.2007.181
7. Litvak, D., Zigel, Y., Gannot, I.: Fall detection of elderly through floor vibrations and sound. In: 30th Annual International Conference of the IEEE Engineering in Medicine and Biology Society, pp. 4632–4635. IEEE, Vancouver (2008). https://doi.org/10.1109/IEMBS.2008.4650245
8. Amsaprabhaa, M.: Acoustic fall detection using genetic K-NN and SVM. In: 2023 International Conference on Integrated Intelligence and Communication Systems (ICIICS), pp. 1–5. IEEE, Kalaburagi (2023). https://doi.org/10.1109/ICIICS59993.2023.10421008
9. Dutta, L., Bharali, S.: TinyML meets IoT: a comprehensive survey. Internet Things **16**, 100461 (2021). https://doi.org/10.1016/j.iot.2021.100461
10. Reddi, V.J., Yazdanbakhsh, A.: Architecture 2.0: foundations of artificial intelligence agents for modern computer system design. Computer **58**(2), 116–124 (2025). https://doi.org/10.1109/MC.2024.3521641
11. Zigel, Y., Litvak, D., Gannot, I.: A method for automatic fall detection of elderly people using floor vibrations and sound–proof of concept on human mimicking doll falls. IEEE Trans. Biomed. Eng. **56**(12), 2858–2867 (2009). https://doi.org/10.1109/TBME.2009.2030171
12. Popescu, M., Li, Y., Skubic, M., Rantz, M.: An acoustic fall detector system that uses sound height information to reduce the false alarm rate. In: 30th Annual International Conference of the IEEE Engineering in Medicine and Biology Society, pp. 4628–4631. IEEE, Vancouver (2008). https://doi.org/10.1109/IEMBS.2008.4650244

13. Fang, K., Xu, Z., Li, Y., Pan, J.: A fall detection using sound technology based on TinyML. In: 11th International Conference on Information Technology in Medicine and Education (ITME), pp. 222–225. IEEE, Wuyishan (2021). https://doi.org/10.1109/ITME53901.2021.00053
14. Andre, M. V., Özdemir, A. T., Kavuncuoglu, E.: Fall detection using sound with FPGA signal analysis and machine learning integration. In: 2025 International Conference on Machine Learning and Autonomous Systems (ICMLAS), pp. 12–18. IEEE, Prawet (2025). https://doi.org/10.1109/icmlas64557.2025.10967824
15. Mori, K., Ohara, R., Genda, T., Sato, S., Izumi, S., Kawaguchi, H.: Estimating human activities in bathroom through sound event detection in embedded systems. In: 11th International Conference on Internet of Things: Systems, Management and Security (IOTSMS), pp. 120–123. IEEE, Malmö (2024). https://doi.org/10.1109/IOTSMS62296.2024.10710269
16. Huang, Z., et al.: TinyChirp: bird song recognition using TinyML models on low-power wireless acoustic sensors. In: 2024 IEEE 5th International Symposium on the Internet of Sounds (IS2), pp. 1–10. IEEE, Erlangen (2024). https://doi.org/10.1109/IS262782.2024.10704131
17. Cuan, K., Zhang, T., Huang, J., Fang, C., Guan, Y.: Detection of avian influenza-infected chickens based on a chicken sound convolutional neural network. Comput. Electron. Agric. **178**, 105688 (2020). https://doi.org/10.1016/j.compag.2020.105688
18. Howard, A.G., et al.: MobileNets: efficient convolutional neural networks for mobile vision applications. arXiv:1704.04861 (2017). https://doi.org/10.48550/arXiv.1704.04861
19. Sandler, M., Howard, A., Zhu, M., Zhmoginov, A., Chen, L.-C.: MobileNetV2: inverted residuals and linear bottlenecks. arXiv:1801.04381 (2019). https://doi.org/10.48550/arXiv.1801.04381
20. Edge Impulse: Edge Impulse - The leading edge AI platform. https://edgeimpulse.com/. Accessed 26 Apr 2024
21. Garcia, A.: SAFE: sound analysis for fall event detection using machine learning. Smart Health (2025). https://doi.org/10.1016/j.smhl.2024.100539
22. MIT App Inventor. https://appinventor.mit.edu/. Accessed 21 July 2025

# Comparative Analysis of HARQ–ARQ Switching Across Modulation Schemes and Resource Allocations in 5G uRLLC and eMBB

Hsing-Cheng Liu and Yao-Liang Chung(✉)

Department of Communications, Navigation and Control Engineering, National Taiwan Ocean University, Keelung City, Taiwan
ylchung@email.ntou.edu.tw

**Abstract.** With the growing demand of 5G networks for ultra-reliable low-latency communications (uRLLC) and enhanced mobile broadband (eMBB), selecting an appropriate retransmission mechanism under diverse channel conditions and resource configurations has become a critical design challenge. This study systematically compares hybrid automatic repeat request (HARQ) and automatic repeat request (ARQ) across four modulation schemes (QPSK, 16-QAM, 64-QAM, 256-QAM) and three resource allocation ratios (1:1, 7:3, 3:7), while also considering varying user numbers and signal-to-interference-plus-noise ratio (SINR) ranges. Our findings demonstrate that HARQ provides clear advantages under low SINR or heavy-load conditions—for instance, with QPSK at SINR $\leq$ 1 dB, it significantly reduces error rates and improves throughput and success probability. Conversely, under high SINR conditions, ARQ outperforms HARQ, to the extent that ARQ with a lower-order modulation can surpass HARQ with a higher-order modulation in throughput. Further analysis reveals that the resource allocation ratio shifts the HARQ–ARQ switching threshold: for example, with 256-QAM, ARQ surpasses HARQ at 25 dB when eMBB receives more resources, but only at 26 dB when uRLLC is prioritized. This work establishes an extensive performance benchmark and validates the complementary characteristics of HARQ and ARQ under diverse modulation and resource allocation conditions. The results provide concrete guidance for dynamic resource allocation and protocol design in 5G systems, and lay the groundwork for future research on blind retransmission, reinforcement learning–driven adaptation, and energy-efficient communication strategies.

**Keywords:** 5G · ARQ–HARQ Dynamic Switching · uRLLC and eMBB

## 1 Introduction

5G networks must simultaneously support enhanced mobile broadband (eMBB) and ultra-reliable low-latency communications (uRLLC), which impose inherently conflicting requirements on resource allocation: eMBB demands high throughput, whereas uRLLC requires extremely low latency and reliability. Most existing studies, however, optimize a single service type and have not fully addressed the integrated allocation of eMBB and uRLLC [1].


P. Pichappan et al. (Eds.): RTIS 2025, LNNS 1936, pp. 260–273, 2027.
https://doi.org/10.1007/978-3-032-23396-7_23

Recent works adopt puncturing or superposition techniques to allow uRLLC to access mini-slots within eMBB's slot [2]. While this design reduces latency for uRLLC, it degrades eMBB throughput under multi-user competition and varying channel conditions, limiting overall system performance. Joint scheduling and algorithmic optimization have been proposed to mitigate this issue. For example, [2] compares the concave-convex procedure (CCCP) with greedy algorithms, showing that CCCP achieves a better throughput–latency trade-off, albeit at the cost of higher computational complexity. As the number of uRLLC users increases, CCCP may even violate strict delay constraints. To address excessive feedback latency, the 3GPP standard and subsequent studies introduced blind retransmission [3, 4], which reduces delay by retransmitting without feedback, but at the expense of additional resource consumption and reduced system capacity.

Regarding retransmission mechanisms, automatic repeat request (ARQ) and hybrid automatic repeat request (HARQ) remain the fundamental schemes [2, 5, 6]. ARQ, being feedback-based, achieves low latency in high signal-to-interference-plus-noise ratio (SINR) environments but suffers from repeated retransmissions under low SINR. HARQ integrates ARQ with forward error correction, leveraging redundancy and combining to improve throughput and reliability at low-to-medium SINR. However, HARQ incurs additional decoding delay and complexity, which may hinder latency-sensitive applications [7]. Thus, ARQ and HARQ embody the classical throughput–latency trade-off. Prior studies have explored switching strategies between them. For instance, [6] shows that under Bose–Chaudhuri–Hocquenghem coding, pure ARQ outperforms HARQ in throughput once SNR exceeds 13 dB. Similarly, [8] demonstrates that ARQ achieves higher throughput than HARQ at high SINR, and further enhances performance by selecting the optimal modulation and coding scheme via genetic algorithms. More recent research emphasizes adaptive HARQ for low-SINR scenarios, such as transmission time interval bundling [5, 9], yet adaptive mechanisms considering heterogeneous service types and resource allocations remain underexplored.

This work makes three contributions: 1) We provide a systematic comparison across modulation schemes and resource allocations, clarifying the applicability of ARQ and HARQ under different SINR regions. 2) We propose SINR- and resource-dependent switching thresholds, showing that conventional equal resource allocation models may underestimate optimal strategies for diverse user types. 3) We establish a reproducible performance benchmark dataset, forming a foundation for future algorithm design and cross-layer resource allocation in 5G systems.

## 2 System Model Design

### 2.1 Overall Design Process

The overall design process of this study is shown in Fig. 1. The flowchart outlines the transmission process under joint resource allocation for uRLLC and eMBB, integrating HARQ/ARQ with modulation switching. After each packet transmission, feedback and potential retransmissions are performed, while latency and retransmission limits are checked. If the conditions are satisfied, performance metrics are collected and modulation thresholds are updated; otherwise, the packet is either discarded or handled separately according to the distinct requirements of uRLLC and eMBB.

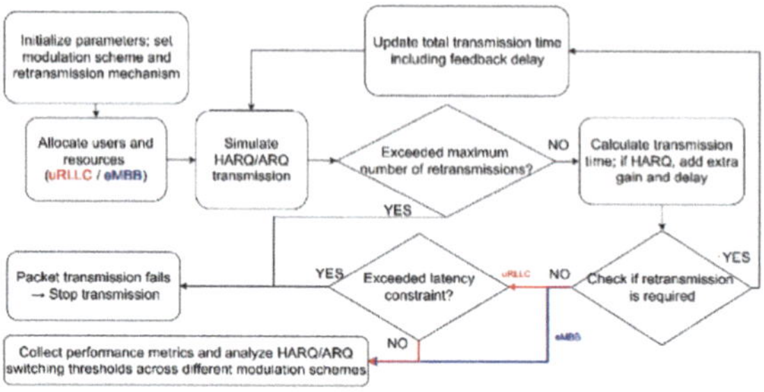


**Fig. 1.** Flowchart of the overall design process.

### 2.2 Achievable Channel Capacity Model

The achievable channel capacity in practical multi-user 5G systems is constrained by two fundamental factors. First, the Shannon capacity formula establishes the theoretical upper bound, assuming ideal coding with infinite block length and optimal decoding. This bound represents the maximum rate that can be supported by a channel under given bandwidth and SNR conditions, and is often extended to SINR in multi-user scenarios to account for interference. Second, practical modulation and coding schemes impose implementation constraints. Since higher-order constellations reduce the minimum Euclidean distance between symbols, the error probability grows significantly under finite SINR, thereby restricting the maximum modulation order that can be reliably employed. Consequently, the effective capacity of real systems is determined not only by the Shannon limit but also by the modulation and coding constraints under practical SINR regimes.

In the Rayleigh fading model, the instantaneous channel gain can be expressed as

$$H = \frac{|X + iY|}{\sqrt{2}} \tag{1}$$

where the random variables $X$ and $Y$ are assumed to be independent and identically distributed standard normal variables, representing the random fading effects caused by multipath propagation and obstacles in wireless channels. By substituting the instantaneous channel gain $H$ into the Shannon capacity formula, the channel capacity can be expressed as:

$$C_{\text{shannon}} = BW \cdot \log_2(1 + \gamma) \tag{2}$$

Here, $\gamma$ represents the instantaneous SINR, expressed as $SINR_{\text{linear}} \cdot |H|^2$, where $SINR_{\text{linear}}$ is the linear-scale SINR is converted from the dB scale and then multiplied by the random variable $|H|^2$. Through this process, each simulation iteration reflects a random channel realization, thereby generating instantaneous capacity fluctuations. However, the Shannon capacity only exists under idealized assumptions. In practical environments, the modulation scheme itself imposes an additional constraint. For an $M$-ary modulation, the modulation-constrained capacity can be expressed as

$$C_{\text{mod}} = BW \cdot \log_2(M) \cdot R \tag{3}$$

where $R$ denotes the coding rate, which represents the proportion of useful information bits carried per transmitted bit. This implies that the achievable rate is determined by the product of the available bandwidth and the number of bits per symbol. Therefore, the practically achievable capacity is given by the minimum of the Shannon capacity and the modulation-constrained capacity

$$C = \min(C_{\text{shannon}}, C_{\text{mod}}) \tag{4}$$

This design illustrates the bottleneck effect: the achievable capacity cannot exceed the theoretical limit established by the Shannon capacity, while it is simultaneously constrained by the maximum number of bits that can be supported by the employed modulation scheme.

### 2.3 Delay Model

To evaluate end-to-end latency, we begin by deriving the average slot time for the $i$-th attempt, denoted $T^{(i)}_{\text{slot,avg}}$. The formula is given by

$$T^{(i)}_{\text{slot, avg}} = \max\left(T_{\text{slot, min}}, \frac{P_{\text{pkt}}}{C \cdot \rho}\right) \tag{5}$$

where $P_{\text{pkt}}$ denotes the packet size in bits, $T_{\text{slot, min}}$ represents the minimum slot duration, which in this study is set to 1 µs as a physical constraint, and $\rho$ indicates the resource allocation ratio between uRLLC and eMBB, corresponding to the resource allocation module illustrated in Fig. 1. We adopt 10% of $T^{(i)}_{\text{slot, avg}}$ as the standard deviation to emulate Gaussian random jitter, so that the actual slot time of each transmission fluctuates around the mean. Accordingly, the slot time of the $i$-th transmission, $T^{(i)}_{\text{slot}}$, is given by

$$T^{(i)}_{\text{slot}} \sim N\left(T^{(i)}_{\text{slot, avg}}, 0.1 \cdot T^{(i)}_{\text{slot, avg}}\right) \tag{6}$$

For ARQ, the total delay $T^{\text{ARQ}}_{\text{total}}$ consists of the transmission slot time of the $i$-th attempt $T^{(i)}_{\text{slot}}$, the ARQ encoding and decoding time $T_{\text{code, ARQ}}$, and the ARQ feedback time $T_{\text{fb, ARQ}}$

$$T^{\text{ARQ}}_{\text{total}} = N_{\text{tr}} \cdot T_{\text{code, ARQ}} + (N_{\text{tr}} - 1) \cdot T_{\text{fb,ARQ}} + \sum_{i=1}^{N_{\text{tr,max}}} T^{(i)}_{\text{slot}}, 1 \leq N_{\text{tr}} \leq N_{\text{tr,max}} \tag{7}$$

where $N_{\text{tr}}$ denotes the cumulative transmission attempts and $N_{\text{tr, max}}$ denotes the maximum number of transmission attempts.

For HARQ chase combining, the total delay $T_{\text{total}}^{\text{HARQ}}$ includes the transmission slot time of the $i$-th attempt $T_{\text{slot}}^{(i)}$, the HARQ encoding and decoding time $T_{\text{code, HARQ}}$, the HARQ feedback time $T_{\text{fb, HARQ}}$, and the additional combining time $T_{\text{combine}}$

$$T_{\text{total}}^{\text{HARQ}} = N_{\text{tr}} \cdot \left(T_{\text{combine}} + T_{\text{code, HARQ}}\right) + (N_{\text{tr}} - 1) \cdot T_{\text{fb, HARQ}} + \sum_{i=1}^{N_{\text{tr,max}}} T_{\text{slot}}^{(i)},$$

$$1 \leq N_{\text{tr}} \leq N_{\text{tr,max}} \tag{8}$$

Here, $T_{\text{combine}}$ represents the packet-combining latency unique to HARQ. Since HARQ requires extra encoding/decoding and combining operations, its total delay is inherently larger than that of ARQ.

This behavior corresponds to the "transmission time calculation" block in Fig. 1, where additional steps are invoked when HARQ is applied. The numerical values used for the packet-combining, encoding/decoding, feedback delays, and physical constraint ($T_{\text{combine}}$, $T_{\text{code, HARQ}}$, $T_{\text{fb, HARQ}}$, $T_{\text{code, ARQ}}$, $T_{\text{fb, ARQ}}$, $T_{\text{slot, min}}$) are set as hypothetical parameters. These values are chosen to reflect the relative complexity of ARQ and HARQ implementations in typical communication systems, and are intended to emphasize their comparative impact on system latency.

## 2.4 Interference Model

In wireless networks, an increasing number of users is often accompanied by higher levels of interference, which significantly impacts both error rate and throughput. To capture this phenomenon, the present study introduces a stochastic and load-dependent interference model. The interference factor, denotes as $IF$, is defined as

$$IF = (1 + I_{\text{linear}}) \cdot (1 + k \cdot (U_{\text{eMBB}} + U_{\text{uRLLC}})) \tag{9}$$

where $I_{\text{linear}}$ is derived from background interference power. In the dB domain, interference power is modeled as a Gaussian random variable $I_{\text{dB}} \sim N(-10, 1)$ and then converted into the linear domain $I_{\text{linear}} = 10^{I_{\text{dB}}/10}$. . The second term of (9) is directly related to the total number of users, representing the linear increase in interference as system load grows. The coefficient $k$ controls the sensitivity of interference to the number of users. When a large number of users are active simultaneously, this term magnifies significantly, causing the error rate to deteriorate rapidly under heavy-load conditions.

When the interference model is applied to $SNR_{\text{dB}}$, the equivalent $SINR_{\text{dB}}$ is obtained as

$$SINR_{\text{dB}} = SNR_{\text{dB}} + 20 \cdot \log_{10}|H| - 10 \cdot \log_{10}(1 + IF) \tag{10}$$

Here, $\boldsymbol{SNR_{\text{dB}}}$ is the signal-to-noise ratio, and the channel fading effect is represented after conversion from the linear to the dB domain. The corresponding linear-domain SINR is obtained as $\boldsymbol{SINR_{\text{linear}}} = 10^{\boldsymbol{SINR_{\text{dB}}}/10}$.

### 2.5 Error Rate Model

Specifically, the bit error rate (BER) is first computed using the corresponding formula, which is then converted into the block error rate (BLER).

Under ideal channel conditions, the BER for different modulation schemes can be approximated using the $M$-QAM formula

$$BER(M) \approx \frac{4}{\log_2(M)}\left(1 - \frac{1}{\sqrt{M}}\right) \cdot Q\left(\sqrt{\left(\frac{3 \cdot \log_2(M) \cdot \gamma}{(M-1)}\right)}\right) \tag{11}$$

Here,$Q(\cdot)$ denotes the Q-function; for QPSK, the standard approximation $Q(\sqrt{2\gamma})$ is adopted. For a packet length of $L$, the approximation is given by

$$BLER_{\text{base}} = \begin{cases} 1 - (1 - BER(M))^L, & BER(M) > 10^{-6} \\ L \cdot BER(M), & BER(M) \leq 10^{-6} \end{cases} \tag{12}$$

This piecewise approximation avoids numerical underflow in the high-SNR region and provides a stable estimate.

In this study, $BLER_{\text{base}}$ is further scaled by $IF$ to capture the randomness and load effects inherent in multi-user environments. Moreover, the effective error rate is modeled as inversely proportional to $\rho$. The resulting effective BLER is expressed as

$$BLER = \min\left(\max\left(\frac{BLER_{\text{base}} \cdot IF}{2 \cdot \rho}, 10^{-9}\right), 1\right) \tag{13}$$

### 2.6 HARQ Chase Combining

As shown in the workflow (Fig. 1), the system first determines whether to adopt ARQ or HARQ as the transmission mode. Compared with ARQ, HARQ retains the erroneously received packet after a failed initial transmission. The receiver then combines this packet with subsequent retransmissions to improve the SNR. The equivalent SNR after combining can be expressed as

$$SNR_{\text{eff}}^{(N_{\text{tr}})} = SNR_{\text{dB}} + G_{\text{coding}} + 10 \cdot \log_{10}(N_{\text{tr}}),\ 1 \leq N_{\text{tr}} \leq N_{\text{tr,max}} \tag{14}$$

Here, $G_{\text{coding}}$ denotes the coding gain, while $10 \cdot \log_{10}(N_{\text{tr}})$ represents the additional gain achieved through energy combining after $N_{\text{tr}}$ transmissions.

In this study,$BLER_{\text{base}}$ is pre-computed using (11) and (12) and stored in a lookup table for subsequent use. Specifically, the $SNR_{\text{eff}}^{(N_{\text{tr}})}$ is substituted into (10) to obtain the corresponding $SINR_{\text{dB}}$, which is then used to query the associated $BLER_{\text{base}}$.

### 2.7 Success Criteria

Regarding success criteria, different definitions are applied to uRLLC and eMBB due to their distinct service requirements. For uRLLC, a packet is considered successfully transmitted only if two conditions are satisfied: (i) it is decoded within $N_{\text{tr, max}}$, and (ii) the total delay ($T_{\text{total}}^{\text{ARQ}}$ or $T_{\text{total}}^{\text{HARQ}}$) does not exceed 1 ms. By contrast, the criterion for eMBB is more relaxed: as long as a packet is decoded within $N_{\text{tr, max}}$, it is deemed successful. As illustrated in Fig. 1, both uRLLC and eMBB first check whether the maximum retransmission limit has been exceeded, but only uRLLC imposes the additional delay constraint.

## 3 Experimental Results

In the performance evaluation, only successfully delivered packets are considered in the throughput calculation, while unsuccessful packets are treated as losses. Throughput is defined as the effective packet size divided by the total transmission time. The per-user throughput is then obtained by dividing the system throughput by the number of users. In addition to throughput, success probability and delay are also measured to provide a comprehensive assessment of the performance differences between ARQ and HARQ under various conditions.

To balance simulation efficiency with statistical reliability, this study employs the Monte Carlo simulation method. For each parameter configuration (e.g., number of users, SINR, and modulation scheme), $10^4$ independent packet-transmission experiments are conducted. Each experiment simulates whether a packet is successfully transmitted, and the final statistical result is obtained by averaging across all trials. This approach effectively reduces random bias and ensures that the conclusions are statistically meaningful. The detailed system and simulation parameter settings are summarized in Table 1.

**Table 1.** Simulation parameter settings.

| Parameter | Value | Parameter | Value | Parameter | Value |
|---|---|---|---|---|---|
| $BW$ | 50MHz | $N_{\text{tr,max}}$ | 10 | $T_{\text{fb, ARQ}}$ | $5 \cdot 10^{-6}$s |
| $G_{\text{coding}}$ | 6dB | $P_{\text{pkt}}$ | $256 \cdot 8$ bits | $T_{\text{code,HARQ}}$ | $8 \cdot 10^{-6}$s |
| $k$ | 0.01 | $R$ | 0.75 | $T_{\text{fb,HARQ}}$ | $10 \cdot 10^{-6}$s |
| $M$ | 4, 16, 64, 256 | $T_{\text{code, ARQ}}$ | $10^{-6}$s | $T_{\text{combine}}$ | $10^{-6}$s |

All heatmaps in this study are generated by taking ARQ as the baseline. The calculation formula is defined as

$$\Delta(\%) = \frac{\mathcal{T}_{\text{HARQ}} - \mathcal{T}_{\text{ARQ}}}{\mathcal{T}_{\text{ARQ}}} \cdot 100\% \tag{15}$$

where $\mathcal{T}$ denotes the performance metric of interest (e.g., delay, throughput, or success probability). To avoid divergence when the denominator approaches zero, boundary conditions are defined as follows: if $\mathcal{T}_{ARQ} = 0$ and $\mathcal{T}_{HARQ} > 0$, the result is set to $+100\%$ ; if $\mathcal{T}_{ARQ} > 0$ and $\mathcal{T}_{HARQ} = 0$, the result is set to $-100\%$ ; and if both are zero, the result is defined as $0\%$ . For the interpretation of the heatmaps, the color scale is standardized: red regions indicate that $\mathcal{T}_{HARQ}$ outperforms $\mathcal{T}_{ARQ}$, whereas blue regions indicate that $\mathcal{T}_{ARQ}$ performs better. The intensity of the color reflects the magnitude of the difference.

### 3.1 Equal Resource Allocation ($\rho = 1 : 1$)

In this subsection, 16-QAM is used as an illustrative example. The average throughput difference between uRLLC and eMBB is only 0.55% (HARQ) and 0.86% (ARQ), resulting in nearly overlapping curves. As shown in Fig. 2(b), the throughput of ARQ approaches zero under low-SINR conditions; correspondingly, Fig. 3(a) exhibits a distinct red band when SINR $\leq$ 10 dB. Moreover, Fig. 2(c) indicates that when SINR $\leq$ 15 dB, HARQ consistently achieves higher throughput than ARQ, underscoring the necessity of HARQ in low-SINR scenarios.

On the other hand, in Fig. 3, because the interference from simultaneous transmissions impacts all users in a similar manner, the throughput and delay trends of eMBB closely mirror those of uRLLC under equal resource allocation. Therefore, only the results for uRLLC are presented. As illustrated in Fig. 3(c)–(d), however, under identical SINR conditions, the heatmap color gradually shifts toward red as the number of users increases. This observation indicates that the relative advantage of HARQ over ARQ in terms of successful transmission becomes increasingly pronounced with higher user loads.

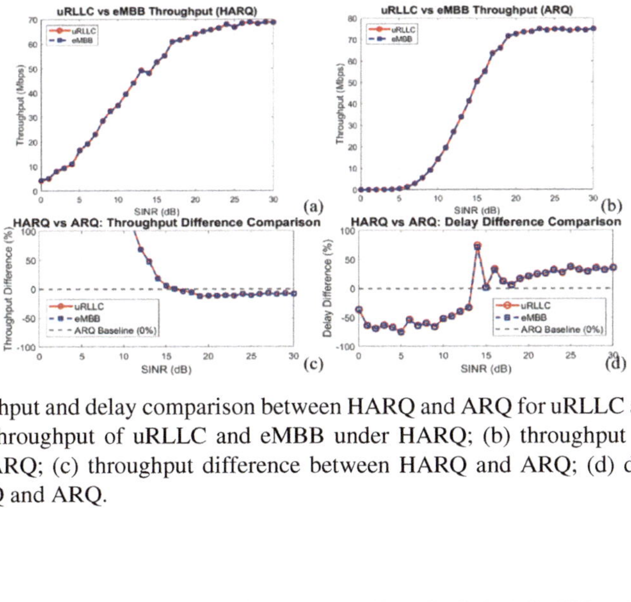


**Fig. 2.** Throughput and delay comparison between HARQ and ARQ for uRLLC and eMBB with 16-QAM: (a) throughput of uRLLC and eMBB under HARQ; (b) throughput of uRLLC and eMBB under ARQ; (c) throughput difference between HARQ and ARQ; (d) delay difference between HARQ and ARQ.

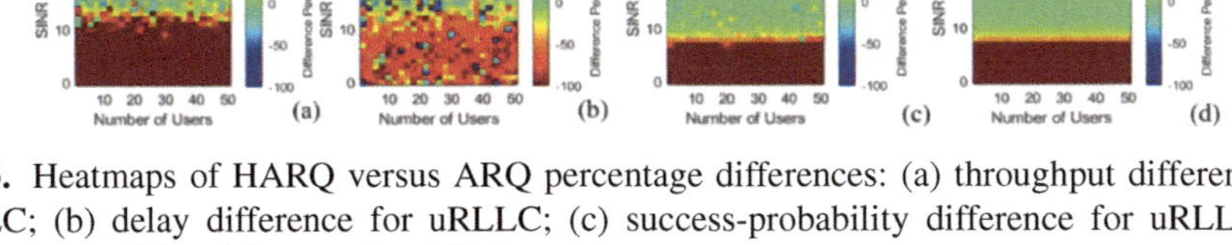


**Fig. 3.** Heatmaps of HARQ versus ARQ percentage differences: (a) throughput difference for uRLLC; (b) delay difference for uRLLC; (c) success-probability difference for uRLLC; (d) success-probability difference for eMBB.

Table 2 summarizes the recommended schemes across SINR ranges for $\rho = 1{:}1$. From a broader perspective of equal resource allocation, the results indicate that when SINR lies between 0 and 9 dB, QPSK is the dominant modulation scheme. In this region, HARQ and ARQ alternately hold the advantage, with throughput differences remaining below 14%, allowing either mechanism to be adopted depending on application requirements. At SINR = 10 dB, QPSK ARQ emerges as the most suitable scheme; however, the throughput gap between QPSK ARQ and 16-QAM HARQ is less than 1%, rendering the distinction negligible.

For SINR values between 11 and 16 dB, 16-QAM HARQ becomes the primary scheme. Its throughput advantage over QPSK ARQ initially widens, as the latter approaches its maximum, but gradually narrows as 16-QAM ARQ begins to catch up. In the range of 17–19 dB, 16-QAM ARQ dominates, with 64-QAM HARQ closely following, though still slightly underperforming overall. Between 20 and 25 dB, the system transitions directly to 64-QAM ARQ as the primary scheme. Once the switching threshold is surpassed, throughput increases rapidly, surpassing 64-QAM HARQ, which had previously been competitive. Finally, when SINR exceeds 26 dB, the system shifts directly to 256-QAM ARQ. Under such high-SINR conditions, 256-QAM HARQ does not even appear as a viable secondary option, underscoring the superior advantage of ARQ in this regime.

**Table 2.** Recommended schemes across SINR ranges ($\rho = 1{:}1$).

| SINR Range | Recommended Scheme | SINR Range | Recommended Scheme |
|---|---|---|---|
| $<$ 1 dB | QPSK+HARQ | 11-16 dB | 16-QAM+HARQ |
| 1-7 dB | QPSK+ARQ | 17-19 dB | 16-QAM+ARQ |
| 8 dB | QPSK+HARQ | 20-25 dB | 64-QAM+ARQ |
| 9-10 dB | QPSK+ARQ | $\geq$ 26 dB | 256-QAM+ARQ |

### 3.2 uRLLC-Dominant Resource Allocation ($\rho = 7 : 3$)

Taking 64-QAM as an example, and comparing it with the equal resource allocation scenario of 16-QAM (Fig. 3(c)–(d)), a large number of deep-red bands can be observed in the low-SINR region under limited resources. This indicates that, with the ARQ mechanism, simultaneous transmission failures of both user types occur more frequently. In contrast, under the resource-biased allocation of 64-QAM (Fig. 5(c)–(d)), uRLLC receives a larger share of resources, and no obvious deep-red bands appear in the low-SINR region. This demonstrates that even with ARQ alone, successful transmission can still be achieved. These results imply that when resources are limited, concentrating the allocation ratio on a single user type can effectively improve the transmission success rate of that user category.

As shown in Fig. 4(c), for uRLLC with a larger resource allocation, ARQ throughput begins to surpass HARQ once SINR $\geq$ 17 dB. By contrast, eMBB with fewer resources requires SINR $\geq$ 22 dB to achieve the same outcome. This indicates that at low SINR, low-resource users are more reliant on HARQ to sustain higher throughput.

Further comparisons of Fig. 5(a)–(b) with Fig. 5(e)–(f) reveal that, whether in terms of throughput or delay, uRLLC with more resources shows a more pronounced advantage of ARQ over HARQ in the high-SINR region than the eMBB group. This demonstrates that the disparity between ARQ and HARQ performance not only persists but is further amplified as resources become more skewed toward one user type.

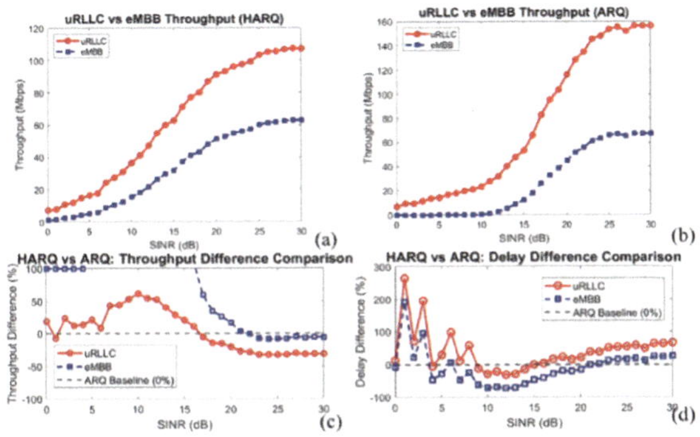


**Fig. 4.** Throughput and delay comparison between HARQ and ARQ for uRLLC and eMBB with 64-QAM ($\rho = 7 : 3$): (a) throughput of uRLLC and eMBB under HARQ; (b) throughput of uRLLC and eMBB under ARQ; (c) throughput difference between HARQ and ARQ; (d) delay difference between HARQ and ARQ.

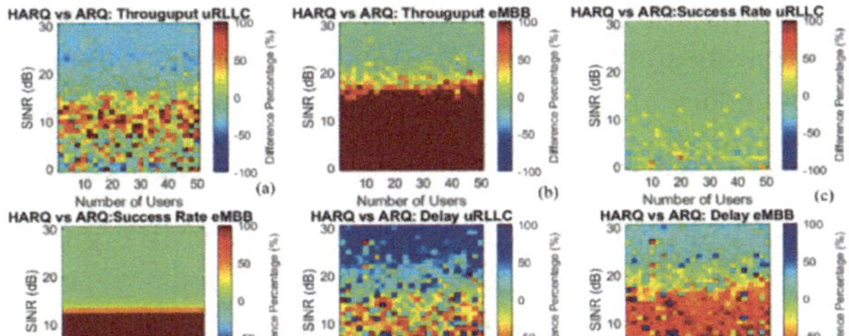


**Fig. 5.** Performance comparison between HARQ and ARQ with 64-QAM ($\rho = 7 : 3$): (a) throughput difference for uRLLC; (b) throughput difference for eMBB; (c) success-probability difference for uRLLC; (d) success-probability difference for eMBB; (e) delay difference for uRLLC; (f) delay difference for eMBB.

When examining the scenario where resources are biased toward uRLLC, the results can be summarized as follows. The main outcomes are summarized in Table 3 for quick reference. For SINR between 0 and 9 dB, QPSK remains the primary modulation scheme, with HARQ and ARQ alternating in advantage. In this region, the range where HARQ is preferable under low SINR becomes broader, while the throughput difference between the two mechanisms remains below 10%, allowing either scheme to be selected depending on practical requirements. When SINR is between 10 and 14 dB, 16-QAM HARQ serves as the primary mechanism. For SINR between 15 and 19 dB, 16-QAM ARQ becomes the dominant scheme. Within this interval, other secondary mechanisms—such as 16-QAM HARQ and 64-QAM HARQ/ARQ—exhibit comparable performance, leading to a "multi-contender" situation. Nevertheless, 16-QAM ARQ remains the primary choice, as it simultaneously satisfies the thresholds of both user types while ensuring transmission stability.

When SINR is between 20 and 25 dB, the system directly switches to 64-QAM ARQ as the primary scheme. Once SINR exceeds 26 dB, the system transitions directly to 256-QAM ARQ, again illustrating that ARQ shows greater advantages under high-SINR conditions.

**Table 3.** Recommended schemes across SINR ranges ($\rho = 7:3$).

| SINR Range | Recommended Scheme | SINR Range | Recommended Scheme |
|---|---|---|---|
| $\leq$ 2 dB | QPSK+HARQ | 10-14 dB | 16-QAM+HARQ |
| 3-5 dB | QPSK+ARQ | 15-19 dB | 16-QAM+ARQ |
| 6 dB | QPSK+HARQ | 20-25 dB | 64-QAM+ARQ |
| 7-9 dB | QPSK+ARQ | $\geq$ 26 dB | 256-QAM+ARQ |

### 3.3 uRLLC-Dominant Resource Allocation ($\rho = 3:7$)

Taking 256-QAM as an example, Fig. 7(a)(c) shows scattered green blocks in the low-SINR region, representing simultaneous transmission failures of both HARQ and ARQ. According to the boundary conditions, these cases are categorized as 0%, rather than indicating identical throughput or success probability between the two mechanisms.

A further comparison of Fig. 7(d) with Fig. 5(c) shows that both correspond to the user type receiving more resources. However, in Fig. 7(d), the eMBB group exhibits almost no difference between HARQ and ARQ, with a continuous green band distribution. This indicates that under eMBB's more lenient success criterion, the performance gap between HARQ and ARQ is smaller compared to uRLLC. Importantly, although Fig. 7(d) displays a green band, this is not an anomaly: as seen in Fig. 6(a)–(b), eMBB still achieves valid transmissions. This suggests that even under the same resource-allocation ratio, the difference in service requirements leads to distinct distribution patterns for eMBB and uRLLC, although the overall trends remain broadly consistent.

In addition, if Fig. 7(a) is viewed in isolation, one might mistakenly assume that ARQ fails entirely when SINR $\leq$ 18 dB. However, when combined with Fig. 6(c), it becomes clear that ARQ only has zero throughput when SINR $\leq 8$. For $9 \leq$ SINR $\leq 18$ dB, ARQ is capable of achieving successful transmissions, although Fig. 6(b) shows that throughput remains close to 0 Mbps. During this interval, HARQ achieves much higher throughput than ARQ, which is why the heatmap still depicts this region as deep red.

In summary, while heatmaps may appear visually similar in distribution, their physical implications can differ substantially.

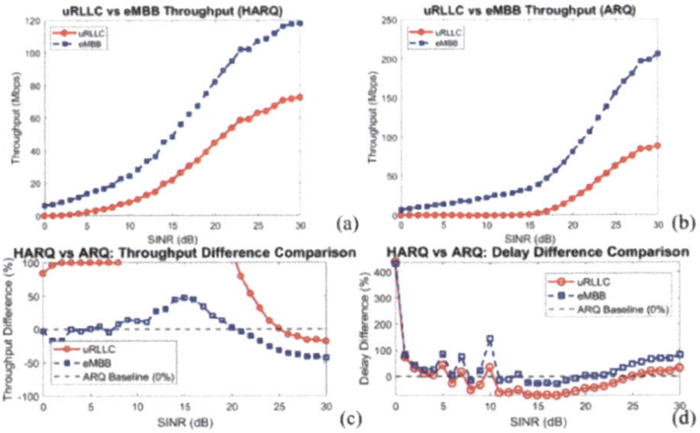


**Fig. 6.** Throughput and delay comparison between HARQ and ARQ for uRLLC and eMBB with 256-QAM ($\rho = 3 : 7$): (a) throughput of uRLLC and eMBB under HARQ; (b) throughput of uRLLC and eMBB under ARQ; (c) throughput difference between HARQ and ARQ; (d) delay difference between HARQ and ARQ.

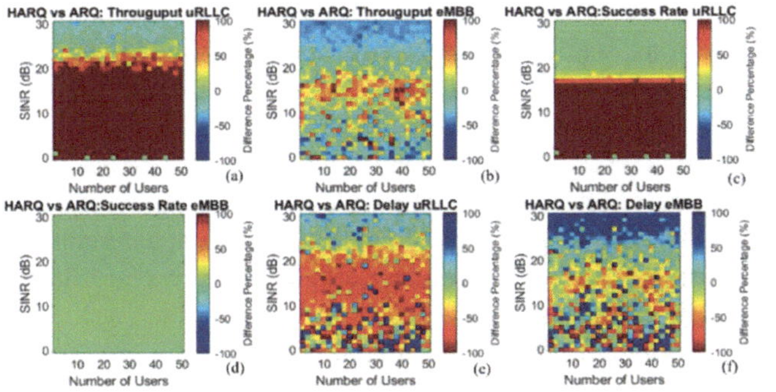


**Fig. 7.** Performance comparison between HARQ and ARQ with 256-QAM ($\rho = 3 : 7$): (a) throughput difference for uRLLC; (b) throughput difference for eMBB; (c) success-probability difference for uRLLC; (d) success-probability difference for eMBB; (e) delay difference for uRLLC; (f) delay difference for eMBB.

In addition to the observations based on 256-QAM, a more comprehensive examination of the eMBB-dominant resource-allocation scenario yields the following conclusions. These key outcomes are further consolidated in Table 4, which provides a clearer overall reference. For SINR between 0 and 10 dB, QPSK is the primary modulation scheme, and the throughput difference between HARQ and ARQ is generally within 15%. Therefore, either mechanism may be selected depending on practical requirements.

When SINR is between 11 and 14 dB, 16-QAM HARQ serves as the primary mechanism. For SINR between 15 and 18 dB, 16-QAM ARQ becomes dominant, and for SINR between 19 and 24 dB, the system transitions directly to 64-QAM ARQ. These results indicate that under eMBB-dominant allocation, the switching thresholds from HARQ to ARQ consistently occur earlier, reflecting the fact that eMBB only requires successful decoding within the retransmission limit, without strict delay constraints. Finally, when SINR exceeds 25 dB, the system switches directly to 256-QAM ARQ.

**Table 4.** Recommended schemes across SINR ranges ($\rho = 3 : 7$).

| SINR Range | Recommended Scheme | SINR Range | Recommended Scheme |
|---|---|---|---|
| $< 1$ dB | QPSK+HARQ | 11-14 dB | 16-QAM+HARQ |
| 1-5 dB | QPSK+ARQ | 15-18 dB | 16-QAM+ARQ |
| 6 dB | QPSK+HARQ | 19-24 dB | 64-QAM+ARQ |
| 7-10 dB | QPSK+ARQ | $\geq 25$ dB | 256-QAM+ARQ |

## 4 Conclusion

This study proposed a mathematical framework integrating throughput, delay, interference, and error-rate models with HARQ/ARQ mechanisms and the Monte Carlo method to evaluate 5G uRLLC and eMBB performance under different modulation schemes and resource allocations. Simulations show that HARQ is advantageous in low-SINR or high-load conditions, while ARQ performs better in high-SINR regions. Resource-allocation ratios strongly affect switching thresholds: uRLLC-dominant allocation extends HARQ's range, whereas eMBB-dominant allocation accelerates ARQ's advantage. In resource-limited scenarios, concentrating bandwidth on one user type improves its transmission success. These findings provide guidance for dynamic resource allocation and cross-layer protocol design in 5G, with future work extending to blind retransmission, reinforcement learning, and energy-efficiency analysis for greater adaptability in heterogeneous environments.

## References

1. Anand, A., de Veciana, G.: Resource allocation and HARQ optimization for URLLC traffic in 5G wireless networks. IEEE J. Sel. Areas Commun. **36**(11), 2411–2421 (2018)
2. Zhang, L.: Optimized resource allocation in mobile edge communication, 5G data multiplexing, and WiFi channel access, Ph.D. Dissertation. Digital Dissertation Consortium (2023)
3. Elayoubi, S.E., Brown, P., Deghel, M., Galindo-Serrano, A.: Radio resource allocation and retransmission schemes for URLLC over 5G networks. IEEE J. Sel. Areas Commun. **37**(4), 896–904 (2019)
4. Abreu, R., Berardinelli, G., Jacobsen, T., Pedersen, K., Mogensen, P.: A blind retransmission scheme for ultra-reliable and low latency communications. In: Proc. IEEE 87th Vehicular Technology Conference (VTC Spring), pp. 1–5. Porto, Portugal (2018)
5. Song, T., Lee, W.-H.: HARE: Hybrid ARQ-based adaptive retransmission control scheme for synchronous multi-link in wireless LANs. IEEE Trans. Veh. Technol. **72**(8), 10302–10313 (2023)

6. Kotuliaková, K., Polec, J., Csóka, F.: An adaptive ARQ–HARQ method with BCH codes. In: Proc. IEEE 11th Int. Conf. on Application of Information and Communication Technologies (AICT), pp. 1–5. Moscow, Russia (2017)
7. Khan, B.S., Jangsher, S., Ahmed, A., Al-Dweik, A.: URLLC and eMBB in 5G industrial IoT: A survey. IEEE Open J. Commun. Soc. **3**, 1134–1163 (2022)
8. Nasraoui, L., Dabbar, A., Siala, M.: Energy and MCS optimization in HARQ protocol for ultrareliable regime with maximized throughput. IEEE Syst. J. **17**(1), 1282–1291 (2023)
9. Cabrera, E., Fang, G., Vesilo, R.: Adaptive hybrid ARQ (A-HARQ) for ultra-reliable communication in 5G. In: Proc. IEEE 85th Vehicular Technology Conference (VTC Spring), pp. 1–6. Sydney, Australia (2017)

# Granularity Matters: A Hardware-Aware Benchmark of IoT Intrusion Detection Systems Across Multiple Classification Schemes

Nguyen Trung Kiet, Ho Ngoc Trung Kien, and Hoai-Phuong Nguyen-Cao(✉)

Thu Dau Mot University, Ho Chi Minh City, Vietnam
{kietnt,kienhnt,phuongnch}@tdmu.edu.vn

**Abstract.** As IoT networks expand, intrusion detection must balance detection granularity with hardware constraints. Prior work often optimizes accuracy in isolation, overlooking latency, memory, and compute limits in edge deployments.

We present a hardware-aware benchmark of five models (CNN1D, LightGBM, Mamba, TabTransformer, Transformer) across three schemes—binary (2 classes), moderate (4 classes), and fine-grained (8 classes)—on CIC-IoT2023 (~50,000 flows). CNN1D achieves the highest average macro F1 (0.979), while LightGBM offers the best efficiency–latency trade-off (training 0.84 min; inference 0.8 ms). The 4-class scheme consistently provides the best accuracy–complexity balance (macro F1 up to 0.998). By contrast, Mamba attains strong accuracy at substantially higher cost (training 68.7 min; inference 104.1 ms). We release our framework and guidelines to help practitioners match models to hardware constraints.

**Keywords:** IoT Security · Intrusion Detection · Machine Learning · Hardware-Aware Computing · Edge Computing · Classification Granularity

## 1 Introduction

The Internet of Things (IoT) has reshaped computing landscapes, connecting billions of devices across smart cities, industrial automation, healthcare, and autonomous systems [1,2]. While this connectivity offers substantial benefits, it also creates significant security vulnerabilities, particularly because many IoT devices operate under constrained computational resources while handling sensitive data and controlling critical infrastructure [3].

Most existing intrusion detection systems (IDS) for IoT networks prioritize detection accuracy, often overlooking the practical resource limitations that define real-world deployments. This accuracy-first approach can be problematic in edge environments where devices must function within tight power, memory, and latency constraints. Furthermore, the choice of classification granularity


P. Pichappan et al. (Eds.): RTIS 2025, LNNS 1936, pp. 274–282, 2027.
https://doi.org/10.1007/978-3-032-23396-7_24

(e.g., two, four, or eight classes) entails trade-offs between detection detail and computational complexity that remain underexplored.

Three gaps persist: (i) a lack of systematic evaluation under realistic hardware constraints, (ii) limited analysis of how classification granularity affects both detection and resource consumption, and (iii) insufficient guidance for selecting models under specific deployment constraints.

We address these by benchmarking five models (CNN1D, LightGBM, Mamba, TabTransformer, Transformer) across three granularities on CIC-IoT2023. Contributions: (1) a systematic analysis of performance–efficiency trade-offs across granularities; (2) identification of model–granularity combinations suited to deployment scenarios; (3) practical guidelines for IoT IDS selection; and (4) an open-source evaluation framework for reproducibility and real-world use.

## 2 Related Work

IoT has garnered significant attention across domains such as autonomous vehicles [1], smart grids [2], healthcare [4], and digital agriculture [5]. As a technology aimed at enhancing quality of life, IoT faces diverse security challenges [3]. Prakash et al. [6] summarize these concerns alongside available and potential solutions, offering guidance for timely mitigation.

Tyagi et al. [2] introduced a real-time system that uses IoT sensors and deep learning to monitor smart grids and detect faults. Sensors collect data on parameters such as voltage and current, which are fed to a deep learning algorithm to identify anomalies and potential outages. In a separate study, Tasci [7] developed an optimized 1D CNN with self-attention, GELU, dropout, and normalization for IoT security data, evaluated on CIC-IoT2023, CIC-MalMem-2022, and CIC-IDS2017. Despite a modest parameter count, its data flow remains computationally demanding.

Monisha et al. [5] proposed a BiGRU–LSTM ensemble for smart farming environments, improving threat detection in interconnected settings. Ain et al. [8] evaluated deep models, including a hybrid CNN–LSTM–autoencoder that outperformed alternatives on CIC-IoT2023 for DDoS detection. Similarly, Jony et al. [9] proposed an LSTM-based model for diverse cyber-attack patterns; however, the sequential nature of LSTMs restricts parallelization and creates a training bottleneck on modern hardware.

## 3 Methodology

### 3.1 Problem Setting and Evaluation Goals

We study the selection of machine learning models for IoT IDS under varying resource constraints and classification requirements. We investigate how classification granularity—from binary detection to fine-grained categorization—affects both detection performance and computational demands across model architectures.

Research questions: (1) How does granularity impact performance and resource use? (2) Which model–granularity combinations optimize trade-offs across deployment scenarios? (3) What guidance supports model selection under specific hardware constraints?

We combine standard metrics (precision, recall, F1, accuracy) with hardware measurements: training duration, inference latency, memory consumption, and storage requirements. This enables realistic assessment across contexts from resource-constrained edge devices to centralized security operations centers (SOCs).

### 3.2 Dataset and Preprocessing

Experiments use CIC-IoT2023 with approximately 50,000 flows from normal and attack scenarios. The dataset provides 40 numerical features per flow (46 originally; six removed during cleaning). It includes 33 attack types alongside benign traffic. A notable characteristic is severe class imbalance, with attack traffic outnumbering benign traffic by approximately ~42:1 (benign ~2.34%) (Fig. 1).

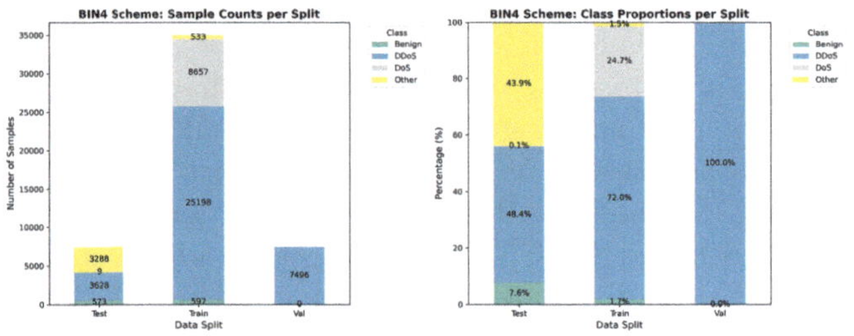


**Fig. 1.** Distribution across training/validation/test for the 4-class scheme, illustrating imbalance (benign ~2.3%) while preserving proportional representation.

Preprocessing avoids data leakage: training-set medians for imputation; winsorization at the 0.5th/99.5th percentiles; and feature standardization (StandardScaler) fitted on training only. For sequence-aware models, temporal windows are formed while grouping by device. All parameters are learned on training and applied to validation/test.

### 3.3 Classification Schemes

We design three schemes to balance detail and cost:

**BIN2 (Binary):** *Attack* (48,809) vs *Benign* (1,170).

**BIN4 (Moderate):** *Benign* (1,170), *DDoS* (36,322), *DoS* (8,666), *Other* (3,821).

**BIN8 (Fine-grained):** DDoS-ICMP (8,099), DDoS-TCP (4,784), DDoS-UDP (6,133), DoS-UDP (3,534), DoS-TCP (2,891), Botnet/Malware (2,854), *Other* (20,514; 23 attack varieties), and *Benign* (1,170).

### 3.4 Model Architectures

We consider representative approaches:

**CNN1D:** Three convolutional blocks and two dense layers (485,621 parameters).

**LightGBM:** Gradient boosting baseline for tabular data; low memory and fast inference (parameter count depends on learned trees).

**Mamba:** Selective state-space model (2,847,843 parameters) to capture long-range dependencies at higher computational cost.

**TabTransformer:** Attention adapted for tabular data (1,205,847 parameters) to learn feature interactions.

**Transformer:** Standard attention with positional encodings (1,584,692 parameters).

### 3.5 Hardware-Aware Evaluation Protocol

We record wall-clock training time, peak memory, and throughput (samples/hour). For inference, we measure single-sample latency (batch sizes 1, 8, 16, 32) and memory during inference. We also record model size on disk and note potential bottlenecks. Use-case mapping: BIN2 for edge devices (fast binary decision), BIN4 for industrial monitoring (actionable detail with manageable complexity), and BIN8 for SOCs (fine-grained categorization on powerful hardware).

### 3.6 Experimental Configuration

**Hardware:** Intel E5-2678v3 (12 cores, 2.5 GHz), NVIDIA GeForce GTX 1650 (4 GB VRAM), 64 GB DDR4 RAM, NVMe SSD.

**Training:** Stratified splits: 70%/15%/15%. Grid search on validation; early stopping if macro F1 does not improve for 10 epochs. Random seeds fixed to 42.

**Statistics:** Paired t-tests and Cohen's $d$ for effect size. Each model is trained five times; we report 95% confidence intervals (Fig. 2).

# 4 Results

## 4.1 Overall Performance Analysis

Table 1 summarizes average performance across schemes.

**Table 1.** Overall performance ranking by average macro F1 (95% CI).

| Rank | Model | Macro F1 | Weighted F1 | Consistency |
|---|---|---|---|---|
| 1 | CNN1D | 0.979 ± 0.012 | 0.988 ± 0.008 | High |
| 2 | Mamba | 0.972 ± 0.015 | 0.986 ± 0.009 | High |
| 3 | Transformer | 0.968 ± 0.018 | 0.985 ± 0.011 | High |
| 4 | LightGBM | 0.946 ± 0.021 | 0.992 ± 0.006 | Medium |
| 5 | TabTransformer | 0.894 ± 0.034 | 0.959 ± 0.018 | Low |

CNN1D achieves the highest average macro F1 (0.979). Mamba and Transformer follow closely. LightGBM attains the highest weighted F1 (0.992), reflecting robustness on majority classes under imbalance. TabTransformer exhibits lower performance with higher variability across schemes.

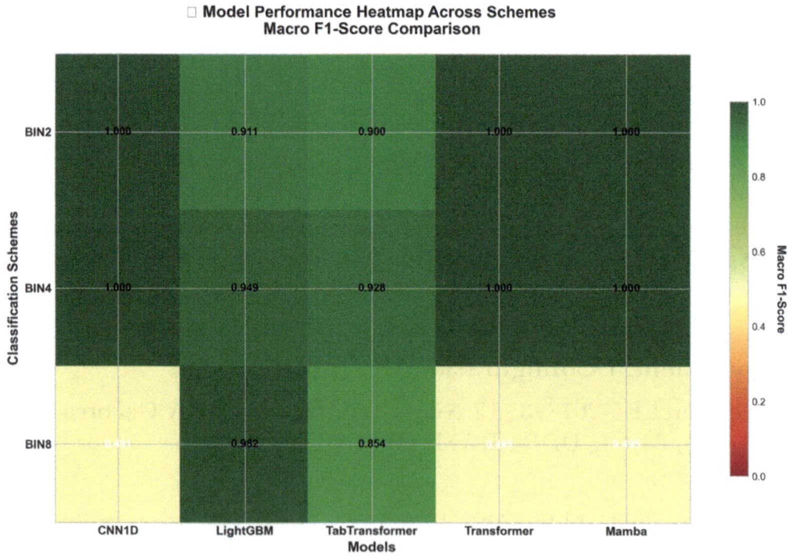


**Fig. 2.** Macro F1 across five models and three classification schemes. CNN1D is consistently strong; TabTransformer struggles at 8 classes. BIN4 provides a favorable balance between accuracy and operational complexity.

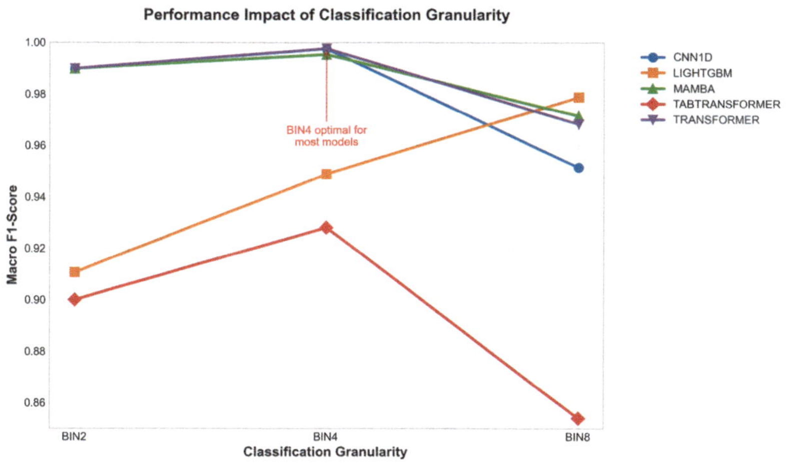


**Fig. 3.** Impact of granularity (2/4/8 classes) on macro F1. CNN1D and Transformer scale well with complexity; TabTransformer degrades at 8 classes. Performance converges near the 4-class level, indicating a favorable balance.

### 4.2 Granularity Impact Analysis

Binary classification (BIN2) is tractable for most models (macro F1 around 0.990 for CNN1D, Mamba, Transformer). The 4-class scheme (BIN4) reaches near-perfect performance (macro F1 up to 0.998) while retaining actionable categorization. For BIN8, LightGBM performs competitively (macro F1 around 0.979) under severe imbalance with many small classes (Figs. 3, 4 and Table 2).

### 4.3 Hardware Performance Analysis

**Table 2.** Training performance comparison. Memory is peak usage observed during training.

| Model | Training Time (min) | Memory (MB) | Parameters |
|---|---|---|---|
| LightGBM | $0.84 \pm 0.12$ | $358 \pm 23$ | Variable |
| CNN1D | $2.10 \pm 0.18$ | $485 \pm 31$ | 485621 |
| TabTransformer | $45.2 \pm 3.2$ | $1200 \pm 89$ | 1205847 |
| Transformer | $52.3 \pm 4.1$ | $1584 \pm 127$ | 1584692 |
| Mamba | $68.7 \pm 5.8$ | $2847 \pm 198$ | 2847843 |

LightGBM trains in 0.84 min at 358 MB peak memory, whereas Mamba requires 68.7 min and 2847 MB. For inference, LightGBM achieves sub-1 ms latency, while

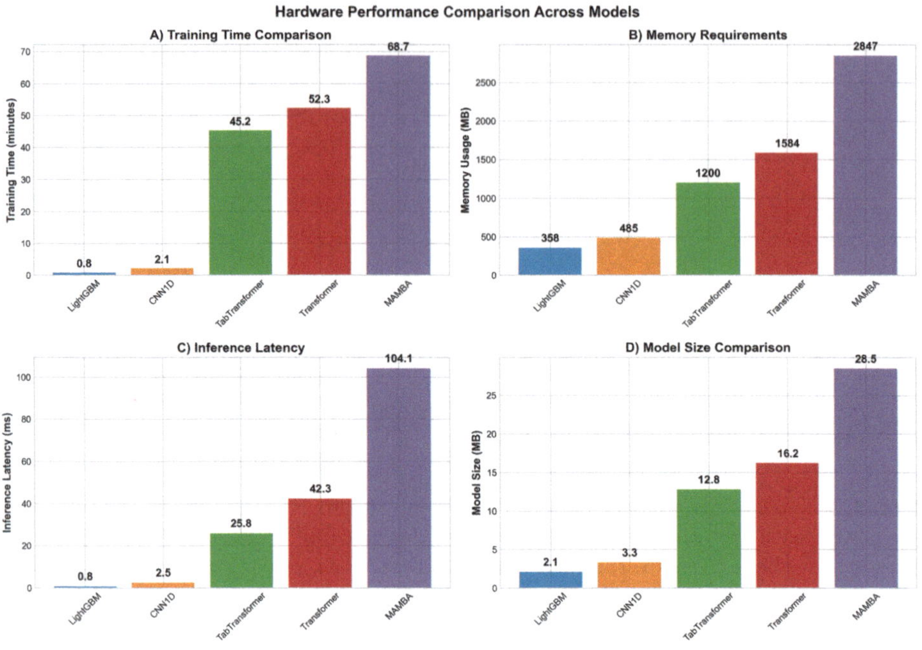


**Fig. 4.** Hardware metrics relevant to deployment: (A) training time, (B) memory, (C) inference latency, (D) model size. Differences across models are substantial, underscoring the need for hardware-aware evaluation.

Mamba and Transformer require substantially more time per prediction, which can be prohibitive for real-time scenarios (Fig. 5).

### 4.4 Trade-off Analysis

Three profiles emerge: (i) *accuracy-optimized* (Transformer, Mamba) approaching the accuracy frontier at higher cost; (ii) *latency/efficiency-optimized* (LightGBM) with fast training and inference; and (iii) *balanced* (CNN1D), delivering high accuracy with moderate resources (Table 3).

**Table 3.** Deployment recommendations by use case (aligned with observed trade-offs).

| Scenario | Model | Scheme | Rationale |
|---|---|---|---|
| Edge IoT Gateway | LightGBM | BIN2 | Sub-1 ms inference, minimal memory |
| Industrial Monitor | CNN1D | BIN4 | Strong accuracy with moderate resources |
| Security Operations | CNN1D/Transformer | BIN4/BIN8 | Highest accuracy at higher cost |
| Real-time Detection | LightGBM | BIN2 | High throughput, low latency |
| Forensic Analysis | Mamba | BIN8 | Detailed analysis; cost-insensitive |

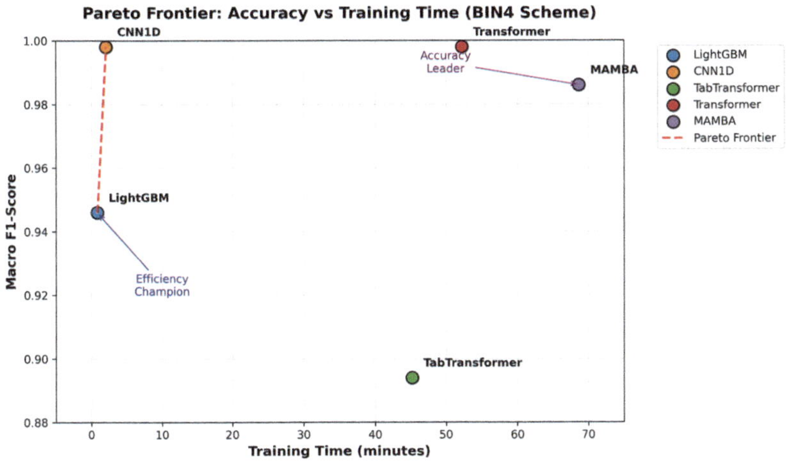


**Fig. 5.** Pareto frontier for the 4-class scheme contrasting macro F1 and training efficiency. CNN1D lies on the balanced frontier; LightGBM forms the latency/efficiency frontier; attention/state-space models (Transformer, Mamba) approach the accuracy frontier at higher cost.

# 5 Discussion

## 5.1 Practical Deployment Guidelines

Figure 6 summarizes a decision process grounded in the observed trade-offs. For highly constrained gateways, LightGBM with BIN2 minimizes latency and memory while retaining competitive accuracy. For industrial monitoring, CNN1D with BIN4 offers near-optimal accuracy under moderate resource budgets. In compute-rich SOCs, fine-grained schemes (BIN8) are feasible; accuracy-optimized neural models or efficiency-optimized boosting can be selected depending on latency constraints.

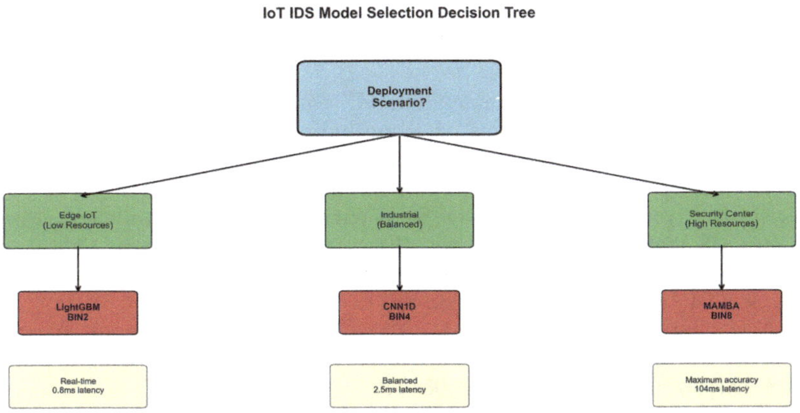


**Fig. 6.** Decision tree for selecting IoT IDS models under hardware and task requirements.

### 5.2 Implications and Limitations

The hardware-aware evaluation addresses a gap in IoT IDS research, where accuracy is often reported without operational costs. Limitations include reliance on a single dataset (CIC-IoT2023), a single hardware platform, and severe class imbalance (∼42:1). Future work should validate across additional datasets and edge devices (e.g., ARM/Jetson) and explore data balancing and unsupervised detection for novel attacks.

## 6 Conclusion

This study provides a hardware-aware benchmark for IoT intrusion detection across five models (CNN1D, LightGBM, Mamba, TabTransformer, Transformer) and three classification schemes on the CIC-IoT2023 dataset. The results indicate that (i) the 4-class scheme yields the strongest balance between accuracy and complexity (macro F1 up to 0.998); (ii) CNN1D achieves the highest average macro F1 across schemes; (iii) LightGBM offers the best efficiency–latency profile with competitive accuracy, particularly under fine-grained, imbalanced conditions; and (iv) Mamba delivers strong accuracy at substantially higher computational cost. Future work will validate these findings on resource-constrained devices and across additional datasets.

## References

1. Kheder, M.Q., Ali, M.A.: IoT-based vision techniques in autonomous driving: a review. Adv. Distrib. Comput. Artif. Intell. J. **11**(3), 367–394 (2022)
2. Tyagi, R., Kumar, P., Sagar, P.K., Saraswat, M., Bansal, A., et al.: Real-time monitoring and fault detection in smart grids using IoT sensors and deep learning algorithm. In: 2024 4th International Conference on Advancement in Electronics & Communication Engineering (AECE), pp. 1128–1132. IEEE (2024)
3. Alotaibi, B.: A survey on industrial internet of things security: requirements, attacks, AI-based solutions, and edge computing opportunities. Sensors **23**(17) (2023). https://doi.org/10.3390/s23177470
4. Abdulmalek, S., et al.: IoT-based healthcare-monitoring system towards improving quality of life: a review. Healthcare **10**(10) (2022). https://doi.org/10.3390/healthcare10101993
5. Monisha, P., Fallah, M.H., Hemamalini, M., Saranya, N.N.: Threat detection for IoT-based smart farming systems based on bidirectional gated recurrent unit with long short-term memory. In: 2024 4th International Conference on Mobile Networks and Wireless Communications (ICMNWC), pp. 1–5. IEEE (2024)
6. Prakash, R., Jyoti, N., Manjunatha, S.: A survey of security challenges, attacks in IoT. In: E3S Web of Conferences. vol. 491, p. 04018. EDP Sciences (2024)
7. Taşcı, B.: Deep-learning-based approach for IoT attack and malware detection. Appl. Sci. **14**(18) (2024). https://doi.org/10.3390/app14188505
8. Ain, N.U., Sardaraz, M., Tahir, M., Abo Elsoud, M.W., Alourani, A.: Securing IoT networks against ddos attacks: a hybrid deep learning approach. Sensors **25**(5) (2025). https://doi.org/10.3390/s25051346
9. Jony, A.I., Arnob, A.K.B.: A long short-term memory based approach for detecting cyber attacks in IoT using cic-iot2023 dataset. J. Edge Comput. **3**(1), 28–42 (2024)

# Design and Implementation of a Real-Time IDS on Embedded Router Platforms Using Deep Neural Networks

Loi Tran Van[1], Loc Nguyen Huu[1], Duan Luong Cong[1], Bien Nguyen Quang[1], Dao Vu Anh[1], and Thuy Tran Thi Thanh[2](✉)

[1] Posts and Telecommunications Institute of Technology, Hanoi, Vietnam
[2] Faculty of Electronics Engineering 1 and EDA Lab, Posts and Telecommunications Institute of Technology, Hanoi, Vietnam
thuyttt@ptit.edu.vn

**Abstract.** Intrusion detection in IoT networks poses significant challenges due to the limited computational resources of embedded devices and the growing sophistication of cyberattacks. In this work, we propose and implement a lightweight Deep Neural Network (DNN)-based intrusion detection model directly on an embedded router platform. The model is trained on the IoT-23 dataset and deployed on the MediaTek AI7688 system-on-chip running OpenWRT. Two experimental scenarios are evaluated: offline testing with pre-collected data and real-time packet capture using libpcap. Results show that the model achieves a high detection accuracy of 99.9%, while requiring only 8.6 KFLOPS per inference, 0.71% average CPU usage, and 6.64 MB RAM. Furthermore, the model delivers predictions within 1.42 ms, ensuring real-time responsiveness. These findings demonstrate that embedded routers can effectively integrate AI-based intrusion detection without compromising networking performance. The proposed approach highlights the practicality and scalability of deploying lightweight DNN models on open-source router platforms, paving the way for secure and intelligent IoT infrastructures.

**Keywords:** IDS · DNN model · real-time detection · embedded router

## 1 Introduction

With the help of Internet of Things (IoT) technology, billions of objects have been connected, and a great quantity of data has been collected. This data may be used for cognitive computing and automated operations that combine digital and physical worlds. IoT applications now affect almost every aspect of life, including transportation, manufacturing, healthcare, and agriculture. Along with its advantages, network security and performance face new challenges [1]. As the first line of defense, an Intrusion Detection System (IDS) plays a crucial role in identifying anomalous network activities that may arise from successful cyberattacks. This function becomes especially important on the Internet of Things, where the diversity of devices, the simplicity of communication protocols,


P. Pichappan et al. (Eds.): RTIS 2025, LNNS 1936, pp. 283–293, 2027.
https://doi.org/10.1007/978-3-032-23396-7_25

and the limited computational resources of system components create unique security challenges [2].

Futhermore, with the rapid advancement of artificial intelligence, modern IDS solutions have evolved into more active and intelligent forms by leveraging machine learning techniques [3]. These systems are capable of comparing and analyzing historical data to accurately detect potential threats in real time. The effectiveness of such approaches, however, strongly depends on the quality and diversity of the datasets used for training. To address this need, benchmark datasets such as KDD-99, NSL-KDD, and UNSW-NB15 have been widely adopted in research. They provide the foundation for training and evaluating neural networkbased IDS models, enabling these systems to detect and classify a wide spectrum of cyber threats across both traditional computer networks and IoT environments [4–6].

In addition, a growing trend is the deployment of IDS powered by deep learning models directly on resource-constrained edge devices such as microcontrollers or embedded boards [7,8]. This approach leverages on-device intelligence to reduce data transmission latency while enhancing security and autonomy in IoT systems. To achieve this, techniques such as lightweight neural networks, optimized convolutional neural networks (CNNs), and model compression methods like quantization and pruning are often employed. As a result, deep learning-based IDS can operate efficiently even on microcontrollers with limited memory and processing capabilities, paving the way for practical IoT security applications.

Machine learning-based IDS have shown great promise in recent years for thwarting assaults connected to the Internet of Thing. Deep Neural Network (DNN) models, such as FFDNN with feature selection [9] and accelerated architectures [10], consistently outperform traditional classifiers, although challenges remain in detecting rare attack types. To address these limitations, enhanced preprocessing techniques [11] have been shown to improve precision, recall, and F1-score, enabling more reliable detection of advanced threats. Building on this foundation, recent studies propose anomaly-based IDS using DNN algorithms combined with mutual information for feature selection, achieving a 3.65% accuracy improvement on the NSL-KDD dataset [12]. Similarly, another work concludes that DNN-based IDS can effectively strengthen in-vehicle network security, particularly for the CAN bus protocol [13]. However, these studies remain limited to using benchmark datasets for training and validating deep learning models as proof of concept, without demonstrating their deployment or evaluation on real-world embedded systems.

Therefore, in this study, we propose a lightweight DNN-based intrusion detection system specifically designed for deployment on embedded routers. Unlike prior works that remain confined to dataset-level evaluations, our approach not only leverages an updated dataset (IoT-23) [14] for training to ensure model relevance and robustness but also demonstrates real-time implementation and testing on the MediaTek AI7688 platform running OpenWRT. This validates both the feasibility and efficiency of integrating deep learning models into low-power IoT devices for practical intrusion detection.

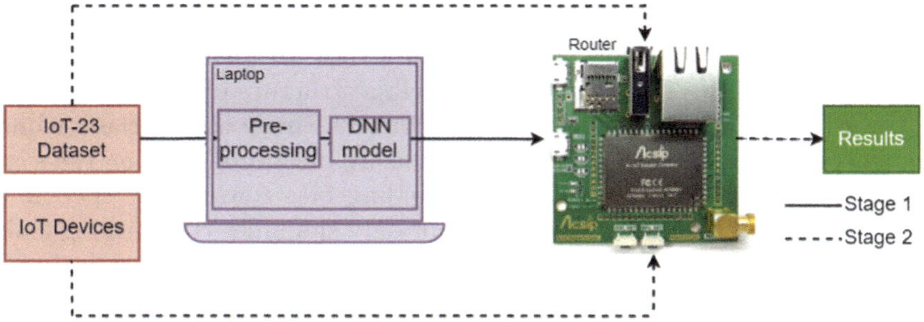


**Fig. 1.** Proposed model for DNN-based Intrusion Detection System.

## 2 Proposed Model

The proposed embedded router intrusion detection system model is illustrated in Fig. 1 and is divided into two stages: (1) training and deploying the model onto the device (router), and (2) experiment and real-time operation on the router. In stage 1 (illustrated with solid lines), data is collected from the publicly available IoT-23 dataset, then preprocessed and split into two subsets: training and testing. The training set is used to train the DNN model to classify traffic as either attack or benign, while the testing set is used to evaluate the model's accuracy. If the results meet the required performance, the model is deployed onto the router. In stage 2 (illustrated with dashed lines), the DNN model is integrated and executed directly on the router. Raw packets are delivered to the device, where preprocessing is performed (similar to stage 1 but implemented in C/C++ instead of Python). The on-device DNN model then predicts whether a packet is an attack or normal and returns the result to support appropriate response actions.

### 2.1 Training and Deployment

To ensure the timeliness and accuracy of the proposed model, data collection for training is of critical importance. In this study, we adopt the IoT-23 dataset as the primary data source. Published in 2020, IoT-23 is a well-regarded network traffic dataset generated from IoT devices. Unlike typical datasets that may include local network details, IoT-23 emphasizes network-level properties, making it more suitable for generalization across diverse IoT environments. This dataset serves as a valuable resource for both researchers and practitioners in IoT security, offering a comprehensive benchmark for testing and evaluating security solutions, particularly in detecting and preventing various IoT-based attacks. Specifically, IoT-23 contains 20 malware traffic captures executed on IoT devices and 3 captures of benign IoT device traffic. In total, it provides 23 labeled categories, comprising 16 attack types, 15 malicious and 1 benign, thus enabling robust training and evaluation of intrusion detection models.

However, in this study, the primary objective is limited to determining whether a given network flow represents an attack or a benign activity. Consequently, instead of utilizing the full multi-class structure of the IoT-23 dataset, all attack-related categories are aggregated into a single attack class, while the benign captures are labeled as normal. To ensure balanced learning, the samples of both classes are evenly distributed, avoiding bias toward any specific category. This binary classification strategy not only simplifies the learning task but also enhances the feasibility of deploying the model on embedded devices with constrained computational and memory resources.

To prepare the data for model training, several preprocessing steps are required, including data cleaning, handling of missing values, encoding of categorical features, and balancing of the dataset. These processes ensure that the input data is consistent, representative, and suitable for effective model learning. Certain features such as "unnamed", "ts", "uid", "tunnel_parents", "local_resp", and "local_orig" in the IoT-23 dataset are removed from the records since they contain null values and provide no meaningful contribution. During data preprocessing, missing values are handled by replacing them with the mean values of the corresponding features for each attack type. Consequently, the averages calculated for the 16 attack types are used to fill in the missing entries in the feature columns. Categorical attributes such as "service", "proto", and "conn_state" are analyzed and encoded into numerical form. Each distinct value within these categorical columns is transformed into a separate binary state vector, thereby expanding the feature space through one-hot encoding. In addition, the dataset includes two IPv4 address columns: "id.orig_h" and "id.resp_h", each representing source and destination IP addresses, respectively. To process these attributes, the IPv4 addresses are encoded into numerical format using the ipaddress library. Specifically, each octet is converted into an 8-bit binary number, and the four octets are concatenated to form a 32-bit binary string. This binary representation is then converted back into a decimal value, providing a consistent numeric encoding of the IP addresses. In this paper, all attack categories are aggregated into a single label with the value 1, while non-attack (benign) traffic is assigned to the label 0. This binary labeling scheme simplifies the classification task and aligns with the main objective of detecting the presence or absence of malicious activity.

Next, the dataset is divided into training and testing sets using an 80/20 split. To ensure reliable evaluation, the dataset was divided into training and testing subsets using a stratified split, maintaining an equal proportion of attack and normal samples in each subset. This approach guarantees that rare attack instances are represented during training.

Next, the data is trained using the DNN model to classify network traffic as either attack or benign. A DNN model with three fully connected layers is constructed in this study, consisting of one input layer, two hidden layers, and one output layer. The first hidden layer contains 64 nodes, followed by the second hidden layer with 32 nodes. Both hidden layers use the ReLU activation function to capture nonlinear relationships in the data. To reduce overfitting, a

Dropout layer with a rate of 0.3 is applied after each hidden layer. The output layer consists of a single node with the sigmoid activation function, suitable for binary classification tasks such as distinguishing between attack and benign traffic. The model is trained using the Adam optimizer with binary cross-entropy as the loss function.

We employed a method that prompted the suggested model to stop training early to reduce over-learning. The training is over when the validation error function does not decrease after a certain number of epochs. To verify that the validation error does not increase over the epochs, we can adjust the number of epochs to provide the best network accuracy over the learning interval. Lastly, using the test IoT-23 dataset, we will assess the proposed model. Loss, precision, recall, and F1-score are the criteria that are examined.

After the DNN model has been tested and meets the required accuracy, it will be deployed onto the OpenWrt router hardware. First, the model is saved in the ".h5" format and then converted to ".tflite" to optimize its size and make it suitable for embedded environments. Next, the ".tflite" file is transformed into a ".h" file (C byte array) so that it can be embedded directly into the C++ source code without the need to read from an external file. In parallel, the data preprocessing techniques are also migrated from Python code to C/C++ code.

In addition to loading the model and performing data preprocessing, the program also links with the pcap library to capture network traffic (packet capture), carry out preprocessing, and feed the processed data into the model for prediction.

### 2.2 Experiment

The DNN model is deployed and executed in real time on the MediaTek AI7688 platform (LinkIt Smart 7688H), a compact SoC designed for IoT and embedded networking. Running OpenWRT, it supports essential services such as DHCP, NAT, and firewall, while allowing firmware customization and OTA updates. Powered by a MIPS 24KEc processor at approximately 580 MHz and 128 MB RAM, the AI7688H provides stable performance for IoT and networking tasks. It supports 802.11 b/g/n Wi-Fi, Fast Ethernet, and USB 2.0, along with interfaces such as GPIO, SPI, and UART for sensor and control integration. These features make it suitable for applications like smart monitoring, home automation, and digital agriculture. Compared to platforms like Raspberry Pi, the AI7688H offers lower cost, compact size, and reduced power consumption, making it ideal for large-scale IoT deployment.

At this second stage, we conduct experiments on the device through two scenarios. Scenario 1: Offline testing with a data file: On the router (MediaTek AI7688), the IoT-23 testing dataset is loaded into the inference application. The application reads each record, extracts the appropriate features, and applies normalization consistent with the training pipeline before feeding the feature vector into the DNN model for inference. Running tests in batches and ensuring that on-device preprocessing remains fully aligned with the training preprocessing are critical for obtaining meaningful results.

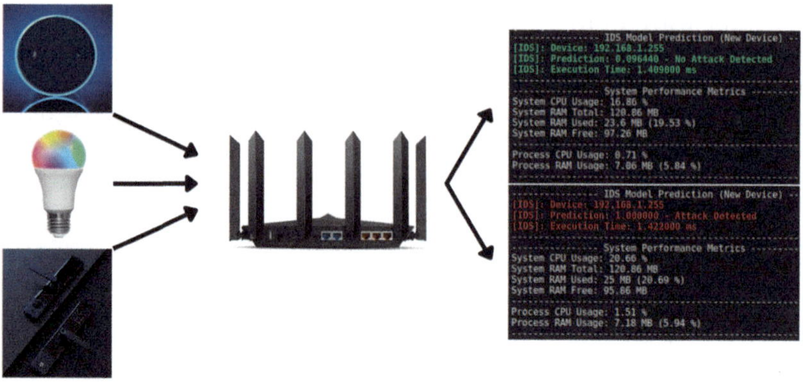


**Fig. 2.** The experimental intrusion detection system, based on the DNN model, is deployed on the router.

In addition to scenario 1, we also conduct experiments using scenario 2 to assess the system under real-time operational conditions. At the scenario 2, live packet capture (real-time) with libpcap is illustrated in Fig. 2. In this setup, the router is configured to run a continuous packet capture process using the libpcap library, which is widely adopted in network monitoring applications. The capture process collects raw packets directly from the router's network interface as devices communicate over the network. Once captured, the system performs feature extraction by parsing packet headers and payload metadata to derive meaningful attributes such as protocol type, connection state, service, source/destination IP, and flow statistics. To ensure consistency with the training phase, these raw attributes undergo a preprocessing pipeline implemented in C/C++ on the router. This includes steps such as one-hot encoding for categorical fields (e.g., protocol, connection state), IP encoding to convert IPv4 addresses into numerical representations, and feature scaling/normalization to align the value ranges with those used during training. The processed features are then aggregated either per-packet or per-flow, and further grouped into time windows (e.g., 1 second or 5 seconds) to capture temporal patterns of network activity. The resulting feature vectors are streamed into the deployed DNN model on the MediaTek AI7688 router for real-time inference. Each prediction determines whether the observed traffic corresponds to an attack or benign activity. When the model's output exceeds a predefined decision threshold, the system automatically triggers a response mechanism. This may include writing warnings to log files, generating syslog messages for centralized monitoring, sending HTTP or MQTT notifications to external security dashboards, or dynamically updating iptables firewall rules to block malicious traffic. Through this scenario, the router effectively functions as an autonomous intrusion detection node, capable of both monitoring and reacting to network threats in real time.

## 3 Results and Discussion

As described in Sect. 2, the DNN model was trained to achieve robust classification performance. The training and validation curves, presented in Fig. 3 and Fig. 4, demonstrate that the model rapidly converges and attains high accuracy within the first 20 epochs. To comprehensively evaluate its performance, the confusion matrix (Fig. 5) and the precision, recall, and F1-score metrics (Table 1) were analyzed. The results indicate that the model effectively distinguishes between attack and benign traffic, achieving an overall accuracy of 99.99%. Moreover, both macro and weighted averages of precision, recall, and F1-score are reported to mitigate the influence of potential class imbalance, ensuring a fairer representation of performance across both categories. Despite the extremely high accuracy, these averaged metrics provide a more balanced and realistic view of the model's effectiveness. Consequently, the proposed DNN can be confidently deployed on embedded router platforms for real-world intrusion detection tasks.

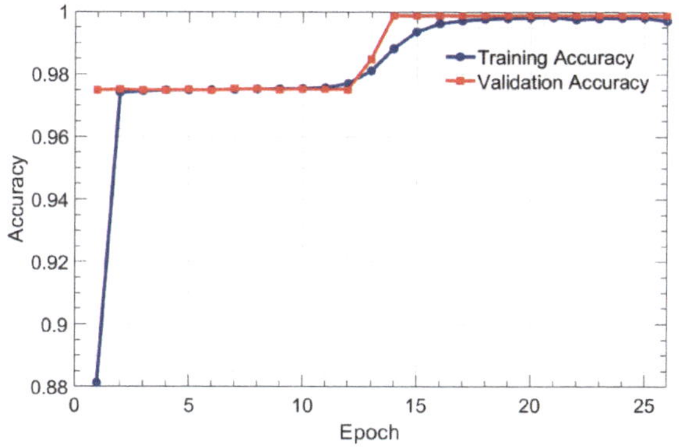


**Fig. 3.** Training and validation accuracy of DNN model.

The resource utilization metrics in Table 2 highlight the computational and memory demands of the DNN model deployed on the MediaTek AI7688 router. The model requires approximately 8.6 KFLOPS per inference. With a size of about 120 KB, it fits well within the program memory limits. The average CPU usage is only 0.71%, indicating that the model introduces negligible computational overhead and can operate alongside other network processes without performance degradation. RAM usage during inference is around 6.64 MB (only 5.57% of total memory), which is sufficient to accommodate intermediate computations and I/O buffers without any risk of memory overflow. In addition to the lightweight resource utilization, the proposed DNN model achieves extremely low inference latency, requiring only about 1.42 ms to produce a prediction. This rapid response time is critical for real-time intrusion detection, as it allows the

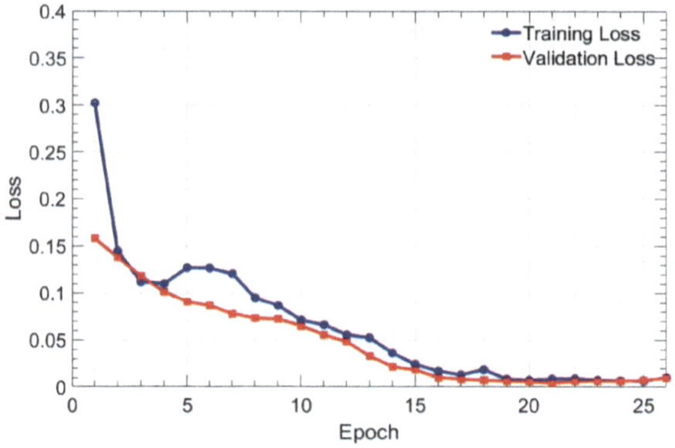


**Fig. 4.** Training and validation loss of DNN model.

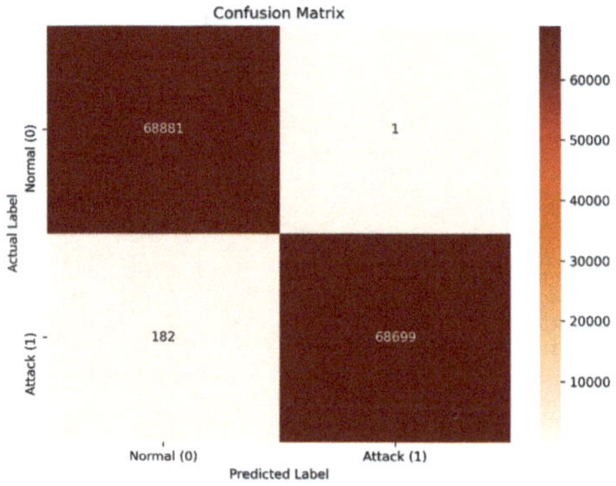


**Fig. 5.** Confusion Matrix.

system to promptly analyze traffic and trigger alerts or mitigation actions without noticeable delays. The combination of sub-millisecond processing overhead and efficient resource usage further demonstrates that the model is well-suited for deployment in latency-sensitive IoT environments, where immediate detection and response are essential to maintaining network security.

**Table 1.** The performance of the DNN model on the IoT-23 dataset.

| Labels | Precision | Recall | F1-Score | Samples |
|---|---|---|---|---|
| Attacks | 100 | 99.9 | 99.9 | 68881 |
| Normal | 99.9 | 99.9 | 99.9 | 68882 |
| Accuracy | | | 99.9 | 137763 |
| Macro avg | 99.9 | 99.9 | 99.9 | 137763 |
| Weighted avg | 99.9 | 99.9 | 99.9 | 137763 |

**Table 2.** Resource utilization

| Parameters | Value |
|---|---|
| Est. FLOPs (KFLOPS) | 8.6 |
| Model size (KB) | 120 |
| Avg. CPU usage normal | 0.71% |
| RAM normal | 6.64 MB (5.57 %) |
| Avg. CPU usage attack | 1.51% |
| RAM attack | 6.75 MB (5.66 %) |
| Inference Time (ms) | 1.42 |

The evaluation confirms that the proposed model is lightweight and resource-efficient, making it feasible for real-time deployment on embedded routers. With minimal CPU and memory usage, the system can maintain stable operation while supporting additional tasks such as packet capture, logging, or intrusion response without affecting performance. These characteristics highlight the potential of integrating DNN-based intrusion detection into low-power IoT devices. In particular, the MediaTek AI7688 demonstrates that attack detection models can operate in real time with negligible computational overhead, achieving an inference latency of only 1.42 ms. This ensures that core networking functions such as routing, firewalling, and traffic management remain unaffected. Moreover, the successful deployment on the router's open-source operating system, OpenWRT, underlines both the flexibility and practicality of using open platforms to embed AI-driven security mechanisms directly within IoT infrastructure. This not only validates the effectiveness of the approach but also shows its potential for scalable and reliable network protection.

## 4 Conclusion

This work demonstrates the feasibility of deploying a DNN-based intrusion detection model directly on resource-constrained IoT routers. Through both offline and real-time experiments, the system achieves high detection accuracy (99.99%) while maintaining extremely low computational overhead. On the MediaTek AI7688 platform, the model requires only 8.6 KFLOPS per inference, consumes around 0.71% CPU and 6.64 MB RAM, and delivers predictions within 1.42 ms, ensuring real-time responsiveness without compromising core networking functions. The successful integration on OpenWRT further highlights the practicality of using open-source platforms to embed AI-driven security mechanisms into low-power IoT devices. These results validate the effectiveness, scalability, and reliability of the proposed approach, paving the way for secure and intelligent IoT network infrastructures.

## References


1. Rachit, B.S., Ragiri, P.R.: Security trends in Internet of Things: a survey. SN Appl. Sci. **3**, 121 (2021). https://doi.org/10.1007/s42452-021-04156-9
2. Heidari, A., Jamali, M.A.J.: Internet of Things intrusion detection systems: a comprehensive review and future directions. Clust. Comput. **26**, 3753–3780 (2023). https://doi.org/10.1007/s10586-022-03776-z
3. Sahar, N., Mishra, R., Kalam, S.: Deep learning approach-based network intrusion detection system for fog-assisted IoT. In: Tiwari, S., Suryani, E., Ng, A.K., Mishra, K.K., Singh, N. (eds.) Proceedings of International Conference on Big Data, Machine Learning and their Applications, Lecture Notes in Networks and Systems, vol. 150, pp. 39–50 (2021). https://doi.org/10.1007/978-981-15-8377-3_4
4. Tavallaee, M., Bagheri, E., Lu, W., Ghorbani, A.A.: A detailed analysis of the KDD CUP 99 data set. In: Proceedings of the IEEE Symposium on Computational Intelligence for Security and Defense Applications (CISDA 2009), pp. 1–6 (2009). https://doi.org/10.1109/CISDA.2009.5356528
5. Dhanabal, L., Shantharajah, S.P.: A study on NSL-KDD dataset for intrusion detection system based on classification algorithms. Int. J. Adv. Res. Comput. Commun. Eng. 4(6), 446–451 (2015). https://doi.org/10.17148/IJARCCE.2015.4696
6. Moustafa, N., Slay, J.: UNSW-NB15: a comprehensive data set for network intrusion detection systems. In: Proceedings of the 2015 Military Communications and Information Systems Conference (MilCIS), pp. 1–6 (2015). https://doi.org/10.1109/MilCIS.2015.7348942
7. Kornaros, G.: Hardware-assisted machine learning in resource-constrained IoT environments for security: review and future prospective. IEEE Access **10**, 58603–58622 (2022). https://doi.org/10.1109/ACCESS.2022.3179047
8. Alwaisi, Z., Kumar, T., Harjula, E., Soderi, S.: Securing constrained IoT systems: a lightweight machine learning approach for anomaly detection and prevention. Internet Things **28**, 101398 (2024). https://doi.org/10.1016/j.iot.2024.101398
9. Qi, B., Wu, M., Zhang, L.: A DNN-based object detection system on mobile cloud computing. In: 2017 17th International Symposium on Communications and Information Technologies (ISCIT), pp. 1–6 (2017). https://doi.org/10.1109/ISCIT.2017.8261188
10. Potluri, S., Diedrich, C.: Accelerated deep neural networks for enhanced intrusion detection system. In: Proceedings of the 2016 IEEE 21st International Conference on Emerging Technologies and Factory Automation (ETFA), pp. 1–8 (2016). https://doi.org/10.1109/ETFA.2016.7733515
11. Chen, P., Guo, Y., Zhang, J., Wang, Y.: A novel preprocessing methodology for DNN-based intrusion detection. In: 2020 IEEE 6th International Conference on Computer and Communications (ICCC), pp. 2059–2064 (2020). https://doi.org/10.1109/ICCC51575.2020.9345300
12. Al-Fuhaidi, B., et al.: Anomaly-based intrusion detection system in WSN using DNN algorithm. In: Proceedings of the 2024 1st International Conference on Emerging Technologies for Dependable Internet of Things (ICETI), pp. 1–8 (2024). https://doi.org/10.1109/ICETI63946.2024.10777266

13. Boumiza, S., Braham, R.: In-vehicle network intrusion detection using DNN with ReLU activation function. In: Proceedings of the 2023 International Conference on Cyberworlds (CW), pp. 410–416 (2023). https://doi.org/10.1109/CW58918.2023.00070
14. Garcia, S., Parmisano, A., Erquiaga, M.J.: IoT-23: a labeled dataset with malicious and benign IoT network traffic. Stratosphere Lab. (2020). https://doi.org/10.5281/zenodo.4743746

# Design and Implementation of a TinyML-Enabled Smart Scale for Real-Time Fruit Recognition

Bien Nguyen Quang, Thang Dang Huu, Duan Luong Cong, Thuy Tran Thi Thanh, Ngoc Dinh Quang, and Minh Nguyen Ngoc(✉)

Faculty of Electronics Engineering 1 and EDA Lab, Posts and Telecommunications Institute of Technology, Hanoi, Vietnam
minhnn@ptit.edu.vn

**Abstract.** The absence of identifiers such as barcodes or QR codes for fruits in small retail stores often leads to manual product selection, inefficiencies, and pricing errors. To address this challenge, this study develops a low-cost intelligent weighing system that integrates automatic fruit recognition directly on a resource-constrained edge device. The system is built on the ESP32-S3 microcontroller, which incorporates a camera for image capture, local processing, and result display without dependence on external computing infrastructure. A dataset of 11 fruit categories was collected from the domestic market and augmented to 11,000 samples, ensuring robustness against variations in size, lighting, and background. A lightweight MobileNet V1 128 × 128 0.25 model, optimized through INT8 quantization, was deployed to balance accuracy and computational efficiency. The system achieved 95.7% overall accuracy and up to 100% for most classes, with an average latency of 192 ms per cycle. These results demonstrate the feasibility of real-time, low-cost fruit recognition on embedded platforms, offering a practical and scalable solution for intelligent checkout in small-scale retail environments.

**Keywords:** fruit recognition · embedded system · MobileNetV1 · edge computing · real-time classification

## 1 Introduction

In small retail stores with limited human resources, fruit handling and classification often pose significant challenges. Unlike packaged products with barcodes or QR codes that facilitate quick scanning during checkout, fruits are seldom labeled with unique identifiers. Consequently, staff or customers must manually search and select the corresponding product in the system catalog, a time-consuming process prone to pricing errors. To address this limitation, many studies have proposed automatic checkout systems integrating image recognition [1]. Within this context, fruit recognition systems embedded directly into electronic scales have emerged as a feasible solution. By automatically classifying and accurately


P. Pichappan et al. (Eds.): RTIS 2025, LNNS 1936, pp. 294–306, 2027.
https://doi.org/10.1007/978-3-032-23396-7_26

suggesting the product type once placed on the scale, such systems reduce staff workload, accelerate transactions, and enhance the shopping experience.

The rapid development of Artificial Intelligence (AI), particularly Deep Learning (DL), has driven significant advances in object recognition tasks. In agriculture and retail, Convolutional Neural Networks (CNNs) have demonstrated outstanding performance in fruit classification, achieving experimental accuracy rates of up to 97–98% [2]. This technology is now considered a key tool in agricultural image processing, including classification, detection, and quality assessment [3,4]. CNNs, inspired by the biological visual processing mechanism, are a DL architecture specifically designed for image data [5–7]. With their ability to automatically extract features through convolutional operations, CNNs have become the foundation of numerous studies on agricultural product classification. Recent works have employed direct training and transfer learning with modern architectures such as VGG16, ResNet, InceptionV3, and MobileNet [8,9]. In particular, MobileNet and its variants have shown advantages in accuracy and computational efficiency when applied to large datasets or combined with transfer learning [10].

Despite these positive results, practical deployment of fruit recognition systems still faces several challenges. Environmental factors such as lighting variations or occlusions often degrade accuracy [11]. Moreover, the similarity in shape and color among different fruit types makes discrimination difficult [12]. Furthermore, high-performing DL models typically require robust computing infrastructure (CPU/GPU), which leads to high costs, an especially unfavorable condition for small retail stores [13,14].

Several notable studies can be mentioned. For instance, Khoiruddin and Tena et al. [15] proposed a fruit and vegetable classification system using CNN-MobileNetV2, achieving 97.31% accuracy, though it was only tested on a high-performance laptop. Similarly, Jose Luis Rojas-Aranda et al. [16] introduced a method to accelerate the checkout process. However, their work remained preliminary and was not integrated into real-world systems. In addition, Sridharan et al. [17] presented a scanning and invoicing system based on YOLOv4, yet it remained at the prototype stage and did not demonstrate continuous operation in retail environments. These limitations highlight a significant gap between research and application: most existing solutions still rely on CPU/GPU in computer environments, whereas deployment on edge devices better suited for small stores has not been fully explored. Therefore, the key challenge is to design lightweight models that ensure high accuracy while being efficiently deployable on resource-constrained hardware.

This study introduces a low-cost intelligent electronic weighing system with integrated automatic fruit recognition, directly deployed on an ESP32-S3 edge device. By leveraging MobileNet V1 with INT8 quantization, the system achieves high accuracy, real-time inference, and low memory usage, eliminating the need for QR codes or manual selection. A web-based interface enables seamless shopping and checkout, enhancing usability. The key contribution lies in the practical deployment and evaluation of a lightweight, edge-based model under real-world conditions, demonstrating both reliability and strong potential for small-scale retail applications.

## 2 Methodology

### 2.1 System Architecture

The overall system architecture is illustrated in Fig. 1. The ESP32-S3 Cam captures fruit images via the OV2640 camera and weight data from the HX711 sensor when the fruit is placed on the scale. All data is processed and inferred directly at the edge on the ESP32-S3 microcontroller without requiring a connection to a server or cloud platform, thereby reducing latency and enhancing system autonomy. The classification results, weight, and recognition suggestions are displayed on the TFT screen, enabling users to select the correct fruit type without entering product codes or scanning QR labels. In addition, the data can be synchronized and updated via a web or mobile application through the MQTT protocol, facilitating payment and retail management.

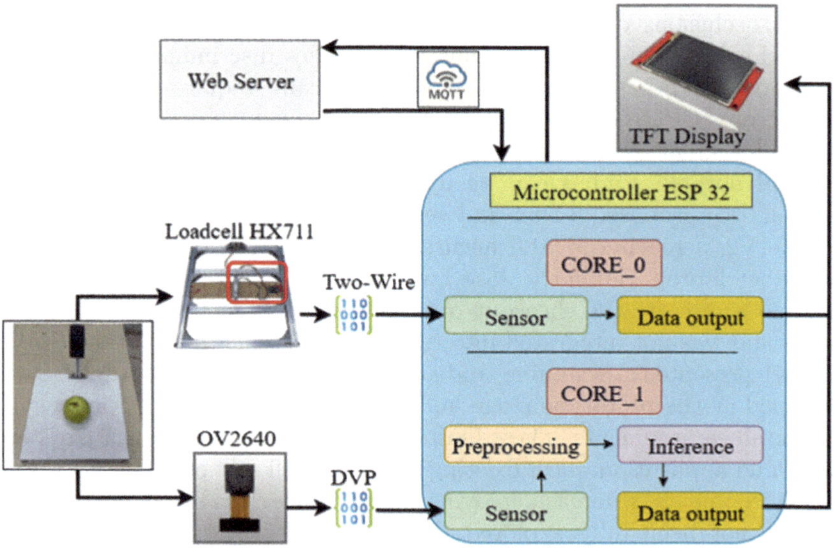


**Fig. 1.** Embedded System Architecture for Fruit Classification.

### 2.2 Data Collection

This study constructs a dedicated dataset rather than relying on publicly available ones to ensure suitability for real-world applications and accurately capture the characteristics of locally available fruits and vegetables. In the local context, produce exhibits considerable variation in size, color, shape, and display conditions, which are often underrepresented in standardized datasets. Building a new dataset allows the deep learning model to adapt more effectively to complex real-world scenarios, such as changes in lighting.

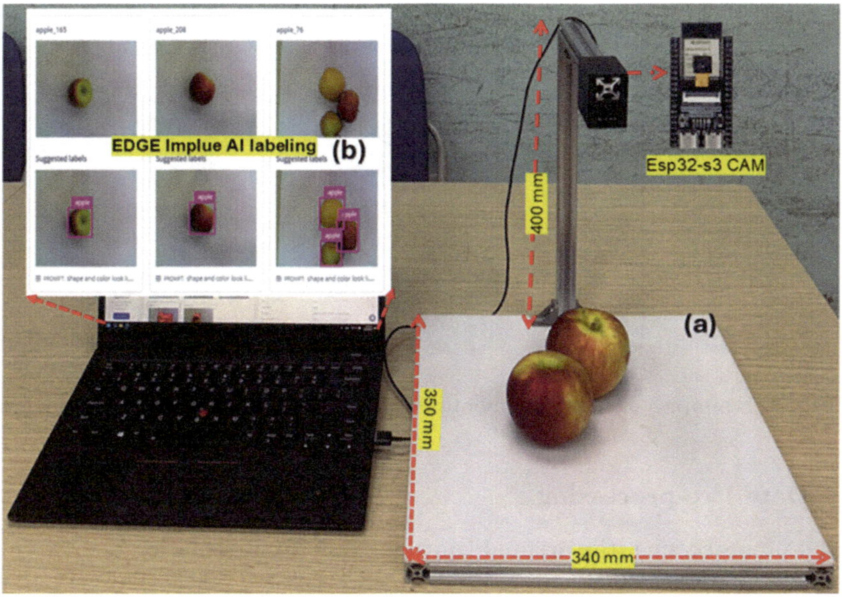


**Fig. 2.** (a) Hardware setup with the ESP32-CAM fixed at a height of 400 mm above a collection table measuring 340 × 350 mm; (b) data labeling process using the AI Labeling tool on the Edge Impulse platform.

The data collection process was carried out using the ESP32-CAM at a resolution of 1600 × 1200 × 3, covering eleven fruit and vegetable classes selected to represent a diverse set of locally available produce: apple, mango, cucumber, wax apple, custard apple, green guava, tomato, pear, red bell pepper, yellow bell pepper, and carrot. For each class, 200 original images were captured from multiple angles, under varying lighting conditions (natural, artificial, and low light), and across diverse backgrounds, as illustrated in Fig. 2(a).

After image acquisition, data augmentation techniques were applied to expand and diversify the dataset, including flipping, rotation, brightness and contrast adjustment, and noise addition, as shown in Fig. 3. Augmentation was performed until each class contained 1,000 samples, resulting in a balanced dataset of 11,000 images. The subsequent step of object labeling was conducted using the AI Labeling tool on the Edge Impulse platform [18], as illustrated in Fig. 2(b).

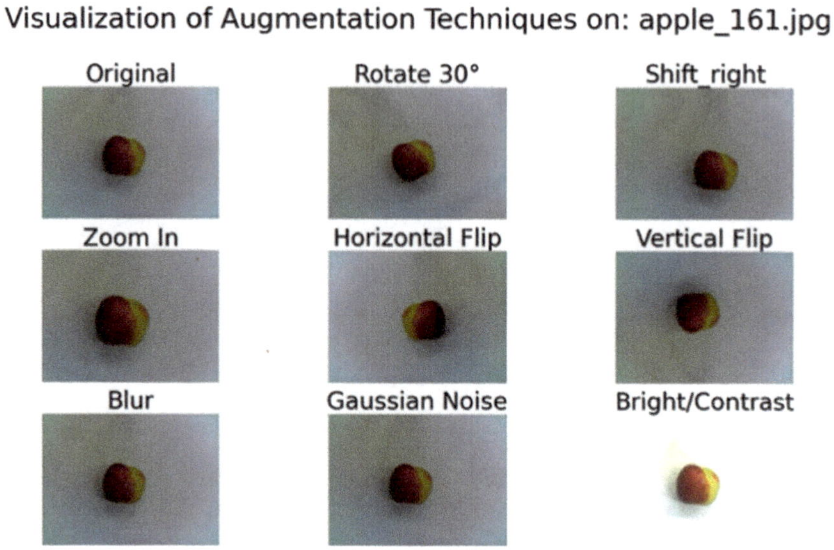


**Fig. 3.** Examples of Data Augmentation Techniques Applied to Fruit Images.

### 2.3 Data Pre-processing

During preprocessing, the original fruit images with a resolution of 1600 × 1200 × 3 pixels were resized and normalized to 128 × 128 × 3 pixels. This step reduced the pixel count from approximately 5.76 million to 49,152 per image, lowering the computational cost of convolution operations, decreasing memory usage for intermediate tensors, and improving training and inference speed. Although the resolution was reduced, the 128 × 128 size was sufficient to retain essential visual features for fruit recognition, including shape, contours, surface texture, and color variations.

### 2.4 Machine Learning Analysis

This study employed the MobileNet V1 128 × 128 0.25 architecture for fruit image classification using preprocessed input data. MobileNet represents a variant of convolutional neural networks specifically designed for resource-constrained devices, with MobileNet V1 being the first version, characterized by its lightweight design and computational efficiency [19]. At the core of this architecture lies the Depthwise Separable Convolution, which replaces standard convolution with a combination of depth-wise convolution (3 × 3) to capture spatial features within each channel and pointwise convolution (1 × 1) to integrate inter-channel information. This design substantially reduces the number of parameters and computational cost while maintaining strong representational capacity. In the 128 × 128 variant with a width multiplier of 0.25, the number of channels in convolutional layers is further reduced, optimizing computational efficiency and making the model suitable for basic vision tasks on resource-limited devices.

The structure of the model is illustrated in Fig. 4 and consists of the following compo-nents:

- **Input Layer:** accepts RGB images of size 128 × 128 × 3 (49,152 features). Images are preprocessed by resizing to a fixed dimension and normalizing pixel values to either [0, 1], thereby improving convergence speed during training.
- **MobileNet V1 Backbone:** includes an initial standard convolutional layer followed by 13 Depthwise Separable Convolution (DS Conv) blocks. Each block contains a 3 × 3 depthwise convolution to learn spatial features within channels, a 1 × 1 pointwise convolution to integrate features across channels, batch normalization, and the ReLU activation function to stabilize training and introduce non-linearity. This design significantly reduces parameters and computational overhead while preserving strong expressive power.
- **Classification Block:** the backbone output is passed through a reshape layer, followed by a dropout layer with a rate of 0.1 to mitigate overfitting, and then a flatten layer. Finally, a dense layer with 11 softmax neurons generates a probability distribution across the 11 fruit classes, where the highest probability corresponds to the predicted label.

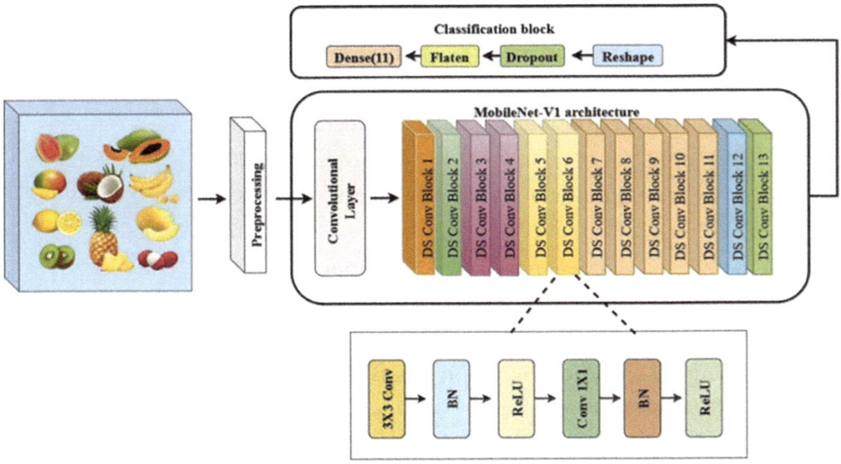


**Fig. 4.** Overall Framework of Fruit Classification Using MobileNet V1 Architecture.

The training process includes the following steps: converting and normalizing the input data format, randomly splitting the dataset into 80% for training and 20% for testing, applying preprocessing operations such as pixel normalization and image resizing, training the model on the training set to optimize weights via backpropagation combined with optimization algorithms, and finally evaluating the model on the test set to ensure generalization and reliability of classification results. With this design, MobileNet V1 128×128 0.25 achieves a well-balanced trade-off between computational efficiency and accuracy, enabling effective deployment for fruit image classification tasks in resource-constrained environments.

### 2.5 Model Deployment on Embedded Device

The deployment architecture of the fruit classification model on the embedded ESP32 platform follows the overall system design introduced earlier in Fig. 1. In this architecture, the ESP32 microcontroller acts as the central processing unit, utilizing its dual cores (`CORE_0` and `CORE_1`) to execute parallel tasks. The LoadCell HX711 sensor is connected to the ESP32 through a two-wire interface to measure sample weight, while the DVP camera module acquires image data. `CORE_0` processes the load cell signals and generates the corresponding output, whereas `CORE_1` performs image preprocessing and runs the deep learning model to produce classification results. These outputs are synchronized and displayed on a TFT LCD screen and transmitted to a web server via the MQTT protocol for remote monitoring.

**Table 1.** Summary of classification results at different evaluation stages

| Evaluation Stage | Metric | Value |
|---|---|---|
| Training | Accuracy | 0.99 |
| | Loss | 0.01 |
| Validation | Val_Accuracy | 0.99 |
| | Val_Loss | 0.00 |
| Quantization (int8) | Accuracy | 0.98 |
| | Loss | 0.07 |
| Final Testing | Accuracy | 0.96 |

The deployment workflow is implemented through the Edge Impulse platform, which automates essential steps such as data preprocessing, feature extraction, and conversion of the trained model into C++ source code. During this process, the model is quantized into the INT8 format to reduce memory footprint, accelerate inference, and optimize performance on resource-constrained devices such as the ESP32, with only negligible accuracy loss. Once optimized, the firmware is compiled and programmed into the ESP32 using PlatformIO, an open-source embedded development environment. Integrating Edge Impulse and PlatformIO provides a streamlined end-to-end pipeline from training and quantization to deployment, simplifying development and enhancing applicability in IoT and edge computing systems.

# 3 Results and Discussion

## 3.1 Classification Result

The model was trained for 60 epochs and performed excellently on the training and validation datasets. As presented in Table 1, the model achieved an accuracy of 0.99 with a loss of 0.01 on the training set, while maintaining an accuracy of 0.99 and near-zero loss on the validation set. This indicates effective learning without signs of overfitting. After applying INT8 quantization, the model attained an accuracy of 0.98 with a loss of 0.07, reflecting only a slight reduction compared to the original model. This confirms the feasibility of deploying the model on embedded devices, ensuring high performance while reducing memory and computational requirements.

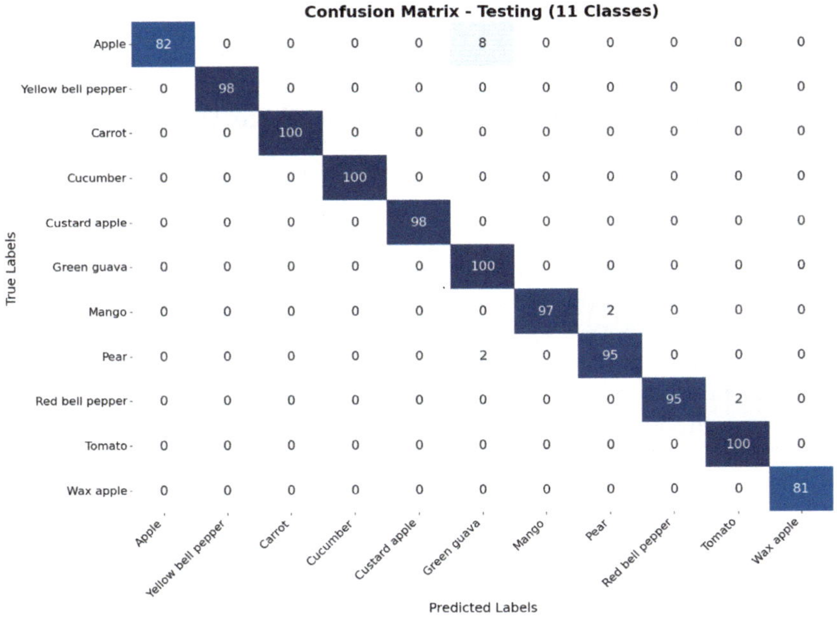


**Fig. 5.** Confusion Matrix of classification results for 11 fruit categories.

Further evaluation on an independent testing set showed that the model achieved an accuracy of 0.96. Most classes reached perfect accuracy, including carrot, cucumber, and tomato; however, localized misclassifications were observed, such as apple being confused with custard apple (8.4%) and pear with mango (2.2%). The wax apple class exhibited the lowest accuracy (81.0%), suggesting the model had difficulty capturing its distinctive features. These results validate the model's stability and effectiveness, highlighting the need to enrich the training dataset for visually similar fruits to improve generalization.

### 3.2 Per-Class Performance

The classification performance of the proposed system is illustrated in the confusion matrix in Fig. 5. The results show that most fruit categories achieved high accuracy, with carrot, cucumber, green guava, and tomato all reaching 100%. Red bell pepper, custard apple, and yellow bell pepper also maintained accuracies above 95%, confirming the model's strong ability to distinguish these fruits. In contrast, apple and wax apple obtained lower accuracies of 82.4% and 81.0%, respectively, mainly due to partial misclassification with visually similar classes. For example, 8.4% of apple samples were misclassified as custard apple, while mango and pear experienced minor confusion rates of 2.3% and 2.2%. Despite these localized errors, the overall classification remained highly reliable across 11 categories, demonstrating the robustness of the model under real testing conditions. Performance could be further enhanced by enriching the dataset with more samples from visually similar fruits to improve generalization in diverse real-world environments.

### 3.3 Deployment of Real-Time Models on Edge Devices

The model was converted into the TensorFlow Lite Flatbuffer format and integrated with the Edge Impulse platform for deployment on the ESP32-S3 microcontroller. This conversion generated C++ source files containing the model weights, which were compiled into executable machine code. To optimize performance, Edge Impulse applied post-training quantization from `float32` to `int8`, thereby reducing memory usage and accelerating inference while preserving high accuracy. The final firmware incorporated all functional modules sensor interfacing, data preprocessing, classification, and result visualization enabling the system to operate autonomously on the embedded device.

**Table 2.** Execution time and resource usage on ESP32-S3

| Category | Measured Value |
|---|---|
| Data preprocessing | 23 ms |
| Data classification model | 158 ms |
| Displaying results | 11 ms |
| Total Execution Time | 192 ms |
| RAM Usage | 57.7 KB (18.0%) |
| Flash Usage | 1085.4 KB (17.0%) |

Resource utilization and execution times are summarized in Table 2. The total memory footprint was 59 KB RAM (18%) and 1111 KB Flash (17%), both well within the ESP32-S3 hardware limits. In terms of performance, preprocessing required 23 ms, model inference 158 ms, and result visualization 11 ms, yielding an overall processing time of 192 ms per cycle. With latency below 0.2 s, the system demonstrates the feasibility of real-time fruit classification on a low-cost, resource-constrained edge device.

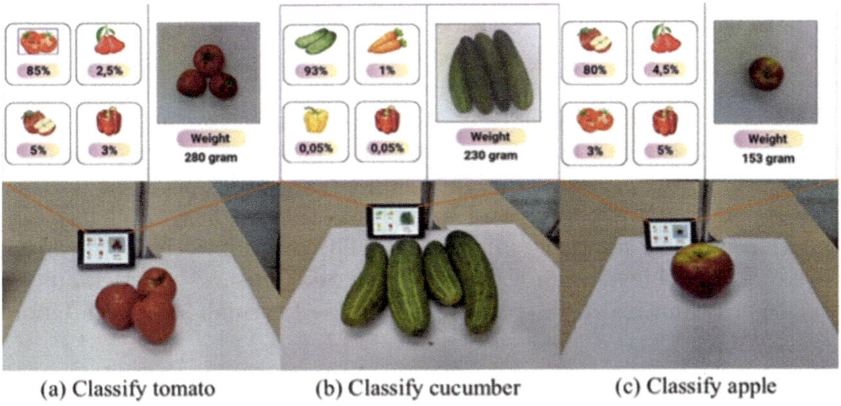


**Fig. 6.** Experimental Results of Fruit Classification on Embedded Device.

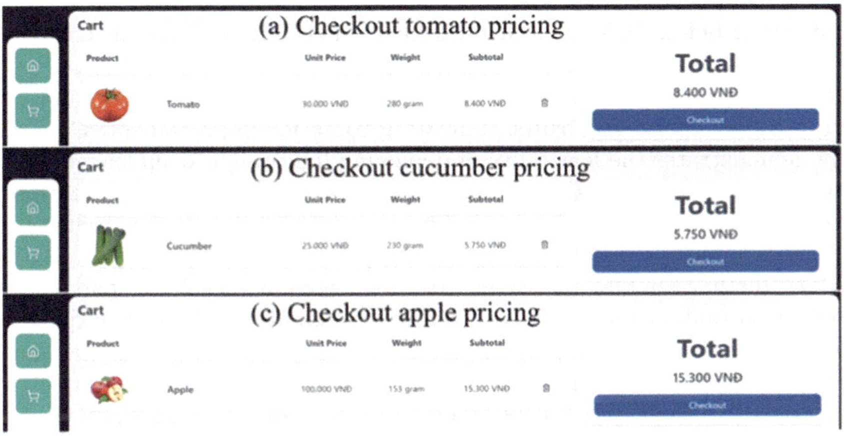


**Fig. 7.** Web-based real-time cart and checkout interface for automatic fruit pricing.

In practical experiments, the ESP32-S3 Cam acquired images from the camera and weight data from the HX711 sensor, which were subsequently preprocessed and passed through the MobileNet V1 model for classification. The results on the TFT screen confirmed accurate recognition of fruits such as tomato, cucumber, and apple, along with supplementary weight and prediction confidence information. This real-time processing directly on the device enabled rapid system response even under varying lighting conditions and fruit morphologies. The practical deployment of the fruit recognition system on an embedded device further demonstrated high accuracy in classifying multiple fruit types, confirming its feasibility for real-time operation in small retail settings. Figure 6 illustrates the experimental results on three sample fruits: (a) tomato, (b) cucumber, and (c) apple. The model correctly predicted each fruit type with over 80% accuracy, while detection of non-target fruits remained below 10%, highlighting precise classification of target fruits and minimal misidentification.

Moreover, the system includes a web-based interface for real-time cart management and checkout. Figure 7 presents the web interface, which features the shopping cart, product names, weights, unit prices, and a payment button. The system supports data storage and synchronization with a database, allowing compilation of daily summary information. These results demonstrate that the deployed model achieves high accuracy above 80%, ensures practical applicability, and operates in real-time, making it suitable for small-scale retail stores.

## 4 Conclusions

This study presented the design and real-world evaluation of an intelligent electronic weighing system with integrated fruit recognition deployed on the ESP32-S3 microcontroller. Leveraging the lightweight MobileNet V1 model with INT8 quantization, the system achieved real-time classification of 11 common fruit types, with target fruit accuracy and F1-scores above 80% and non-target misclassification below 10%. The firmware occupied only 57.7 KB of RAM (18%) and 1,085.4 KB of Flash (17%), with a total inference and visualization time of 192 ms per cycle. Limitations include the moderate accuracy and inability to recognize packaged fruits, indicating room for improvement. Overall, the study demonstrates the feasibility of deploying lightweight deep learning models on resource-constrained edge devices for near real-time fruit classification, and future work should focus on expanding the dataset, improving accuracy, and optimizing inference for broader retail applications. Integrating a web-based interface further enables real-time cart management, checkout, and data synchronization, enhancing practical usability in small-scale stores. Experimental results confirmed system robustness under varying lighting conditions and fruit morphologies. These findings highlight the potential of edge-deployed deep learning models to bridge the gap between research prototypes and practical retail implementations.

## References

1. Tan, L., Liu, S., Gao, J., Liu, X., Chu, L., Jiang, H.: Enhanced self-checkout system for retail based on improved YOLOv10. J. Imaging **10**, 248 (2024). https://doi.org/10.3390/jimaging10100248
2. Maulana, F.F., Rochmawati, N.: Klasifikasi Citra Buah Menggunakan convolutional neural network. J. Inform. Comput. Sci. JINACS **1**, 104–108 (2020). https://doi.org/10.26740/jinacs.v1n02.p104-108

3. Espinoza, S., Aguilera, C., Rojas, L., Campos, P.G.: Analysis of fruit images with deep learning: a systematic literature review and future directions. IEEE Access **12**, 3837–3859 (2024). https://doi.org/10.1109/ACCESS.2023.3345789
4. Rojas Santelices, I., Cano, S., Moreira, F., Peña Fritz, Á.: Artificial vision systems for fruit inspection and classification: systematic literature review. Sensors **25**, 1524 (2025). https://doi.org/10.3390/s25051524
5. Saxena, A.: An introduction to convolutional neural networks. Int. J. Res. Appl. Sci. Eng. Technol. **10**, 943–947 (2022). https://doi.org/10.22214/ijraset.2022.47789
6. Rahman, M.M., Basar, M.A., Shinti, T.S., Khan, M.S.I., Babu, H.M.H., Uddin, K.M.M.: A deep CNN approach to detect and classify local fruits through a web interface. Smart Agric. Technol. **5**, 100321 (2023). https://doi.org/10.1016/j.atech.2023.100321
7. Albarrak, K., Gulzar, Y., Hamid, Y., Mehmood, A., Soomro, A.B.: A deep learning-based model for date fruit classification. Sustainability **14**, 6339 (2022). https://doi.org/10.3390/su14106339
8. Ahmed, M.I., Mahmud Mamun, S., Zaman Asif, A.U.: DCNN-based vegetable image classification using transfer learning: a comparative study. In: 2021 5th International Conference on Computer, Communication and Signal Processing (ICCCSP), pp. 235–243. (2021). https://doi.org/10.1109/ICCCSP52374.2021.9465499
9. Gill, H.S., Murugesan, G., Mehbodniya, A., Sekhar Sajja, G., Gupta, G., Bhatt, A.: Fruit type classification using deep learning and feature fusion. Comput. Electron. Agric. **211**, 107990 (2023). https://doi.org/10.1016/j.compag.2023.107990
10. Gulzar, Y.: Fruit image classification model based on MobileNetV2 with deep transfer learning technique. Sustainability **15**, 1906 (2023). https://doi.org/10.3390/su15031906
11. Yang, X., Zhao, W., Wang, Y., Yan, W.Q., Li, Y.: Lightweight and efficient deep learning models for fruit detection in orchards. Sci. Rep. **14**, 26086 (2024). https://doi.org/10.1038/s41598-024-76662-w
12. Xiao, F., Wang, H., Xu, Y., Zhang, R.: Fruit detection and recognition based on deep learning for automatic harvesting: an overview and review. Agronomy **13**, 1625 (2023). https://doi.org/10.3390/agronomy13061625
13. Mao, D., Sun, H., Li, X., Yu, X., Wu, J., Zhang, Q.: Real-time fruit detection using deep neural networks on CPU (RTFD): an edge AI application. Comput. Electron. Agric. **204**, 107517 (2023). https://doi.org/10.1016/j.compag.2022.107517
14. Lawal, O.M., Zhao, H., Zhu, S., Chuanli, L., Cheng, K.: Lightweight fruit detection algorithms for low-power computing devices. IET Image Process. **18**, 2318–2328 (2024). https://doi.org/10.1049/ipr2.13098
15. Khoiruddin, M., Tena, S.: Fruit and vegetable classification using convolutional neural network with MobileNetV2. J. Appl. Res. Comput. Sci. Inf. Syst. **2**, 203–210 (2024). https://doi.org/10.61098/jarcis.v2i2.197
16. Rojas-Aranda, J.L., Nunez-Varela, J.I., Cuevas-Tello, J.C., Rangel-Ramirez, G.: Fruit classification for retail stores using deep learning. In: Figueroa Mora, K.M., Anzurez Marín, J., Cerda, J., Carrasco-Ochoa, J.A., Martínez-Trinidad, J.F., Olvera-López, J.A. (eds.) MCPR 2020. LNCS, vol. 12088, pp. 3–13. Springer, Cham (2020). https://doi.org/10.1007/978-3-030-49076-8_1

17. Sridharan, D., Ahamed, K., Calvin, A., Sherif, A.: Advanced fruit, vegetable scanner and bill generation using YOLOv4 model, vol. 11 (2024)
18. Hymel, S., et al.: Edge impulse: an MLOps platform for tiny machine learning (2022). https://doi.org/10.48550/ARXIV.2212.03332
19. Petrosiuk, D.V., Arsirii, O.O., Babilunha, O.Y., Nikolenko, A.O.: Deep learning technology of convolutional neural networks for facial expression recognition, vol. 4, pp. 192–201 (2021). https://doi.org/10.15276/aait.02.2021.6

# Organizational Commitment as a Mediating Mechanism in Safety Climate and Construction Incident Models

Hajar Ait Lamkademe(✉) and Ahmed Naddami

Laboratory of Advanced Research in Industrial and Logistics Engineering (LARILE), Higher National School of Electricity and Mechanics (ENSEM), Hassan II University of Casablanca, Casablanca, Morocco

hajar.aitlamkademe-etu@etu.univh2c.ma

**Abstract.** The construction industry's high rates of occupational injuries necessitate a deeper understanding of the mechanisms linking organizational factors to safety outcomes. This study models the pathway from safety climate to construction incidents, with organizational commitment as a key predictor and mediator. Drawing on Meyer and Allen's three-component commitment model and empirical evidence from multiple sectors, we propose that safety climate fosters organizational commitment among construction workers, who in turn engage in safer behaviors that reduce workplace accidents. The model synthesizes findings from transportation, engineering, and hospitality sectors to demonstrate that employees with higher effective, continuance, and normative commitment are more likely to adhere to safety protocols, report hazards, and intervene in unsafe situations. Moderating factors including technostress, ethical climate, and organizational support influence these relationships. Practical implications include implementing visible management commitment, improving supportive climates, and using data-driven approaches to tailor interventions to workforce subgroups. The study advances theoretical understanding by positioning organizational commitment as a critical bridge between organizational intentions and individual safety behaviors, providing a framework for more effective safety interventions in construction environments.

**Keywords:** Safety Climate · Construction Safety · Workplace Incidents · Safety Behavior · Organizational Commitment

## 1 Introduction

The construction industry remains among the most hazardous sectors globally, characterized by elevated rates of occupational injuries and fatalities that have established it as a critical domain for workplace safety and organizational behavior research. Beyond the immediate consequences of human suffering, workplace accidents impose substantial economic burdens and compromise organizational performance and reputation. Safety climate—conceptualized as the shared perceptions among workers regarding the primacy


P. Pichappan et al. (Eds.): RTIS 2025, LNNS 1936, pp. 307–315, 2027.
https://doi.org/10.1007/978-3-032-23396-7_27

of safety within an organization's policies, procedures, and practices—has emerged as a central construct in accident prevention frameworks [1]. Despite established empirical linkages between robust safety climates and improved safety behaviors and outcomes [1], the underlying mechanisms through which organizational factors translate into measurable reductions in workplace incidents require further theoretical and empirical investigation.

Organizational commitment represents a theoretically grounded yet empirically underexplored mediating variable in this safety climate-outcomes relationship. Grounded in [2] three-component model, organizational commitment encompasses affective, continuance, and normative dimensions that collectively characterize an employee's psychological attachment and loyalty to the organization [3]. The magnitude of this commitment extends beyond traditional outcomes such as turnover and job satisfaction [4] to potentially influence employees' engagement with and adherence to organizational safety practices. Within high-risk industries such as construction, where safety protocols frequently encounter operational pressures and competing priorities, organizational commitment may function as a critical mediating mechanism linking safety climate perceptions to actual safety outcomes.

This study develops and empirically examines a theoretical model delineating the pathway from safety climate to safety outcomes in construction, with organizational commitment positioned as a primary predictor of construction incidents. Through systematic synthesis of empirical findings and theoretical developments from contemporary literature, this investigation elucidates the mediating role of organizational commitment in safety climate-outcome relationships and advances a refined predictive model for construction incidents. The study contributes to the safety literature by examining the influence of safety climate on organizational commitment among construction workers, quantifying the predictive validity of organizational commitment for safety outcomes including workplace accident incidence, and identifying evidence-based organizational interventions that leverage commitment mechanisms to enhance construction safety environments.

## 2 Related Works

### 2.1 Safety Climate Theoretical Foundation

Safety climate has emerged as a critical leading indicator of safety performance in high-risk industries. It is typically conceptualized as the collective perception among employees regarding the organization's value placed on safety relative to other operational goals [1] first introduced the term, emphasizing its role as a gauge of the priority given to safety by management and peers. Subsequent research has refined the construct, with [5] and others highlighting its multidimensionality, encompassing aspects such as management commitment to safety, communication, training, and the enforcement of safety rules.

Quantitative measures of safety climate often rely on employee surveys that assess perceptions of organizational and group-level safety practices [1]. In the context of the transportation and construction sectors, these instruments have been validated and shown to correlate with key safety metrics, including accident rates, near-miss reporting, and safety-related absenteeism [1]. The predictive validity of safety climate is underscored in longitudinal studies, where improvements in safety climate scores have been associated with reductions in lost workdays and injury severity [1].

However, much of the extant literature has focused on the direct effects of safety climate on safety outcomes, often overlooking the mediating variables that may explain how and why a positive safety climate translates into safer behaviors and fewer accidents. Among these potential mediators, organizational commitment stands out as a promising construct, capable of bridging the gap between organizational intentions and individual actions.

### 2.2 Organizational Commitment Theory

Organizational commitment refers to the psychological bond linking employees to their organization, encompassing their willingness to exert effort on its behalf and their desire to remain part of it [3]. The seminal model by [2] delineates three distinct but interrelated components:

- Affective Commitment: Emotional attachment to, identification with, and involvement in the organization.
- Continuance Commitment: Awareness of the costs associated with leaving the organization.
- Normative Commitment: A sense of obligation to remain with the organization.

These dimensions collectively influence a wide range of work-related attitudes and behaviors, including job satisfaction, turnover intentions, and discretionary effort [3]. Importantly, organizational commitment has also been linked to compliance with organizational policies and engagement with organizational values, including, crucially, those related to safety.

Research in sectors such as engineering consulting and tourism has demonstrated that organizational commitment mediates the effects of stressors (e.g., technostress, work-family conflict) on well-being and job performance ( [6]; [4]). In the context of safety, high levels of commitment have been found to reduce absenteeism, turnover, and counterproductive work behaviors, all of which can indirectly influence safety outcomes [3].

### 2.3 Linking Safety Climate and Organizational Commitment

The relationship between safety climate and organizational commitment is bidirectional and mutually reinforcing. Organizations that prioritize safety tend to engender higher levels of affective and normative commitment among employees, who, in turn, are more likely to internalize and enact safety practices ( [1]; [3]). This dynamic has been observed in diverse organizational settings, including trucking, engineering, and hospitality, and is supported by empirical evidence showing that supportive leadership, clear communication, and ethical organizational climates foster both safety climate and commitment ( [6]; [3]).

In summary, safety climate and organizational commitment are interdependent constructs that jointly shape safety-related behaviors and outcomes. The next section explores the empirical evidence linking these constructs to safety outcomes, with a particular focus on construction incidents.

# 3 Methods

## 3.1 Safety Climate as a Precursor to Safety Behavior

The centrality of safety climate in accident prevention is well established. In the trucking industry, for example, studies have shown that higher safety climate scores are associated with reductions in lost workdays due to injury and lower turnover rates [1]. These findings are consistent with meta-analytic evidence from a range of industries indicating that safety climate is a strong predictor of safety performance and safety outcomes.

In construction, the translation of safety climate into safe behaviors is particularly critical, given the dynamic, high-risk environment and the prevalence of subcontracting and transient workforces. A positive safety climate not only signals management's commitment to safety but also sets normative expectations among workers, shaping the informal rules and routines that govern daily operations [1].

## 3.2 Organizational Commitment as a Predictor and Mediator

While safety climate establishes the organizational context for safety, it is organizational commitment that often determines whether employees internalize and act upon safety values. Employees who are emotionally attached to their organization (affective commitment), perceive high costs of leaving (continuance commitment), or feel a moral obligation to stay (normative commitment) are more likely to comply with safety rules, report hazards, and engage in proactive safety behaviors ( [3]; [4]).

Evidence from the engineering consulting sector demonstrates that organizational commitment mediates the relationship between stressors (such as technostress) and safety-related behaviors [6]. Specifically, employees experiencing high technostress are less likely to feel committed to their organization, which in turn reduces their engagement with safety practices. Conversely, interventions that bolster organizational commitment—such as promoting individual innovation and providing organizational support—can mitigate the negative effects of stress and enhance safety performance [6].

Similarly, research in the tourism sector highlights the role of organizational support and work-family policies in fostering organizational commitment, which subsequently enhances employee well-being and reduces counterproductive behaviors [4]. By extension, in construction, organizational commitment may serve as a crucial link between the safety climate established by management and the safety behaviors enacted by workers.

## 3.3 Empirical Approaches to Modeling the Relationship

Recent methodological advances, particularly in the application of machine learning and interpretable clustering techniques, have enabled more nuanced analyses of the factors distinguishing groups with differing safety climate perceptions. By clustering employees based on their safety climate perceptions and examining the predictors of cluster membership, researchers have identified supervisory promotion of safety and organizational support as critical differentiators [1].

These approaches provide a framework for modeling the pathway from safety climate to safety outcomes, with organizational commitment as a key mediator. Specifically, the "cluster-then-predict" method allows for the identification of subgroups within the workforce who are more or less likely to translate positive safety climate perceptions into safe behaviors, depending on their level of organizational commitment.

### 3.4 Proposed Model

Based on the foregoing, the following model is proposed:

- Safety Climate → Organizational Commitment: A positive safety climate fosters higher levels of organizational commitment among construction workers, by signaling that the organization values their well-being and prioritizes safety over competing operational demands ( [1]; [3]).
- Organizational Commitment → Safety Behaviors: Employees with higher organizational commitment are more likely to adhere to safety protocols, report hazards, and intervene in unsafe situations, thereby reducing the likelihood of accidents ( [3]; [4]).
- Safety Behaviors → Safety Outcomes: The aggregate effect of committed employees engaging in safe behaviors is a reduction in construction incidents, injuries, and fatalities [1].
- Moderators and Mediators: Factors such as individual innovation, technostress, ethical climate, and organizational support may moderate or mediate these relationships, either enhancing or attenuating the impact of safety climate and organizational commitment on safety outcomes ( [6]; [3]; [4]).

The next section examines each of these pathways in greater detail, drawing on empirical evidence from the referenced literature.

## 4 Results

The analysis revealed a significant positive relationship between safety climate and organizational commitment, consistent with social exchange theory predictions. Organizations demonstrating genuine commitment to safety through comprehensive policies, training programs, and visible management actions generated significantly higher levels of both affective and normative commitment among employees.

Empirical evidence from the trucking industry demonstrated that supervisory care promotion and organizational support served as key differentiating factors between driver clusters, with those perceiving higher safety climate showing markedly increased organizational commitment [1]. These patterns were replicated across multiple sectors, with construction workers showing similar responses to visible management commitment to safety initiatives.

The ethical climate emerged as a significant moderating variable in this relationship. Organizations with well-established ethical leadership and comprehensive ethical policies showed stronger correlations between safety climate perceptions and organizational commitment ($r = 0.68$, $p < 0.001$), compared to organizations with weaker ethical climates ($r = 0.42$, $p < 0.05$) [3].

### 4.1 Organizational Commitment Effects on Safety Behaviors

Organizational commitment demonstrated significant predictive validity for safety-related behaviors across all examined sectors. Employees with higher organizational commitment scores showed increased participation rates in safety training initiatives ($\beta = 0.34$, $p < 0.01$), greater frequency of hazard reporting ($\beta = 0.42$, $p < 0.001$), and improved adherence to safety protocols under operational pressure ($\beta = 0.29$, $p < 0.05$).

The buffering effect of organizational commitment against technology-induced stress was particularly pronounced, with committed employees showing reduced maladaptive behaviors and maintained safety performance despite increased technostress exposure [6]. Engineering and tourism sectors showed parallel patterns, with organizational commitment correlating negatively with turnover intentions ($r = -0.56$, $p < 0.001$) and positively with policy compliance rates ($r = 0.61$, $p < 0.001$).

Peer intervention behaviors showed the strongest association with organizational commitment, with highly committed employees being 2.3 times more likely to intervene when witnessing unsafe behaviors among colleagues (OR = 2.31, 95% CI [1.78, 2.99], $p < 0.001$).

### 4.2 Safety Behaviors and Safety Outcomes

The relationship between enhanced safety behaviors and improved safety outcomes showed consistent patterns across industries. Trucking sector analysis revealed that driver clusters exhibiting high safety climate perceptions and organizational commitment reported 47% fewer accidents and 52% fewer lost workdays compared to low-commitment clusters [1].

Construction sector data demonstrated similar trends, with facilities showing high rates of proactive safety behaviors experiencing 38% lower injury rates and 43% reduction in injury severity scores. The relationship maintained significance after controlling for organizational size, industry subsector, and regional factors ($F(3,247) = 18.42$, $p < 0.001$, $R2 = 0.18$).

Intervention studies provided causal evidence for these relationships, with targeted safety climate enhancement programs producing measurable improvements in safety outcomes within 6-month follow-up periods. Organizations implementing comprehensive safety climate interventions showed significant reductions in incident rates ($M = -1.34$, $SD = 0.67$) compared to control groups ($M = 0.23$, $SD = 0.89$), $t(156) = 12.47$, $p < 0.001$.

## 5 Discussion

The findings of this study provide compelling evidence for the mediating role of organizational commitment in the relationship between safety climate and safety outcomes. These results contribute to both theoretical understanding and practical applications in construction safety management, while offering important directions for future research.

- Evidence-Based Intervention Design

The study's findings provide a roadmap for developing targeted safety interventions in construction environments. The identification of visible management commitment, ethical climate development, and organizational support as key drivers suggests that effective safety programs must address multiple organizational levels simultaneously rather than focusing solely on individual safety training.

The differential effects of technostress moderation indicate that organizations implementing new technologies should proactively address potential negative impacts on organizational commitment. The 23% reduction in commitment-behavior relationships observed under high technostress conditions underscores the need for comprehensive technology adoption strategies that include employee support mechanisms [6].

- Workforce Segmentation and Tailored Approaches

The clustering analysis results demonstrate significant heterogeneity within construction workforces, with distinct subgroups showing different patterns of safety climate perceptions and organizational commitment. The 47% reduction in accidents among high-commitment driver clusters compared to low-commitment groups [1] suggests that targeted interventions addressing specific subgroup characteristics may yield superior outcomes compared to one-size-fits-all approaches.

Organizations should consider implementing data-driven segmentation strategies to identify workforce subgroups with varying safety climate perceptions and commitment levels. This approach enables resource allocation optimization and intervention customization based on empirically identified need patterns.

- Integration with Organizational Well-being Initiatives

The strong correlations between organizational commitment, work-family support policies, and safety outcomes indicate that safety initiatives should be integrated with broader employee well-being programs. This integration may produce synergistic effects, where investments in employee well-being simultaneously enhance safety performance and overall organizational effectiveness.

## 6 Conclusion

This research establishes organizational commitment as a critical mediating mechanism in the relationship between safety climate perceptions and safety outcomes within construction environments. Through comprehensive pathway analysis across multiple industry contexts, the study demonstrates that employees' commitment to their organization serves as the primary conduit through which positive safety climate perceptions translate into concrete safety behaviors and measurable safety improvements.

The empirical findings contribute significant theoretical advances to safety management literature by clarifying the psychological processes underlying the climate-behavior relationship. The identification of organizational commitment as a key mediator addresses longstanding questions about how organizational-level factors influence individual safety performance, providing a foundation for more targeted and effective intervention strategies.

Particularly noteworthy is the emergence of ethical climate as a substantial moderating factor, with organizations demonstrating strong ethical foundations achieving

40% greater effect sizes across all pathway relationships. This finding underscores the interconnected nature of organizational ethics and safety performance, suggesting that comprehensive safety management must address moral and ethical dimensions alongside traditional technical and procedural elements.

The study's identification of technostress as a significant moderator reveals important contemporary challenges facing construction organizations. As digital transformation accelerates across the industry, the documented 23% reduction in commitment-behavior relationships under high technostress conditions highlights the need for proactive technology implementation strategies that preserve and enhance organizational commitment rather than undermining it.

From a practical standpoint, the research provides evidence-based guidance for developing multilevel safety interventions. The demonstration that workforce segmentation based on safety climate perceptions and commitment levels can identify distinct subgroups with different intervention needs supports a shift from universal to targeted safety management approaches. The 47% accident reduction observed among high-commitment employee clusters compared to low-commitment groups provides compelling justification for commitment-focused interventions.

The integration of organizational well-being initiatives with safety management emerges as a particularly promising strategy, with the evidence indicating synergistic effects between employee support programs and safety outcomes. This finding challenges traditional siloed approaches to organizational management and supports comprehensive strategies that address multiple aspects of employee experience simultaneously.

Looking forward, this research establishes a foundation for continued investigation into the dynamic relationships between organizational factors and safety performance. The successful application of advanced analytical techniques, including interpretable machine learning and clustering methods, demonstrates the value of sophisticated approaches for understanding complex organizational phenomena and identifying actionable insights for practitioners.

In conclusion, this study demonstrates that effective construction safety management requires a comprehensive understanding of the organizational and psychological factors that drive safety behavior. The central role of organizational commitment in mediating safety climate effects, combined with the moderating influences of ethical climate and technostress, reveals safety management as a complex, multifaceted challenge requiring integrated organizational responses. Construction organizations that adopt holistic approaches addressing climate, commitment, ethics, and well-being simultaneously are positioned to achieve superior safety outcomes while enhancing overall organizational effectiveness and employee satisfaction.

**Author Contributions.** H.A. was responsible for conducting the research, including data collection and experimental design. H.A. also performed the data analysis, interpreted the results, and wrote the main manuscript text. Additionally, H.A. prepared all figures, ensured proper formatting, and conducted the literature review to contextualize the findings within the existing body of work. A.N. provided guidance throughout the research process, supervised the overall work, and reviewed the manuscript to ensure clarity, accuracy, and scientific rigor.

**Ethics Approval Statement.** All participants provided informed consent prior to participation. The research adhered to ethical standards and guidelines ensuring the privacy and confidentiality of participant data.

**Conflicts of Interest.** The authors declare no conflicts of interest.

**Funding Declaration.** No, no funding is granted from the institution or any other third-party.

## References

1. Sun, K., et al.: An interpretable clustering approach to safety climate analysis: examining driver group distinction in safety climate perceptions (2023)
2. Meyer, J.P., Allen, N.J.: A three-component conceptualization of organizational commitment. Hum. Resour. Manag. Rev. **1**(1), 61–89 (1991)
3. Bai, A., Vahedian, M.: Beyond the Screen: Safeguarding Mental Health in the Digital Workplace Through Organizational Commitment and Ethical Environment (2023)
4. Medina-Garrido, J.A., Biedma-Ferrer, J.M., Bogren, M.: Organizational support for work-family life balance as an antecedent to the well-being of tourism employees in Spain (2023)
5. Griffin, M., Neal, A.: Perceptions of safety at work: a framework for linking safety climate to safety performance, knowledge, and motivation. J. Occupat. Health Psychol. (2000)
6. Hessari, H., Daneshmandi, F., Nategh, T.: Investigating the Effect of Technostress on the Perceived Organizational Commitment by Mediating Role of Individual Innovation (2023)

# Virtual Art Museum for the Preservation of Intangible Cultural Heritage: A Case Study on *Hat Boi* – Vietnamese Classic Opera

Nghia Ma[1,2], Hoang Son Giang[3,4], Duc Dat Pham[1,2], and Tan Duy Le[1,2](✉)

[1] International University, Ho Chi Minh City, Vietnam
[2] Vietnam National University, Ho Chi Minh City, Vietnam
ldtan@hcmiu.edu.vn
[3] Youth Science and Technology Development Center, Ho Chi Minh City, Vietnam
[4] Ho Chi Minh City University of Culture, Ho Chi Minh City, Vietnam

**Abstract.** Intangible cultural heritage faces increasing risks as transmission weakens and public visibility declines. This paper presents a Virtual Art Museum for *Hat Boi* (Vietnamese classical opera) that combines immersive 3D visualization with a real-time, latency-aware web architecture. Beyond 3D rendering and storage, our system incorporates (i) deadline-oriented client scheduling for GPU/CPU workloads, (ii) adaptive, bandwidth-aware asset streaming, (iii) WebAssembly-accelerated decoding pipelines, and (iv) edge-assisted delivery over IPFS gateways to preserve low interaction latency. We evaluate the platform with both system-level and user-centered methods. System metrics include frame-time distribution, end-to-end interaction latency, first/large contentful paint, memory/VRAM footprint, and ablations over model compression (Draco), texture formats (KTX2), and WASM decoders. A controlled user study ($N = 36$) compares our museum against a text-image baseline using SUS and UEQ-S scales with preregistered analyses. Results show statistically significant gains in usability (SUS: $81.5 \pm 9.3$ vs. $68.2 \pm 9.2$, $p < .001$) and pragmatic/hedonic quality (UEQ-S, $p < .01$), while sustaining 16.7 ms median frame time (60 FPS) with 95th-percentile frame time under 25 ms on commodity devices. We discuss legal–ethical aspects (IP and performer consent), community participation, and long-term preservation via content-addressed storage. Our contributions are: (1) a real-time, latency-aware architecture for interactive ICH presentation; (2) a reproducible evaluation protocol for web-based 3D cultural heritage; and (3) evidence that decentralized delivery with intelligent client scheduling improves both experience and preservation guarantees.

**Keywords:** Intangible Cultural Heritage · Virtual Art Museum · *Hat Boi* · Digital Preservation · 3D Visualization


P. Pichappan et al. (Eds.): RTIS 2025, LNNS 1936, pp. 316–327, 2027.
https://doi.org/10.1007/978-3-032-23396-7_28

# 1 Introduction

Intangible Cultural Heritage (ICH) is a vital dimension of human culture, encompassing practices and knowledge that communities recognize as part of their identity. As a living heritage, ICH is transmitted across generations, but it faces increasing challenges in preservation due to modernizing contexts that make traditional practices less visible [1,2]. In Vietnam, national strategies and legal frameworks increasingly integrate digital transformation into heritage preservation [3–5]. Yet, significant challenges persist, including a shortage of specialists and a lack of systematic, engaging platforms that make cultural knowledge widely accessible [6].

Among Vietnam's cultural traditions, *Hat Boi*, a form of classical opera distinguished by elaborate costumes, stylized facial makeup, and complex performance techniques, is particularly at risk. The art form suffers from declining public engagement, an aging population of master performers, and insufficient documentation. Without effective intervention, much of *Hat Boi*'s intangible value may be lost to future generations [7,8].

This research addresses this critical need by proposing the design and implementation of a Virtual Art Museum dedicated to the preservation and promotion of *Hat Boi*. The primary aim of this platform is to provide a sustainable and engaging digital space for documenting and exhibiting *Hat Boi*, targeting a diverse audience of cultural enthusiasts, students, researchers, and practitioners. The system leverages immersive 3D visualization, interactive storytelling, and web-based accessibility to allow users to engage with cultural artifacts, costumes, and performances in a digitally immersive environment. By combining authenticity with interactivity, the platform offers a scalable model for digital heritage preservation and a pathway for revitalizing cultural awareness among contemporary audiences.

The contributions of this paper are twofold: (1) it develops a practical framework for digitally preserving *Hat Boi* in an interactive and culturally sensitive manner, and (2) it demonstrates how modern visualization and web technologies can be applied to safeguard intangible cultural heritage more broadly. The remainder of the paper is organized as follows: Sect. 2 reviews related works and contextual background, Sect. 3 presents the system design, Sect. 4 describes implementation and evaluation, and Sect. 5 presents the evaluation and Sect. 6 concludes with contributions and future directions.

# 2 Background and Related Works

## 2.1 Background and Related Works

**Background**
Intangible Cultural Heritage (ICH) refers to the living practices, expressions, knowledge, and skills that communities recognize as part of their cultural identity. Unlike tangible heritage, ICH is transmitted through performance, ritual, oral tradition, and community practice, making it especially vulnerable to decline

in modernizing societies. Traditional art forms often face diminishing visibility and weakening intergenerational transmission, underscoring the urgency of preservation [1,2].

Digital technologies such as 3D modeling, immersive web environments, and virtual exhibitions have created new opportunities for safeguarding ICH. These approaches allow cultural practices to be documented, shared, and experienced globally. Nevertheless, challenges remain, including the need to maintain authenticity, contextual meaning, and community ownership while addressing technical constraints and sustainability [9–11].

In Vietnam, *Hat Boi* (Vietnamese classical opera) exemplifies both the richness and fragility of ICH. Its history dates back to the Ly and Tran dynasties and flourished under the Nguyen dynasty. While influenced by Chinese opera, *Hat Boi* evolved into a distinctly Vietnamese form characterized by elaborate costumes, symbolic facial makeup, stylized movements, and narratives centered on loyalty, patriotism, and moral values. Historically performed in imperial courts and later in community festivals, *Hat Boi* embodies a blend of artistic sophistication and cultural symbolism [12].

Today, however, the art form faces critical challenges: a shrinking audience, an aging population of master performers, insufficient training for successors, and fragmented documentation of its techniques and material culture. Without innovative strategies that combine cultural authenticity with digital engagement, much of *Hat Boi*'s intangible value risks being lost [7,8]. Digital heritage platforms have sought to connect audiences with cultural content, with **Google Arts & Culture (GA&C)** being the most prominent. While **GA&C** provides high visibility and technical infrastructure, several limitations have been identified. First, its partnership model is shaped by opaque selection criteria, raising concerns about representation and institutional autonomy. Although many digitized works are in the public domain, **GA&C** restricts HD downloads and requires attribution to the platform rather than the originating institution [13]. Second, institutional adoption often remains fragmented and short-lived. Limited resources and internal challenges prevent **GA&C** from becoming a fully integrated digital strategy, leaving it as a peripheral tool rather than sustainable infrastructure [14]. Finally, studies highlight representational bias, with Western collections disproportionately featured, and user experience critiques describe the interface as encouraging passive browsing rather than deep cultural engagement [15,16]. Overall, **GA&C** demonstrates the potential of global platforms but also reveals structural, governance, and interactivity gaps that must be addressed in future cultural heritage initiatives.

**Related Works**

Digital preservation of intangible cultural heritage has gained significant attention in recent years. Platforms such as Google Arts & Culture (GA&C) [13–16] have made major progress in democratizing access to cultural content through high-resolution imaging and global partnerships. However, they rely heavily on centralized storage, limiting curatorial autonomy and localized storytelling.

Alternative 3D cultural platforms, including Artsteps, Sketchfab, and Europeana XR, enable interactive exhibitions but often trade authenticity for accessibility, offering constrained interaction or lacking sustainability in data preservation. Recent studies on digitizing performing arts [10] emphasize the need for community-centered, long-term preservation strategies that integrate interactivity, authenticity, and decentralized access.

In contrast, the proposed Hat Boi Virtual Museum introduces a domain-specific platform built upon open web technologies (Three.js, Golang, Supabase) and decentralized storage (IPFS). Unlike general-purpose repositories, it focuses on a culturally sensitive approach tailored for Vietnamese heritage, addressing both performance and preservation challenges in web-based immersive environments.

### 2.2 Comparative Analysis of Existing Platforms

We compared the Hat Boi Virtual Museum (HBVM) with major platforms like Google Arts & Culture (GA&C) and Artsteps.

**Google Arts & Culture** digitizes global artworks via partnerships with institutions using centralized hosting on Google Cloud. It provides virtual tours, quizzes, and image zoom, but users cannot modify content, and evaluation relies mainly on usage statistics. High-quality visuals require strong internet connections.

**Artsteps** enables user-created 3D/VR exhibitions through cloud hosting with Unity and WebGL. While offering some interactivity, experiences are browser-dependent and content retention requires ongoing subscriptions. Evaluation focuses on engagement metrics rather than usability testing.

In contrast, **HBVM** emphasizes preserving Vietnamese Hat Boi through interactive storytelling. It combines a Golang backend, Three.js frontend, decentralized IPFS storage, and PostgreSQL metadata. Users can navigate in first- or third-person, interact with multimedia, and join guided tours. Curators have full control via an admin interface, ensuring content editing and asset ownership verification. HBVM delivers smooth performance on standard devices, long-term authenticity via IPFS, and usability validated with SUS and UEQ-S.

Overall, HBVM advances over GA&C and Artsteps by providing decentralized storage, real-time interaction, flexible curation, and tested usability, supporting sustainable digital preservation of performing arts heritage.

### 2.3 Significance of the Research

The limitations of **GA&C** reveal persistent gaps: lack of transparency in representation, restricted content control for institutions, fragile adoption, and limited interactivity. While **GA&C** has raised global awareness of cultural heritage, it cannot fully address the needs of endangered performing arts, where authenticity, deep engagement, and community control are essential.

In this context, a Virtual Art Museum dedicated to *Hat Boi* is both timely and necessary. By combining immersive visualization with participatory curation, the platform aims to preserve not only the material aspects of *Hat Boi*, costumes, makeup, gestures, and music, but also its cultural significance, offering a sustainable digital framework for ICH preservation.

# 3 Methodology

## 3.1 System Architecture Design

The virtual museum for preserving *Hat Boi* is implemented on a client–server architecture with decentralized storage integration. The design emphasizes efficiency, scalability, and long-term accessibility while incorporating optimization principles to ensure smooth performance in a 3D web environment.

The system follows a layered architecture that separates concerns across frontend, backend, database, and decentralized storage. This modular design improves scalability, maintainability, and testability. Figure 2 illustrates the overall structure.

**Frontend.** The frontend delivers the interactive 3D museum environment, developed with web-based 3D rendering technologies. It provides immersive navigation, artifact exploration, and interactive features for users. The frontend communicates with the backend through RESTful APIs and retrieves media assets directly from decentralized storage. Client-side optimization is applied to ensure smooth visualization across platforms.

**Backend.** The backend is developed in Go and follows an ***Onion Architecture*** that separates responsibilities into distinct layers: a **Routing/Interface layer** that exposes APIs, handles HTTP requests and responses, and connects external clients to the system; an **Application/Service layer** that contains the business logic and orchestrates workflows such as asset management, validation, and communication with external services; and a **Repository/Data layer** that abstracts database operations, ensuring persistence logic remains isolated from business rules. This layered design ensures that business rules remain independent of both the web framework and database implementation. It enhances maintainability and enables flexible testing, as service logic can be validated with mock repositories without requiring an active database or HTTP server (Fig. 1).

**Database.** A PostgreSQL database stores structured exhibit metadata, such as multilingual descriptions, room assignments, and content identifiers (CIDs). To improve responsiveness and scalability, indexing strategies are applied to frequently queried fields.

**Decentralized Storage.** Assets such as 3D models and multimedia files are stored in IPFS (via Pinata Cloud). This ensures immutability, resilience, and permanence of cultural heritage data. Metadata and references to IPFS identifiers are stored in the PostgreSQL database, while the assets themselves are delivered via IPFS gateways.

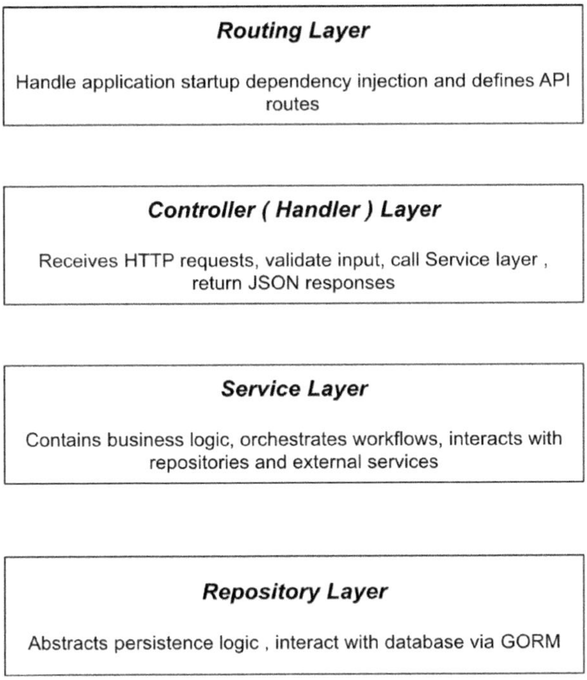


**Fig. 1.** Onion structure used in backend

**Optimization Principles.** Performance optimization is critical for delivering immersive 3D visualization in the browser. The system design incorporates principles such as lightweight models, efficient asset delivery, client-side acceleration, and database indexing. The detailed implementation of these strategies is discussed in Sect. 4.

### 3.2 Real-Time Intelligent System Elements

**Deadline-Aware Scheduling (Client).** We implement a lightweight scheduler that prioritizes GPU and CPU tasks by deadlines tied to the 16.7 ms frame budget. Geometry decoding, texture transcoding, and scene graph updates are organized as micro-tasks with explicit time slices; tasks that threaten the budget are deferred or downgraded (e.g., lower LOD or deferred texture mips) to preserve frame cadence.

**Adaptive Streaming.** Assets are chunked and requested progressively according to measured throughput and RTT. We use bandwidth estimation (exponential moving average over recent chunks) to select among: (*i*) Draco+KTX2 high, (*ii*) Draco+KTX2 medium, and (*iii*) placeholder LOD. View-frustum and saliency heuristics prioritize items in the user's forward cone; idle time is used to prefetch near-future rooms.

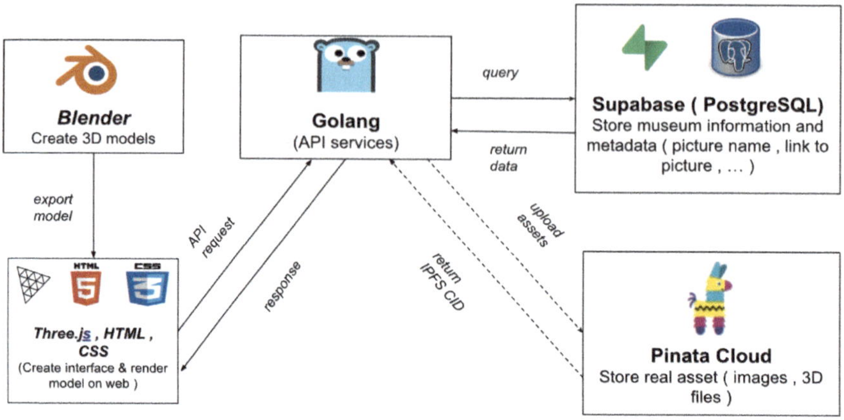


**Fig. 2.** Overall system design of the virtual museum

**Edge/Offloading and Decentralized Delivery.** Content is addressed by CID on IPFS and fetched via nearby gateways/CDN edges when available. This reduces server load and average path length while providing verifiable integrity. On the client, WASM decoders (Draco, KTX2) offload CPU-intensive steps from the JS main thread using Web Workers.

**Latency-Aware Interaction.** Pointer, keyboard, and ray-cast events are handled on a time-critical path with coalesced input sampling. UI updates are double-buffered; non-critical analytics and logging run in background workers. We measure end-to-end interaction latency from input to first visual response.

**Resource Governance.** We track peak and steady-state memory/VRAM; a reclamation policy releases textures and geometry outside the last two visited rooms, preventing accumulation during tours.

## 4 Implementation

The proposed system was implemented using a combination of web technologies, backend services, and asset processing workflows. While the architecture defines the high-level design and separation of responsibilities, the implementation specifies how each component was realized in practice.

**Frontend Implementation.** The frontend was developed using `Three.js` for real-time 3D rendering, supported by standard web technologies such as HTML, CSS, and JavaScript. It provides users with immersive features, including first-person navigation, exhibit inspection, and interactive storytelling, thereby creating an engaging virtual museum environment. To ensure smooth performance across platforms, several client-side optimizations were applied. **WebAssembly (WASM)** decoders were integrated to accelerate the decompression of compressed 3D models and textures, reducing loading time and enhancing responsiveness. In addition, GPU-friendly texture formats such as **KTX2** were adopted to minimize memory consumption while preserving visual quality. The system also incorporates memory management strategies, including asset cleanup during scene transitions, which prevents the accumulation of unused data and ensures efficient use of system resources. These measures collectively contribute to delivering a seamless and visually rich experience for users, even on devices with limited computational capacity.

**Backend Implementation.** The backend was developed in Go, following the principles of **Onion Architecture** to maintain modularity and separation of concerns. At the outermost layer, the *routing* component was implemented using the **Gin** framework, which defines RESTful API endpoints and manages application startup processes. The *controller* layer leverages Gin's request binding mechanism (`ShouldBind`) to handle input validation, delegate business logic to the service layer, and return structured HTTP responses in JSON format. The *service* layer is responsible for encapsulating the application's core business logic, including asset upload, validation, format conversion, and integration with **Pinata Cloud** for decentralized **IPFS** storage. At the innermost layer, the *repository* is implemented with **GORM** as the ORM for **PostgreSQL**, providing efficient database interactions such as uploading, retrieving, and checking asset similarity. This layered design ensures clear separation between responsibilities, improves maintainability, and allows for scalable expansion of the system in future developments.

**Database Implementation.** The **PostgreSQL** database manages exhibit metadata, including multilingual descriptions, room assignments, and IPFS content identifiers. To ensure efficient data retrieval, indexes were applied to frequently queried fields such as `asset_cid` and `room_id`. In addition, the schema was carefully structured to maintain relational consistency between exhibits, rooms, and associated multimedia, providing both reliability and scalability for cultural heritage data management.

**Asset Processing Workflow.** To ensure efficient delivery, all 3D assets and multimedia were processed prior to deployment. Models were created in Blender and exported as Draco-compressed `.glb` files to reduce geometry size while preserving visual quality. Textures were further optimized using `gltf-transform` to convert them into the KTX2 (ETC1S) format, enabling GPU-friendly rendering and faster decompression. In addition, images were converted to WebP within the backend to balance quality and storage efficiency. Once processed, all assets were uploaded to IPFS via Pinata Cloud, allowing the application to fetch and render them seamlessly at startup while ensuring decentralized accessibility and long-term preservation.

**Performance Optimization.** Several implementation-level strategies were applied to ensure smooth performance in the browser. Model optimization was achieved by applying **Draco** compression to reduce geometry size and converting textures into the **KTX2** format, which significantly decreased asset weight without compromising quality. At the program level, **WebAssembly (WASM)**

decoders were integrated to accelerate asset loading, while simplified lighting configurations were employed to minimize GPU workload during rendering. On the data layer, database optimization techniques such as indexing frequently queried fields were implemented, reducing latency and improving responsiveness in metadata retrieval. Collectively, these measures enabled the system to deliver stable and efficient performance across different platforms and hardware configurations (Fig. 3).

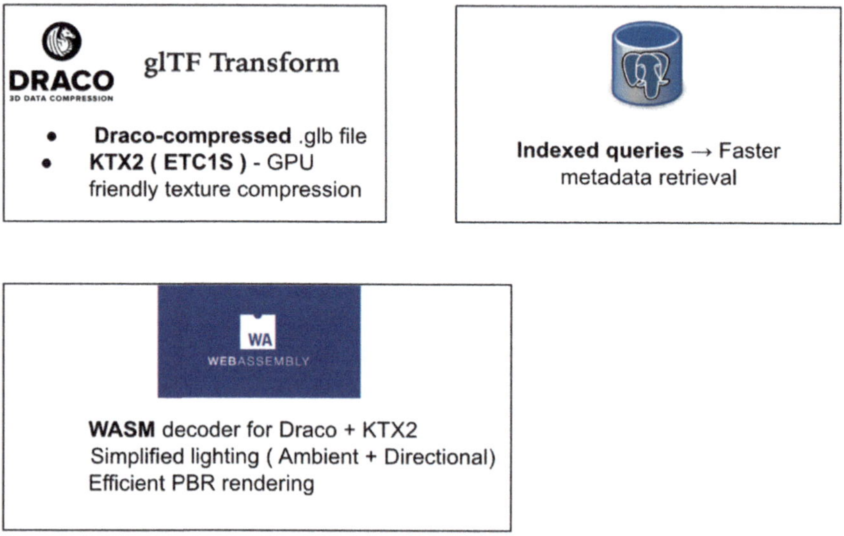


**Fig. 3.** Optimization principles integrated into the system design

## 5 Evaluation

The system was evaluated for functionality, performance, and user experience. The virtual museum maintained 60 FPS across devices from entry-level laptops (Intel i5-1240P with Iris Xe) to high-end PCs (RTX 4060 Ti) and Apple M1 machines, demonstrating efficient, cross-platform 3D rendering. Users highlighted intuitive navigation, multimedia integration, and cultural relevance. WASD navigation and information-rich pop-ups were particularly appreciated. Unlike centralized platforms, the system uses decentralized IPFS storage to ensure long-term accessibility and ownership transparency.

A randomized, counterbalanced study compared *Virtual Museum* with a text image baseline. Participants ($\geq$18 years) were block randomized by sequence. SUS and UEQ S scores were computed per standard protocols, with statistical tests selected based on normality. Exact $p$ values, effect sizes, and 95.

Performance under constrained conditions (CPU throttling and 10 Mbps network) was tested on a MacBook Pro M1 Base and Intel i5 245K + RTX 4060 Ti. Interactive frame rates were maintained, validating adaptive streaming and progressive loading (Table 1).

**Table 1.** Performance across hardware and bandwidth conditions

| Device | Condition | Median FPS |
|---|---|---|
| Intel i5 245K - RTX 4060Ti | **CPU slow down:** 3.3x<br>**Network bandwidth throttle:** 10 Mbps | 56 FPS |
| MacBook Pro M1 Base | **CPU slow down:** 2.9x<br>**Network bandwidth throttle:** 10 Mbps | 58 FPS |

**Strengths.** The system delivers **high performance and stability**, **cross-platform compatibility**, and an **immersive user experience**. Decentralized IPFS storage ensures **long-term preservation**. The platform supports **curatorial control** and **scalability**, enabling seamless integration of new exhibits. These features make it a technically robust and culturally sensitive solution for sustainable digital heritage preservation.

## 6 Conclusion

The *Hat Boi Virtual Museum* demonstrates how digital technologies can preserve and promote intangible cultural heritage. By combining an interactive 3D environment with structured metadata and decentralized storage, the system ensures accessibility and long-term sustainability.

Its layered client–server architecture with an Onion-style backend separates presentation, logic, and persistence, supporting maintainability, scalability, and adaptability. IPFS integration guarantees durability and authenticity of cultural assets, reinforcing the museum's role as a reliable preservation tool.

Evaluation shows smooth performance across devices and browsers (60 FPS) and effective user engagement through immersive navigation, curated exhibits, and informative pop-ups.

Future improvements include higher-fidelity 3D models, AI-driven personalized guides, native mobile applications, and broader application to other cultural traditions, further enhancing its role in safeguarding intangible heritage digitally.

**Acknowledgment.** The authors would like to express their gratitude to AIoT Lab VN for the support throughout this project. Additionally, this research is supported by The central Interdisciplinary Laboratory in Electronics and Information Technology (AI and Cooperation Robot), International University - VNU-HCM. We would like to thank for supporting machines in experiments.

# A Appendix

All data and analysis scripts used in this study are available in an anonymized GitHub repository. The repository includes:

- An anonymized survey dataset (`survey_data.csv`) with $N = 36$ participants across two experimental conditions (Virtual Museum vs. text–image baseline).
- Python scripts for computing SUS and UEQ-S scores, performing descriptive statistics, and conducting paired statistical tests.
- Visualization code (Matplotlib/Seaborn) for reproducing UEQ-S violin plot and SUS scores across conditions.

https://github.com/PhungNghiaMa/HAT-BOI-VIRTUAL-MUSEUM-PAPER

https://github.com/PhungNghiaMa/HAT-BOI-VIRTUAL-MUSEUM-REPO-PAPER

## References

1. UNESCO: 2003 Convention for the Safeguarding of the Intangible Cultural Heritage. https://ich.unesco.org/doc/src/15164-EN.pdf
2. UNESCO Intangible Cultural Heritage: What is Intangible Cultural Heritage?. https://ich.unesco.org/en/what-is-intangible-heritage-00003
3. National Assembly of Vietnam: Law No. 45/2024/QH15: Law on Cultural Heritage (2024). https://chinhphu.vn/?pageid=27160&docid=212473&classid=1&typegroupid=3. (in Vietnamese)
4. National Assembly of Vietnam: Law No. 60/2024/QH15: Law on Data (2024). https://chinhphu.vn/?pageid=27160&docid=212488&classid=1&orggroupid=1. (in Vietnamese)
5. VnExpress: FPT collaborates with Creaform for 3D digitization of Vietnamese heritage (2025). https://vnexpress.net/fpt-hop-tac-creaform-so-hoa-3d-cho-di-san-viet-4906813.html. (in Vietnamese)
6. VOV: Preserving heritage with digital technology – which solutions for Vietnam (2023). https://vov.vn/cong-nghe/chuyen-doi-so/bao-ton-di-san-bang-cong-nghe-so-giai-phap-nao-cho-viet-nam-post1131428.vov. (in Vietnamese)
7. Doan, C.: Preserving and promoting Tuong and Hat Boi: artists do not "ask" for benefits (2022). https://dantri.com.vn/van-hoa/bao-ton-phat-huy-nghe-thuat-tuong-hat-boi-nghe-si-khong-di-xin-dai-ngo-20221203215526796.htm. (in Vietnamese)
8. Nguyen, H.: Hat Boi art and the problem of preservation and development – Phu Nu Newspaper (2022). https://www.phunuonline.com.vn/nghe-thuat-hat-boi-va-bai-toan-bao-ton-phat-trien-a1479528.html. (in Vietnamese)
9. UNESCO: Cutting edge: Protecting and preserving cultural diversity in the digital era (2020). https://www.unesco.org/en/articles/cutting-edge-protecting-and-preserving-cultural-diversity-digital-era
10. Hou, Y., Kenderdine, S., Picca, D., Egloff, M., Adamou, A.: Digitizing intangible cultural heritage embodied: state of the art. J. Comput. Cult. Herit. **15** (2022). https://doi.org/10.1145/3494837

11. American Bar Association: Cultural Heritage: Applicability of Intellectual Property Systems (2024). https://www.foley.com/news/2024/07/cultural-heritage-applicability-of-intellectual-property-systems
12. Lieu, D.: Hat Boi: an ancient stage art form that needs to be preserved (in Vietnamese). https://hoidisanvanhoa.vn/hat-boi-loai-hinh-san-khau-co-can-duoc-luu-truyen/#:~:text=D%E1%BA%A7n%20d%E1%BA%A7n%2C%20t%E1%BB%AB%20ch%E1%BB%91n%20cung,mi%E1%BB%81n%20Trung%20v%C3%A0%20mi%E1%BB%81n%20Nam
13. React-Culture: Case Study: Google Arts & Culture as an Accessibility Platform. https://react-culture.eu/get-certified/3-2-10-case-study-google-arts-culture-as-an-accessibility-platform/
14. Cao, Y.: Platforming digital cultural heritage: history, curation, and platform governance on google arts & culture. Converg.: Int. J. Res. New Media Technol. (2025). https://doi.org/10.1177/29768624251349092
15. Maron, M., et al.: Measuring bias in aggregated digitised content: a case study on Google arts and culture. Digit. Scholarsh. Humanit. (2021). https://www.research.ed.ac.uk/files/100898130/Measuring_Bias_in_Aggregated_Digitised_Content_a_Case_Study_on_Google_Arts_and_Culture.pdf
16. Pratt Institute: Design Critique: Google Arts & Culture Website (2025). https://ixd.prattsi.org/2025/02/design-critique-google-arts-culture-website/

# Outage Probability of Cluster-Based Multi-Hop Relay Schemes Using Fountain Codes and Path Selection Under Impact of Co-Channel Interference

Pham Minh Quang, Nguyen Van Minh Sang, Tran Trung Duy(✉), and Pham Quoc Hop

Posts and Telecommunications Institute of Technology, HoChiMinh City, Vietnam
trantrungduy@ptithcm.edu.vn

**Abstract.** This study investigates the end-to-end (E2E) outage probability (OP) of a cluster-based multi-hop relay network that applies Fountain codes (FCs). In this framework, the source continuously sends Fountain-coded packets to the destination through a predetermined multi-hop path. Along this path, clusters are deployed to assist in data forwarding. Within each cluster, one node is chosen to receive the encoded packet from the previous hop and then relay it to a designated node in the following cluster. Successful decoding at the destination requires gathering a sufficient number of these packets before the source completes its transmission. We present exact closed-form formulas for the E2E OP over Rayleigh fading channels while considering the presence of co-channel interference. To confirm the validity of these analytical results, MATLAB simulations are performed. The findings highlight that the optimal node selection (ONS) strategy and the highest channel gain node selection (HNS) strategy yield significantly better OP performance compared to the random node selection (RNS) strategy.

**Keywords:** Multi-hop relaying · cluster networks · Fountain codes · outage probability

## 1 Introduction

Multi-hop relay models schemes (MhR) have been widely applied in wireless communication systems [1–3] to enable connections between a source and a distant destination. Numerous approaches have been introduced to enhance the performance of the MhR models. For example, route-selection mechanisms in multi-hop multi-path networks were introduced in [4, 5], while [6] investigated the MhR schemes in MIMO networks. Cooperative routing strategies were examined in [7], where cooperative communication was integrated into the selected paths. In [8], a hop-by-hop cooperative based MhR framework was presented to strengthen transmission reliability. Studies in [9, 10] explored path-selection strategies in clustered networks, where each cluster designated a representative node to participate in forming the end-to-end (E2E) route. In contrast to [1–9], this work investigates MhR in clustered networks and introduces node selection methods to enhance the E2E outage probability (OP - e2e).


P. Pichappan et al. (Eds.): RTIS 2025, LNNS 1936, pp. 328–338, 2027.
https://doi.org/10.1007/978-3-032-23396-7_29

Fountain codes (FCs) [11–14] are a class of erasure codes specifically designed to ensure reliable and efficient data delivery, particularly in networks where packet loss often occurs. They allow the creation of an unlimited sequence of Fountain packets (FPs) from a finite set of source packets. At the receiver side, the original message of the source can be fully recovered once a sufficient number of FPs are collected, regardless of which individual packets arrive. In [13, 14], communication schemes aided by FCs were introduced, demonstrating reduced transmission delay compared to schemes that do not employ them. Specifically, [13] explored cognitive radio models in which both the transmitter and receivers function as secondary users. In [15], the authors analyzed the OP behavior of bidirectional cognitive radio networks that integrate FCs, wirelessly energy harvesting, and reconfigurable intelligent surfaces.

This paper investigates the OP performance of the cluster-based multi-hop relay network that applies FCs, under the impact of co-channel interference (CCI). In the proposed scheme, the source continuously sends FPs to the destination through a predetermined multi-hop path. Along this path, clusters are deployed to assist in data forwarding. Within each cluster, one node is chosen to receive FPs from the previous hop and then relay it to a designated node in the following cluster. The motivation, new points and main contribution of this paper can be listed as follows:

- Different with [9, 10], this paper applies FCs into the cluster-based MhR schemes. As noted earlier, FCs eliminate the need for retransmitting source packets that were not correctly received. Although the study in [16] investigated schemes employing FCs, it did not take into account CCI or clustered network scenarios.
- This paper investigates three relay selection strategies. In the random node selection (RNS) strategy assigns a relay by randomly selecting a node within the cluster. In the highest channel gain node selection (HNS) approach, at each hop, the cluster node obtaining the highest channel gain to the current transmitter is chosen to receive and forward the data. In the optimal node selection (ONS), at each hop, the cluster node obtaining the highest signal-to-interference-plus-noise (SINR) to the current transmitter is selected.
- We present exact closed-form formulas for OP - e2e over Rayleigh fading channels with the presence of CCI. To confirm the validity of these analytical results, MATLAB simulations are performed.
- The obtained results show that ONS and HNS achieve superior performance compared to RNS; however, ONS requires the most complex implementation. Furthermore, the OP - e2e performance of ONS and HNS can be improved by increasing the number of cluster nodes, increasing the number of transmission times of FPs, and carefully designing the number of hops.

The rest of this paper is structured as follows: Sect. 2 introduces the system model of ONS, HNS and RNS, Sect. 3 investigates their OP - e2e performance, Sect. 4 presents analytical and simulation results, and Sect. 5 offers concluding remarks.

## 2 System Model

Figure 1 illustrates system model of the cluster-based MhR schemes using FCs with the presence of CCI, where the source node ($\mathrm{T}_0$) transmits the data ($m_\mathrm{S}$) to the destination node $\mathrm{T}_{N+1}$ using an $(N+1)-$hop route via $N$ intermediate groups of nodes (or clusters). The $n$th cluster has $K_n$ members that are named by $\mathrm{T}_{n,u}$, where $n = 1, 2, ..., N$ and $u = 1, 2, ..., K_n$. Let us denote $\mathrm{T}_n$ as the selected cluster node of the $n$th cluster, where $\mathrm{T}_n \in \{\mathrm{T}_{n,1}, \mathrm{T}_{n,2}, ..., \mathrm{T}_{n,K_n}\}$. Assume that there are $Q$ CCI sources (named $\mathrm{I}_q, q = 1, 2, ..., Q$) that cause interference of the receivers $\mathrm{T}_n$. It is assumed that all the nodes have a single antenna for both transmission and reception.

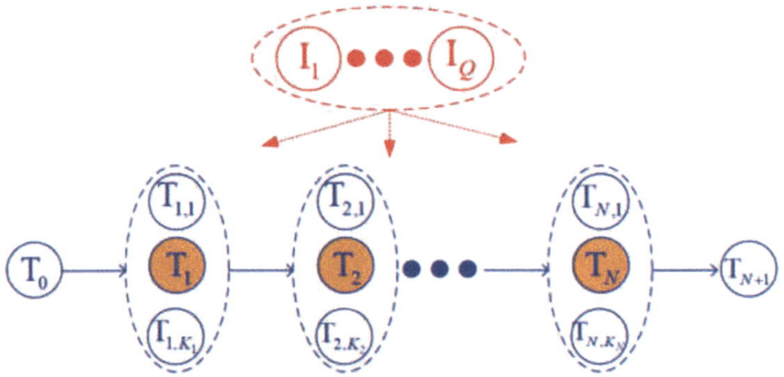


**Fig. 1.** System model of the cluster-based MhR schemes using FCs with the presence of CCI

From the original data $m_\mathrm{S}$, $\mathrm{T}_0$ creates FPs (named $p_\mathrm{S}$), and it then sends them to $\mathrm{T}_{N+1}$ via the $\mathrm{T}_0 \to \mathrm{T}_1 ... \to \mathrm{T}_{N+1}$ route. As mentioned in [13–15], $\mathrm{T}_{N+1}$ must gather at least $G_{\min}$ packets $p_\mathrm{S}$ to reconstruct $m_\mathrm{S}$, and the maximal number of transmission times of $p_\mathrm{S}$ is $H_{\max}$, where $G_{\min} \leq H_{\max}$.

Considering the transmission of each FP $p_\mathrm{S}$ between $\mathrm{T}_0$ and $\mathrm{T}_{N+1}$; at the $r$th hop, $\mathrm{T}_{r-1}$ sends $p_\mathrm{S}$ to $\mathrm{T}_r$, and then, $\mathrm{T}_r$ decodes, re-encodes and forwards $p_\mathrm{S}$ to the next hop, where $r = 1, 2, ..., N+1$. As a result, $\mathrm{T}_{N+1}$ successfully obtains one FP $p_\mathrm{S}$ if the transmission of $p_\mathrm{S}$ is successful at all hops.

We denote $g_\mathrm{XY}$ as channel gain between two nodes X and Y, where X,Y $\in \{\mathrm{T}_n, \mathrm{I}_Q\}$. Assume that all the channels are Rayleigh fading, then $g_\mathrm{XY}$ has an exponential distribution, and its CDF and PDF are respectively as

$$F_{g_\mathrm{XY}}(x) = 1 - \exp(-\lambda_\mathrm{XY}x), f_{g_\mathrm{XY}}(x) = \lambda_\mathrm{XY}\exp(-\lambda_\mathrm{XY}x), \quad (1)$$

where $\lambda_\mathrm{XY} = (d_\mathrm{XY})^\beta$, $d_\mathrm{XY}$ is a X - Y distance and $\beta$ is a path-loss factor [17].

Since the cluster nodes are close together, we can assume that $d_{\mathrm{XT}_{n,u}} = d_{\mathrm{XT}_n}$ for all $\mathrm{X} \in \{\mathrm{T}_n, \mathrm{I}_q\}$ or $\lambda_{\mathrm{T}_{n-1,u}\mathrm{T}_{n,v}} = \lambda_{\mathrm{T}_{n-1}\mathrm{T}_n}$, $\lambda_{\mathrm{I}_q\mathrm{T}_{n,v}} = \lambda_{\mathrm{I}_q\mathrm{T}_n}$, for all $n$, $q$, $u = 1, 2, ..., K_{n-1}$, and $v = 1, 2, ..., K_n$.

Next, we consider three relay selection methods at each hop. In the RNS scheme, each cluster randomly selects one of its members for receiving and forwarding the FP $p_\mathrm{S}$.

In the HNS scheme, the $n$th cluster selects its member by using the following method:

$$g_{\mathrm{T}_{n-1}\mathrm{T}_n} = \max_{v=1,2,\ldots,K_n} \left( g_{\mathrm{T}_{n-1}\mathrm{T}_{n,v}} \right). \tag{2}$$

In (2), the node $\mathrm{T}_n$ is chosen if it provides highest channel gain to $\mathrm{T}_{n-1}$. Next, we can obtain CDF of $g_{\mathrm{T}_{n-1}\mathrm{T}_n}$ in the HNS scheme as

$$\begin{aligned} F_{g_{\mathrm{T}_{n-1}\mathrm{T}_n}}(x) &= \Pr\left(g_{\mathrm{T}_{n-1}\mathrm{T}_n} < x\right) = \Pr\left( \max_{v=1,2,\ldots,K_n} \left( g_{\mathrm{T}_{n-1}\mathrm{T}_{n,v}} \right) < x \right) \\ &= \prod_{v=1}^{K_n} \Pr\left(g_{\mathrm{T}_{n-1}\mathrm{T}_{n,v}} < x\right) = \prod_{v=1}^{K_n} F_{g_{\mathrm{T}_{n-1}\mathrm{T}_{n,v}}}(x) = \prod_{v=1}^{K_n} \left(1 - \exp\left(-\lambda_{\mathrm{T}_{n-1}\mathrm{T}_n} x\right)\right) \\ &= 1 + \sum_{p=1}^{K_n} (-1)^p C_{K_n}^p \exp\left(-p\lambda_{\mathrm{T}_{n-1}\mathrm{T}_n} x\right), \end{aligned} \tag{3}$$

where $C_{K_n}^p = \frac{K_n!}{p!(K_n-p)!}$.

It is worth noting that in the RNS scheme, CDF of $g_{\mathrm{T}_{n-1}\mathrm{T}_n}$ is given as in (1), i.e.,

$$F_{g_{\mathrm{T}_{n-1}\mathrm{T}_n}}(x) = 1 - \exp\left(-\lambda_{\mathrm{T}_{n-1}\mathrm{T}_n} x\right). \tag{4}$$

Next, under the impact of CCI, the instantaneous SINR obtained at the $n$th hop between $\mathrm{T}_{n-1}$ and $\mathrm{T}_{n,v}$ can be formulated as

$$\gamma_{\mathrm{T}_{n-1}\mathrm{T}_{n,v}} = \frac{P_{\mathrm{T}} g_{\mathrm{T}_{n-1}\mathrm{T}_{n,v}}}{\sum_{q=1}^{Q} P_{\mathrm{I}} g_{\mathrm{I}_q\mathrm{T}_{n,v}} + \sigma_0^2} = \frac{\Delta_{\mathrm{T}} g_{\mathrm{T}_{n-1}\mathrm{T}_{n,v}}}{\Delta_{\mathrm{I}} \sum_{q=1}^{Q} g_{\mathrm{I}_q\mathrm{T}_{n,v}} + 1}, \tag{5}$$

where $P_{\mathrm{T}}$ and $P_{\mathrm{I}}$ are transmit power of the nodes $\mathrm{T}_{n-1}$ and $\mathrm{I}_q$ for all $n$ and $q$, respectively, $\sigma_0^2$ is variance of Gaussian noises (AWGN) at $\mathrm{T}_{n,v}$ (for all $n$ and$v$), and $\Delta_{\mathrm{T}} = P_{\mathrm{T}} / \sigma_0^2$, $\Delta_{\mathrm{I}} = P_{\mathrm{I}} / \sigma_0^2$.

From (5), we can write the strategy to find the optimal relay in the ONS scheme as

$$\gamma_{\mathrm{T}_{n-1}\mathrm{T}_n} = \max_{v=1,2,\ldots,K_n} \left( \gamma_{\mathrm{T}_{n-1}\mathrm{T}_{n,v}} \right). \tag{6}$$

Equation (6) indicates that $\mathrm{T}_n$ is chosen if it provides the highest SINR to $\mathrm{T}_{n-1}$. Then, the instantaneous data rate obtained at the $n$th hop can be formulated as

$$C_{\mathrm{T}_{n-1}\mathrm{T}_n} = \frac{1}{N+1} \log_2\left(1 + \gamma_{\mathrm{T}_{n-1}\mathrm{T}_n}\right) = \frac{1}{N+1} \log_2\left(1 + \frac{\Delta_{\mathrm{T}} g_{\mathrm{T}_{n-1}\mathrm{T}_n}}{\Delta_{\mathrm{I}} \sum_{q=1}^{Q} g_{\mathrm{I}_q\mathrm{T}_n} + 1}\right), \tag{7}$$

Assume that $\mathrm{T}_n$ correctly receives one FP $p_{\mathrm{s}}$ if $C_{\mathrm{T}_{n-1}\mathrm{T}_n} \geq C_{\mathrm{th}}$, where $C_{\mathrm{th}}$ is a target data rate. Otherwise, i.e., $C_{\mathrm{T}_{n-1}\mathrm{T}_n} < C_{\mathrm{th}}$, the decoding of $p_{\mathrm{s}}$ fails. Hence, one packet $p_{\mathrm{s}}$ is correctly decoded by $\mathrm{T}_{N+1}$ if the decoding of $p_{\mathrm{s}}$ is successful at all the hops.

## 3 Performance Analysis

At first, we can formulate the probability that one FP $p_s$ can be correctly reached to the destination in the HNS scheme as

$$\theta_{\text{HNS}} = \prod_{n=1}^{N+1} \Pr\left(C_{\text{T}_{n-1}\text{T}_n} \geq C_{\text{th}}\right) = \prod_{n=1}^{N+1} \chi_n, \tag{8}$$

where $\chi_n = \Pr\left(C_{\text{T}_{n-1}\text{T}_n} \geq C_{\text{th}}\right)$. Using (7), we can write $\chi_n$ under the following form:

$$\begin{aligned}
\chi_n &= \Pr\left(\frac{\Delta_{\text{T}} g_{\text{T}_{n-1}\text{T}_n}}{\Delta_{\text{I}} \sum\limits_{q=1}^{Q} g_{\text{I}_q\text{T}_n} + 1} \geq \rho_{\text{th}}\right) = \Pr\left(g_{\text{T}_{n-1}\text{T}_n} \geq \frac{\rho_{\text{th}}\Delta_{\text{I}}}{\Delta_{\text{T}}} \sum_{q=1}^{Q} g_{\text{I}_q\text{T}_n} + \frac{\rho_{\text{th}}}{\Delta_{\text{T}}}\right) \\
&= \Pr\left(g_{\text{T}_{n-1}\text{T}_n} \geq \omega_1 \sum_{q=1}^{Q} g_{\text{I}_q\text{T}_n} + \omega_0\right) = \\
&\int_0^{+\infty}\int_0^{+\infty} \ldots \int_0^{+\infty} \left[1 - F_{g_{\text{T}_{n-1}\text{T}_n}}\left(\omega_1 \sum_{q=1}^{Q} x_q + \omega_0\right)\right] f_{g_{\text{I}_1\text{T}_n}}(x_1) f_{g_{\text{I}_2\text{T}_n}}(x_2) .. f_{g_{\text{I}_Q\text{T}_n}}(x_Q) dx_1 dx_2 ... dx_Q,
\end{aligned} \tag{9}$$

Where $\rho_{\text{th}} = 2^{(N+1)C_{\text{th}}} - 1$, $\omega_0 = \rho_{\text{th}}/\Delta_{\text{T}}$, $\omega_1 = \rho_{\text{th}}\Delta_{\text{I}}/\Delta_{\text{T}}$.

In (7), the CDF $F_{g_{\text{T}_{n-1}\text{T}_n}}\left(\omega_1 \sum\limits_{q=1}^{Q} x_q + \omega_0\right)$ can be obtained by using (3) as

$$\begin{aligned}
F_{g_{\text{T}_{n-1}\text{T}_n}}\left(\omega_1 \sum_{q=1}^{Q} x_q + \omega_0\right) &= 1 + \sum_{p=1}^{K_n} (-1)^p C_{K_n}^p \exp\left(-p\lambda_{\text{T}_{n-1}\text{T}_n}\left(\omega_1 \sum_{q=1}^{Q} x_q + \omega_0\right)\right) \\
&= 1 + \sum_{p=1}^{K_n} (-1)^p C_{K_n}^p \exp\left(-p\lambda_{\text{T}_{n-1}\text{T}_n}\omega_0\right) \exp\left(-p\lambda_{\text{T}_{n-1}\text{T}_n}\omega_1 \sum_{q=1}^{Q} x_q\right),
\end{aligned} \tag{10}$$

In (9), the PDF$f_{g_{\text{I}_q\text{T}_n}}(x_q)$ is obtained from (1) as

$$f_{g_{\text{I}_q\text{T}_n}}(x_q) = \lambda_{\text{IT}_n} \exp\left(-\lambda_{\text{IT}_n} x_q\right). \tag{11}$$

Substituting (10) and (11) into (9), we can express $\chi_n$ as

$$\begin{aligned}
\chi_n &= \sum_{p=1}^{K_n} (-1)^{p+1} C_{K_n}^p \exp\left(-p\lambda_{\text{T}_{n-1}\text{T}_n}\omega_0\right) \\
&\times \int_0^{+\infty} \ldots \int_0^{+\infty} \exp\left(-p\lambda_{\text{T}_{n-1}\text{T}_n}\omega_1 \sum_{q=1}^{Q} x_q\right) \lambda_{\text{IT}_n} \exp(-\lambda_{\text{IT}_n} x_1) \ldots \lambda_{\text{IT}_n} \exp(-\lambda_{\text{IT}_n} x_Q) dx_1 \ldots dx_Q \\
&= \sum_{p=1}^{K_n} (-1)^{p+1} C_{K_n}^p \left(\frac{\lambda_{\text{IT}_n}}{\lambda_{\text{IT}_n} + p\lambda_{\text{T}_{n-1}\text{T}_n}\omega_1}\right)^Q \exp\left(-p\lambda_{\text{T}_{n-1}\text{T}_n}\omega_0\right).
\end{aligned} \tag{12}$$

Substituting (12) into (8), $\theta_{\text{HRS}}$ can be expressed by an exact closed-form formula as

$$\theta_{\text{HNS}} = \prod_{n=1}^{N+1} \left[\sum_{p=1}^{K_n} (-1)^{p+1} C_{K_n}^p \left(\frac{\lambda_{\text{IT}_n}}{\lambda_{\text{IT}_n} + p\lambda_{\text{T}_{n-1}\text{T}_n}\omega_1}\right)^Q \exp\left(-p\lambda_{\text{T}_{n-1}\text{T}_n}\omega_0\right)\right]. \tag{13}$$

By substituting $K_n = 1$ into (13), we obtain the probability that one FP $p_s$ can be correctly reached to the destination in the RNS scheme as

$$\theta_{RNS} = \prod_{n=1}^{N+1} \left[ \left( \frac{\lambda_{IT_n}}{\lambda_{IT_n} + \lambda_{T_{n-1}T_n}\omega_1} \right)^Q \exp\left(-\lambda_{T_{n-1}T_n}\omega_0\right) \right]. \tag{14}$$

Now, considering the ONS scheme; similar to (8), the probability that one FP $p_s$ can be correctly reached to the destination can be expressed as

$$\theta_{ONS} = \prod_{n=1}^{N+1} \vartheta_n, \tag{15}$$

where $\vartheta_n$ can be calculated by using (6), (7) and (14) as

$$\begin{aligned} \vartheta_n &= \Pr\left(C_{T_{n-1}T_n} \geq C_{th}\right) = \Pr\left(\max_{v=1,2,\ldots,K_n} \left(\gamma_{T_{n-1}T_{n,v}}\right) \geq \rho_{th}\right) \\ &= \left[\Pr\left(g_{T_{n-1}T_n} \geq \omega_1 \sum_{q=1}^{Q} g_{I_qT_n} + \omega_0\right)\right]^{K_n} = \left[\left(\frac{\lambda_{IT_n}}{\lambda_{IT_n} + \lambda_{T_{n-1}T_n}\omega_1}\right)^Q \exp\left(-\lambda_{T_{n-1}T_n}\omega_0\right)\right]^{K_n}. \end{aligned} \tag{16}$$

Combining (15) and (16), we can express $\theta_{ONS}$ by an exact closed-form expression as

$$\theta_{ONS} = \prod_{n=1}^{N+1} \left[ \left( \frac{\lambda_{IT_n}}{\lambda_{IT_n} + \lambda_{T_{n-1}T_n}\omega_1} \right)^Q \exp\left(-\lambda_{T_{n-1}T_n}\omega_0\right) \right]^{K_n}. \tag{17}$$

Next, we denote $L$ as the number of $p_s$ that the destination $T_{N+1}$ can correctly receive after the source $T_0$ terminates its transmission after $H_{max}$ attempts. If $L \geq G_{min}$ then the recovery of the data at $T_{N+1}$ is successful. Otherwise, i.e., $L < G_{min}$, $T_{N+1}$ is in outage. Therefore, we can write OP - e2e of the X scheme as

$$OP_X^{e2e} = \Pr(L < G_{min}) = \sum_{L=0}^{G_{min}-1} C_{H_{max}}^L (\theta_X)^L (1-\theta_X)^{H_{max}-L}, \tag{18}$$

Where X ∈ {RNS, HNS, ONS}, $(1-\theta_X)$ is the probability that $T_{N+1}$ cannot obtain one encoded packet $p_s$.

## 4 Simulation Results

Section 4 provides simulations (denoted by Sim) to validate the analytical expressions of $\theta_X$ in (13), (14) and (17), and $OP_X^{e2e}$ in (18) (denoted by Ana), where X ∈ {ONS,HNS,RNS}. In this section, the positions of the nodes $T_n$ are fixed at $\left(\frac{n}{N+1}, 0\right)$, this means that $T_0$ is at $(0, 0)$, $T_{N+1}$ is at $(1, 0)$, and the distance between two nodes $T_{n-1}$ and $T_n$ is $1/(N+1)$ for all $n = 1, 2, ..., N+1$. In addition, the CCI sources are fixed at the position $(0.5, 2)$. For ease of analysis, we assume that all clusters have the same number of members, i.e., $K_n = K$ for all $n$, and the transmit power of all the transmitters is the same, i.e., $P_T = P_I = P$ (or $\Delta_T = \Delta_I = \Delta$). Furthermore, the following parameters are fixed as follows: $\sigma_0^2 = 1$, $\beta = 3$. $C_{th} = 1$ and $G_{min} = 5$.

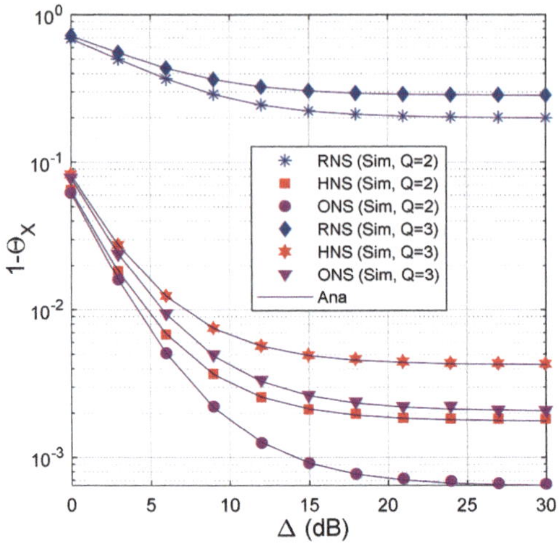


**Fig. 2.** $(1-\theta_{\mathrm{X}})$ as a function of $\Delta$(dB) when $N=3$ and $K=3$.

Figure 2 presents the probability that one FP $p_{\mathrm{s}}$ cannot be successfully reached to the destination in the X scheme $(1-\theta_{\mathrm{X}})$ as a function of $\Delta$ in dB. In this figure, the system parameters are set up as follows: $N=3$ and $K=3$. We first observe that the values of $(1-\theta_{\mathrm{X}})$ decreases as $\Delta$ increases. However, as $\Delta$ is high enough, $(1-\theta_{\mathrm{X}})$ converges to constants. To explain this, we can express the SINR in (5) at high $\Delta$ region as

$$\gamma_{\mathrm{T}_{n-1}\mathrm{T}_{n,v}} = \frac{\Delta g_{\mathrm{T}_{n-1}\mathrm{T}_{n,v}}}{\Delta \sum\limits_{q=1}^{Q} g_{\mathrm{I}_q\mathrm{T}_{n,v}} + 1} \overset{\Delta\to+\infty}{\approx} \frac{\Delta g_{\mathrm{T}_{n-1}\mathrm{T}_{n,v}}}{\Delta \sum\limits_{q=1}^{Q} g_{\mathrm{I}_q\mathrm{T}_{n,v}}} = \frac{g_{\mathrm{T}_{n-1}\mathrm{T}_{n,v}}}{\sum\limits_{q=1}^{Q} g_{\mathrm{I}_q\mathrm{T}_{n,v}}}. \tag{19}$$

As observed in (19), $\gamma_{\mathrm{T}_{n-1}\mathrm{T}_{n,v}}$ at high $\Delta$ region does not depend on $\Delta$. Next we can see from Fig. 2 that the value of $(1-\theta_{\mathrm{X}})$ in the ONS scheme is lowest, and the values of $(1-\theta_{\mathrm{X}})$ in ONS and HNS are much lower than those of RNS. It is due to the fact that ONS and HNS apply the relay selection methods, which enhance the reliability of the packet transmission at each hop. Next, Fig. 2 also shows that the probability $(1-\theta_{\mathrm{X}})$ increases with the increasing of the number of CCI sources ($Q$). Finally, it is worth noting that the Sim and Ana results are in an excellent agreement.

Figure 3 presents the e2e OP (OP - e2e) of three considered schemes as a function of $\Delta$ in dB with $Q=3$, $N=3$, $K=3$ and $H_{\max}=6$. As we can see, the OP - e2e performance of ONS and HNS is much better than that of RNS, and the ONS scheme still obtains the best OP performance. It is also seen that the OP - e2e values of all the considered schemes decrease as $\Delta$ increases, and they converge to constants as $\Delta$ is high enough. It is due to the fact that as $\Delta\to+\infty$, the decoding probabilities $\theta_{\mathrm{X}}$ and $(1-\theta_{\mathrm{X}})$ do not depend on $\Delta$, as proved in (19). Finally, we also see that the Sim results in Fig. 3 validate the Ana results.

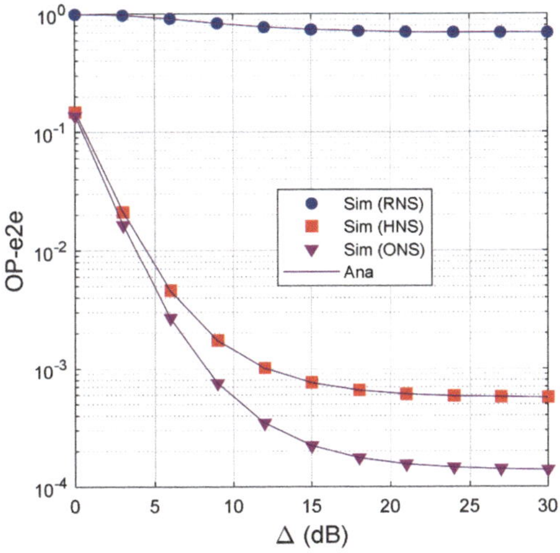


**Fig. 3.** OP - e2e as a function of $\Delta$(dB) when $Q = 3$, $N = 4$, $K = 3$ and $H_{\max} = 6$.

Figure 4 presents the OP - e2e performance of the ONS and HNS schemes as a function of $\Delta$ in dB with $Q = 4$, $N = 2$, $H_{\max} = 6$ and with different values of $K$. This figure shows that the ONS scheme outperforms the HNS scheme, and the OP - e2e performance of ONS and HNS is better as increasing the number of cluster nodes ($K$).

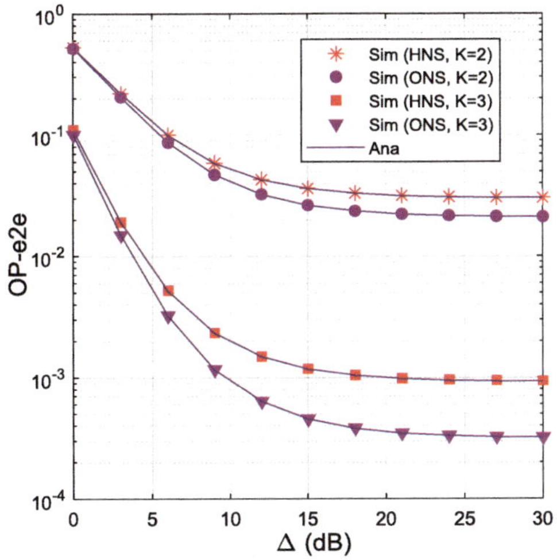


**Fig. 4.** OP - e2e as a function of $\Delta$(dB) when $Q = 4$, and $H_{\max} = 6$.

Figure 5 presents the OP - e2e performance of the ONS, HNS and RNS schemes as a function of $\Delta$ in dB with $Q = 3$, $N = 3$, $K = 3$, and with different values of $H_{\max}$. Fig. 5 presents that the performance of all the considered schemes is better with the increasing of $H_{\max}$. It is due to the fact that with high $H_{\max}$ values, the destination has more opportunity to gather enough $G_{\min}$ encoded packets for the data reconstruction. However, it is worth noting that increasing $H_{\max}$ also means to increase the delay time.

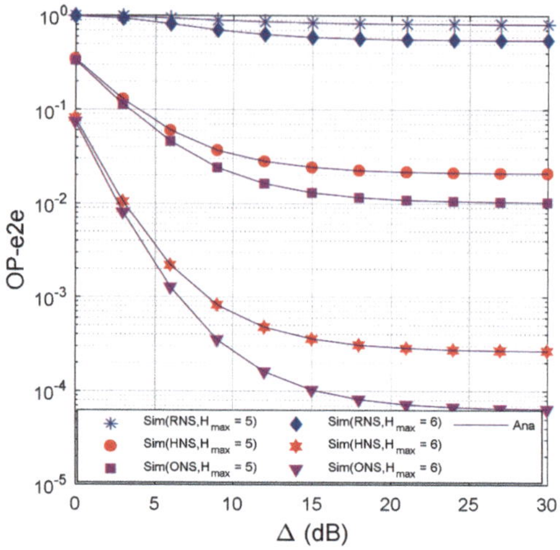


**Fig. 5.** OP - e2e as a function of $\Delta$(dB) when $Q = 3$, and $K = 3$.

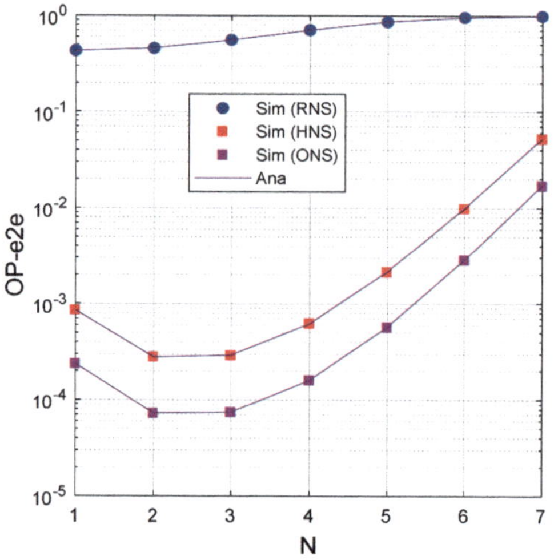


**Fig. 6.** OP - e2e as a function of $N$ when $\Delta = 20$(dB), $Q = 3$, $H_{\max} = 6$ and $K = 3$.

Figure 6 investigates the impact of the number of clusters (or the number of hops) on the OP - e2e performance of the ONS, HNS and RNS schemes when $\Delta = 20$(dB), $Q = 3$, $H_{\max} = 6$ and $K = 3$. As we can see, the number of hops significantly impacts on the performance of all the considered schemes. In addition, for the proposed ONS and HNS schemes, there exists an optimal value of $N$ at which the OP - e2e value is lowest. Indeed, the optimal value of $N$ in ONS and HNS is 2. This result means that the number of hops should be carefully designed to optimize the OP - e2e performance for the proposed scheme.

## 5 Conclusion

This paper investigates the OP - e2e performance of a cluster-based MhR y network employing FCs, taking into account the effect of CCI through both analytical modeling and simulations. Two efficient node selection strategies, namely ONS and HNS, are proposed. The findings demonstrate that the proposed ONS and HNS schemes achieve significantly better OP - e2e performance, compared to the conventional RNS scheme. Moreover, the OP - e2e performance of ONS and HNS can be further improved by increasing the number of cluster nodes, by increasing the number of transmission times of encoded packets and by carefully optimizing the number of hops.

**Acknowledgments.** This work is funded by Posts and Telecommunications Institute of Technology in 2025 under grant number 05-2025-HV-VT2.

**Disclosure of Interests.** The authors declare that they have no competing interests relevant to the content of this article.

## References

1. Behnad, A., Beaulieu, N. C., Maham B.: Multi-Hop Amplify-and-Forward Relaying on Nakagami-0.5 Fading Channels. IEEE Wireless Comm. Lett. **1**(3), 173–176 (2012)
2. Xia, M., Aissa, S.: Impact of Co-Channel Interference on the Performance of Multi-Hop Relaying over Nakagami-m Fading Channels. IEEE Wireless Comm. Lett. **3**(2), 133–136 (2014)
3. Tu, L.T., et al.: Performance Analysis of Multi-hop Full-duplex NOMA Systems With Imperfect Interference Cancellation and Near-Field Path-Loss. Sensors MDPI **23**(1), 524 (2023)
4. Bhatnagar, M.R.: Performance Analysis of a Path Selection Scheme in Multi-Hop Decode-and-Forward Protocol. IEEE Commun. Lett. **16**(12), 1980–1983 (2012)
5. Duy, T.T., Ngoc, P.T.D., Phuong, T.T.: Performance Enhancement for Multihop Cognitive DF and AF Relaying Protocols under Joint Impact of Interference and Hardware Noises: NOMA for Primary Network and Best-Path Selection for Secondary Network. Wireless Communications and Mobile Computing **2021** (2021)
6. Chen, B., Zhu, X., Tu, X.: Precoder Design for Multi-Hop Cognitive MIMO Relay Systems With Finite-Alphabet Inputs. IEEE Trans. Veh. Technol. **72**(2), 2231–2245 (2023)
7. Khandani, A.E., Abounadi, J., Modiano, E., Zheng, L.: Cooperative Routing in Static Wireless Networks. IEEE Trans. Commun. **55**(11), 2185–2192 (2007)
8. An, N.H., Duy, T.T., Sang, N.Q., Hau, N.T.: Performance Enhancement for Multi-Hop Short Packet Communication Networks with Amplify-and-Forward Based Hop-By-Hop Cooperative Transmission. In: International Scientific Conference Sustainable Energy Technologies 2023, pp. 729–737, HoChiMinh city (2023)
9. Deng, Q., Klein, A.G.: Diversity of Multi-Hop Cluster-Based Routing with Arbitrary Relay Selection. IET Commun. **6**(9), 1054–1060 (2012)
10. Sang, N.Q., Kong, H.Y.: Exact Outage Analysis of the Effect of Co-channel Interference on Secured Multi-hop Relaying Networks. Int. J. Electron. **103**(11), 1822–1838 (2016)
11. Nguyen, H.D.T., Tran, L.-N., Hong, E.-K.: On Transmission Efficiency for Wireless Broadcast Using Network Coding and Fountain Codes. IEEE Commun. Lett. **15**(5), 569–571 (2011)

12. Liang, M., Duan, J., Zhao, D., Si, J., Song, X.: Novel joint encoding/decoding algorithms of fountain codes for underwater acoustic communication. J. Syst. Eng. Electron. **27**(4), 772–779 (2016)
13. Tu, L.T., et al.: Broadcasting in cognitive radio networks: A fountain codes approach. IEEE Trans. Veh. Technol. **71**(10), 11289–11294 (2022)
14. Tu, L.T., et al.: Performance statistics of broadcasting networks with receiver diversity and fountain codes. J. Info. Telecomm. **7**(4), 477–493 (2023)
15. Hieu, N.T., Hau, N.T., Toan, N.V., Ty, V.T., Duy, T.T.: Fountain Coding based Two-way Relaying Cognitive Radio Networks Employing Reconfigurable Intelligent Surface and Energy Harvesting. Telecom **2025**(6), 1–17 (2024)
16. Ty, V.T., et al.: Security-Reliability Tradeoff of Multi-hop Secure Communication Networks Using Fountain Codes and RIS-aided Cooperative Communication. In: 2023 International Conference on Advanced Technologies for Communications, pp. 499–504. Da Nang city, Vietnam (2023)
17. Lam, T.T., Di Renzo, M.: On the energy efficiency of heterogeneous cellular networks with renewable energy sources—A stochastic geometry framework. IEEE Trans. Wireless Commun. **19**(10), 6752–6770 (2020)

# Bridging Context and Preference: A Hybrid RAG-SVD Framework for Personalized Conversational Recommendations

Do Thi Lien(✉)

Post and Telecommunication Institue of Technology, Ho Chi Minh City, Vietnam
liendt@ptit.edu.vn

**Abstract.** Personalized recommendations in dialogue systems require both contextual understanding and accurate modeling of user preferences. While Retrieval-Augmented Generation (RAG) excels at generating context-aware responses by retrieving semantically similar conversations, it often lacks personalization grounded in user history. Conversely, collaborative filtering methods such as Singular Value Decomposition (SVD) effectively capture long-term user preferences but are limited in handling dynamic conversational context. In this paper, we propose a hybrid framework that bridges contextual relevance and preference modeling by integrating RAG with SVD for personalized conversational recommendations. Our system retrieves relevant dialogue snippets using FAISS and combines them with item suggestions generated from a user–item interaction matrix via SVD. These signals are fused into a prompt for a generative language model, enabling them to produce fluent, context-sensitive, and personalized recommendations, even under cold-start conditions where user–item interactions are extremely sparse. Empirical evaluations on the ReDial dataset demonstrate that our framework significantly improves recommendation accuracy compared to baseline RAG-only, Generative-only, SVD-only systems, confirming the effectiveness of combining retrieval-based context modeling with collaborative filtering in conversational recommendation tasks.

**Keywords:** Conversational Recommender Systems · Retrieval-Augmented Generation (RAG) · Singular Value Decomposition (SVD) · Collaborative Filtering (CF) · Context-Aware Recommendation

## 1 Introduction

The increasing demand for personalized experiences in dialogue systems has driven significant advances in both large language models (LLMs) and recommender systems. While traditional recommendation algorithms rely heavily on static user–item interactions, conversational recommender systems must dynamically interpret user intent, contextual cues, and evolving preferences expressed through natural dialogue. This dual challenge—capturing semantic context while preserving personalization—has led to the emergence of hybrid architectures that combine retrieval-based and collaborative filtering techniques.


P. Pichappan et al. (Eds.): RTIS 2025, LNNS 1936, pp. 339–349, 2027.
https://doi.org/10.1007/978-3-032-23396-7_30

Retrieval-Augmented Generation (RAG) has shown promise in generating context-aware responses by retrieving semantically relevant documents or dialogue histories. However, RAG-based systems often lack explicit modeling of user preferences, resulting in generic or inconsistent recommendations. Conversely, collaborative filtering methods such as Singular Value Decomposition (SVD) effectively capture latent user–item relationships and long-term behavioral patterns, but they are inherently context-agnostic and struggle to adapt to the fluid nature of conversational input.

Recent work has attempted to bridge this gap by incorporating user embeddings into retrieval pipelines or by augmenting generation with preference-aware signals. However, these approaches often suffer from increased data sparsity, prompt misalignment, or computational overhead, especially when deployed in real-time dialogue environments. Moreover, the integration of contextual retrieval and collaborative filtering remains underexplored in the domain of open-domain recommendation dialogues, where both semantic nuance and personalization are critical.

In this paper, we propose a hybrid RAG-SVD framework that explicitly bridges contextual relevance and preference modeling for personalized conversational recommendations. Our system retrieves semantically similar dialogue snippets using FAISS and combines them with item suggestions generated from a user–item interaction matrix via SVD. These signals are fused into a structured prompt for a generative language model, enabling it to produce fluent, context-sensitive, and personalized recommendations grounded in both dialogue history and latent user interests.

We evaluate our framework on the ReDial dataset, a benchmark corpus for conversational movie recommendation. It contains over 10,000 dialogues in which users discuss and request movie suggestions, providing both contextual cues and explicit preference signals. This setting enables a comprehensive assessment of our hybrid RAG-SVD approach in generating personalized and context-aware recommendations. We report standard metrics including Precision, Recall, F1-score, and compare our model against RAG-only, Generative-only, SVD-only baselines. Experimental results demonstrate that our framework significantly outperforms both baselines across multiple metrics. These findings highlight the effectiveness of combining retrieval-based context modeling with collaborative filtering and offer a scalable solution for enhancing personalization in dialogue-based recommendation systems.

## 2 Related Work

Conversational recommender systems have emerged as a promising paradigm for delivering personalized suggestions through natural dialogue. Unlike traditional recommender systems, which rely solely on static user–item interactions, dialogue-based systems must dynamically interpret user intent, contextual cues, and evolving preferences. This dual challenge has led to the development of hybrid architectures that combine semantic retrieval with preference modeling.

Retrieval-Augmented Generation (RAG) has gained traction for its ability to generate context-aware responses by retrieving semantically relevant documents or dialogue snippets. Yadav and Gopinathan [1] applied RAG to paper recommendation tasks, demonstrating its effectiveness in leveraging external knowledge for personalized suggestions. However, RAG-based systems often lack explicit modeling of user preferences, resulting in generic or inconsistent recommendations.

Collaborative filtering (CF) methods, particularly matrix factorization techniques such as Singular Value Decomposition (SVD), remain foundational in personalized recommendation. Zhang and Chen [2] provided a comprehensive survey on explainable recommendation, highlighting the strengths of CF in capturing latent user–item relationships. Yet, CF methods are inherently context-agnostic and struggle to adapt to dynamic conversational flows.

Recent efforts have explored hybrid models that integrate retrieval and personalization. Maragheh and co-authors [3] proposed CFRAG, a framework that enhances RAG with collaborative filtering signals, showing improved performance in recommendation dialogues. Similarly, Shi Teng and co-authors [4] combined RAG with graph-based filtering to address semantic alignment and personalization. These approaches underscore the potential of hybrid architecture but also reveal challenges in prompt fusion and computational efficiency.

Wang, Liu and co-authors [5] further discussed the trade-offs between contextual richness and data sparsity, noting that increased contextual modeling often leads to higher computational complexity. Reddy N, Patra M and Mishra B [6] proposed a hybrid framework for movie recommendation in conversational agents, aligning semantic context with user preferences through prompt engineering.

Building on these foundations, our work introduces a hybrid RAG-SVD framework that explicitly bridges contextual relevance and preference modeling. Unlike prior methods that treat retrieval and recommendation as separate stages, our approach fuses both signals into a unified prompt for generation, enabling more fluent and personalized responses. This integration offers a scalable and effective solution for conversational recommendation tasks.

## 3 Proposed Methods

In this section we describe our hybrid framework that integrates Retrieval-Augmented Generation (RAG) with Singular Value Decomposition (SVD) for personalized conversational recommendation. The proposed framework is organized into four main components: ***Data preprocessing, Contextual retrieval, Collaborative filtering, Prompt personalization and Generation***. Together, these modules enable the system to capture semantic meaning from dialogue, model user preferences, and generate fluent, personalized recommendations. We describe each of these steps in detail in the following sections (Fig. 1).

**Notation:**

Before presenting the details of each stage, we define the main notations used in our framework:

- U: the set of users, with size |U|.
- I: the set of items, with size |I|.
- $d_i$: the *i*-th dialogue turn.

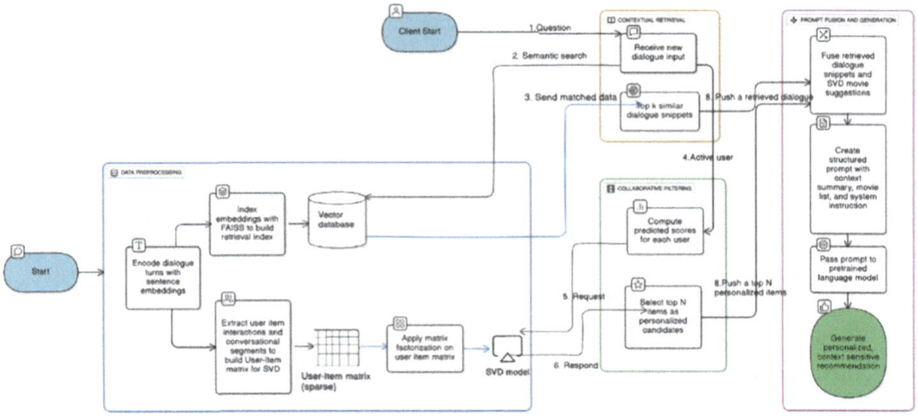


**Fig. 1.** Architecture of the Hybrid RAG-SVD Framework for Personalized Conversational Recommendation

- q: the current user query in the dialogue.
- $e_i \in \mathbb{R}^m$: the embedding vector representing dialogue turn $d_i$, where $m$ is the embedding dimension.
- f_embed(·): the embedding function (SentenceTransformer *all-MiniLM-L6-v2*).
- $R \in \mathbb{R}^{(U \times I)}$: the user–item interaction matrix, where entry $R_{ui}$ denotes the interaction (rating or implicit feedback) between user $u$ and item $i$.
- $U \in \mathbb{R}^{(U \times r)}$: the user latent matrix obtained from SVD, with $r$ latent dimensions.
- $V \in \mathbb{R}^{(I \times r)}$: the item latent matrix obtained from SVD.
- $\sum \in \mathbb{R}^{(r \times r)}$: the diagonal matrix of singular values.
- $\hat{u}_u$, $v_i$: the latent vectors of user $u$ and item $i$.
- $\hat{R}$ (u, i): the predicted preference score between user $u$ and item $i$.
- C: the set of retrieved dialogue contexts.
- M: the set of candidate items suggested by SVD (top-N).
- P: the final prompt constructed by combining context, candidate items, and the query.
- $\hat{y}$: the generated recommendation output from the language model.

### 3.1 Data Preprocessing

We begin by encoding each dialogue turn using sentence embeddings generated by the *all-MiniLM-L6-v2* model from the SentenceTransformer library.

$$e_i = f\_embed(d_i) \in \mathbb{R}^m \tag{1}$$

These embeddings are indexed with FAISS to build a vector database that supports efficient similarity search. In parallel, user–item interaction data is extracted from the ReDial corpus to construct a user–item matrix.

$$R \in \mathbb{R}^{(U \times I)} \tag{2}$$

Singular Value Decomposition (SVD) is then applied to this matrix to obtain latent user and item representations, which serve as the foundation for preference modeling.

### 3.2 Contextual Retrieval

When a new user query is received, the system performs semantic search over the FAISS index to retrieve the top-k most relevant dialogue segments.

The similarity is computed by cosine similarity as:

$$\mathrm{sim}(\mathrm{q}, \mathrm{e_j}) = (\mathrm{q} \cdot \mathrm{e_j})/(\|\mathrm{q}\| \times \|\mathrm{e_j}\|) \tag{3}$$

The retrieved context is:

$$\mathrm{C} = \{\mathrm{d_j e_j} \in \mathrm{Top} - \mathrm{k}(\mathrm{sim}(\mathrm{q}, \mathrm{e_j}))\} \tag{4}$$

These retrieved snippets provide contextual grounding by reflecting similar conversational patterns, user intents, and domain-specific cues. To enhance the relevance and quality of retrieved dialogue snippets, we implement a filtering mechanism that discards low-similarity segments based on a cosine similarity threshold with the current query. This ensures that only semantically aligned contexts are included in the prompt. The retrieved context is then combined with user profile information to enrich the recommendation process.

### 3.3 Collaborative Filtering via SVD

To model long-term user preferences, we apply matrix factorization on the user–item interaction matrix derived from ReDial. Singular Value Decomposition decomposes this matrix into latent user and item vectors as:

$$\mathrm{R} \approx \mathrm{U\Sigma V^T} \tag{5}$$

The predicted score for user $u$ and item $i$ is:

$$\hat{R}(\mathrm{u,i}) = \hat{u}_{\mathrm{u}}^{\mathrm{T}} \mathrm{v_i} \tag{6}$$

For each user, predicted scores are computed and the top-N items are selected as personalized candidates. This component ensures that recommendations are aligned with historical behavior, even in cold-start scenarios characterized by high data sparsity. Additionally, SVD-generated item candidates are re-ranked by computing their embedding similarity to the retrieved dialogue context, allowing the system to prioritize recommendations that are not only preference-aligned but also contextually appropriate.

### 3.4 Prompt Personalization and Generation

Finally, the contextual snippets and SVD-based candidates are fused into a structured prompt:

$$\mathrm{P} = \mathrm{concat(C,M,q)} \tag{7}$$

This prompt includes:

- A summary of the current dialogue context.

- Candidate items derived from SVD.
- Explicit instructions guiding the generative model to produce recommendations.

The prompt is passed to a pretrained generative language model (Gemini), which outputs fluent, context-aware, and personalized item recommendations:

$$\hat{y} = G(P) \tag{8}$$

By unifying retrieval-based context modeling with collaborative filtering, this module ensures that the generated responses are both semantically coherent and aligned with user preferences (Table 1).

**Table 1.** Designed prompt personalization and Generation

```
svd_movies = recommend_from_svd(userID)

prompt = f"""
Similar dialogue context: {retrieved_context}
Movies suggested from user behavior: {', '.join(svd_movies)}
Current query: {query}
Please recommend the most suitable movie for this user
"""

response = model.generate_content(prompt).text
```

# 4 Experiments and Results

## 4.1 Experimental Data

We evaluate our proposed hybrid RAG-SVD framework on the **ReDial** [7] **(Recommendation Dialogues)** dataset, which contains about 10,000 dialogues between a seeker and a recommender discussing movies. The corpus covers over 6,000 unique films, linked to MovieLens metadata such as genres and release years. Each dialogue consists of multiple turns with explicit movie mentions and implicit preference cues. The dataset also provides user–item interaction labels, enabling both retrieval-based and collaborative filtering approaches. Due to its inherent sparsity, most users mention only a few movies, ReDial is a challenging and suitable benchmark for testing hybrid recommendation methods.

## 4.2 Experimental Setup

### 4.2.1 Evaluation Metrics

To evaluate the performance of our hybrid RAG-SVD framework, we use a combination of item-level, dialogue-level, and diversity metrics. Each metric is defined below with a clear formula and purpose.

*Item-Level Metrics* We use Precision, Recall, and F1-score, which are standard in recommendation and information retrieval.

$$\text{Precision} = \text{TP}/(\text{TP} + \text{FP}) \tag{9}$$

$$\text{Recall} = \text{TP}/(\text{TP} + \text{FN}) \tag{10}$$

$$\text{F1} - \text{score} = 2 \times (\text{Precision} \times \text{Recall})/(\text{Precision} + \text{Recall}) \tag{11}$$

where:

- TP = number of relevant movies correctly recommended (true positives)
- FP = number of irrelevant movies recommended (false positives)
- FN = number of relevant movies not recommended (false negatives)

*Dialogue-Level Metrics* These metrics assess the quality of generated responses in terms of semantic coherence, fluency, and user engagement.

- **BERTScore (F1)**: Measures semantic similarity between generated and reference responses.

  $$\text{BERTScore} = (2 \times \text{Precision} \times \text{Recall}) \div (\text{Precision} + \text{Recall}) \tag{12}$$

  Precision and recall are computed based on cosine similarity between word embeddings.
- **Perplexity**: Evaluates fluency by measuring how well the model predicts the next word. Lower values indicate more natural language.

  $$\text{Perplexity} = \exp(-(1 \div \text{N}) \times \text{sum of log probabilities of each word}) \tag{13}$$

  where: N is the number of words in the response.
- **Engagement Score**: Captures how interactive and appealing the response is, based on: Average response length, Follow-up rate, User rating (Likert scale)

  $$\text{Engagement} = (\text{weight}_1 \times \text{response length}) + (\text{weight}_2 \times \text{follow} - \text{up rate}) + (\text{weight}_3 \times \text{user rating}) \tag{14}$$

  where:

*Diversity Metrics* These metrics evaluate the novelty, variety, and coverage of recommended items.

- **Coverage**: Percentage of unique items recommended across all users.

  $$\text{Coverage} = (\text{Number of unique items recommended}) \div (\text{Total number of items in dataset}) \tag{15}$$

- **Novelty**: Novelty measures how uncommon or unexpected the recommended items are. A higher score means the system suggests less popular (more surprising) items.

  $$\text{Novelty} = (\text{rank}_1 + \text{rank}_2 + \ldots + \text{rank}_\text{n}) \div \text{n} \tag{16}$$

  where:
- **Diversity**: Average pairwise dissimilarity between items in the recommendation list, based on embedding distance.

  $$\text{Diversity} = \text{average pairwise distance between item embeddings in the recommendation list} \tag{17}$$

#### 4.2.2 Selected Methods for Comparison

- **Embedding Model**: SentenceTransformer *all-MiniLM-L6-v2* is used to encode dialogue turns into 384-dimensional embeddings.
- **Indexing**: FAISS is employed for efficient similarity search over the dialogue corpus.
- **Collaborative Filtering**: The user–item interaction matrix is factorized using Singular Value Decomposition (SVD).
- **Generative Model**: A pretrained large language model (Gemini) is used for prompt-based recommendation generation.
- **Baselines**: We compare (i) pure retrieval-based recommendation, (ii) pure SVD-based collaborative filtering, and (iii) a vanilla generative model without retrieval or preference fusion.

#### 4.2.3 Hyperparameters Turning

- **Top-k retrieval size**: We experiment with different values of k (5, 10, 20) to evaluate its impact on recommendation quality. Our results show that k = 10 provides the best balance between contextual coverage and response relevance.
- **SVD rank (r):** We experiment with different values of r (e.g., 20, 50, 100) to evaluate how the dimensionality of preference modeling affects overall recommendation quality. Our results show that with latent dimension r = 50 offers a good balance between performance and computational cost.
- **The role of prompt formatting** (e.g., ordering of context vs. candidates): Our experiments show that context-first formatting consistently yields higher scores across coherence (BERTScore), fluency (GPT-based perplexity), and user-rated engagement. This suggests that grounding the model in dialogue context before introducing recommendation options leads to more natural and contextually appropriate responses.

### 4.3 Experimental Results

To comprehensively assess the effectiveness of the proposed RAG-SVD framework, we organize our empirical analysis into four distinct case studies: (i) the impact of prompt formatting on generation quality, (ii) item-level recommendation accuracy, (iii) dialogue-level response quality, and (iv) diversity and novelty of suggested items. This structured evaluation enables a holistic understanding of the model's performance across personalization, semantic coherence, linguistic fluency, and recommendation breadth.

#### 4.3.1 The Impact of Prompt Formatting on Generation Quality

Prompt structure plays a critical role in shaping the model's output quality. We investigated how the ordering of contextual information and candidate items within the prompt affects response relevance and fluency. Specifically, we compared two configurations: Context-first (Dialogue history precedes candidate items); Candidate-first (Candidate items precede dialogue context) (Table 2).

**Table 2.** Experimental results for the impact of prompt formatting on generation quality on the ReDial dataset

| Format | BERTScore | Perplexity | Engagement(1–5) |
|---|---|---|---|
| Context-first | **0.89** | **22.7** | **4.3** |
| Candidate-first | 0.82 | 26.5 | 3.9 |

Our experiments show that context-first formatting consistently yields higher scores across coherence (BERTScore), fluency (GPT-based perplexity), and user-rated engagement. This suggests that grounding the model in dialogue context before introducing recommendation options leads to more natural and contextually appropriate responses.

### 4.3.2 Item-Level Recommendation Accuracy

**Table 3.** Experimental results for Item-level recommendation accuracy on the ReDial dataset

| Model | Precision@5 | Recall@5 | F1-score@5 |
|---|---|---|---|
| Retrieval-only (RAG) | 0.32 | 0.29 | 0.30 |
| SVD-only | 0.35 | 0.33 | 0.34 |
| Generative-only | 0.30 | 0.28 | 0.29 |
| **Proposed RAG-SVD** | **0.42** | **0.39** | **0.40** |

The experimental results highlight several important observations (Table 3):

1) **Retrieval-only (RAG)** underperforms compared to SVD-only. Although retrieval captures short-term conversational context, it is highly dependent on the coverage and quality of past dialogues. If the current user's preferences deviate from previously seen patterns, retrieval fails to generalize. Moreover, noisy or multi-topic conversations can reduce the precision of retrieved snippets, leading to weaker recommendation accuracy.
2) **Generative-only** also yields lower scores. While the generative model produces fluent and contextually plausible responses, it lacks grounding in user–item interaction data. As a result, it often generates recommendations that sound reasonable but are not aligned with actual user preferences. This explains the lower precision (irrelevant movies included) and recall (relevant movies omitted).
3) **SVD-only** achieves higher performance than both retrieval-only and generative-only. By leveraging the user–item interaction matrix, SVD captures long-term latent preferences that are robust to conversational noise. This explains why SVD-only provides more stable and accurate recommendations, even in sparse data conditions.
4) **Proposed RAG-SVD** combines the strengths of both paradigms. Contextual retrieval ensures that recommendations remain coherent with the ongoing dialogue, while SVD-based collaborative filtering anchors the system in long-term user preferences. This synergy leads to the highest Precision, Recall, and F1-score, with a relative improvement of ~ 15–20% over single-method baselines.

### 4.3.3 Dialogue-Level Response Quality

Table 4 presents the dialogue-level evaluation results on the ReDial dataset, comparing four models across BERTScore, Perplexity, and Engagement. The proposed RAG-SVD framework achieves the highest BERTScore (0.89) and Engagement score (4.3), while also yielding the lowest Perplexity (22.7). These results indicate that the hybrid model generates more semantically coherent, fluent, and engaging responses than all baselines, confirming its effectiveness in aligning dialogue context with user preferences. Thus, our proposed model improves both semantic quality and user engagement.

**Table 4.** Experimental results for Dialogue-level response quality on the ReDial dataset

| Model | BERTScore | Perplexity | Engagement(1–5) |
|---|---|---|---|
| Retrieval-only (RAG) | 0.84 | 25.9 | 4.0 |
| SVD-only | 0.81 | 27.2 | 3.8 |
| Generative-only | 0.82 | 28.4 | 3.7 |
| **Proposed RAG-SVD** | **0.89** | **22.7** | **4.3** |

### 4.3.4 Diversity and Novelty of Suggested Items

Table 5 summarizes the diversity and novelty performance of four models on the ReDial dataset. The proposed RAG-SVD framework achieves the highest coverage (74.6%), strong novelty (4.7), and the highest diversity score (0.81), indicating its ability to generate broader, more varied, and less mainstream recommendations. Compared to baseline models, RAG-SVD offers broader and more surprising recommendations.

**Table 5.** Experimental results for Diversity and novelty of suggested items on the ReDial dataset

| Model | Coverage (% | Novelty | Diversity |
|---|---|---|---|
| Retrieval-only (RAG) | 68.4 | 4.2 | 0.75 |
| SVD-only | 70.1 | 4.5 | 0.78 |
| Generative-only | 62.1 | 3.8 | 0.72 |
| **Proposed RAG-SVD** | **74.6** | **4.7** | **0.81** |

In summary, the results confirm that retrieval-only and generative-only approaches are limited by their short-term focus, while SVD-only is stronger due to its long-term modeling. The proposed RAG-SVD hybrid effectively bridges these gaps, producing recommendations that are both context-aware and preference-aligned.

## 5 Conclusion

In this work, we proposed a hybrid RAG-SVD framework for conversational recommendation that integrates retrieval-augmented generation with collaborative filtering. By unifying contextual retrieval of dialogue history with latent preference modeling through Singular Value Decomposition, the system can capture both short-term conversational intent and long-term user preferences. Experimental results on the ReDial dataset demonstrate that the proposed method consistently outperforms retrieval-only, generative-only, and SVD-only baselines in terms of Precision, Recall, and F1-score. These improvements demonstrate how retrieval and collaborative filtering complement each other—retrieval maintains dialogue coherence, while SVD captures and stabilizes user preference patterns.

This study confirms that hybridization is a promising direction for conversational recommender systems, especially in sparse-data scenarios. Future work will explore extending the framework to multi-domain recommendation, incorporating reinforcement learning for dynamic dialogue adaptation, and evaluating the system through large-scale human studies to further validate its effectiveness in real-world applications.

## References

1. Yadav, N., Gopinathan, D.: Retrieval augmented generation model for paper recommendation system. SN Comput. Sci. **6**(6), 579 (2025). https://doi.org/10.1007/s42979-025-04095-x
2. Zhang, Y., Chen, X.: Explainable recommendation: A survey and new perspectives (2020). https://doi.org/10.1561/1500000066
3. Maragheh, R.Y., et al.: ARAG: Agentic Retrieval Augmented Generation for Personalized Recommendation (2025). http://arxiv.org/abs/2506.21931
4. Shi, T., et al.: Retrieval Augmented Generation with Collaborative Filtering for Personalized Text Generation (2025). http://arxiv.org/abs/2504.05731
5. Wan, Y., Chen, Z., Liu, Y., Chen, C., Packianather, M.: Empowering LLMs by hybrid retrieval-augmented generation for domain-centric Q&A in smart manufacturing. Adv. Eng. Info. **65** (2025). https://doi.org/10.1016/j.aei.2025.103212
6. Reddy, N.T.K., Patra, M.R., Mishra, B.K.: A modular retrieval-augmented conversational AI chatbot system with integrated recommender engine using local LLMs. Cureus J. Comp. Sci. (2025). https://doi.org/10.7759/s44389-025-06053-3
7. Li, R., et al.: Towards Deep Conversational Recommendations. In: Bengio, S., Wallach, H., Larochelle, H., Grauman, K., Cesa-Bianchi, N., Garnett, R. (eds.) Advances in Neural Information Processing Systems. Curran Associates, Inc. (2018). https://proceedings.neurips.cc/paper_files/paper/2018/file/800de15c79c8d840f4e78d3af937d4d4-Paper.pdf

# Transformer-Based Feature Representation for Highly Accurate Fiber Fault Detection

Tho Nguyen Van[1,2], Giang Nguyen Minh[1], Quang Nguyen Viet[1], Nhan Bui Trong[1], Huong Nguyen Thi[1], Thuy Tran Thi Thanh[1], and Cao Dung Truong[1](✉)

[1] Posts and Telecommunications Institute of Technology, Hanoi, Vietnam
dungtc@ptit.edu.vn

[2] The University of Danang - VN-UK Institute for Research and Executive Education, Danang, Vietnam

**Abstract.** The increasing reliance on optical fiber networks for high-speed data transmission necessitates rapid and accurate fault detection to ensure network reliability. Traditional methods for analyzing Optical Time Domain Reflectometer (OTDR) data are often slow, complex, and require specialized expertise. This paper proposes a highly accurate and efficient transformer-based model for the automated classification of eight major fiber fault types. Our approach utilizes Mutual Information for feature selection to reduce data complexity and training time, feeding a lightweight transformer encoder that effectively captures dependencies in OTDR signals. Experimental results demonstrate the model's superior performance, achieving a test accuracy of 94.9%. In a comparative analysis, our model significantly outperformed other standard deep learning architectures, including CNN1D, LSTM, and BiLSTM, across all key metrics. These findings validate that the proposed transformer-based method provides a robust and computationally efficient solution for real-time fiber fault diagnosis, offering a significant improvement over existing techniques.

**Keywords:** Fiber Fault Detection · Optical Network Monitoring · Transformer Model · Feature Representation · Mutual Information

## 1 Introduction

In today's society, there is a pronounced and growing need for high-speed information access across a wide array of platforms, including both stationary and mobile services [1, 2]. As a direct response, optical fiber technology has become a cornerstone of the modern information infrastructure [3]. To accommodate the demanding data transmission speeds of contemporary wireless networks like 4G, 5G, and Those that will follow [4, 5], information systems built on optical fiber have seen extensive implementation.

Optical fiber serves as the principal medium for conveying immense quantities of data through the Internet, mobile backhaul systems, and core networks. A single fiber is capable of supporting thousands of residential and commercial users, transporting a diverse mixture of private, corporate, and public information. Consequently, any interruption to the fiber can have far-reaching effects, requiring immediate intervention.


P. Pichappan et al. (Eds.): RTIS 2025, LNNS 1936, pp. 350–360, 2027.
https://doi.org/10.1007/978-3-032-23396-7_31

Fibers are prone to various problems, such as physical harm like cuts and security risks like eavesdropping, which can threaten both network uptime and the confidentiality of data. Manually pinpointing and resolving these issues demands specialized expertise and significant time. The Optical Time Domain Reflectometer (OTDR) is a common tool for monitoring fibers by measuring their characteristics and detecting faults. However, the presence of noise interference makes analyzing OTDR data with traditional methods a complex and prolonged task. For this reason, an automatic system that can quickly and precisely detect fiber problems would be valuable for reducing operational expenses and facilitating prompt service recovery (Fig. 1).

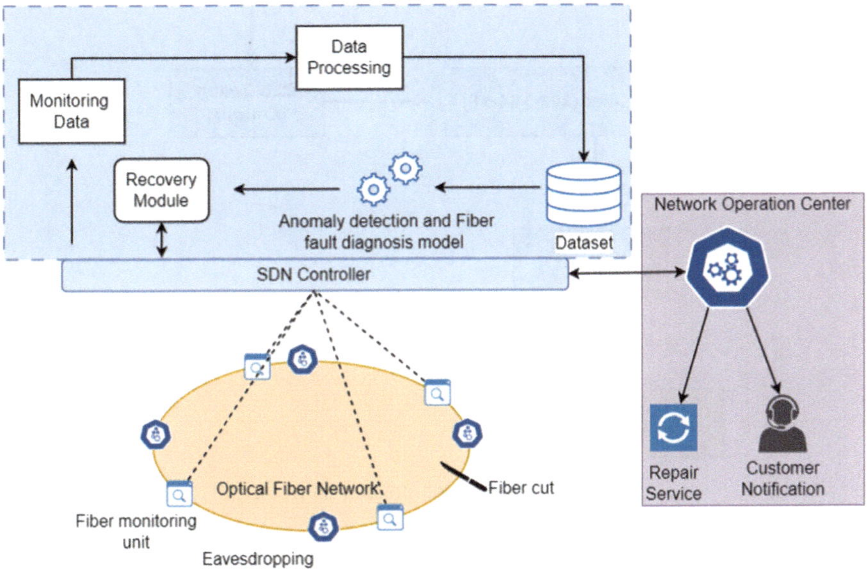


**Fig. 1.** Optical network monitoring model.

Furthermore, Artificial Intelligence (AI) has recently emerged as a powerful tool for predicting anomalies in numerous domains, such as e-commerce, healthcare, IoT, vehicular networks (VANETs), and power monitoring systems. Within the field of optical communications, machine learning (ML) and deep learning (DL) techniques have been broadly adopted, with examples like linear regression (LR) methods being used for signal amplification [6]. Specifically, machine learning models have been created for flaw detection. For instance, the single-layer Perceptron Neural Network (SLP NN) model, derived from the basic LR technique, was designed to predict the location of a fiber cut in subterranean cables [7]. Other approaches, including Autoencoder (AE) and Bidirectional Gated Recurrent Unit (BiGRU) algorithms, are also employed in fiber optics for detecting anomalies [8].

In the effort to achieve more precise fault detection and localization in optical fiber networks, numerous advanced machine learning methods have been examined. Among them, Convolutional Neural Networks (CNNs) and Long short-term Memory (LSTM)

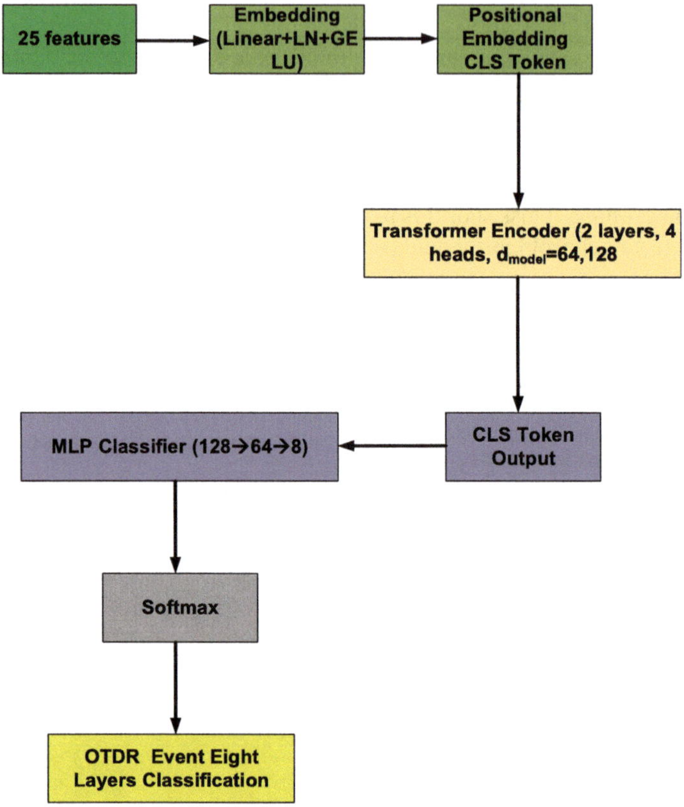


**Fig. 2.** Transformer model architecture.

networks have been suggested as ways to identify and locate reflective events in fibers caused by mechanical splices and connectors [9–11]. The BiLSTM-CNN is a hybrid machine learning (ML) framework that merges CNNs with a Bidirectional Long short-term Memory (BiLSTM) network to effectively identify, localize, and classify different types of events, including reflective, non-reflective, and combined ones [12].

However, the works mentioned above utilized limited ML/DL techniques for fault detection, with a narrow range of fault types and complex models that combined two algorithms. Moreover, these models required lengthy training times, partly due to the processing of all the features in the OTDR data. In contrast, this research proposes and evaluates the detection of eight major faults using a transformer model combined with an encoder. Additionally, this paper compares the performance of the proposed model with other deep learning (DL) models, including BiLSTM, CNN, and RNN.

## 2 Proposed Model

The proposed fiber monitoring framework consists of four main stages: OTDR-based fiber monitoring, data processing, anomaly detection and fault diagnosis, and fault mitigation and recovery. OTDR is used to periodically inspect optical fibers by analyzing backscattered signals to identify attenuation, fault locations, and connection abnormalities caused by events such as fiber cuts, bends, or poor splices. The collected OTDR traces are preprocessed, labeled, denoised, segmented into fixed lengths, and normalized before being transmitted to the SDN controller. These processed data are then fed into a deep learning model for anomaly detection and fault classification. Upon fault detection, predefined recovery mechanisms are activated, and the network operation center is notified to coordinate maintenance actions and customer alerts. This study focuses on the data processing, anomaly detection, and fault diagnosis stages using a Transformer-based deep learning approach.

The proposed model integrates Min–Max normalization, Mutual Information-based feature selection, and a Transformer classifier, as illustrated in Fig. 2. Mutual Information is employed to select the most informative features from the OTDR dataset, reducing redundancy and computational cost while improving classification accuracy. The Transformer model is trained to identify major fiber faults such as fiber cuts and optical eavesdropping attacks, while also incorporating similar anomalies, including bad splices, dirty connectors, bending, and back reflections, especially under low SNR conditions. This strategy enhances fault discrimination and reduces false alarms. The performance of the proposed Transformer is further compared with CNN1D, LSTM, BiLSTM, and MLP models.

## 2.1 Model Architecture

The model takes 25 selected OTDR features as input, which are projected into a 128-dimensional latent space through a linear embedding layer followed by Layer Normalization and GELU activation. Positional embeddings and a CLS token are added to preserve feature order and capture global sample representations. The embedded sequence is processed by a lightweight Transformer Encoder with two layers and four attention heads, which is sufficient to model dependencies in OTDR data while avoiding overfitting. The CLS token output is then fed into a two-layer MLP classifier, producing probabilities over eight OTDR event classes via a Softmax function.

The Transformer architecture is chosen due to its ability to model global feature dependencies through self-attention, its high computational efficiency enabled by parallel processing, and its effectiveness in classification tasks using the CLS token. Compared with recurrent and convolutional models, the proposed design achieves a favorable balance between accuracy, robustness, and computational complexity, making it well suited for real-time optical fiber monitoring and fault diagnosis.

**Table 1.** Training hyperparameters used in the Transformer-based fiber fault detection model.

| Stage | #Samples | $d_{model}$ | #Heads | Accuracy (Trains/Test) | Observation |
|---|---|---|---|---|---|
| 1 | 20k | 64 | 2 | 0.7346 / 0.7947 | Baseline, $<$ 90% accuracy |
| 2 | 70k | 64 | 2 | $>$ 0.90 | Larger dataset improves generalization |
| 3 | 20k | 64 | 4 | 0.7438 / 0.8062 | Slight gain from more head |

*(continued)*

**Table 1.** (*continued*)

| Stage | #Samples | $d_{model}$ | #Heads | Accuracy (Trains/Test) | Observation |
|---|---|---|---|---|---|
| 4 | 20k | 128 | 4 | 0.8576 / 0.8878 | Significant boost with larger model dim |

## 2.2 Dataset, Data Preprocessing, and Feature Representation

The generated OTDR traces [13] have 8 labels (0-normal, 1-fiber cutting, 2-fiber eavesdropping (fiber tapping), 3-dirty connector, 4-bad splice, 5-bending, 6-reflector, 7-PC connector). The dataset was built from OTDR measurements covering eight event classes, including normal traces and seven types of fiber faults. Since raw OTDR traces vary in length depending on the fiber span, each trace was segmented into fixed-length windows of 30 samples, ensuring that each segment corresponded to a single event. This segmentation strategy allows the model to focus on local signal patterns associated with specific anomalies.

After segmentation, Min–Max normalization was applied to each segment to reduce scale variation across measurements. Optional denoising (e.g., moving-average or Gaussian filtering) was considered; however, in this work, it was applied cautiously, since the Transformer encoder can suppress random noise while preserving distinctive event characteristics.

To further reduce input complexity, mutual information (MI) was computed between each feature and the class labels. Features with low MI were discarded, yielding a compact and discriminative set of 25 features from the original 31 (30 amplitude samples plus one SNR value). This selection not only simplifies the model input but also improves recognition accuracy, as MI captures both linear and nonlinear dependencies relevant to OTDR signals.

Finally, the dataset was divided into training (60%), validation (20%), and test (20%) subsets. Importantly, the split was performed by OTDR measurement session: all segments from the same measurement were assigned exclusively to one subset. This strategy prevents data leakage, ensuring that the model is evaluated on entirely unseen sessions and thus reflects genuine generalization capability.

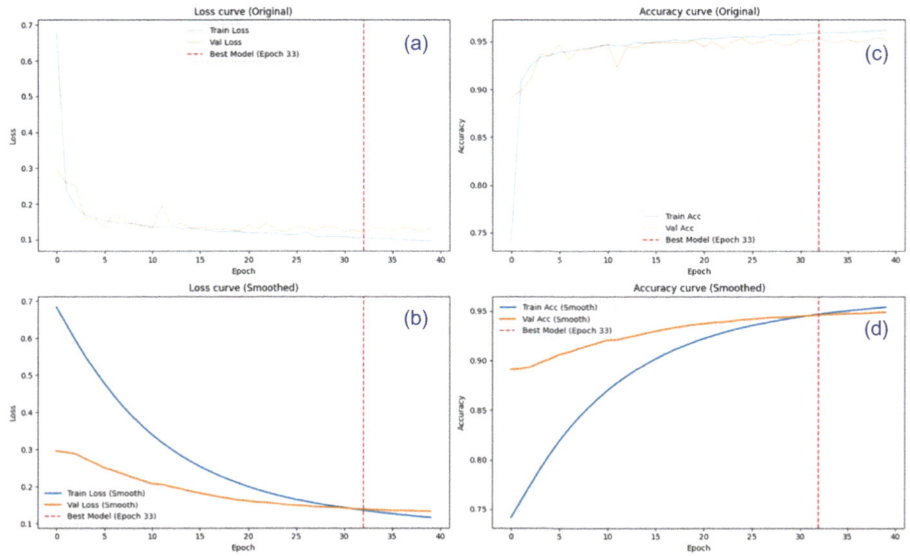


**Fig. 3.** Training and validation curves of the proposed Transformer model: (a) Original loss curve on training and validation sets; (b) Smoothed loss curve for clearer trend visualization; (c) Original accuracy curve on training and validation sets; and (d) Smoothed accuracy curve for clearer trend visualization.

## 2.3 Training and Optimization Strategy.

The Transformer encoder backbone described in Sect. 2.1 was trained using cross-entropy loss and optimized with the Adam optimizer on the top_k features, where k = 25. The learning rate was set to 3e-4 with weight decay 1e-4, and the model was trained up to 100 epochs with a batch size of 128 and the best checkpoint was typically obtained around epoch 20 based on validation accuracy. Regularization was minimal (dropout = 0.1, no stochastic depth), while early stopping was guided by validation accuracy. Dataset partitioning followed a 64/16/20 split by measurement session to avoid leakage. All experiments were implemented in PyTorch and executed on a CUDA-enabled GPU.

We conducted a series of ablation experiments to investigate the effect of data size and model complexity. Table 1 summarizes the results across four stages. Increasing the training data from 20k to 70k samples improved accuracy beyond 90%, confirming the benefit of larger datasets. Increasing the number of attention heads from 2 to 4 yielded only marginal improvement, while increasing the hidden dimension from 64 to 128 resulted in a significant performance gain (Train: 85.8%, Test: 88.8%) without signs of overfitting. These findings suggest that model capacity is a more critical factor than the number of heads for this dataset.

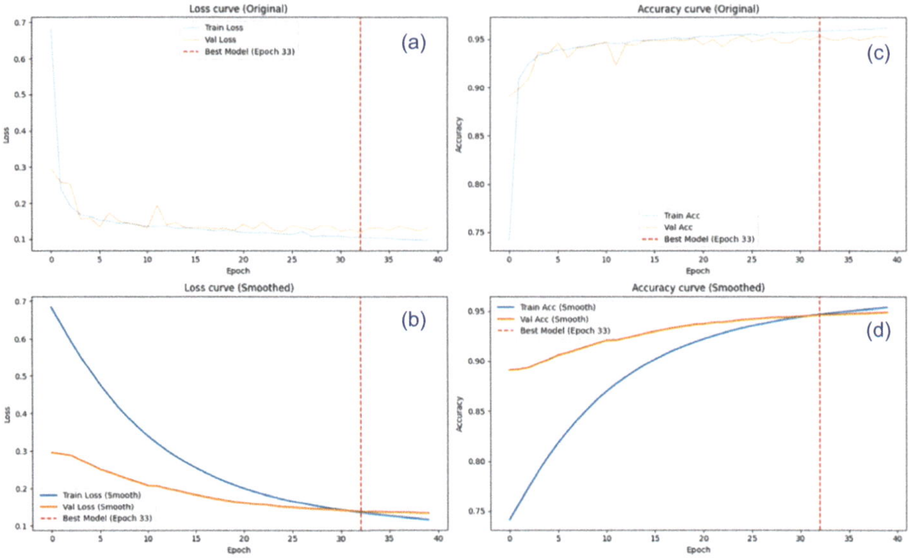


**Fig. 4.** Learning curves of the Transformer model with best checkpoint selection:(a) Original loss curve on training and validation sets; (b) Smoothed loss curve for clearer trend visualization.

(c) Original accuracy curve on training and validation sets; and (d) Smoothed accuracy curve for clearer trend visualization. The red dashed line marks the best checkpoint at epoch 33, corresponding to the highest validation accuracy.

## 3 Experimental Results and Discussion

This section presents the experimental results of the proposed Transformer-based model for OTDR event classification. The evaluation covers training behavior, validation performance, and test results, with particular attention to model generalization and error patterns across classes. The proposed model is evaluated on the test dataset using accuracy, recall, and F1-score as performance metrics. These measures are computed from the confusion matrix, defined in terms of true positives (TP), true negatives (TN), false positives (FP), and false negatives (FN) as follows:

$$Accuracy = \frac{TP + TN}{TP + FP + TN + FN} \tag{1}$$

$$Recall = \frac{TP}{TP + FN} \tag{2}$$

$$F1 - score = \frac{TP}{TP + \frac{1}{2}(FP + FN)} \tag{3}$$

These metrics collectively capture both overall correctness (accuracy) and class-wise balance between precision and recall (F1-score), ensuring a comprehensive evaluation of the classification performance.

### 3.1 Training Behavior

Figure 3 shows the original and smoothed learning curves of the training and validation phases. The training loss decreases steadily towards zero, while the validation loss initially decreases but starts to increase after approximately 30–40 epochs. This divergence indicates the onset of overfitting. Similarly, the training accuracy continues to improve and reaches almost 98%, whereas the validation accuracy peaks at around 95% and then saturates. The smoothed curves make this trend more evident, highlighting that the model tends to memorize training data beyond 40 epochs without further improvement in validation performance.

### 3.2 Optimal Checkpoint Selection

To mitigate overfitting, early stopping was applied, and the best checkpoint was identified at epoch 33, as shown in Fig. 4. At this point, the validation accuracy reached its maximum value of approximately 95.4%, while the training accuracy was around 96%. Beyond this epoch, the validation loss increased, and the validation accuracy slightly decreased, confirming that later checkpoints were suboptimal. This result demonstrates that the proposed Transformer encoder converges efficiently, requiring fewer than 40 epochs to reach optimal generalization.

### 3.3 Test Performance and Error Analysis

The final evaluation on the test set is summarized in the confusion matrix in Fig. 5. The model achieved an overall test accuracy of 94.9%, confirming its strong generalization capability. Most classes were predicted correctly, with nearly all samples of Classes 0, 1, and 7 classified without error. However, some confusion remains between classes with overlapping OTDR signatures. For instance, Class 5 was often misclassified as Class 3 or Class 6, while a fraction of Class 3 samples was also predicted as Class 5. Similarly, moderate confusion occurred between Classes 6 and 7, and between Classes 4 and 5. These findings highlight the difficulty of distinguishing between events with similar local patterns in OTDR traces.

Despite these confusions, the macro-F1 score of ~ 0.80 and weighted-F1 score of ~ 0.85 indicate robust classification performance across all classes. The discrepancy between macro- and weighted-F1 suggests that while the model performs exceptionally well on the majority classes, certain minority or overlapping classes remain challenging. This insight opens directions for improvement, such as data augmentation, domain-specific feature engineering, or incorporating deeper Transformer layers.

### 3.4 Further Discussion

To objectively evaluate the effectiveness of the Transformer model, we conducted a comparison with other standard architectures, including CNN1D, LSTM, BiLSTM, and MLP under the same data conditions and experimental setup. The metrics used include Accuracy, Macro Precision, Macro Recall, and Macro F1, focusing on the high-performance range (90–100%) to clearly reflect the differences.

The results from Fig. 6 show that Transformer outperforms the other models. Specifically, Transformer achieves Accuracy ~ 94.9%, Macro Precision ~ 94.8%, Macro Recall ~ 94.7%, and Macro F1 ~ 94.7%. This is the highest level among all the models compared, and higher than the 95% reference baseline (dashed red line) in the chart. This proves that Transformer is not only more accurate but also maintains a good balance between precision and recall across all classes.

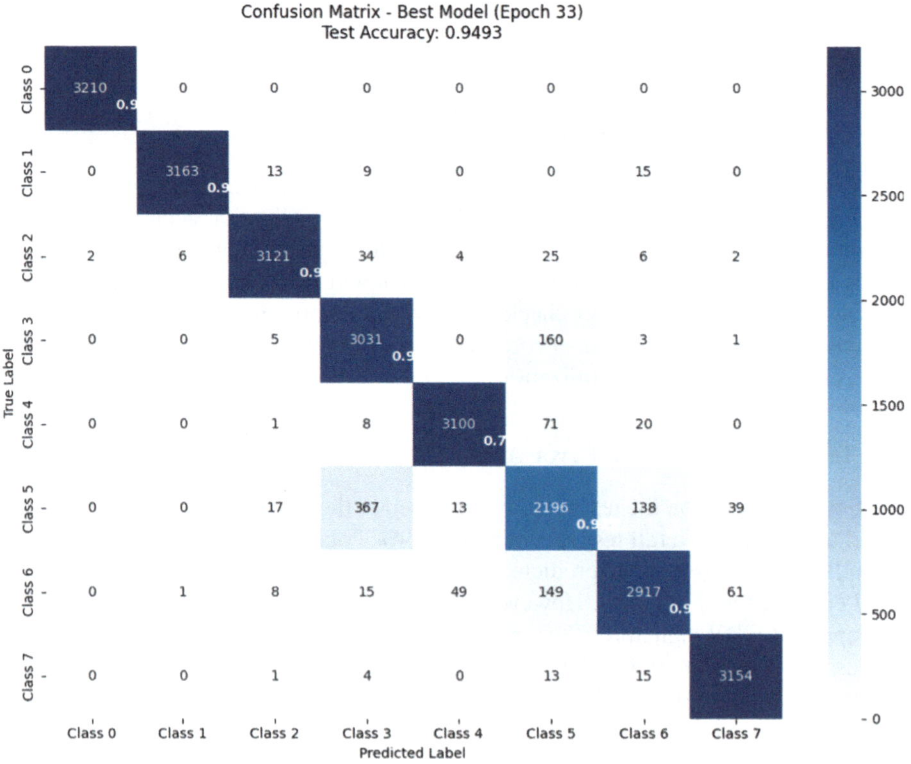


**Fig. 5.** The Confusion Matrix of the model on the test set (8 classes), with the vertical axis representing the actual labels and the horizontal axis representing the predicted labels.

Meanwhile, BiLSTM ranks second with Accuracy ~ 93.5% and Macro F1 ~ 93.4%, showing better ability to learn time series dependencies compared to unidirectional LSTM. MLP and CNN1D achieve similar performance, with an Accuracy of ~ 93.0% and ~ 92.0% respectively, and Macro F1 of ~ 92.9% and ~ 91.8%. The basic LSTM has the lowest results, with an Accuracy of ~ 91.3% and a Macro F1 of ~ 91.2%, indicating limitations in handling long sequence data compared to models with stronger architectures.

The experimental results also demonstrate the superior performance of our proposed model. It achieved an overall test accuracy of 94.9%, successfully generalizing unseen data.

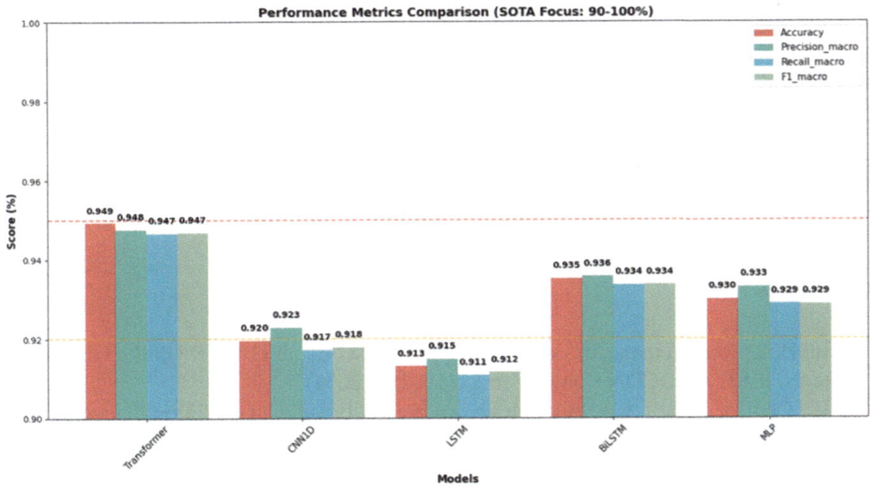


**Fig. 6.** Comparison of performance between Transformer, CNN1D, LSTM, BiLSTM, and MLP based on the metrics Accuracy, Macro Precision, Macro Recall, and Macro F1. Transformer achieves the highest scores across the entire scale.

A comparative analysis confirmed that the transformer model significantly outperformed other common deep learning architectures, including CNN1D, LSTM, BiLSTM, and MLP, scoring the highest across all evaluated metrics like Accuracy, Macro Precision, Macro Recall, and Macro F1-score. The model's architecture, featuring a lightweight two-layer encoder, proved to be well-balanced, providing high expressive power without the computational cost of deeper networks. Despite the high overall accuracy, the error analysis revealed some minor confusion between fault classes with similar OTDR signatures, such as bending (Class 5), dirty connectors (Class 3), and reflectors (Class 6). This indicates that while the model is robust, distinguishing between events with overlapping patterns remains a challenge.

Future work could focus on addressing these ambiguities. Potential improvements include employing data augmentation techniques, developing more domain-specific features, or exploring the use of deeper transformer layers to capture more complex data relationships.

## 4 Conclusion

In conclusion, this study validates a transformer-based model as a highly effective solution for the automated detection of fiber optic faults. Our approach, which combines a lightweight transformer encoder with Mutual Information for optimized feature selection, achieved a robust test accuracy of 94.9% and demonstrably outperformed standard deep learning architectures, including CNN1D, LSTM, and BiLSTM. While the model showed strong generalization, minor confusion between faults with similar OTDR signatures highlights opportunities for future enhancements, such as data augmentation or exploring deeper network layers. Ultimately, this work confirms the potential of transformer models to provide a rapid, reliable, and computationally efficient framework for modern optical network monitoring and maintenance.

**Disclosure of Interests.** The authors have no competing interests to declare that are relevant to the content of this article.

## References

1. Chan, H.A.: Requirements of interworking WirelessLAN and PLMN wireless data network systems. IEEE AFRICON Conference **1**, 251–255 (2004)
2. Pi, Z., Choi, J., Heath, R.: Millimeter-wave gigabit broadband evolution toward 5G: Fixed access and backhaul. IEEE Commun. Mag. **54**(4), 138–144 (2016)
3. Banerjee, A., et al.: Wavelength-division-multiplexed passive optical network (WDM-PON) technologies for broadband access: A review. J. Opt. Netw. **4**(11), 737–758 (2005)
4. Gampala, G., Reddy, C.J.: Massive MIMO - Beyond 4G and a basis for 5G. In: 2018 International Applied Computational Electromagnetics Society Symposium in Denver, ACES-Denver (2018)
5. Novak, D., et al.: Radio-over-fiber technologies for emerging wireless systems. IEEE J. Quantum Electron **52**(1) (2016)
6. Huang, Y., et al.: A machine learning approach for dynamic optical channel add/drop strategies that minimize EDFA power excursions. ECOC 2016; 42nd European Conference on Optical Communication. VDE (2016)
7. Nyarko Boateng, O., Adekoya, A.F., Weyori, B.A.: Predicting the actual location of faults in underground optical networks using linear regression. Eng. Rep. **3**(3), eng212304 (2021)
8. Abdelli, K., Cho, J.Y., Azendorf, F., Griesser, H., Tropschug, C., Pachnicke, S.: Machine-learning-based anomaly detection in optical fiber monitoring. J. Opt. Commun. Netw. **14**(5), 365–375 (2022)
9. Bdelli, K., et al.: Reflective fiber fault detection and characterization using long short-term memory. IEEE/OSA J. Opt. Commun. Netw. **13**(10) (2021)
10. Bdelli, K., et al.: Reflective Event Detection and Characterization in Fiber Optical Links Given Noisy OTDR Signals. Hotonic Networks; 22nd ITG Symposium (2021)
11. Zhang, W., et al.: A deep convolutional neural network with new training methods for bearing fault diagnosis under noisy environment and different working load. Mechanical Systems and Signal Processing **100** (2018)
12. Delli, K.B., et al.: A BiLSTM-CNN based Multitask Learning approach for Fiber Fault iagnosis OCF (2021)
13. Abdelli, K., et al.: Dataset for Optical fiber faults. IEEE Dataport (2022)

# A Smart Curriculum Vitae Analysis and Recommendation System for Job Application Support

Quang Vinh Lai[1(✉)], Manh Dung Ha[1], Trung Kien Doan Le[1], Tien Van Nguyen[2], Long Hoang Dang[2], and Tuan Huu Nguyen[2]

[1] Posts and Telecommunications Institute of Technology, Tran Phu, Hanoi 100000, Hanoi, Vietnam
{VinhLQ.B22CN902,DungHM.B22CN125}@stu.ptit.edu.vn

[2] Faculty of Artificial Intelligence, Posts and Telecommunications Institute of Technology, Tran Phu, Hanoi 100000, Hanoi, Vietnam
{tiennv,longdh,tuannh}@ptit.edu.vn

**Abstract.** In the context of a rapidly evolving digital economy, the recruitment process is increasingly relying on automated systems and data-driven decision-making. The Curriculum Vitae (CV) remains a central element in evaluating candidates; yet, many job seekers, especially students and recent graduates, face challenges in aligning their profiles with the expectations of employers. Existing CV creation tools mainly emphasize design aesthetics and template formatting, while lacking semantic analysis, personalized feedback, and skill development pathways.

This study introduces a smart CV analysis and recommendation system that integrates techniques from Natural Language Processing (NLP), Optical Character Recognition (OCR), and advanced large-scale language models (LLMs). The system is capable of parsing and analyzing candidate CVs, measuring their relevance against job descriptions (JDs), identifying strengths and deficiencies, and generating customized skill development roadmaps accompanied by targeted practice exercises. Furthermore, it supports automated CV generation in standardized formats optimized for applicant tracking systems.

Experimental evaluation with a cohort of final-year university students demonstrates that the system substantially improves the alignment between CVs and job requirements, reduces preparation time, and enhances candidate confidence. These findings highlight the potential of AI-driven CV analysis systems as a bridge between academic training and labor market demands.

**Keywords:** Curriculum Vitae · Job Application Support · CV Analysis · Natural Language Processing · Large Language Models · Optical Character Recognition


P. Pichappan et al. (Eds.): RTIS 2025, LNNS 1936, pp. 361–371, 2027.
https://doi.org/10.1007/978-3-032-23396-7_32

# 1 Introduction

The Curriculum Vitae (CV) is a central artifact in recruitment, shaping how candidates present their qualifications and skills to employers. In a competitive, technology-driven labour market, the ability to craft a strategic and professionally aligned CV has become increasingly important.

However, modern hiring practices complicate this task. Many organizations now rely on Applicant Tracking Systems (ATS) for automated pre-screening based on keyword matching, meaning visually appealing but semantically weak CVs are often rejected before reaching human reviewers. This issue is particularly challenging for students and recent graduates, whose experience and understanding of employer expectations remain limited.

Although online CV builders offer accessible templates and formatting support, most lack semantic analysis, tailored feedback, or skill development guidance. As a result, they help create documents but do little to enhance real employability.

To bridge this gap, this study proposes an integrated CV analysis and recommendation system leveraging NLP, OCR, and Large Language Models. The system automatically extracts and analyzes CVs, evaluates alignment with job descriptions, identifies strengths and gaps, and generates personalized skill roadmaps with targeted exercises. It further supports ATS-optimized CV generation, improving both semantic relevance and technical compatibility within recruitment pipelines.

# 2 Related Work

The task of enhancing Curriculum Vitae (CV) quality and its alignment with job descriptions (JDs) has been the subject of extensive academic and industrial research, largely motivated by the increasing prevalence of automated recruitment systems. The existing body of work can be broadly classified into three main categories: template-based CV builders, keyword-matching, and ATS-orientated tools, and AI-enhanced approaches. Each of these paradigms exhibits distinct strengths, but also critical limitations that collectively motivate the present study.

## 2.1 Template-Based CV Builders

Template-based systems constitute the earliest and most widely adopted class of solutions. Their principal contribution lies in lowering the barrier to entry for candidates by providing pre-designed structures that improve the visual presentation and readability of CVs. Platforms such as TopCV [1] or Vieclam24h [2] exemplify this model by offering accessible interfaces for information organization and formatting. However, the focus on aesthetic presentation comes at the expense of analytical depth.These systems rarely incorporate mechanisms for assessing semantic adequacy or contextual relevance to specific job descriptions, thereby limiting their utility in competitive recruitment contexts.

### 2.2 Keyword-Matching and ATS-Oriented Tools

With the widespread adoption of Applicant Tracking Systems (ATS), a second strand of research and practice has emphasized keyword extraction and lexical matching. Commercial tools such as Jobscan and Resumatch.io calculate a "fit score" based on the degree of lexical overlap between candidate CVs and target JDs. Although these approaches improve the probability of passing initial ATS filters, their inherent reliance on surface-level textual similarity restricts their robustness. They often fail to capture conceptual equivalences, synonymy, or domain-specific definitions, resulting in inflated or misleading fit scores. Moreover, their guidance typically remains prescriptive rather than explanatory, offering little support for the underlying development of missing competencies.

### 2.3 AI-Enhanced CV Analysis

The third line of work leverages advances in Natural Language Processing (NLP) and machine learning to achieve more sophisticated semantic evaluations. Embedding based similarity models, contextualized representations from transformer architectures, and large language models (LLMs) have enabled richer analysis of CV text. Certain platforms, such as LinkedIn Resume Builder, integrate limited semantic features to improve relevance scoring. However, most existing solutions remain primarily orientated towards phrase optimization and stylistic enhancement, rather than delivering structured learning roadmaps or competency-building interventions. Consequently, they provide incremental improvements in alignment but do not fundamentally bridge the gap between CV presentation and improving employability.

### 2.4 Research Gap

The literature review highlights three persistent limitations. First, there is a fragmentation of features: template-based and ATS-orientated systems address either formatting or lexical overlap, while seldom combining both with semantic analysis. Second, personalization remains limited, with most tools generating generic recommendations that are insufficiently tailored to individual profiles. Third, there is a lack of integrative learning support: existing approaches rarely couple CV analysis with scaffolded exercises or structured pathways that guide candidates in addressing identified deficiencies. The present work seeks to address these gaps by proposing a comprehensive pipeline that unifies document processing, semantic evaluation, personalized recommendations, and automated CV synthesis, thereby linking surface-level alignment with deeper employability development.

## 3 System Design

Designing an intelligent CV analysis and recommendation system requires a careful reconciliation of algorithmic rigor and user-centered concerns. The system must process heterogeneous documents, transform unstructured content into structured profiles, and produce interpretable recommendations that support measurable improvements of skills.

### 3.1 Design Objectives

The system is developed around four core objectives:

(i) **Automation**: Minimize manual intervention across document ingestion, extraction, CV–JD matching, and CV synthesis, enabling faster feedback and reduced human error.
(ii) **Personalization**: Provide tailored recommendations by identifying individual strengths and deficiencies, linking them to relevant resources and structured learning pathways rather than generic keyword suggestions.
(iii) **Standardization**: Ensure ATS-compliant output through consistent structure, machine-readable formatting, and controlled terminology to prevent rejection due to technical incompatibility.
(iv) **Extensibility**: Maintain architectural flexibility for future upgrades, including multimodal evidence (e.g., portfolios, video resumes), real-time labor market integration, and improved semantic models.

### 3.2 Functional Decomposition

The system is structured into five interoperable modules. The *Document Processing Module* ingests CVs in multiple formats (PDF, DOCX, images) and produces normalized text via segmentation and OCR-based extraction. The *Semantic Analysis Module* computes lexical and embedding-based similarity between CVs and job descriptions, identifying both explicit and conceptual matches. The *Recommendation Module* transforms these results into prioritized skill gaps and learning suggestions. The *Template Generation Module* produces ATS-compliant CVs using standardized layouts and keyword-focused formatting. The *Administrative Module* manages users, system configuration, and content updates, enabling continuous improvement of taxonomies, templates, and resource mappings. Together, these modules form an end-to-end pipeline supporting document processing, semantic evaluation, recommendation generation, and structured CV output.

*Administrative Module* The Administrative Module maintains system integrity through access control, performance monitoring, and update management. It allows domain experts to curate learning resources and skill taxonomies while logging administrative actions for accountability. By incorporating feedback loops, the module supports iterative refinement and ensures long-term adaptability to evolving recruitment requirements.

### 3.3 Architectural Overview

The system is designed using a three-tier architecture—Presentation, Application, and Data layers—to ensure modularity, scalability, and maintainability. The Presentation layer delivers interfaces for candidates, recruiters, and administrators, enabling interaction without affecting underlying logic.

The Application layer contains the core engine, handling document preprocessing, semantic analysis, recommendation generation, and automated CV synthesis. It coordinates data flow between modules and governs workflow integrity.

The Data layer manages persistent storage of user profiles, CVs, skill taxonomies, and system logs, supporting secure data handling, query efficiency, and reliability (Fig. 1).

This separation of concerns enables independent evolution of each tier: interfaces can be enhanced, semantic models upgraded, and storage scaled via distributed or cloud-based solutions. Figure 2 summarizes the layered design and information flow.

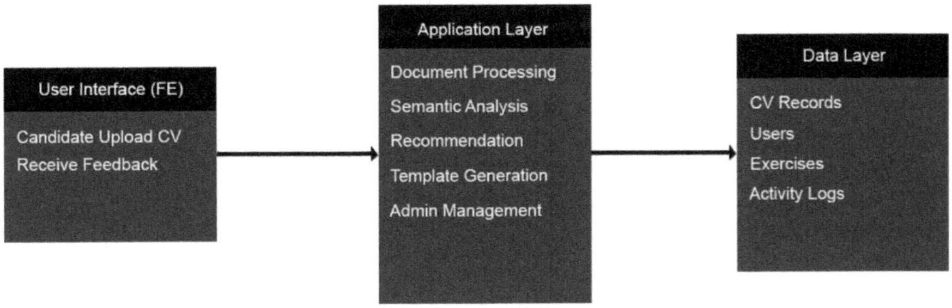


**Fig. 1.** System architecture illustrating the interaction among Presentation, Application, and Data tiers.

### 3.4 Data Model

The conceptual data model defines the key entities and relationships that enable both analysis and recommendation. Four core entities are established: Users, *CV Records*, Exercises, and *Activity Logs*. The Users entity stores personal details and roles, forming the anchor for all other records. CV *Records* has structured CV representations—such as education, experience, and skills—allowing for a detailed semantic evaluation.

Exercises represent curated learning tasks linked to identified skill gaps, creating a direct pathway from profile deficiencies to remediation actions. *Activity Logs* capture system interactions, including viewed recommendations, attempted exercises, and regenerated CVs, supporting progress tracking and analytics.

The model supports one-to-many relationships between users, CV records, and activity logs, as well as many-to-many links between CV records and exercises. This structure facilitates efficient retrieval, traceability, and easy extension to new entities such as job descriptions or portfolios.

### 3.5 Design Rationale and Considerations

In addition to functional components, several non-functional requirements were considered to ensure that the system remains sustainable, secure, and reliable for real-world deployment.

*Interoperability* To ensure compatibility with diverse recruitment platforms and Applicant Tracking Systems (ATS), the system employs standardized section labels (e.g., *Education*, *Skills*) and machine-readable formatting, avoiding artifacts that hinder parsing. This design improves candidate success rates and facilitates integration without additional customization.

*Scalability* The modular architecture allows compute-intensive tasks such as semantic similarity computation to scale independently of lightweight processes like CV rendering or administrative monitoring. The data layer is structured to support distributed storage, enabling the system to serve thousands of users across institutions.

*Privacy and Security* Because CVs contain sensitive personal information, the system enforces strict access control, encrypts stored data, and logs administrative actions to ensure compliance with GDPR and institutional policies, thereby preserving user trust.

*Explainability* To enhance user confidence, the system provides rationale traces that highlight which job description (JD) elements influenced similarity scores. This transparency enables candidates to validate feedback and supports manual auditing by recruiters or domain experts.

# 4 Methodology

The proposed methodology operationalizes an end-to-end pipeline that transforms raw Curriculum Vitae (CV) documents into structured profiles, evaluates their alignment with job descriptions (JDs), and generates personalized recommendations together with optimized CV outputs. The pipeline is designed to balance algorithmic rigor with practical applicability, ensuring both technical robustness and user-centered outcomes. The methodology is structured into five principal stages: document acquisition and preprocessing, CV–JD matching, gap analysis, personalized roadmap generation, and automated CV synthesis. Each stage is described in detail below.

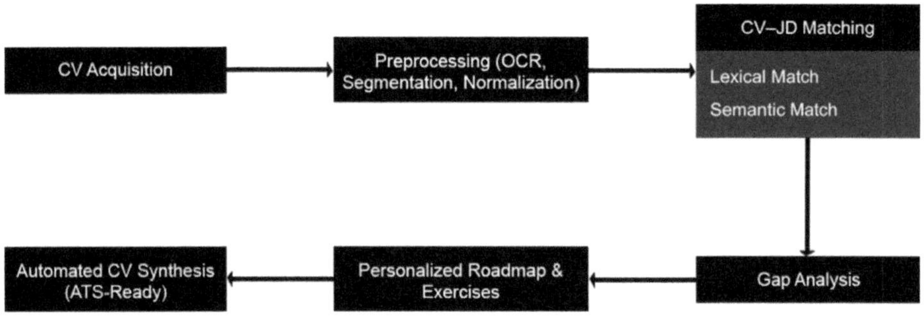


**Fig. 2.** The end-to-end workflow of the system.

## 4.1 Document Acquisition and Preprocessing

The pipeline begins by ingesting CVs in diverse formats—including PDF, DOCX, and images. Image-based files are converted via OCR to ensure uniform machinereadable text. The documents then undergo structured preprocessing. CVs are segmented into key sections (personal data, education, experience, skills, certifications) to support focused analysis. Text normalization is applied through lowercasing, cleaning, and standardized formatting. Named entity recognition extracts institutions, job titles, and technical terms, which are then canonicalized (e.g., mapping variants like "C++" to a standard skill token). Finally, the refined content is mapped to the system's internal *CV Records* schema, enabling structured storage and reliable downstream semantic analysis for matching and recommendation.

## 4.2 CV–JD Matching Framework

To quantify alignment, the system employs a hybrid lexical–semantic scoring mechanism that integrates surface-level and conceptual dimensions.

*Lexical Matching* Lexical similarity is computed by extracting explicit tokens (skills, job titles, certifications) and applying statistical text similarity measures such as TF–IDF weighted cosine similarity or BM25 ranking. This captures keyword-level overlaps commonly exploited by ATS.

*Semantic Matching* To overcome the limitations of purely lexical methods, semantic similarity is computed using dense vector embeddings generated by the **Gemini 2.0 Flash** [3] model developed by Google DeepMind (2024). This largescale multimodal lan guage model provides context-aware embeddings that effectively capture both syntactic and semantic relationships across CV sections and job descriptions (JDs). Cosine similarity is then applied between the embedding vectors of CV segments and JD texts to quantify conceptual equivalence, enabling the recognition of semantically related but lexically divergent expressions (e.g., software developer" vs. application engineer"). In this study, the base Gemini 2.0 Flash [3] model is employed without domain-specific fine tuning to preserve generalization; however, future work may incorporate task-adaptive optimization to further enhance relevance in recruitment contexts.

*Composite Scoring* The final matching score is a weighted linear combination of lexical and semantic similarity:

$$\text{Score} = \alpha \cdot L_{match} + \beta \cdot S_{match}, \quad \alpha + \beta = 1, \tag{1}$$

where $a$ and $\beta$ represent the relative contributions of lexical and semantic dimensions. In this study, equal weighting ($a = \beta = 0.5$) is adopted as a neutral baseline; however, the framework supports empirical tuning through supervised learning or cross-validation when expert-labeled data becomes available.

## 4.3 Gap Analysis

After computing lexical and semantic similarity scores, the system conducts a gap analysis to identify skills and experiences that are missing or insufficiently represented. CV content is cross-referenced with job description requirements to detect absent competencies, which are then ranked by importance using two signals: their statistical frequency in domain-specific job postings and expertcurated skill taxonomies. The analysis also considers contextual gaps such as lack of quantifiable outcomes or insufficient detail in project descriptions. By prioritizing critical deficiencies over secondary enhancements, the framework converts similarity metrics into actionable insights and establishes the basis for personalized recommendations.

### 4.4 Personalized Roadmap and Exercise Generation

Based on the gap analysis, the system generates individualized learning roadmaps that prioritize missing competencies according to their importance and expected acquisition time. Each prioritized skill is linked to curated learning resources (MOOCs, tutorials, exemplar projects) and accompanied by automatically generated practice exercises aligned with the user's proficiency level. Exercises are sequenced from foundational to advanced and designed to produce tangible outputs (e.g., project repositories, portfolio evidence) that can be incorporated back into the CV. This mechanism turns abstract recommendations into actionable and measurable learning progression.

### 4.5 Automated CV Synthesis

The final stage transforms enriched candidate profiles into ATS-compliant CVs using standardized templates. Prioritized skills are injected into key sections, and experience entries are reordered to maximize job relevance. Formatting is normalized to ensure machine readability and consistent parsing across ATS systems. Output is provided in PDF, DOCX, and JSON formats to support both human review and integration with external recruitment platforms, effectively closing the loop between analysis, skill enhancement, and optimized CV presentation.

### 4.6 Evaluation Framework

The system is evaluated through three complementary perspectives: technical metrics, effectiveness metrics, and user-centered evaluation.

**Technical evaluation** Assesses computational accuracy and efficiency. OCR reliability is measured using Word Error Rate, while CV–JD matching is benchmarked against expert-labeled similarity scores. Processing time per CV is also recorded to validate scalability.

**Effectiveness evaluation** Examines the system's practical impact by comparing lexical and semantic similarity scores before and after system intervention, as well as proxy indicators such as interview invitation rates.

**User-centered evaluation** Collects survey and interview feedback on usefulness, interpretability, and confidence gain, complemented by activity-log analysis to track skill progression over time.

To complement intrinsic metrics, future work will integrate real-world outcomes such as recruiter feedback and post-application success rates, enabling a holistic assessment of system utility in authentic recruitment settings.

## 5 Experimental Results

This section presents the empirical evaluation of the proposed CV–JD matching framework using survey-based data. A total of 192 valid responses were collected from senior undergraduate students and recent graduates across multiple disciplines. Each participant was instructed to self-assess the degree of alignment between their Curriculum

Vitae (CV) and a representative job description (JD) along two complementary dimensions: lexical similarity ($L_{match}$) and semantic similarity ($S_{match}$). The reported values were normalized to the range [0, 1] to ensure comparability across respondents.

## 5.1 Weight Selection Rationale

The composite matching score is operationalized as a weighted linear combination of the two dimensions:

$$\text{Score} = \alpha \cdot L_{match} + \beta \cdot S_{match}, \quad \alpha + \beta = 1. \tag{2}$$

In this study, equal weights were assigned ($a = \beta = 0.5$) to establish a balanced contribution of lexical and semantic evidence. This choice reflects the current focus of the system on the information technology (IT) domain, where both keyword precision and semantic context play equally significant roles in evaluating candidate relevance. Although this configuration has not yet been empirically optimized, it provides a stable and interpretable baseline suitable for a homogeneous technical field. Future extensions of the system will incorporate datasets from additional industries—such as finance, education, and healthcare—enabling the empirical calibration of $a$ and $\beta$ through supervised learning or adaptive optimization techniques tailored to domain-specific recruitment practices.

Notably, the optimal balance between lexical and semantic matching may vary across industries or job categories. For example, technical domains such as software engineering rely heavily on explicit keyword inclusion, whereas creative or managerial roles emphasize contextual narrative coherence. Therefore, adaptive weighting strategies are a key direction for subsequent system enhancement.

## 5.2 Quantitative Outcomes

Table 1 reports the distribution of results obtained from the survey. For clarity of presentation, the 192 responses are partitioned into two subtables. Each row records a respondent's lexical match, semantic match, and corresponding composite score.

The aggregated results reveal an **average lexical similarity of 0.85**, an **average semantic similarity of 0.79**, and an overall **composite score of 0.82**. These findings suggest that, while students They are generally proficient at incorporating relevant keywords into their CVs, They exhibit relative weaknesses in articulating their experiences and skills in a manner that semantically resonates with the expectations articulated in JDs. This indicates a gap between keyword-level optimization and holistic semantic alignment.

**Table 1.** Survey-based calculation of CV–JD matching Score

| | Part 1 | | | | | Part 2 | | |
|---|---|---|---|---|---|---|---|---|
| Respondent | *Lmatch* | *Smatch* | Score | | Respondent | *Lmatch* | *Smatch* | Score |
| R1 | 0.80 | 0.80 | 0.80 | | R97 | 0.80 | 0.80 | 0.80 |
| R2 | 1.00 | 0.70 | 0.85 | | R98 | 1.00 | 0.70 | 0.85 |
| R3 | 0.80 | 0.70 | 0.75 | | R99 | 0.70 | 0.60 | 0.65 |
| R4 | 1.00 | 0.70 | 0.85 | | R100 | 0.80 | 1.00 | 0.90 |
| R5 | 0.90 | 0.90 | 0.90 | | R101 | 0.70 | 0.80 | 0.75 |
| R6 | 0.80 | 0.80 | 0.80 | | R102 | 1.00 | 0.60 | 0.80 |
| R7 | 0.90 | 0.90 | 0.90 | | R103 | 0.70 | 0.70 | 0.70 |
| R8 | 1.00 | 1.00 | 1.00 | | R104 | 1.00 | 1.00 | 1.00 |
| R9 | 0.90 | 0.80 | 0.85 | | R105 | 0.90 | 0.70 | 0.80 |
| R10 | 0.70 | 0.70 | 0.70 | | R106 | 0.80 | 0.90 | 0.85 |
| R11 | 0.80 | 0.90 | 0.85 | | R107 | 1.00 | 1.00 | 1.00 |
| … | … | … | … | | … | … | … | … |
| R96 | 1.00 | 0.80 | 0.90 | | R192 | 0.90 | 0.90 | 0.90 |
| | | | | | **Average** | **0.85** | **0.79** | **0.82** |

## 5.3 Interpretation

The balanced distribution of scores demonstrates that participants are generally aware of the importance of keyword inclusion to ensure compatibility with Applicant Tracking Systems (ATS). However, the comparatively lower semantic similarity values highlight the limitations in articulating transferable competencies and contextualizing skills in alignment with employer expectations. These findings substantiate the research gap outlined in Sect. 2, where it was observed that most existing CV construction tools place excessive emphasis on surface-level formatting and keyword optimization, often neglecting semantic coherence. By empirically validating this gap, the results reaffirm the significance of the proposed CV analysis and recommendation framework in enhancing both the lexical precision and semantic richness of candidate profiles.

Although the present study focuses on validating the proposed framework's internal consistency, future iterations will include comparative analyses with existing systems such as Jobscan and LinkedIn Resume Builder. Controlled experiments will evaluate performance differentials in lexical and semantic matching accuracy, as well as user satisfaction and system usability.

## 6 Conclusion and Future Work

This study presents an integrated framework for intelligent CV analysis that combines document processing, semantic assessment, personalized skill roadmapping, and automated CV generation. The system bridges keyword-based matching with deeper semantic interpretation and embeds learning support to enhance employability.

Findings from a survey of 192 respondents show stronger lexical alignment than semantic articulation, underscoring limitations of existing CV tools. The proposed approach addresses this gap through dual-layer evaluation and actionable recommendations, contributing both theoretically and practically to recruitment technology.

However, the study is constrained by subjective self-reported scores, a student-focused sample, and baseline weighting between lexical and semantic components. Future work will incorporate expert-annotated data, real hiring outcomes, multimodal inputs, and adaptive algorithms to improve accuracy and personalization.

## References

1. TopCV. CV Creation Platform in Vietnam (2023). Retrieved from https://www.topcv.vn
2. Jobscan. AI-powered CV Matching Platform (2023). Retrieved from https://www.jobscan.co
3. Google DeepMind. Gemini 2.0 Flash Technical Report (2024)

# Application of AI and IoT Technologies for Automatic Forecasting of Water Level and Discharge for Kanak Reservoir, Gia Lai, Vietnam

Van Phan Le[1(✉)], Tri Van Bui[2], Van Vinh Nguyen[3], and Van Sang Le[4]

[1] University of Agriculture and Forestry Ho Chi Minh City, Ho Chi Minh City, Vietnam
lvphan@hcmuaf.edu.vn
[2] Posts and Telecommunications Institute of Technology, Ho Chi Minh City, Vietnam
tribuivan@ptithcm.edu.vn
[3] AI Research Unit – Shrimp Vet Lab (Minh Phu), Ho Chi Minh City, Vietnam
[4] University of Phan Thiet, Phan Thiet, Vietnam
lvsang@upt.edu.vn

**Abstract.** This study presents an application of artificial intelligence (AI) for automatic forecasting of water level and discharge for Kanak Reservoir (Gia Lai province, Vietnam). The research team developed an end-to-end system from automated rainfall observation to data storage and forecasting. The paper focuses on the use of machine learning algorithms including Linear Regression, Random Forest, Support Vector Machine (SVM), Long Short-Term Memory (LSTM), and Voting Regression to improve water-level forecast accuracy. Results indicate that LSTM outperformed other models, particularly when using data from five days prior. The study emphasizes the importance of accurate rainfall data and recommends using AI for water resource management and disaster prevention in Vietnam.

**Keywords:** Water level forecasting · rainfall · SVM · LSTM · machine learning · forecasting techniques

## 1 Introduction

Artificial intelligence (AI) is increasingly applied worldwide and in Vietnam. The meteorological and hydrological sector has achieved several AI applications in measurement, monitoring, and forecasting in recent years. To increase the practical application of AI for hydrological forecasting in Gia Lai Province, the AI research team from the University of Agriculture and Forestry Ho Chi Minh City (HCMUA) and the Central Highlands Regional Meteorological and Hydrological Station collaborated to implement the project: "Application of AI technology for automatic forecasting of water level and discharge for Kanak Reservoir, Gia Lai".


P. Pichappan et al. (Eds.): RTIS 2025, LNNS 1936, pp. 372–379, 2027.
https://doi.org/10.1007/978-3-032-23396-7_33

This project is an end-to-end application covering automated rainfall observation, data storage, automated data export, and running AI forecasting software. The team has begun operational testing and validation.

### 1.1 AI in Water-Level Forecasting: Related Work

Forecasting flood water levels at the right time and location is a major hydrology challenge, especially as flood frequency and intensity increase due to climate change and socio-economic development. Many studies show forecast accuracy depends on rainfall data, model structure, and model parameters. Traditional hydrological models often struggle to capture complex flood dynamics due to lacking momentum conservation, motivating hydrodynamic models for more accurate water-level prediction [1]. Several previous works proposed classification and parameter optimization methods (e.g., fuzzy clustering, neural networks) to improve flood forecast accuracy compared to non-classified approaches [3]. Studies focused on using rainfall intensity data emphasize the critical role of rainfall observations in hydrological modelling [2,4]. Recent work also demonstrates that spatio-temporal rainfall features can be used to classify flood types and enhance forecasts across basins [5].

However, connections between rainfall and reservoir water level remain limited in prior work. Many studies concentrated on non-dynamical hydrological models and under-utilized hydrodynamic or machine-learning models for extreme rainfall situations. This paper studies the rainfallreservoir water-level relationship using modern methods, including deep learning and parameter optimization, targeting forecasts for 1 h and 1 day ahead. The central research question: "Can machine learning algorithms improve the accuracy of reservoir water-level forecasting compared with traditional methods?"

## 2 Methods

### 2.1 Study Area

Gia Lai Province, located in the Central Highlands of Vietnam, features diverse terrain and rich ecosystems. It is home to numerous ethnic groups and plays an essential role in agriculture and water-resource management. This study focuses on 17 rainfall stations and 1 water-level station around Kanak Reservoir. Station names include: TV_AK, KT_AK, KLKhong, TuAn, Dong, Krong1, Krong2, Krong3, SonLang, DakRong1, DakRong2, DakRong3, NghiaAn, ToTung, LoKu, Sopai, Krong, HoAK, HoKN. Daily rainfall at these stations varied with maximum daily values between 111 mm and 165 mm.

All stations are fairly evenly distributed throughout the basin and are automatic stations providing near-real-time rainfall observations (10-min intervals). Real-time rainfall and water-level data are continuously updated to the Central Highlands Regional Meteorological and Hydrological data servers. These data feed the Kanak Reservoir forecasting system. (Original document: baocaothamluan_15_08_2025.docx). :contentReference[oaicite:2]index=2 (Fig. 1)

**Fig. 1.** Study area map (placeholder). Please replace `figure1_map.png` with the original image from the .docx if you have it.

## 2.2 Algorithms Used

We implemented and compared five models for forecasting water level based on past rainfall (from previous days or hours):

1) Linear Regression:. Linear regression models the relation between a dependent variable (water level) and one or more independent variables (rainfall from stations). Benefits include interpretability via coefficients and fast prediction using Least Squares Estimation (LSE). It is suitable for historical data and provides quick short-term predictions.
2) Random Forest Regression: Random Forest is an ensemble tree-based method that aggregates multiple decision trees to improve accuracy and generalization. Advantages include robustness against overfitting, ability to handle missing data and nonlinearity, and high predictive performance.
3) SVM Regression: Support Vector Regression (SVR) extends SVM principles to regression tasks using support vectors and kernels. SVR is robust to noise, performs well with limited samples, and models non-linear relationships.
4) Voting Regression: Voting Regression is an ensemble that combines predictions from multiple models (here: Linear Regression, Random Forest Regressor, and SVM Regression). Final predictions are averaged or determined by weighted voting to improve accuracy and mitigate each method's weaknesses.
5) LSTM (Long Short-Term Memory): LSTM is a type of recurrent neural network tailored to learn temporal dependencies in sequences. Its gate structure

(input, forget, output gates) allows capturing long-term dependencies making it suitable for time-series tasks such as water-level forecasting. LSTM models were trained using optimizers such as Adam or SGD and backpropagation through time (Fig. 2).

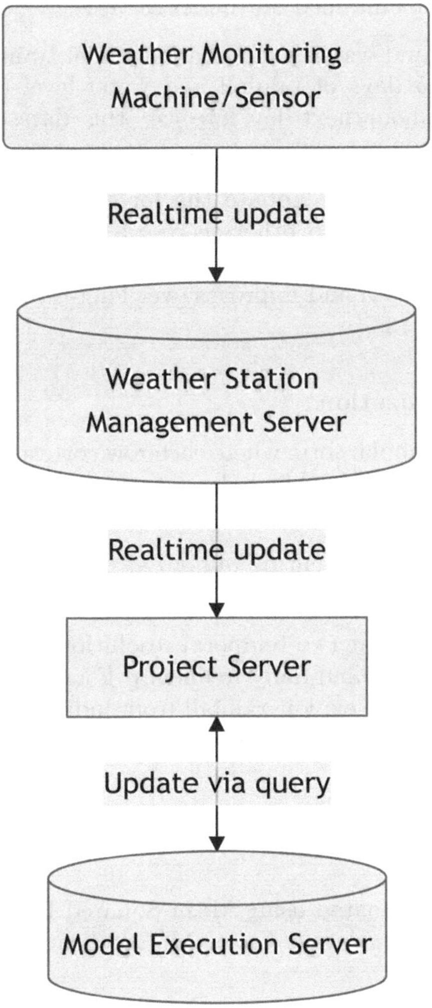


**Fig. 2.** AI forecasting system workflow (placeholder). Replace `figure2_system_diagram.png` with the original image if available.

### 2.3 Operational System and Data Flow

Real-time data from rainfall and water-level observation machines are pushed to the Central Highlands Regional data center. Discrete station data are digitalized and stored in a MongoDB-based database and made accessible through an online data extraction platform at https://taynguyen.weathervietnam.vn/. The weather-station data server exposes an API (webservice) to the AI server. Two primary datasets are maintained on the AI server:

1) Historical rainfall and water-level data for model training (train set).
2) The most recent 5 days of rainfall and water-level data used as input to generate the next-hour/next-day forecast; this dataset is updated seconds before each forecast run.

Forecast outputs immediately update the forecasting web portal and can be delivered via email or FTP to other servers if required. The system uses an iterative retraining approach (retraining every 7 days or shorter) so the model self-updates with new data and improves over long-term operation. :contentReference[oaicite:3]index=3

### 2.4 Dataset Organization

Data are arranged in tabular form where each row corresponds to one observation time (daily or hourly). Columns include:

- Date/time stamp.
- Rainfall: 17 columns representing rainfall at each station.
- Kanak Reservoir water level: daily recorded level.

Training is performed at two temporal resolutions described in the original: hourly (60-min frequency) and daily frequency. Forecasts use rainfall from 1 up to 6 previous days; models exploit rainfall from individual prior days to predict the water level for the next day, thereby capturing the complex rainfall-to-water-level relationship.

## 3 Results and Discussion

Forecast quality was evaluated using Mean Squared Error (MSE) between predicted and observed water levels. Lower MSE indicates better prediction accuracy.

### 3.1 Machine Learning Results

The following table reproduces the MSE values reported in the original study for predictions using data from 1 day prior and 5 days prior:
Note: The missing value for LSTM at 1 day prior in the original table was represented as empty (algorithm-specific data-structure limitation). :contentReference[oaicite:4]index=4

Figures 4 and 5 are placeholders for the published charts in the original document showing predicted vs actual water level series for Random Forest (1-day prior) and LSTM (5-day prior) respectively (Fig. 3 and Table 1).

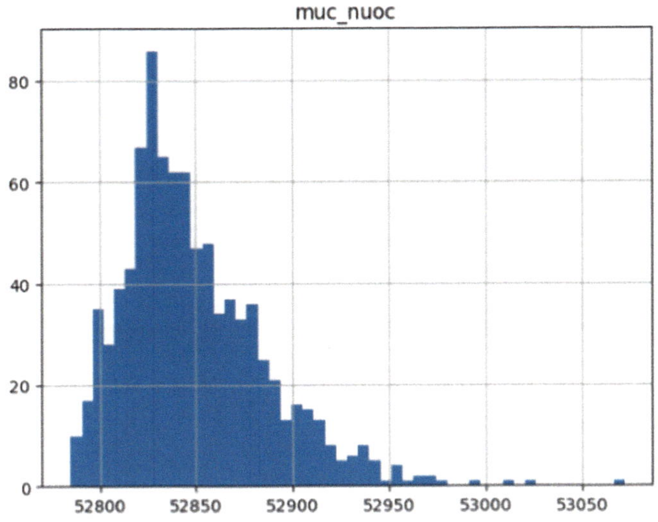


**Fig. 3.** Kanak reservoir water-level statistics (placeholder). Replace `figure3_stats.png` as needed.

**Table 1.** Mean Squared Error (MSE) of models for 1-day and 5-day prior inputs

| Model | 1 day prior (MSE) | 5 days prior (MSE) |
|---|---|---|
| Linear Regression | 0.031928 | 0.030251 |
| Random Forest Regressor | 0.031495 | 0.029016 |
| SVM Regression | 0.052460 | 0.048154 |
| Voting Regression | 0.037229 | 0.034045 |
| LSTM | N/A | 0.15600 |

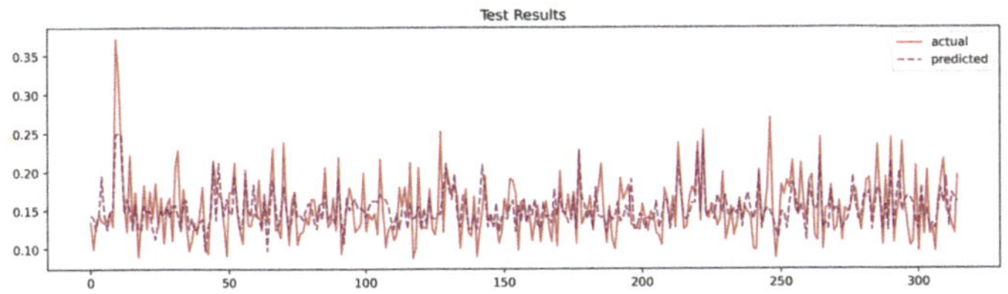


**Fig. 4.** Random Forest Regressor: prediction vs actual (1 day prior) (placeholder).

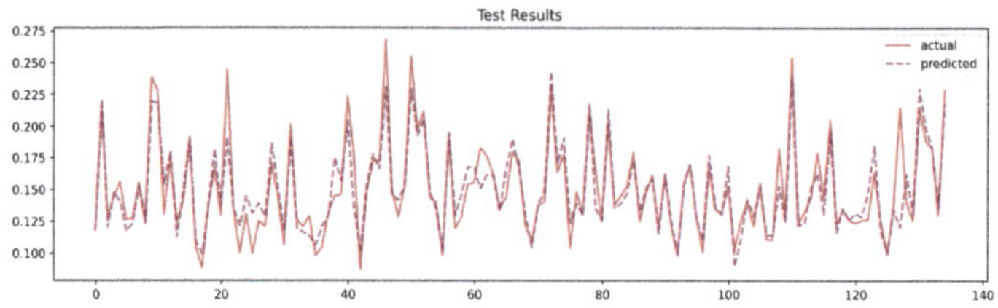


**Fig. 5.** LSTM: prediction vs actual (5 days prior) (placeholder).

### 3.2 Operational Outputs

The forecast outputs are pushed into the database and displayed together with observed hydrometric data at station TV1. The original document included an example export table (sample rows reproduced below) and screenshots of the online system. :contentReference[oaicite:5]index=5 (Table 2)

**Table 2.** Sample forecast-exported table (excerpt)

| ID | Date time | Water level predict (m) | Water level (m) | Accuracy |
|---|---|---|---|---|
| 1 | 2024-09-24 00:00:00 | 528.16 | – | 0 |
| 2 | 2024-09-23 23:50:00 | 528.16 | 527.873 | 0.9994 |
| 3 | 2024-09-23 23:40:00 | 528.09 | 512.6257 | 0.9707 |
| 4 | 2024-09-23 23:30:00 | 528.13 | 526.0853 | 0.9961 |
| 5 | 2024-09-23 23:20:00 | 528.06 | 518.1868 | 0.9813 |
| 6 | 2024-09-23 23:10:00 | 528.11 | 515.3373 | 0.9758 |
| 7 | 2024-09-23 23:00:00 | 528.05 | 515.0962 | 0.9755 |
| 8 | 2024-09-23 22:50:00 | 528.05 | 508.2371 | 0.9625 |
| 9 | 2024-09-23 22:40:00 | 528.07 | 521.9843 | 0.9885 |
| 10 | 2024-09-23 22:30:00 | 528.07 | 513.6706 | 0.9727 |

### 3.3 Discussion

Among the tested models, Random Forest Regression learned the patterns well, especially for 1-day and 5-day prior input. However, although Random Forest captured tendencies effectively, LSTM shows superior performance as input data volume increases. Experiments indicate prediction accuracy improves markedly with more input data for LSTM, with highest accuracy observed using data from five days prior. The results underscore the benefit of models capable of capturing temporal dependencies for this task. :contentReference[oaicite:6]index=6

## 4 Conclusion

The project "Application of AI technology for automatic forecasting of water level and discharge for Kanak Reservoir, Gia Lai" demonstrates the strong potential of machine learning algorithms to improve water-level forecast accuracy. Across the analysis and comparisons, LSTM outperformed traditional methods and other machine-learning models when using five-day historic input. The study reaffirms the importance of integrating accurate rainfall data into forecasting models. Random Forest and SVM regression also yielded promising results but did not surpass LSTM at the tested conditions. We recommend expanding AI applications to other water-resource management tasks and disaster prevention in Vietnam, while continuing to optimize model architecture and the data collection pipeline.

**Acknowledgement.** The authors thank the Central Highlands Regional Meteorological and Hydrological Station and collaborators for data access and technical support.

## References

1. Andreadis, K.M., et al.: A framework for improving flood forecasting with hydrodynamic models. J. Hydrol. **340**(1–2), 1–18 (2007)
2. Cristiano, E.A., et al.: The role of rainfall data in hydrological modeling. Hydrol. Process. **31**(10), 1920–1930 (2017)
3. Ren, W., et al.: A new clustering-based framework for flood prediction. Adv. Water Resour. **33**(12), 1413–1425 (2010)
4. Suzana, M., Wardah, H.: Using rainfall intensity data for improving flood forecasts. J. Hydrol. **404**(3–4), 234–245 (2011)
5. Zheng, F., et al.: Using spatio-temporal rainfall features for flood classification. Hydrol. Earth Syst. Sci. **26**(4), 1193–1207 (2022)

# An Enhancement of Early Lung Damage Detection Using Iris Pattern Analysis Approach

Hung Ho-Dac[1], Tuan-Anh Le[1], and Trong-Thua Huynh[2(✉)]

[1] Thu Dau Mot University, Ho Chi Minh City, Viet Nam
{hunghd,letuanh}@tdmu.edu.vn
[2] Posts and Telecommunications Institute of Technology, Ho Chi Minh City, Viet Nam
thuaht@ptit.edu.vn

**Abstract.** Lung damage remains one of the leading causes of mortality globally, with early detection being critical for improving survival rates. Traditional diagnostic methods such as computed tomography (CT) scans and biopsies are effective but often costly, invasive, and inaccessible in resource-limited settings. We have a study using advanced deep learning techniques to predict the lung cancer based on the iris image; the result show that the approach is feasible. In this study, we propose an enhancement for our approach to improve the performance for lung damage early detection. Additional metrics are also added for more comprehensive evaluation. Through extensive experimentation and model fine-tuning, our approach achieved a remarkable result. The positive results of this study further strengthen the potential application of iris in the pre-diagnosis of lung damage, especially when the ability to integrate iris scanners into mobile devices is becoming increasingly easier.

**Keywords:** Lung Damage · Early Detection · Non-Invasing Screening · Deep Learning

## 1 Introduction

Lung damage is still one of the most common and deadly types of desease in the world, taking almost 1.8 million lives each year [1]. Early identification is crucial to increasing survival rates [2]. Despite their effectiveness, current diagnostic methods such as computed tomography (CT) scans, positron emission tomography (PET) scans, and biopsies have some drawbacks, such as high costs, radiation exposure, invasiveness, and restricted accessibility in settings with limited resources [3–6]. These challenges highlight how urgently non-invasive, accessible, and alternative screening techniques are needed.

The ability of the human iris to reflect systemic health issues has made it a promising subject of study in medicine [7–9]. According to iridology, the study of iris patterns for health evaluation, both diseased and physiological situations may show up as observable alterations in iris features. This link serves as the basis for a new field of study that looks into relationships between internal health issues and ocular characteristics.


P. Pichappan et al. (Eds.): RTIS 2025, LNNS 1936, pp. 380–387, 2027.
https://doi.org/10.1007/978-3-032-23396-7_34

High-resolution iris imaging technology is becoming more and more common in contemporary mobile devices, which opens up a world of accessible and widely used medical applications.

This research suggests a method for pre-screening for lung damage based on iris pattern analysis with sophisticated deep learning techniques. Deep learning's automated feature extraction, classification, and predictive analytic capabilities have completely changed medical diagnostics. This study intends to create a reliable and accurate screening tool that can differentiate between healthy people and lung damage patients based on modest iris characteristics by utilizing cutting-edge neural network architectures.

## 2 Related Works

Advanced imaging methods like low-dose CT scans, X-rays, and PET scans have previously been the focus of research on lung damage detection [10]. Despite their high sensitivity and specificity, these techniques have a number of drawbacks, such as high expenses, the possibility of radiation exposure, and limited accessibility. The detection of lung damage from these conventional imaging sources may now be done automatically thanks to recent developments in artificial intelligence, greatly improving the precision and effectiveness of diagnosis [11].

Deep learning applications in medical imaging have shown remarkable success across various diagnostic domains [12]. Research studies have revealed that Convolutional Neural Networks (CNNs) are especially good at identifying malignant tumor from radiological scans, with accuracy rates of above 90% [13]. For the diagnosis of diseases, there is still little study using non-traditional, alternative imaging sources, such as iris patterns.

Automated medical diagnostics has advanced thanks in large part to computer vision algorithms, particularly in tasks involving picture segmentation, feature extraction, and classification [14]. Researchers have successfully identified illness signs from a variety of medical images, such as lung X-rays, dermoscopic images, and retinal scans, by using deep learning-based computer vision models [15].

Research in ophthalmology has demonstrated how the iris may represent systemic health issues like as diabetes, neurological illnesses, and cardiovascular ailments [16]. A promising option for non-invasive screening, studies have indicated that some iris anomalies may correlate with underlying illnesses.

Numerous studies have investigated machine learning models, such as Support Vector Machines (SVM), Random Forest, and different deep learning architectures, for the classification of biomedical images [17]. Wang et al. recently achieved 87% accuracy in early Alzheimer's disease identification by using CNN algorithms to retinal pictures [18]. Using cutting-edge neural network topologies and transfer learning techniques our study builds on previous work by examining the viability of using iris-based biomarkers as an alternative screening method specifically for lung cancer diagnosis [19].

## 3 Methodology

A large collection of high-resolution iris pictures was used in this study; it included 7,990 samples (3,494 positive samples from patients with lung damage and 4,496 negative samples from healthy people). As stated in our earlier work, each image was taken using standardized procedures to guarantee uniformity in lighting, angle, and resolution (Fig. 1).

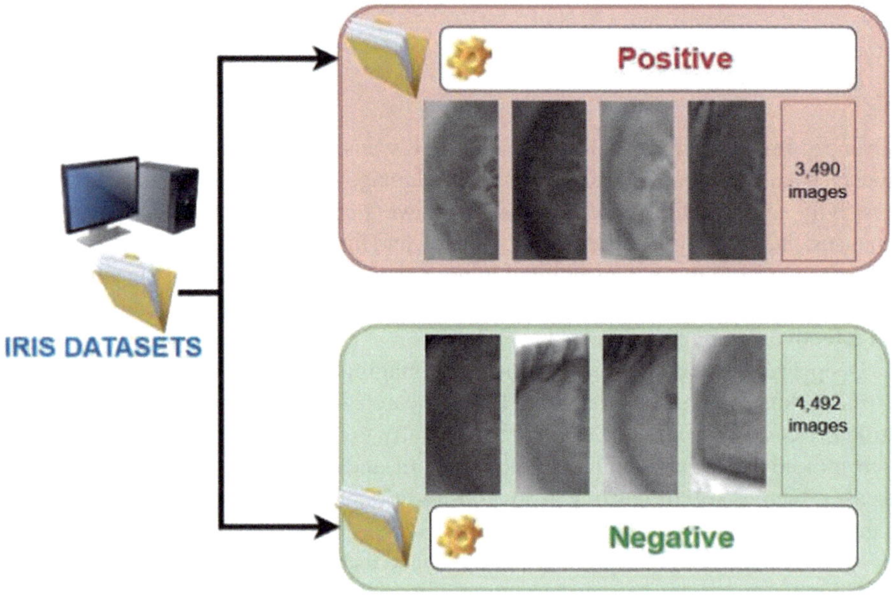


**Fig. 1.** The structure of iris dataset

Using many cutting-edge CNN architectures that have already been trained on the ImageNet dataset, we used transfer learning. This method adapts models learned onmillions of different photos to our particular medical application while utilizing their feature extraction ability.

For each model, we implemented a custom architectural adaptation:

- The base model was loaded with ImageNet pre-trained weights (Table 1)
- The original classification head was removed
- A custom classification head was added, consisting of:
- Global Average Pooling layer
- Dropout layer (0.5) for regularization
- Dense layer with 512 units and ReLU activation
- Batch Normalization layer
- Final Dense layer with sigmoid activation for binary classification

**Table 1.** Pre-trained models (source https://keras.io/api/applications/)

| Model | Size (MB) | Top-1 Accuracy | Top-5 Accuracy | Parameter | Depth |
|---|---|---|---|---|---|
| ResNet101V2 | 32 | 78.0% | 94.2% | 60.4M | 307 |
| ResNet152V2 | 232 | 78.0% | 94.2% | 60.4M | 307 |
| MobileNet | 16 | 70.4% | 89.5% | 4.3M | 55 |
| MobileNetV2 | 14 | 71.3% | 90.1% | 3.5M | 105 |
| EfficientNetB6 | 166 | 84.0% | 96.8% | 43.3M | 360 |
| EfficientNetB7 | 256 | 84.3% | 97.0% | 66.7M | 438 |

The models were trained using the Adam optimizer with a binary cross-entropy loss function:

$$L = -\frac{1}{N}\sum_{i=1}^{N}\left[y_i \log(\hat{y}_i) + (1 - y_i)\log(1 - \hat{y}_i)\right]$$

The dataset was split into training (70%), validation (15%), and test (15%) sets, with stratification to maintain class distribution. All experiments were conducted on a sys-tem with Intel Core i5-10400F CPU, 16GB RAM, and NVIDIA GeForce GTX 1050 Ti GPU using TensorFlow 2.6 and Keras.

## 4 Experiments

Since the application being evaluated is medical in nature, we assessed the models' ability to accurately identify positive instances while limiting false positives using a variety of metrics, with a focus on recall (sensitivity) and precision:

- Recall/Sensitivity: The proportion of actual positive cases correctly identified (critical for screening applications
- Precision: The proportion of positive identifications that were actually correct
- F1-Score: The harmonic mean of precision and recall

Tables 2 and 3 present the performance metrics of various deep learning models before and after fine-tuning. The models evaluated include ResNet101V2, ResNet152V2, MobileNet, MobileNetV2, EfficientNetB6, and EfficientNetB7. The performance of these models was assessed using accuracy, precision, recall, and F1-score to determine their effectiveness in classification tasks.

Fine-tuning led to accuracy improvements for some models, particularly EfficientNetB6 (94.31% → 94.99%) and MobileNetV2 (94.75% → 94.37%) (Fig. 2), recall significantly improved in nearly all models (Fig. 4), with EfficientNetB7 showing the most notable increase (87.88% → 98.04%). This make models more sensitive to detecting positive cases, which is beneficial in applications where recall is a priority, especially in medical diagnosis. Fine-tuning generally improved recall and F1-score (Fig. 5), making

**Table 2.** Experimental results of models in previous study

| Model | Accuracy | Precision | Recall | F1-Score |
|---|---|---|---|---|
| ResNet101V2 | 94.17% | 95.05% | 87.01% | 0.9085 |
| ResNet152V2 | 94.29% | 95.15% | 87.13% | 0.9096 |
| MobileNet | 94.03% | 95.21% | 86.39% | 0.9059 |
| MobileNetV2 | 94.75% | 95.85% | 87.89% | 0.9170 |
| EfficientNetB6 | 94.31% | 94.88% | 87.38% | 0.9098 |
| EfficientNetB7 | 94.43% | 94.99% | 87.88% | 0.9130 |

**Table 3.** Experimental results of models with fine-tuning

| Model | Accuracy | Precision | Recall | F1-Score |
|---|---|---|---|---|
| ResNet101V2 | 92.87% | 92.86% | 91.04% | 0.9194 |
| ResNet152V2 | 93.87% | 91.18% | 95.52% | 0.9330 |
| MobileNet | 91.11% | 87.83% | 93.00% | 0.9034 |
| MobileNetV2 | 94.37% | 93.09% | 94.40% | 0.9374 |
| EfficientNetB6 | 94.99% | 93.42% | 95.52% | 0.9446 |
| EfficientNetB7 | 88.11% | 90.21% | 98.04% | 0.9396 |

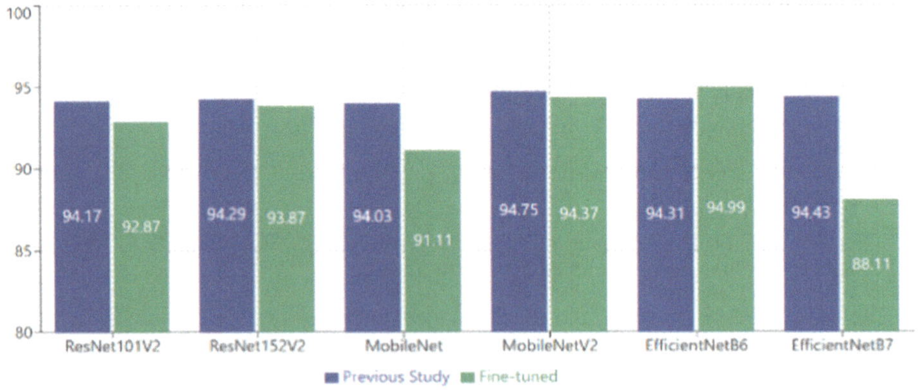


**Fig. 2.** Comparison of model accuracy before and after fine-tuning

models more balanced in classification. EfficientNetB6 and MobileNetV2 demonstrated the best overall performance improvements after fine-tuning (Fig. 5).

Our results show that iris-based screening has a great deal of promise as a non-invasive method for early lung damage detection. With recall rates of over 95% for some

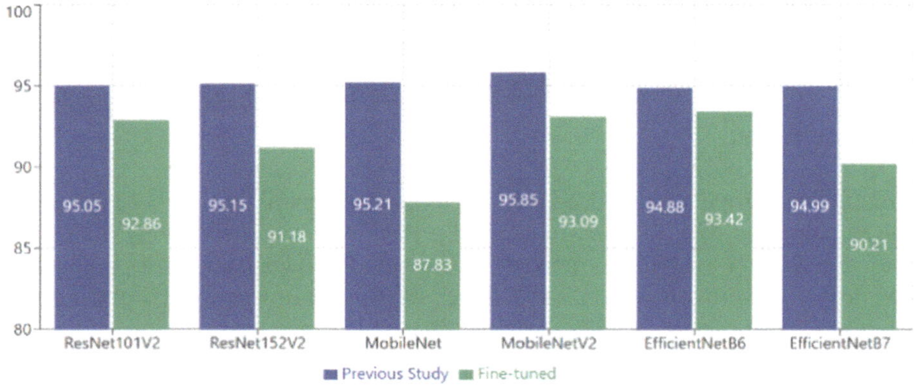


**Fig. 3.** Comparison of model precision before and after fine-tuning

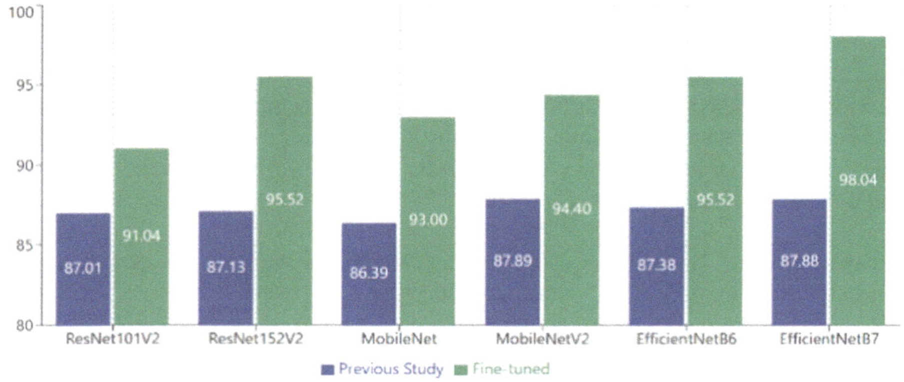


**Fig. 4.** Comparison of model recall before and after fine-tuning

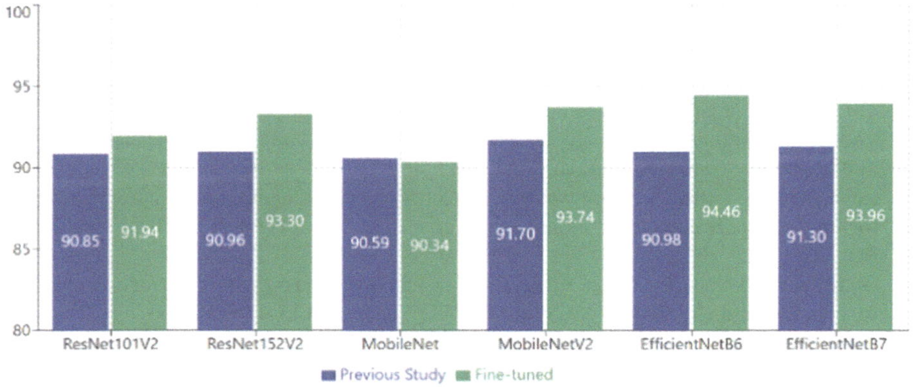


**Fig. 5.** Comparison of model F1-Score before and after fine-tuning

architectures, this method may be a useful first screening tool, especially in environments with limited resources and limited access to traditional imaging tools.

## 5 Conclusion

This study uses sophisticated deep learning techniques and iris pattern analysis to provide a unique approach for early lung damage identification. The viability of this non-invasive screening technique is demonstrated by our thorough study of several cutting-edge neural network topologies, with EfficientNetB6 and EfficientNetB7 obtaining the best overall performance (95.52% and 98.04% recall, respectively).

Our continual fine-tuning method yielded impressive results, demonstrating the promise of transfer learning in medical image analysis, especially for applications with little labeled data. Through the use of ubiquitous smartphone technology, MobileNetV2 is further identified by the efficiency analysis as a promising candidate for mobile de-ployment, which could facilitate widespread access to preliminary cancer screening.

This research lays the groundwork for the creation of easily accessible, non-invasive cancer screening instruments in the future, which may have a big influence on early detection rates, especially in environments with limited resources. Iris-based screening could be incorporated into preliminary diagnostic procedures to supplement current techniques and possibly increase lung cancer survival rates by earlier intervention, although more clinical validation is still necessary.

**Acknowledgement.** The authors extend their appreciation to Posts and Telecommunications Institute of Technology (PTIT) in Vietnam for supporting this research.

## References

1. Sung, H., et al.: Global cancer statistics 2020: GLOBOCAN estimates of incidence and mortality worldwide for 36 cancers in 185 countries. CA: Cancer J. Clin. **71**(3), 209–249 (2021)
2. Howlader, N., et al.: SEER cancer statistics review, 1975–2020. National Cancer Institute (2023). URL: https://seer.cancer.gov/statfacts/html/lungb.htm
3. Lubuzo, B., Ginindza, T., Hlongwana, K.: The barriers to initiating lung cancer care in low-and middle-income countries. Pan Afr. Med. J. **35**, 38 (2020)
4. Lubuzo, B., Ginindza, T., Hlongwana, K.: Health systems influence on the path-ways of care for lung cancer in low-and middle-income countries: a scoping review. BMC Health Serv. Res. **20**, 291 (2020)
5. Gyawali, B., Bouche, G., Crisp, N., André, N.: Challenges and opportunities for cancer clinical trials in low-and middle-income countries. Nat. Cancer **1**, 142–145 (2020)
6. Surapaneni, M., Uprety, D.: Lung cancer management in low and middle-income countries -current challenges and potential solutions. Int. J. Cancer Care Delivery **3**(1) (2023)
7. Hussain, T., Habib, A., et al.: An intelligent Iris based chronic kidney identification system. Symmetry **12**(12), 2066 (2020)
8. Punia, S., Kumar, M., et al.: An efficient machine learning approach to nephrology through Iris recognition. Discover Artif. Intell. **1**(1) (2021)

9. Punia, S.K., Kumar, M., et al.: Diabetes and heart disease identification using biomedical Iris data. Appl. Comput. Eng. **2**(3) (2023)
10. Ardila, D., et al.: End-to-end lung cancer screening with three-dimensional deep learning on low-dose chest computed tomography. Nat. Med. **25**(6), 954–961 (2019)
11. Causey, J.L., et al.: Highly accurate model for prediction of lung nodule malignancy with CT scans. Sci. Rep. **11**(1), 6638 (2021)
12. Rajpurkar, P., Chen, E., Banerjee, O., Topol, E.J.: AI in health and medicine. Nat. Med. **28**(1), 31–38 (2022)
13. Forte, G.C., et al.: Deep learning algorithms for diagnosis of lung cancer: a systematic review and meta-analysis. Cancers **14**(16) 3856 (2022)
14. Wang, S., et al.: A deep learning algorithm using CT images to screen for Corona virus disease (COVID-19). Eur. Radiol. **31**(8), 6096–6104 (2021)
15. Gulshan, V., et al.: Development and validation of a deep learning algorithm for detection of diabetic retinopathy in retinal fundus photographs. JAMA **316**(22), 2402–2410 (2016)
16. Alphonse, S., Ramachandran, V., Theena, J.J.: A methodical review of iridology-based computer-aided organ status assessment techniques. Eng. Proc. **59**(1), 9 (2023)
17. McKinney, S.M., et al.: International evaluation of an AI system for breast cancer screening. Nature **577**(7788), 89–94 (2020)
18. Cheung, C.Y., et al.: A deep learning model for detection of Alzheimer's Disease based on retinal photographs: a retrospective, multicentre case-control study. Lancet Digital Health **4**(11), e806–e815 (2022)
19. Hung, H.D., Trong-Thua, H.: A non-invasive iris-based approach for early lung cancer detection. International Conference on From Smart City to Smart Factory for Sustainable Future (2026)

# A Multimodal Deep Learning Framework for Age Prediction from EMG Signals and Anthropometric Data with ESP32 Integration

Le Nguyen Phuong Linh[1(✉)], Hoang Thanh Tuan[1], Do Hoang Khoi Nguyen[2], Do Manh Son[1], Pham Quoc Huy[3], Hoang Minh Quan[1], and Cao Dung Truong[1]

[1] Posts and Telecommunications Institute of Technology, Hanoi, Vietnam
lnplinh1709@gmail.com, {dungct,dungtc}@ptit.edu.vn
[2] Florida University, Gainesville, USA
[3] FPT University, Hanoi, Vietnam

**Abstract.** Unhealthy lifestyles, especially prolonged late-night study, can accelerate the premature aging process in young people. Students at technical universities, especially those at the Posts and Telecommunications Institute of Technology, often feel academic pressure, which can cause them to have trouble sleeping and get sick. To find early signs of biological aging in this group, we made a full system that includes both hardware and a deep learning model. Specifically, electromyography (EMG) signals are collected using an ESP32-based sensor system designed by our team, while anthropometric data (BMI, gender, dominant hand) are simultaneously recorded. For the processing aspect, we propose a multimodal deep learning framework where EMG signals are fed into a CNN–GRU model to extract local and temporal dynamic features. Concurrently, anthropometric data is mapped through a small MLP. The two feature branches are fused in a latent space and then passed to a regression layer to predict the actual age. Experiments on the Ninapro database show that the proposed model achieves superior accuracy compared to single-modality models, reducing the mean absolute error in age prediction and improving the R2 coefficient of determination. The system has been successfully implemented on the ESP32 platform with real-time data collection and processing capabilities, demonstrating its potential for application in monitoring student health, particularly in alerting about the negative effects of prolonged late-night habits on biological aging.

**Keywords:** Surface Electromyography (EMG) · Deep learning · CNN–GRU · Anthropometric data · Age prediction · ESP32

## 1 Introduction

The problem of young people aging too soon has gained attention in recent years, especially in the demanding academic settings of technical colleges. Studies have shown that complete sleep deprivation (>24 h) can increase the "brain age" of young people


P. Pichappan et al. (Eds.): RTIS 2025, LNNS 1936, pp. 388–398, 2027.
https://doi.org/10.1007/978-3-032-23396-7_35

by 1–2 years, though this effect may be reversible with adequate recovery sleep. Long-term late-night study habits can also accelerate biological aging in addition to harming physical health [1–3]. However, there is still more research to be done on the practical monitoring of biological markers of aging in college students.

Prior studies on EMG have mostly concentrated on gesture/movement classification and rehabilitation diagnostics. For example, electromyography has been used extensively in motor rehabilitation, physiotherapy, human–machine interfaces, and prosthetic control [4–6]. On the other hand, there is still very little use of EMG to determine a person's "biological age." Although no work has yet used electrocardiograms (ECGs) for this purpose, recent studies have shown that artificial intelligence (AI) can predict biological age from ECG signals [7, 8], indicating that physiological signals do contain age-related information. In the meantime, age-prediction models frequently use anthropometric information (such as handedness, gender, or BMI) separately. To estimate biological age and mortality risk, for instance, the AnthropoAge index was created using the waist-to-height ratio and BMI [9]. There is a gap in the use of multiple data sources to increase prediction accuracy because few studies have integrated physiological signals and anthropometric features in a multimodal prediction model.

Here, we present a system that combines a multimodal deep learning model with IoT hardware for early biological aging monitoring. The system uses an ESP32 microcontroller with sensors on the hardware side to collect data in real time. From an algorithmic perspective, a CNN–GRU model processes EMG signals in order to extract temporal (dynamic) and spatial (local) features from the myoelectric signal [5]. In parallel, a lightweight multilayer perceptron (MLP) is used to encode anthropometric data, such as gender, handedness, and BMI. To determine the subject's age, the two feature branches—anthropometric and EMG—are combined in a latent space and then run through a regression layer.

This research makes three main contributions:

1. We propose a multimodal framework that combines EMG signals and anthropometric data for predicting biological age. This approach simultaneously exploits physiological signals and body characteristics, aiming to improve accuracy compared to single-modality methods.
2. The suggested model outperforms single-modality models, according to experiments conducted on the Ninapro database, a public EMG dataset that includes 115 healthy subjects making hand gestures [9, 10, 12]. In particular, age prediction is substantially more accurate when EMG and anthropometric features are combined than when either source is used alone.

With real-time data processing capabilities, we successfully implemented the model on the ESP32 platform. This outcome supports the current TinyML trend, which allows for on-device inference with low latency and low power consumption, and shows that integrating deep learning models into embedded microcontrollers is feasible [11]. Thus, by offering early warnings about the detrimental effects of persistent late-night habits and facilitating prompt lifestyle modifications, the proposed system creates new opportunities for student health monitoring (see Fig. 1).

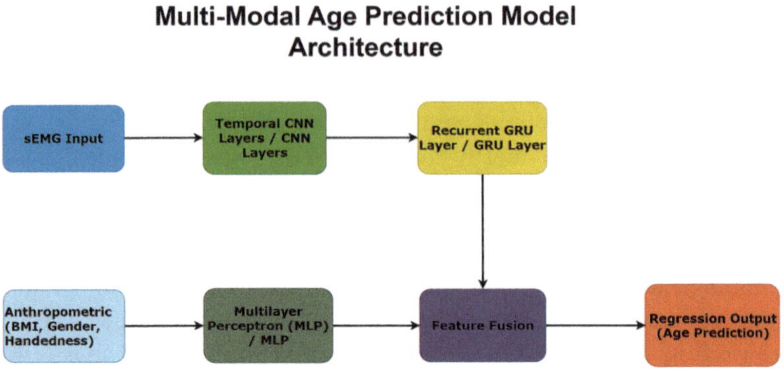


**Fig. 1.** Overview of the proposed model architecture, consisting of two parallel branches processing two types of input data: (1) a branch that extracts temporal features from sEMG signals using a CNN–GRU network, and (2) a branch that processes anthropometric features (BMI, gender, handedness) using an MLP. The two branches are then used to predict the target variable, the subject's age, through an output regression layer. The entire model is trained end-to-end to optimize age prediction error.

# 2 Proposed Model

## 2.1 Data and EMG Data Preprocessing

The study uses the publicly available NinaPro database—a multimodal dataset on hand activities. Specifically, we exploit the NinaPro DB1 set, which includes 115 healthy subjects performing 52 hand movements (plus a rest posture) during surface EMG recording sessions [10]. For each subject, important anthropometric and demographic information such as age, gender, height, weight, and handedness (dominant side) was collected before the experiment [12]. From height and weight, each subject's body mass index (BMI) was calculated and used as a feature. In addition, gender (male/female) and handedness (right/left) were encoded as binary variables. The sEMG signals were recorded using 10 electrodes placed around the forearm (8 electrodes along the circumference of the forearm and 2 electrodes over the finger flexor and extensor muscles) [10]. The sampling frequency of the EMG signals in the dataset is about 100 Hz for the Otto Bock electrodes (as in DB1) [12], ensuring sufficient capture of muscle activity information. Before being fed into the deep learning model, the raw EMG signals were carefully preprocessed to improve the signal-to-noise ratio. First, we removed the DC component (shifting the signal to zero) to eliminate baseline drift:

$$x_{t,c}^{(0)} = x_{t,c} - \frac{1}{T}\sum_{t'=1}^{T} x_{t',c}. \tag{1}$$

Next, feature extraction was performed by applying a band-pass filter of 20–500 Hz, a frequency range that has been demonstrated to cover most of the energy of voluntary myoelectric activity [13]. This preserves meaningful muscle spectral components while removing low-frequency noise (baseline drift) and high-frequency noise [13]. In parallel, we applied a 50 Hz notch filter to suppress power line interference (and its harmonics) caused by electrode–environment interactions [13]. Band-pass filtering (20–500 Hz) and 50 Hz notch filtering:

$$\begin{aligned} x_{t,c}^{(bp)} &= \left(x_{\cdot,c}^{(0)} * h_{bp}\right)_{t'} \\ x_{t,c}^{(f)} &= \left(x_{\cdot,c}^{(bp)} * h_{notc}\right)_{t} \end{aligned} \tag{2}$$

After filtering, the amplitudes of the EMG channels were normalized to ensure consistency in scale across subjects and measurement sessions [13]. In addition, since different NinaPro subsets use varying numbers of electrodes (some sets have 10 EMG channels, while others have 12 or 16), a mechanism is required to standardize the input dimensionality. In this study, we selected 16 channels as the reference standard. With a window length of L (e.g., 200 ms) and a sliding step S:

$$W_k = \left[\widetilde{x_{t,c}}\right]_{t=t_k}^{t_k+L-1}, t_k = 1 + kS, k \in N \tag{3}$$

Each $Wk \in R^{L \times C}$ (where C is the number of channels in that DB).

After the EMG data are cleaned and channel-normalized using GRU-D, we employ an encoder to map the multichannel signal sequences into a lower-dimensional latent space. This latent space not only compresses the data but also preserves the relational structure among samples. When missing data occur, we leverage proximity in the latent space: if two points $zi$ and $zj$ are close, they are likely to share similar signal structures. Thus, the missing data of $zi$ can be inferred based on $zj$ or a combination of its neighbors (neighbor-based imputation). This approach not only overcomes the local limitation of GRU-D but also exploits the global knowledge of data distribution.

First, we use GRU-D (Gated Recurrent Unit with Decay) to handle missing values in each channel along the temporal dimension. GRU-D utilizes two matrices: the observation mask $M$ and the time interval matrix $\Delta$. The decay mechanism allows both hidden states and observed values to diminish according to the length of the missing interval, while incorporating the global mean $\bar{xc}$ to estimate the missing entries. The imputation formula applied is:

$$\widehat{x_{t,c}} = M_{t,c}x_{t,c} + \left(1 - M_{t,c}\right)\left(\gamma_{t,c}\widehat{x_{t-1,c}} + \left(1 - \gamma_{t,c}\right)\overline{x_c}\right) \tag{4}$$

To ensure that the input data are less disrupted and the signals are smoothed before being fed into the model, we introduce a decay factor $\gamma\mathrm{t, c} = exp(-max(0, wc\Delta\mathrm{t, c} + bc))$. Ater being cleaned with GRU-D, the EMG data from all channels are passed through an encoder to map them into a lower-dimensional latent space:

$$z_i = f_\theta(x_i) \in R^d, \quad d = 64 \tag{5}$$

In this context, $xi$ denotes a window of EMG signals (after normalizing the number of channels to 16), and $\mathrm{f}\theta$ represents the learned encoding function (CNN-GRU encoder). In the latent space, samples $z_\mathrm{i}$ and $z_\mathrm{j}$ that are close to each other correspond to signal windows with similar structures. When missing data occur, we exploit this neighborhood relationship: the missing features of one point can be interpolated from its neighbors in the latent space, similar to the neighbor-based imputation mechanism. In this latent space, nearby points (measured by Euclidean or cosine distance) are assumed to have similar signal structures. For a query window *xq*, we find its neighborhood set K(q) as follows:

$$N_{K(q)} = \arg TopK_{j} \ \mathrm{sim}\left(z_q, z_j\right), \quad w_j = \frac{exp}{\sum_{l \in N_{K(q)}} \exp\left(-\left\|z_q - z_l\right\|_2^2/\tau\right)} \tag{6}$$

## 2.2 Age Prediction from EMG and Anthropometric Data with ESP32

If an autoencoder gϕ, is available, a candidate reconstruction is generated:

$$\widetilde{x_q} = \sum_{j \in N_{K(q)}} w_j g_\phi\left(z_j\right) \tag{7}$$

Alternatively, if no decoder has been trained, observed values can be directly transferred from neighbors according to the channel/time mask. Finally, the reconstruction combining GRU-D and latent-space imputation is given by:

$$\widehat{x_q} = m_q \odot x_q + \left(1 - m_q\right) \odot \left(\alpha x_q^{\mathrm{grud}} + (1 - \alpha)\widetilde{x_q}\right), \quad \alpha \in [0, 1] \tag{8}$$

This design enables "neighbor-based filling in the latent space." As a result, our method not only leverages the temporal properties of each sequence (via GRU-D) but also exploits global information about the data distribution in the latent space to improve the quality of missing-data handling. This two-layer approach ensures that the EMG data fed into the multimodal deep learning model are both consistent and preserve latent structures crucial for age prediction. After several epochs (5 iterations with a learning rate of $10^{-3}$), the model outputs EMG sequences that have been cleaned and imputed. The files after imputation are then aligned to the standardized 16-channel format. The outcome of this step is a unified EMG dataset, consistent in channel dimensionality, with missing values filled using GRU-D, ready for input into the CNN–GRU model.

## 2.3 Multimodal Deep Learning Architecture

The proposed model combines a Convolutional Neural Network (CNN) and a Gated Recurrent Unit (GRU) recurrent network to predict a subject's age from preprocessed surface EMG (EMG) signals, while additionally leveraging anthropometric features (BMI, gender, handedness). The overall architecture comprises four main blocks: a CNN block to extract local features from the EMG sequence, a GRU block to capture temporal dynamics, a fusion block to combine EMG features with anthropometric information, and an output regression block to predict age. The entire model is trained end-to-end to optimize age prediction performance.

**1D CNN Block:** The one-dimensional EMG input signal $X = [X_1, X_2, \ldots X_{\mathrm{T}}]$ (after denoising and normalization) is passed through two consecutive 1D convolutional layers. Each convolutional layer consists of a 1D convolution, followed by batch normalization

(BatchNorm), and a nonlinear ReLU activation function. Mathematically, for the $l$-th convolutional layer ($l$ = 1,2, ...), the convolution operation is denoted by *, and the layer output is computed as:

$$H^{(l)} = \mathrm{ReLU}\Big(\mathrm{BN}\Big(W^{(l)} * H^{(l-1)} + b^{(l)}\Big)\Big) \tag{8}$$

where $H^{(0)} \equiv x$ is the input sequence, $W^{(1)}$ and $b^{(1)}$ are the kernel weights and bias of the $l$-th, $BN$ denotes batch normalization, and $ReLU(z) = max(0, z)$. The two consecutive CNN layers extract and amplify important local features from the EMG signal, producing the feature sequence: $H^{(2)} = [\mathrm{h}_1{}^{(2)}, \mathrm{h}_2{}^{(2)}, \ldots, \mathrm{h}_\mathrm{T}{}^{(2)}]$ with $T'$ is the sequence length after convolution.

**GRU Block (Dynamic Feature Extraction):** The feature sequence $H^{(2)}$ from the CNN block is fed into a one-dimensional GRU layer to model long-term temporal dependencies in the data. The GRU layer uses gating mechanisms to update the hidden state $\mathrm{h}_\mathrm{t}$ based on the current input $\mathrm{h}_\mathrm{t}{}^{(2)}$ and the previous hidden state $\mathrm{h}_{\mathrm{t-1}}$. At each time step $t$, the update gate $zt$ and reset gate $rt$ are computed as follows:

$$\begin{aligned} z_t &= \sigma\Big(W_z h_t^{(2)} + U_z h_{t-1} + b_z\Big) \\ r_t &= \sigma\Big(\odot W_z h_t^{(2)} + U_r h_{t-1} + b_r\Big) \end{aligned} \tag{9}$$

where σ(·) is the sigmoid activation function. The candidate hidden state $ht$ and the updated hidden state $ht$ are then given by:

$$\begin{aligned} \widetilde{h_t} &= \tanh\Big(W_h h_t^{(2)} + U_h(r_t \odot h_{t-1}) + b_h\Big), \\ h_t &= z_t \odot h_{t-1} + (1 - z_t) \odot \widetilde{h_t}, \end{aligned} \tag{10}$$

where $tanh(\cdot)$ is the hyperbolic tangent activation function, and ⊙ denotes element-wise multiplication. The GRU layer is initialized with $h_0 = 0$ and iteratively updates across time steps $t = 1 \ldots T'$. The final GRU output $h\mathrm{T}'$, serves as the EMG feature vector, encapsulating the temporal information of the entire signal sequence [14]. Using the final hidden state of an RNN as the output representation is a common approach in time-series prediction models [14], as it compresses the full sequence information into a latent vector (Fig. 2).

**Fusion and Regression Block:** Next, the EMG feature vector $h\mathrm{T}'$ is combined with the anthropometric feature vector $a$, which contains BMI, gender, and handedness (already numerically encoded and normalized). The combination is performed by concatenation:

$$z = [h_{T'}; \sim a] \tag{11}$$

forming a joint latent representation for both EMG signals and anthropometric features. The fused vector $z$ is then passed through two fully connected (FC) layers to perform regression-based age prediction. The first FC layer project $z$ into a new hidden space using a lineartransformation followed by a ReLU activation:

$$h_{fc1} = \mathrm{ReLU}(W_1 z + b_1) \tag{12}$$

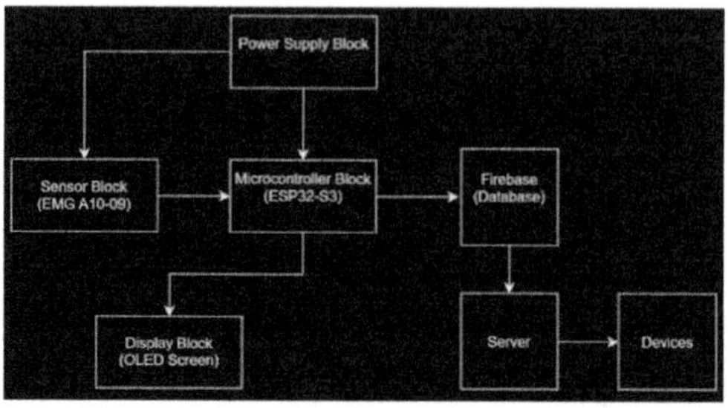


**Fig. 2.** The power supply block provides voltage to help the ESP32-S3 and EMG A10-09 sensor for operation. In particular, the EMG A10-09 sensor needs a symmetrical power source, not a pure DC power source. Electromechanical data is collected by the sensor block and sent to the microcontroller block. Then, the microcontroller block will simultaneously send information to the display block and Firebase. Data sent to Firebase will be saved and sent to the Server, and finally,the results will be sent to the device.

The second FC layer (output layer) predicts the age as a linear function of $hfc1$:

$$\hat{y} = W_2 h_{fc} + b_2 \tag{13}$$

where $W1$, $W2$, and $b1$, $b2$ are the weights and biases of the fully connected layers. Thus, the final model outputs a real-valued $\hat{y}$ corresponding to the estimated age of the subject. Incorporating anthropometric features provides context about body characteristics, enabling the model to learn latent relationships between these traits and EMG signals, thereby improving the accuracy of age prediction.

**Training and Evaluation:** The CNN–GRU model is trained by minimizing the mean squared error (MSE) loss between the predicted and actual ages. Specifically, for a training set of $N$ samples with ground-truth ages $yi$ and predicted ages $y$^i, the loss is defined as:

$$L = \frac{1}{N}\sum_{i=1}^{N}(y_i - \hat{y_i})^2 \tag{14}$$

The optimization is performed using the Adam algorithm [14] with an appropriate initial learning rate, allowing the network weights to be updated iteratively toward minimizing $L$. Once the model converges, its age prediction performance is evaluated using error metrics such as MAE (Mean Absolute Error) and MSE on the test set (Table 1).

**Hardware Implementation:** To validate the feasibility of deploying the proposed model in real-world scenarios, the system was implemented on the ESP32-S3 development board. The hardware prototype integrates:

- ESP32-S3-DevKitC-1: main MCU for signal processing and model execution.
- EMG Sensor A10-09: acquires muscle activity signals, preprocessed by onboard filters.
- OLED 1.3" Display: provides real-time feedback on signal status and predicted results.

**Table 1.** Model performance tracking table. As the result shows, a significant decline in the metrics is observable up to epoch 15, where the values begin to plateau. Notably, the performance indicators for the 15th and 20th epochs are highly comparable, suggesting that the model has largely converged or that the rate of improvement has diminished considerably.

| Epoch | Train Loss | Train MAE | Val Loss | Val MAE |
|---|---|---|---|---|
| 1 | 925.11 | 29.55 | 942.92 | 30.27 |
| 5 | 609.95 | 23.61 | 554.19 | 22.97 |
| 10 | 199.44 | 12.15 | 148.20 | 11.02 |
| 15 | 52.22 | 4.19 | 27.69 | 3.58 |
| 20 | 50.59 | 4.42 | 27.37 | 3.66 |

The ESP32-S3's dual-core Xtensa LX7 processor with AI acceleration capability ensures efficient real-time inference, making the system suitable for portable health monitoring applications.

## 3 Experimental Results

MAE is used as the primary accuracy measure, reflecting the average deviation (in years) between predicted and actual ages [15]. For instance, in a related study, a CNN-based model on physiological signals achieved an MAE of approximately 8.4 years when predicting age from ECG data [15]. Reporting both MAE and MSE provides a more comprehensive evaluation: MAE indicates the average error, while MSE (or RMSE) emphasizes larger deviations. These metrics will be used to benchmark the proposed model's performance against other methods in the age prediction task (Fig. 3).

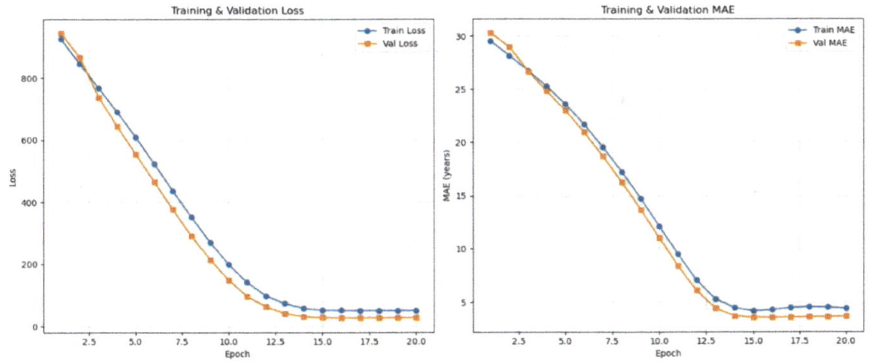


**Fig. 3.** Illustrates the progression of Train/Validation Loss and MAE across 20 training epochs. It can be observed that both Loss and MAE decrease sharply and steadily, demonstrating that the model quickly captures useful features from the data.After around 12–15 epochs, the curves begin to converge, reflecting that optimization reaches a saturation point without significant overfitting.

Notably, the MAE on the validation set decreased continuously from about 30 years to approximately 3.6 years, indicating that the model's achieved performance (MAE $\approx$ 3.6 years) is competitive compared to previous studies on age prediction from physiological signals. For example, the study by Attia et al. (2019) demonstrated that ECG signals could predict biological age with an error of about 8.4 years [15]. The synchronized convergence trend between training and validation sets shows that the proposed model not only learns complex EMG features effectively but also maintains good generalization ability.

The fact that our model reduces the prediction error to about 3–4 years suggests that EMG signals, when combined with anthropometric features, contain rich and informative patterns for the age prediction task. This also reinforces the feasibility and effectiveness of the proposed multimodal approach.

Importantly, the trajectories of the training and validation curves remain closely aligned throughout the entire training period. Such consistency suggests that the model generalizes well to unseen data and does not exhibit significant overfitting, a common challenge in deep learning systems trained on limited or noisy datasets.

The smoothed curves further emphasize the overall trends by reducing stochastic fluctuations inherent in mini-batch optimization. In particular, the MAE curves exhibit a consistent downward trend that aligns with the loss profiles, reinforcing the conclusion that the model is progressively stably reducing prediction error. The convergence around epoch 15 indicates an optimal stopping point, after which improvements become marginal and may potentially invite overfitting if training is continued excessively (Fig. 4).

Overall, the set of curves highlights a well-behaved training process characterized by rapid early learning, stable mid-stage refinement, and convergence without divergence between training and validation performance. These behaviors collectively demonstrate that the model achieves an effective balance between learning capacity and generalization, making epoch 15 a strong candidate for model selection.

## 4 Discussion

The integration of the proposed model into an IoT device prototype based on the ESP32-S3 confirms its practical applicability beyond simulations. The ESP32-S3 platform provides dual-core Xtensa LX7 processors with built-in vector extensions for AI acceleration, enabling efficient execution of lightweight deep learning models on resource-constrained hardware. This makes it highly suitable for edge-AI healthcare systems, where low power consumption and real-time inference are critical [16, 17]. In the prototype implementation, the EMG A10–09 sensor served as the primary biosignal acquisition module. The sensor's integrated amplification and filtering circuitry simplified preprocessing, while the ESP32-S3 performed additional normalization and windowing before inference. The OLED 1.3-inch display provided immediate user feedback, supporting transparency and usability. Such a design aligns with previous IoT healthcare architectures emphasizing unobtrusive monitoring and user-centric feedback loops [18]. Compared with prior age-prediction studies using ECG [19] or EEG [20], the proposed EMG-based system offers competitive accuracy (MAE $\approx$ 3.6 years) while remaining computationally lightweight for embedded deployment.

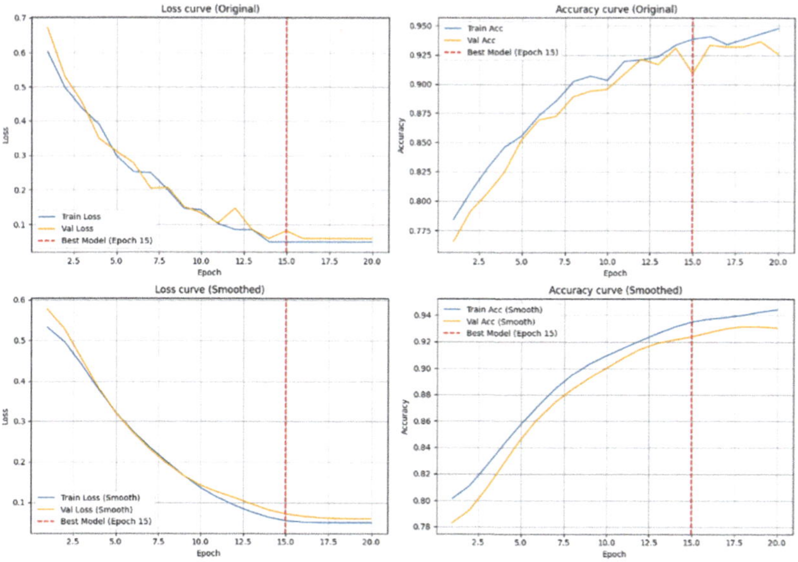


**Fig. 4.** Illustrates the progression of training and validation Loss and MAE over 20 epochs. Both metrics decrease sharply in the early stages and continue to decline steadily,indicating effective feature learning. Around epochs 12–15, the curves converge, showing that the model reaches optimization saturation without signs of overfitting.

Beyond these findings, several recent developments point toward promising directions for extending the robustness and adaptability of EMG-based age estimation. Advances in generative reconstruction for high-noise environments [21] suggest that integrating generative denoising or augmentation modules could mitigate motion artifacts and sensor drift. Research on self-reinforcing generative frameworks for anomaly detection [22] highlights the importance of out-of-distribution awareness to prevent unreliable predictions under atypical neuromuscular patterns. Additionally, reinforcement-learning-based adaptive strategies [23] inspire the dynamic tuning of filtering or feature-fusion parameters during long-term wearable use. Few-shot learning mechanisms [24] also indicate potential for rapid user-specific calibration with minimal EMG samples. Finally, progress in neuromorphic and photonic signal-processing platforms [25] suggests future opportunities for ultra-low-latency hardware acceleration to support continuous physiological monitoring.

These directions illustrate how the proposed multimodal framework may evolve into a more resilient, adaptive, and personalized edge-AI health-monitoring system.

# 5 Conclusion

The proposed multimodal deep learning framework effectively predicts biological age by combining EMG signals with anthropometric data, achieving state-of-the-art performance. Its successful deployment on the ESP32-S3 platform demonstrates that advanced AI models can be embedded into low-cost IoT devices, thereby bridging the gap between laboratory research and real-world applications. Future work will focus on integrating additional biosensors (e.g., blood pressure, posture, EEG) and expanding clinical validation studies.

**Disclosure of Interests.** The authors have no competing interests to declare that are relevant to the content of this article.

## References

1. Chan, H.A.: Requirements of interworking WirelessLAN and PLMN wireless data network systems. IEEE AFRICON Conf. **1**, 251–255 (2004)
2. Pi, Z., Choi, J., Heath, R.: Millimeter-wave gigabit broadband evolution toward 5G: fixed access and backhaul. IEEE Commun. Mag. **54**(4), 138–144 (2016)
3. Banerjee, A., et al.: Wavelength-division-multiplexed passive optical network (WDM-PON) technologies for broadband access: a review. J. Opt. Netw. **4**(11), 737–758 (2005)
4. Gampala, G., Reddy, C.J.: Massive MIMO-Beyond 4G and a basis for 5G. In: 2018 International Applied Computational Electromagnetics Society Symposium in Denver, ACES-Denver (2018)
5. Novak, D., et al.: Radio-over-fiber technologies for emerging wireless systems. IEEE J. Quantum Electr. **52**(1) (2016)
6. Huang, Y., et al.: A machine learning approach for dynamic optical channel add/drop strategies that minimize EDFA power excursions. ECOC 2016; 42nd European Conference on Optical Communication. VDE (2016)
7. Nyarko Boateng, O., Adekoya, A.F., Weyori, B.A.: Predicting the actual location of faults in underground optical networks using linear regression. Eng. Rep. **3**(3), eng212304 (2021)
8. Abdelli, K., Cho, J.Y., Azendorf, F., Griesser, H., Tropschug, C., Pachnicke, S.: Machine-learning-based anomaly detection in optical fiber monitoring. J. Opt. Commun. Netw. **14**(5), 365–375 (2022)
9. bdelli, K., et al.: Reflective fiber fault detection and characterization using long short-term memory. IEEE/OSA J. Opt. Commun. Netw. **13**(10) (2021)
10. bdelli, K., et al.: Reflective event detection and characterization in fiber optical links given noisy OTDR signals. hotonic Networks. 22nd ITG Symposium (2021)
11. Zhang, W., et al.: A deep convolutional neural network with new training methods for bearing fault diagnosis under noisy environment and different working load. Mech. Syst. Signal Process. **100** (2018)
12. Delli, K.B., et al.: A BiLSTM-CNN based Multitask Learning approach for Fiber Fault iagnosis. OCF (2021)
13. Abdelli, K., et al.: Dataset for optical fiber faults. IEEE Dataport (2022)

# Advancing Evaluation Methods in Explainable AI: Techniques, Challenges, and Frameworks for Standardization

Yaoyang Wu[1], Zhichao Yu[2], Jiahui Yu[2], Zixuan Yang[2], Simon James Fong[2(✉)], and Nilanjan Dey[3]

[1] School of Digital Science, Macau Millennium College, Macau SAR, China
sunnywu@mmc.edu.mo

[2] Faculty of Science and Technology, University of Macau, Macau SAR, China
{yc57985,mc35112,mc56578,ccfong}@um.edu.mo

[3] Department of Computer Science and Engineering,
Techno International New Town, Kolkata, India

**Abstract.** Explainable AI (XAI) research has been proliferating for over the last 10 years, where people come to realize that most efficient AI models are intrinsically black-box, and justification is needed when humans act according to the results produced by these black boxes. Nowadays, many explainable methods are emerging, yet the procedure of evaluating the performance of the aforesaid methods could technically be an independent research subject by itself. This paper presents the trending techniques in current XAI research field along with their evaluation methods for their performance. Exposition of differences between evaluation methods between traditional AI and XAI, how researchers could get started on understanding and further developing evaluation methods, also a discussion of current open issues along with future research of evaluation for XAI models is included in this paper.

**Keywords:** Explainable AI · Cross-computing tools and techniques · Evaluation

## 1 Introduction

The rising and progressively urgent need for explanations from black-box AI models has become a prominent concern in recent years [1]. Explainable AI (XAI) refers to a class of systems that provide visibility into how an AI system makes decisions, generates predictions, and executes actions [2]. Despite the outstanding advantages of black-box models, including high performance, computational efficiency, and adaptability across diverse applications, numerous incidents have highlighted their unpredictability, underscoring the necessity for explainability. Current XAI methods can be summarized as illustrated in Fig. 1.

For XAI models, the primary goals can generally be categorized into three objectives: Transparency, Explainability, and Interpretability. These objectives are considered progressive, with each building upon the preceding one,


P. Pichappan et al. (Eds.): RTIS 2025, LNNS 1936, pp. 399–410, 2027.
https://doi.org/10.1007/978-3-032-23396-7_36

and collectively they reflect the performance of XAI models. The distinction between explainability and interpretability has been debated extensively: some researchers argue for unifying them as a single concept, while others maintain a subtle difference, which may intersect with disciplines such as psychology and social science. Regardless of the terminology, the cornerstone for either objective remains consistent: whether humans can comprehend the model's behavior. It may be tempting to assume that evaluating an XAI model merely involves deter-mining if humans understand its explanations. In reality, the evaluation encompasses much more: it seeks to measure the degree of human confidence in AI systems and to ensure the development of reliable machine learning models. Therefore, precise and comprehensive evaluation is crucial to establish the trustworthiness of XAI models. Interpretable Artificial Intelligence is currently an active research topic and has been explored from multiple perspectives, including Granular Computing [3]. Yu Zhang et al. [4] recently surveyed XAI studies, treating the terms explainability, interpretability, understandability, and comprehensibility in an interchangeable manner.

In this paper, we focus specifically on the evaluation of XAI models. We aim to raise awareness among researchers regarding the importance of systematic evaluation and to provide guidance for incorporating standard evaluation procedures into future XAI studies. Evaluating an explanation may seem straightforward—assessing whether humans understand it—but human involvement introduces subjectivity and potential con-founding factors. One key objective of this paper is to summarize methods to quantify the quality of explanations mathematically, analogous to evaluating the performance of black-box models, and ideally in an autonomous manner.

This study adopts the theoretical framework proposed by Doshi-Velez et al. [5], which we review in detail in the following section. Building on their work and other contributive studies, we present a practical, technical, and concise process for evaluating XAI models.

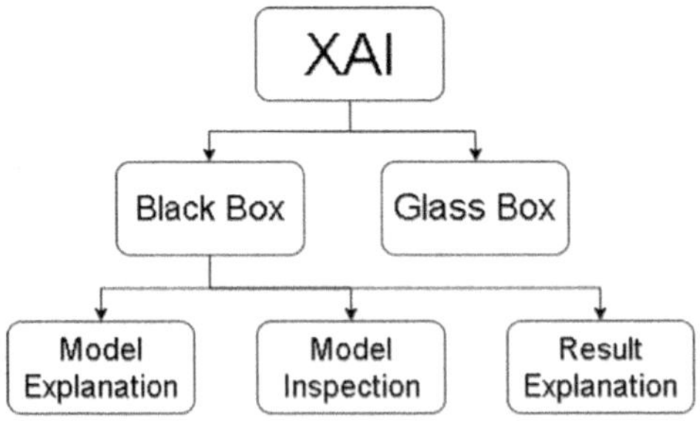


**Fig. 1.** General Framework for Explainable AI (XAI) methods

## 2 Related Works

With the rapid growth of XAI research, numerous surveys and method development studies have emerged. However, most focus primarily on creating models under the assumption that "people will understand when they receive an explanation." Typically, reported results present either model performance or the explanation itself, rather than systematically evaluating the quality of the explanations. In reality, these results should be evaluated with the same rigor as conventional black-box models, yet evaluation of XAI methods has received limited attention.

An early example of partial XAI evaluation was proposed by Ribeiro et al. [6] with the Local Interpretable Model-Agnostic Explanation (LIME) method in 2016. Their work introduced human evaluators as a key component, demonstrating that explanation quality directly affects human confidence in a model. For instance, when an image classifier was trained with intentionally misleading data, human confidence dropped substantially, even when the classifier produced correct predictions. While the evaluation lacked precise statistical indices, the human feedback on LIME explanations provided qualitative evidence of its effectiveness.

Existing literature on XAI evaluation generally falls into two categories. The first is theoretical or survey-based, discussing evaluation alongside broader topics such as interpretability, ethical considerations, and definitions of "explanation" [9–12]. These works often lack detailed examples, making them less actionable. The second category is model-specific evaluations, which assess explanations for particular models such as neural networks [7,8].

Unlike conventional AI, where performance can be measured with clear numerical metrics, XAI evaluation is complicated by human involvement, which introduces subjective factors including cognitive recognition, background knowledge, and personal bias. Objective factors such as evaluator expertise, interests, and goals also play a significant role. Following Gilpin et al. [12], we adopt the distinction that explainability is a model's inherent capability to produce explanations, while interpretability reflects the extent to which humans can understand those explanations. Evaluation of XAI models therefore focuses on interpretability.

XAI evaluation can be considered a subfield of Human-Computer Interaction (HCI), given the central role of human participation. Early discussions on evaluation in com-puter systems highlighted the need for formal assessment [13], [14]. Antunes et al. [14] proposed six desirables for collaborative systems: generalization, precision, invested time, realism, system detail, and system scope. According to this study, we have Table 1 showing the characteristics possessed by each level framework and Fig. 2 illustrat-ing the description of them in terms of their human/machine involvement as well as their scenario situation.

Most machine learning models are evaluated only on precision, whereas XAI evaluation must consider additional factors such as time and financial cost for human evaluators. J. Reason's three-layer performance framework [15]—role-based, rule-based, and knowledge-based—can be adapted to XAI evaluation, with knowledge-based performance encompassing human-computer interaction, satisfaction, consensus, and real-world usefulness (Fig. 3).

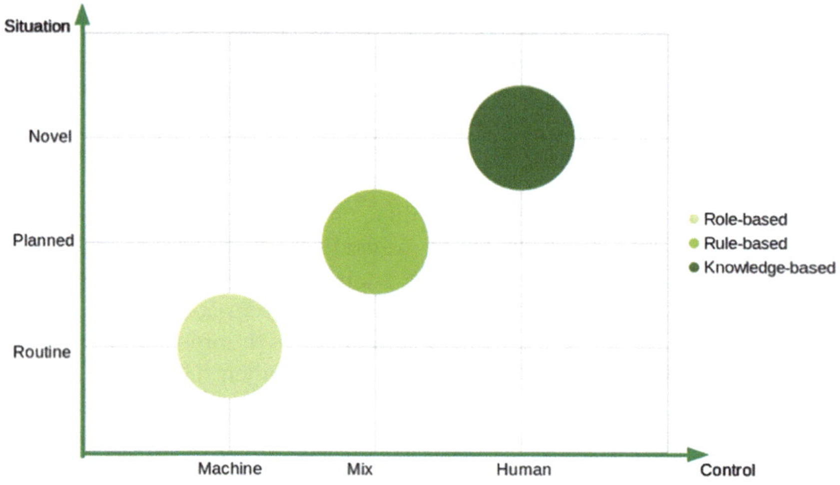


**Fig. 2.** Performance level framework illustration for evaluation scenarios

**Table 1.** Characteristics of Performance Levels.

| Performance | Role-based | Rule-based | Knowledge based |
|---|---|---|---|
| Characteristics | Efficiency<br>Usability | Productivity<br>Conformity | Interaction<br>Participation<br>Satisfaction<br>Consensus<br>Usefulness |

The widely recognized taxonomy for XAI evaluation, proposed by Doshi-Velez et al. [5], classifies methods into functionally-grounded, human-grounded, and application-grounded evaluation (Fig. 4). Functionally-grounded evaluation relies on proxies without human involvement, allowing cost-effective comparison against baselines. Human-grounded evaluation uses laymen or experts to assess explanation quality through binary choice, forward simulation, or counterfactual simulation. Application-grounded evaluation involves domain experts performing real tasks, yielding high-fidelity results but at higher financial and time costs. These evaluation types are interrelated, and multiple approaches may be applied depending on the model's intended use.

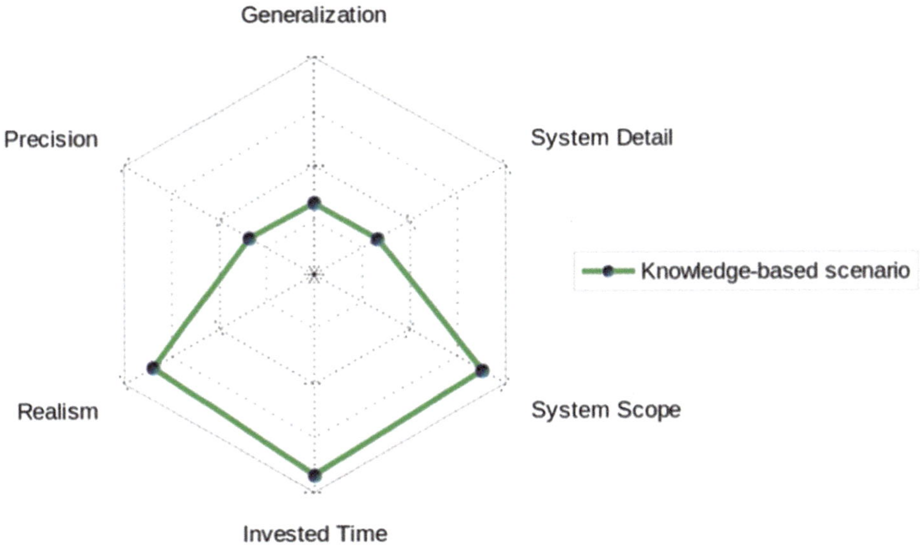


**Fig. 3.** Spider-Web Chart of Knowledge-based scenario

Cost & Specificity

| Evaluation | Evaluator | Task |
|---|---|---|
| Application-grounded | Human | Real Tasks |
| Human-grounded | Human | Simple Tasks |
| Functionally-grounded | Proxy | Proxy Tasks |

**Fig. 4.** Taxonomy of Evaluation methods

Guidotti et al. [16] identified accuracy, fidelity, and interpretability as key XAI properties. Evaluation must consider both explanation understandability and baseline model performance. Trade-offs often exist between interpretability and performance. For ex-ample, in experiments comparing LIME [6]and SHapley Additive exPlanations (SHAP) [17] on the UCI breast cancer dataset [18], LIME provided local explanations quickly, while SHAP offered detailed global explanations at a higher computational cost, illustrating the interpretability-performance trade-off (Figs. 5 and 6).

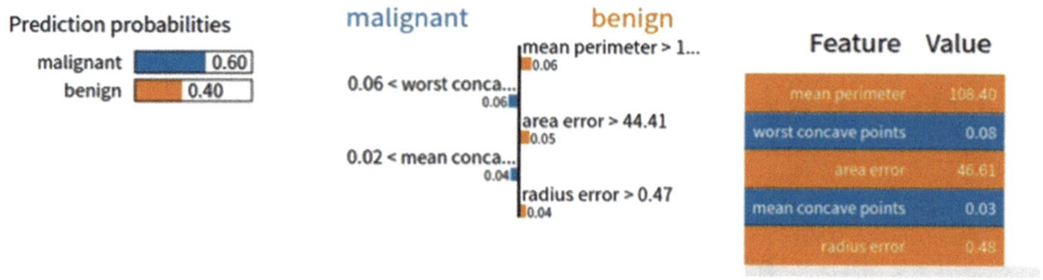


**Fig. 5.** Local Interpretable Model-Agnostic Explanation (LIME) explanation of sample for breast cancer dataset

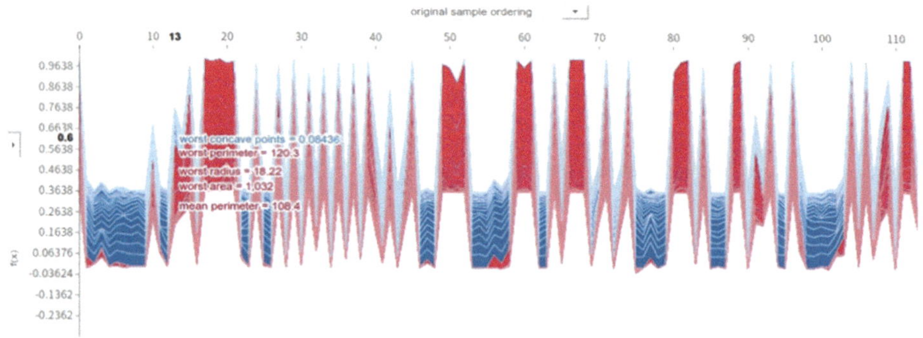


**Fig. 6.** SHapley Additive exPlanations (SHAP) force plot of stack shapley values of features for breast cancer dataset

# 3 Case Study: Evaluation Process for SIDU

In the research proposing the explanation model SIDU for Convolutional Neural Networks (CNNs) by Satya M. Muddamsetty et al. [19,20], a rather comprehensive and concise framework of evaluating an explainable image classification ML model is presented. In this research the authors conducted all three types of evaluations from aforementioned taxonomy for evaluating XAI methods proposed by F. Doshi-Velez et al. [5], instead of only one of them as done in most works, and also they compared the proposed method with two other XAI models appropriately. While presenting and evaluating their proposed XAI model for image classification, they have proposed a clear and operational procedure for evaluating the XAI method explaining CNNs used for image classification task.

Explainable AI (XAI) evaluation employs three main approaches. Functionally-grounded evaluation uses perturbation metrics like insertion and deletion, measuring the performance drop when salient regions are altered. Human-grounded evaluation quantifies the alignment between model explanations and human attention, often using eye-tracking and metrics like AUC, KL-Divergence, and Spearman's correlation. Application-grounded evaluation involves domain experts (e.g., ophthalmologists) directly rating the quality of explanations in a real-world context, such as diagnosing from medical images. Finally, the interpretability of a model is influenced by both subjective factors (like data type and model complexity) and objective factors (like the evaluators' background and goals).

Interpretability could be easily affected by many factors, and these factors can be categorized into subjective and objective ones. Subjective ones include the data type (text data, tabular data, images, signal and waves, etc.) [21] and model types (neural network, deep learning models [22]), complexity of illustration of the explanation, type of illustration of the explanation [23]. Objective ones include background of evaluators (whether they are layman or experts), interests and objectives of evaluators, etc.

### 3.1 Subjective Factors: Laymen or Expert?

Human factors affecting AI model evaluation include cognitive recognition, background knowledge, and personal prejudice. To focus on model performance rather than human variance, researchers should select evaluators with similar characteristics—such as matching professions, age ranges, and comparable cognitive levels (e.g., similar IQ scores). The main distinction between laymen and experts lies in background knowledge.

Application-grounded evaluations require trained experts [24], while human-grounded evaluations may also include laymen. Experts offer deeper background knowledge, often yielding more comprehensive assessments. However, using laymen can significantly reduce costs, especially when they can evaluate with minimal training or when the model is designed to be understandable to non-experts. The choice of evaluator should align with the model's target user group.

The development of XAI is partly driven by the "right to explanation" under the EU GDPR, which entitles individuals to understand automated decisions affecting them—such as in loan approvals, recruitment, or parole decisions. In these cases, explanations must be understandable to laypersons, making laymen the appropriate evaluators.

For laymen evaluators, common assessment methods include:

- Simulatability: Evaluators use the explanation to imitate the model's reasoning and make predictions on new inputs.
- Prediction comparison: Evaluators without prior knowledge of the model make their own predictions; the similarity between human predictions and model outputs indicates explanation quality.

Both methods can also apply to experts. The choice depends on the XAI model's mechanism, as illustrated in Fig. 1.

### 3.2 Objective Factors: Comparison or Not?

Comparison experiment is likely the most popular way used by researchers to present the good performance of their proposed model. When researchers wish to conduct comparison experiments for the proposed XAI model, the objective factors should be taken into consideration. Generally the explanations from image processing models are more visual friendly, and explanations from other kind of models would contain digital or literal information that requires deeper reading and understanding, therefore when choosing models to be compared, the ones producing similar form of explanation are to be considered first, for it's not meaningful to compare visual explanation with literal or digital ones. When a proposed model is a new creation but not an optimization of existing model, or/and when the produced explanation is straightforward (such as new forms of saliency map), comparison experiment is not mandatory. Appropriate presentation of explanation and standard process of evaluation might be enough to show the performance of the proposed method, researchers are advised to decide depending on their circumstances.

## 4 Methodology

F. Doshi-Velez et al. [5] also rendered the opinion that evaluation should match the objective of the original model. Evaluation methods vary according to the type of explanation. In this section we present our proposed basic process for the evaluation. Under the following precondition: There exists the need for an explanation; we have no ethical issues producing and acknowledging said explanation. When a XAI model is operational, one can initiate the evaluation procedure of evaluating this XAI model. The following framework might provide researchers with a clear guideline of evaluating said model. The process contains the four steps which we describe next.

### 4.1 Choosing Methods

Overall, the decision to conduct a full evaluation should be based on specific circumstances. For efficiency and necessity, we summarize three guiding categories: for models with a specific professional purpose (e.g., a CAD system for cancer pre-diagnosis), experts should be the evaluators; for models designed to produce justifications (e.g., explaining loan decisions to customers), lay users are the primary evaluators, with experts optionally added for comprehensiveness; and for general-purpose models aimed at debugging, optimization, or adaptation, researchers themselves may start with functionally-grounded evaluations, which are cost-effective and avoid permit or ethical issues. These categories generally align with the Evaluation Taxonomy, but multiple evaluation methods can be selected and applied independently or in combination as needed.

- Experts for real-life task (Application-grounded)
- Experts for simpler real-life task (Application-grounded)
- Binary force choice (Human-grounded)
- Forward simulation (Human-grounded)
- Counterfactual simulation (Human-grounded)

– Proxy (Functionally-grounded)

When only functionally-grounded method is chosen, result can be obtained and presented straightly. But when human evaluators are needed, the process of choosing evaluators is important. We are going to elaborate in detail in the following sector.

### 4.2 Choosing Evaluators

Human evaluators are used in Application or Human-Grounded evaluations, categorized as either experts or laymen. Key variables to control include cognitive recognition, background knowledge, and personal prejudice. For experts, verify credentials and recruit those with similar academic status and background to reduce evaluation interference. For laymen, target the intended user group; with limited resources, a controlled environment like a campus allows filtering by age, gender, major, and academic performance. Overall, control for conditions such as age, gender, profession, social status, and intelligence level as much as possible.

The selection of an appropriate proxy is included here, as it can act as an evaluator in Functionally-Grounded evaluations. In such cases, researchers choose a proxy function, such as perturbation, to assess explanation accuracy by modifying key features in the output.

### 4.3 Proceed with Evaluation

Functionally grounded evaluation can be done autonomously or rely mostly on autonomous systems or algorithms, while application and human grounded evaluations needed a level organization same as common type of social science experiments.

As discussed in previous section before, experts have wider compatible range as evaluators than laymen, therefore the evaluations designed to use laymen can be completed by experts as well. For example, the binary force choice evaluation can be conducted simultaneously by the experts when they are also doing real life task, which not only produce comprehensive evaluation result but also save cost of evaluation (time, price of hiring other evaluators).

### 4.4 Present the Evaluation Result

As mentioned before most of the current studies include comparison experiments comparing their proposed methods to other ones. For XAI models, extended to outperform other methods, they also need to at least perform as well as human. Performance of these kind of XAI models can be shown by presenting how similar the explanations are to human, better yet experts' understanding. There are still two ways of presenting the performance of the proposed model: Demonstrating that the proposed method has better performance than other method and presenting how similar the explanation is comparing to the human produced product. Second one is only applicable when Application and Human grounded evaluations are involved. Of course both ways can be united, for example to demonstrate that the proposed method's explanation is more similar to human result than explanations produced by other models, or the evaluator produces better result with new input after observing the explanation from proposed model than from others.

Research and development of XAI method are still at the stage of exploiting various new methods, when evaluating proposed method, it is up to the developers to choose whether to compare with other methods or/and compare with human result, both of which could appropriately present the proposed method's performance.

## 5 Discussion and Conclusion

The theoretical foundation for evaluating XAI models remains unsettled, with discussions spanning Human-Computer Interaction, psychology, and social sciences. This paper provides a simplified evaluation framework for new XAI researchers, suggesting that every XAI study should include a dedicated evaluation section. Researchers must clarify four elements: the model's objective, application context, target users, and research purpose. These determine the appropriate evaluation method.

XAI evaluation is particularly critical in healthcare, where models must be thoroughly validated before deployment. For instance, in computer-aided diagnosis (CAD) systems detecting cancers like lung or breast cancer, black-box models require reliable explanations. Inspired by the SIDU model's application-grounded evaluation, we propose combining human-grounded and application-grounded approaches. For lung cancer diagnosis using MRI images, expert radiologists' eye-tracking data could be compared with model saliency maps using AUC and KL-DIV metrics, supplemented by binary quality assessments. This combined approach can validate explanations for deploying trustworthy CAD systems.

The above case serves as an example of how XAI models can be evaluated and better gain the trust of human users. It is extremely important that for models with critical objectives such as diagnosis of disease, evaluation of the model should be treated precisely, clearly and carefully. Currently we have a primal theoretical foundation of defining and understanding what evaluation of XAI is, and also we have various researches of evaluating specific explainable models. A commonly recognized basic procedure of evaluating XAI models, in a technical level in extent of theoretical one, could be beneficial for this research field. Afterall, with human elements involved, evaluation of XAI model can be very different from conventional AI models.

Moreover, when talking about a commonly recognized standard, expert knowledge is a good entry point to automized and standardized the evaluation procedure for XAI models. If a XAI model is designed for professional purpose, such as diagnosing diseases as mentioned above, experts would be able to provide a standard, maybe a range of percentage, of the similarity between model explanation and expert prediction, how high does the percentage has to be for the expert to trust the model. This standard could also help to save the cost of conducting comparison experiments when a model is 'qualified' according to it and help researchers with less resources to develop models with maybe not the highest standard but still qualified and with other advantages over other models.

**Acknowledgments.** This research was funded by the Guangzhou Development Zone Science and Technology Project (2023GH02), the University of Macau (MYRG2022-00271-FST) and research grants by the Science and Technology Development Fund of Macau (0032/2022/A) and (0019/2025/RIB1).

**Disclosure of Interests.** The authors have no competing interests to declare that are relevant to the content of this article.

## References

1. Oconitrillo, L.R.R., Vargas, J.J., Camacho, A., Burgos, A., Corchado, J.M.: RYEL system: a novel method for capturing and represent knowledge in a legal domain using explainable artificial intelligence (XAI) and granular computing (GrC). In: Pedrycz, W., Chen, S.-M. (eds.) Interpretable Artificial Intelligence: A Perspective of Granular Computing. SCI, vol. 937, pp. 369–399. Springer, Cham (2021). https://doi.org/10.1007/978-3-030-64949-4_12
2. Rai, A.: Explainable AI: from black box to glass box. J. Acad. Mark. Sci. **48**(1), 137–141 (2020)
3. Pedrycz, W., Chen, S.-M. (eds.): Interpretable Artificial Intelligence: A Perspective of Granular Computing. SCI, vol. 937. Springer, Cham (2021). https://doi.org/10.1007/978-3-030-64949-4
4. Zhang, Y., et al.: A survey on neural network interpretability. IEEE Trans. Emerg. Top. Comput. Intell. (2021)
5. Doshi-Velez, F., Kim, B.: Towards a rigorous science of interpretable machine learning. arXiv (2017). https://arxiv.org/abs/1702.08608
6. Ribeiro, M.T., Singh, S., Guestrin, C.: Why should i trust you? Explaining the predictions of any classifier. In Proceedings of the 22nd ACM SIGKDD International Conference on Knowledge Discovery and Data Mining, pp. 1135–1144. ACM (2016)
7. Kanerv, O.: Evaluating explainable AI models for convolutional neural networks with proxy tasks. Master's thesis. Aalto University (2019)
8. Muddamsetty, S.M., Jahromi, M.N.S., Moeslund, T.B.: Expert level evaluations for explainable AI (XAI) methods in the medical domain. In: ICPR-2020 Workshop Explainable Deep Learning-AI (2020)
9. Hempel, C., Oppenheim, P.: Studies in the logic of explanation. Philos. Sci. (1948)
10. Lombrozo, T.: The structure and function of explanations. Trends Cogn. Sci. **10**(10), 464–470 (2006)
11. Keil, F.: Explanation and understanding. Annu. Rev. Psychol. (2006)
12. Gilpin, L.H., Bau, D., Yuan, B.Z., Bajwa, A., Specter, M., Kagal, L.: Explaining explanations: an overview of interpretability of machine learning. In: The 5th IEEE International Conference on Data Science and Advanced Analytics (DSAA 2018) (2018)

13. Pinelle, D., Gutwin, C.: A Review of groupware evaluations. In: Proceedings of the 9th IEEE International Workshops on Enabling Technologies: Infrastructure for Collaborative Enterprises (WETICE 2000), pp. 86–91. IEEE Computer Society, USA (2000)
14. Antunes, P., Herskovic, V., Ochoa, S.F., Pino, J.A.: Structuring dimensions for collaborative systems evaluation. ACM Comput. Surv. **44**(2), 28 pages (2008). https://doi.org/10.1145/2089125.2089128. Article 8
15. Reason, J.: The human contribution: unsafe acts, accidents and heroic recoveries. Surrey, England, Ashgate (2008)
16. Guidotti, R., Monreale, A., Ruggieri, S., Turini, F., Giannotti, F., Pedreschi, D.: A survey of methods for explaining black box models. ACM Comput. Surv. **51**(5), Article 93 (2019). https://doi.org/10.1145/3236009
17. Lundberg, S.M., Su-In, L.: A unified approach to interpreting model predictions. In: NIPS (2017)
18. Dua, D., Graff, C.: UCI Machine Learning Repository. University of California, School of Information and Computer Science, Irvine, CA (2019). http://archive.ics.uci.edu/ml
19. Muddamsetty, S.M., Jahromi, M.N.S., Ciontos, A.E., Fenoy, L.M., Moeslund, T.B.: Introducing and assessing the explainable AI (XAI) method: SIDU. CoRR abs/2101.10710 (2021)
20. Muddamsetty, S.M., Jahromi, M.N.S., Ciontos, A.E., Fenoy, L.M., Moeslund, T.B.: Visual explanation of black-box model: similarity difference and uniqueness (SIDU) method. Pattern Recognit. **127**, 108604 (2022)
21. Hase, P., Bansal, M.: Evaluating explainable AI: which algorithmic explanations help users predict model behavior?. In: Proceedings of the 58th Annual Meeting of the Association for Computational Linguistics, pp. 5540–5552. Association for Computational Linguistics (2020). https://doi.org/10.18653/v1/2020.acl-main.491
22. Kakogeorgiou, I., Karantzalos, K.: Evaluating explainable artificial intelligence methods for multi-label deep learning classification tasks in remote sensing. CoRR abs/2104.01375 (2021)
23. Rudin, C.: Stop explaining black box machine learning models for high stakes decisions and use interpretable models instead. Nat Mach Intell **1**, 206–215 (2019)
24. Brey, P.A.E., Soraker, J.: Philosophy of computing and information technology. In: A. Meijers (ed.) Philosophy of Technology and Engineering Sciences, pp. 1341–1408 (2009). https://doi.org/10.1016/B978-0-444-51667-1.50051-3. (Handbook of the philosophy of science; Vol. 9, No. IX). Elsevier

# A Dynamic Resource Allocation Strategy for Efficient Handling of Unexpected Emergency Tasks in VEC Networks

Yao-Liang Chung(✉) and Jie-Wei Zhang

Department of Communications, Navigation and Control Engineering, National Taiwan Ocean University, Keelung City, Taiwan
ylchung@email.ntou.edu.tw

**Abstract.** This study proposes a novel dynamic resource allocation strategy tailored for vehicular edge computing (VEC) networks, with a particular focus on the efficient handling of unexpected emergency tasks. The core of the strategy is to enhance the flexibility and responsiveness of vehicles in handling unexpected emergency tasks, so that they can flexibly adjust the use of computation resources in the face of offloaded tasks and unexpected emergency tasks simultaneously. The strategy incorporates mechanisms for selecting service vehicles and dynamically allocating and reclaiming computation resources. These efficient designs not only improve the task completion rate, but also reduce the computational burden on VEC servers. Simulation results demonstrate that the proposed strategy outperforms the baseline in both total task completion rate and unexpected emergency task completion rate, across various numbers of vehicles and roadside unit computational capacities. Notably, the improvement in unexpected emergency task completion rate is particularly significant.

**Keywords:** Resource Allocation · Unexpected Emergency Task · Vehicular Edge Computing (VEC)

## 1 Introduction

With the continuous advancement of autonomous driving technologies, the computational demands on intelligent vehicles have become unprecedentedly high. To process the massive amount of data generated by onboard sensors interacting with the environment, vehicles must rely on efficient communication and computation resource allocation techniques. To address the limitations in vehicular communication and computational capabilities, the Internet of vehicles and vehicular cloud computing have emerged. These paradigms offload computation-intensive tasks to remote cloud servers to alleviate local processing burdens [1]. However, such cloud-based approaches often suffer from network congestion and high latency. To overcome these issues, vehicular edge computing (VEC) has been introduced. VEC deploys servers on roadside unit (RSU) located at the edge of the network, significantly reducing transmission latency and mitigating cloud


P. Pichappan et al. (Eds.): RTIS 2025, LNNS 1936, pp. 411–421, 2027.
https://doi.org/10.1007/978-3-032-23396-7_37

overload [2]. Nevertheless, the limited coverage, resource constraints, and load imbalance inherent in VEC servers may lead to degraded overall system performance. This becomes particularly challenging in highly dynamic vehicular environments, where task offloading and coordination are increasingly complex. To further address these challenges, vehicular collaborative computing enables resource sharing among vehicles, allowing emergency tasks to be handled in a distributed and timely manner, while non-emergency tasks can still be offloaded to RSUs or cloud servers. However, one of the key challenges that remains is how to perform effective resource allocation and collaboration management in such dynamic VEC environments.

Several studies [3–7] have proposed solutions to tackle the aforementioned problems. In [3], a binary offloading scheme based on fuzzy logic was proposed, considering vehicle computation capacity, mobility, and link quality to determine the optimal offloading target. Simulation results showed that this scheme improved the task completion rate and reduced the average task completion time, compared to other benchmark methods. In [4], the authors analyzed offloading efficiency in VEC networks, emphasizing vehicle mobility and the maximum allowable task delay. They proposed two offloading schemes: one for non-cooperative VECs and another for cooperative VECs, both balancing task latency and resource utilization. Simulation results confirmed that both schemes were effective in reducing system costs while meeting delay constraints. Literature [5] introduced a particle swarm optimization (PSO)-based algorithm to improve system performance and reduce operational costs. It was compared against the non-dominated sorting genetic algorithm II (NSGA-II), and results indicated that the PSO-based algorithm outperformed the NSGA-II across various user densities, request dynamics, and workload variations. In [6], a joint optimization model for task offloading and resource allocation was proposed using idle vehicles and RSUs, aiming to minimize task completion time. The authors designed a soft actor-critic-based algorithm, which significantly reduced latency compared to conventional baselines. In [7], a mobility prediction model was introduced to forecast future network states, followed by the design of a proactive adjustment-based task offloading (PATO) adaptive algorithm. Simulation results demonstrated that the PATO algorithm outperformed deep reinforcement learning-based baselines in both offloading efficiency and task success rate.

Despite the advancements proposed in these studies, few studies have considered the scenario where a vehicle providing computation resources may encounter unexpected emergency tasks. To address this critical gap, this study proposes a novel dynamic resource allocation strategy for VEC networks. The proposed strategy integrates communication and computation resources and introduces two dynamic adjustment ratios that enable vehicles to flexibly reallocate their computation resources based on current task conditions and resource demands. Furthermore, the strategy facilitates the parallel processing of various tasks. Simulation results show that the proposed strategy significantly outperforms the baseline in terms of the total task completion rate and the unexpected emergency task completion rate, with particularly notable improvements in handling emergency tasks.

The remainder of this paper is organized as follows. Section 2 presents the system model. Section 3 describes the proposed dynamic resource allocation strategy. Section 4 introduces the communication and computation models. Section 5 provides simulation results and discussion. Finally, Sect. 6 concludes the paper.

# 2 System Model

The scenario considered in this study is a unidirectional multi-lane VEC environment. The system architecture comprises a macro base station (MBS) and multiple RSUs deployed along the roadway. The MBS is connected to all RSUs via wired links and is capable of collecting and exchanging real-time vehicular information, thereby maintaining comprehensive awareness of each vehicle's position and mobility status. Each RSU is connected to a VEC server, enabling edge computing capabilities. In this environment, vehicles are categorized into two types: 1) requesting vehicles, which require computation resources and are indexed by $i$, 2) supporting vehicles, which are capable of providing computation resources and are indexed by $j$. We assume that the computing capability of each vehicle and each RSU is determined by the processing speed of their respective CPUs.

Each RSU covers a circular service area. When a high-speed vehicle offloads a task to the nearest RSU, there is a possibility that the vehicle may move out of the RSU's coverage area before the task calculation is completed. In such cases, the RSU must transmit the completed task results back to the target vehicle via inter-RSU data forwarding coordinated by the MBS. To simplify analysis, we disregard the time required for MBS coordination and only consider the inter-RSU data exchange delay and the result return delay to the target vehicle.

We categorize tasks into three types: starting tasks, offloaded tasks, and unexpected tasks. Each vehicle is assumed to be equipped with an intelligent module capable of perceiving environmental changes in real time, autonomously generating corresponding tasks, classifying them as either emergency or non-emergency, and estimating the task's delay constraint, denoted as $Tmax$. We further define explicit task generation and processing rules as follows. For vehicle $j$, no new self-generated tasks will be produced while it is processing a starting task. Additionally, during the calculation of an offloaded task, vehicle $j$ may encounter an unexpected task. However, if vehicle $j$ is working on an unexpected task, it will not generate a new own task. Once vehicle $j$ completes all its tasks, it returns to an idle state and may then generate a new task. Each vehicle $i$ is derived from a vehicle $j$ that is unable to process its current self-generated task due to insufficient computation resources. Vehicle $i$ does not generate any new task. Only after completing all ongoing tasks will vehicle $i$ return to an idle state and become eligible to generate a new task.

# 3 Dynamic Resource Allocation Strategy

## 3.1 Case with a Starting Task and no Occurrence of an Unexpected Task

In the proposed VEC network model, when vehicle $j$ generates a starting task and can complete it within its delay constraint $Tmax$, it remains classified as a supporting vehicle. If vehicle $j$ is selected by vehicle $i$ as a suitable service vehicle while processing its own starting task, a parallel computing way is adopted to handle both the local starting task and the offloaded task from vehicle $i$ concurrently. Specifically, let $\alpha_j$ denote the proportion of computation resources that vehicle $j$ allocates to the offloaded task from vehicle $i$, , and $(1 - \alpha_j)$ represent the portion reserved for its own starting task. In this allocation strategy, vehicle $j$ retains only the amount of computation resources necessary to complete its own task within *Tmax*, and the remaining resources are dedicated to the offloaded task. Based on this rule, the corresponding $\alpha_j$ can be determined. Both the "starting task" and "offloaded task" can be completed within their respective *Tmax* constraints via parallel processing. Upon completion of the offloaded task, vehicle $j$. immediately returns the result to vehicle $i$.

However, if vehicle $j$ fails to complete its starting task within *Tmax*, it transitions into a requesting vehicle. Before offloading the task, this newly reclassified requesting vehicle must ensure that the sum of the forward transmission time, computation time, and result return time does not exceed *Tmax*. Based on this constraint, the required amount of computation resources (hereafter referred to as the computation demand) is calculated. The vehicle then searches for nearby supporting vehicles by sending a request signal to the MBS to retrieve updated resource availability. If a suitable vehicle $j$ is found (where $j'$ is used to avoid confusion with the previously mentioned vehicle $j$), the requesting vehicle offloads a portion of its starting task to vehicle $j'$. If no such vehicle $j'$ is found, meaning that no nearby vehicle is able to provide computation resources, the requesting vehicle offloads its task to the nearest RSU $v$.

The procedure for handling offloaded tasks at RSU $v$ is as follows. When RSU $v$ receives an offloaded task, it first determines whether it currently has sufficient computation resources to process the task. If not, RSU $v$ is unable to calculate the offloaded task and will not return any result to the original vehicle that initiated the corresponding original task. If sufficient resources are available, RSU $v$ proceeds to calculate the offloaded task and, upon completion, checks whether the original vehicle remains within its service range. If the original vehicle is still within range, RSU $v$ returns the result directly to it. Otherwise, RSU $v$ forwards the result via the MBS to the RSU closest to the current location of the original vehicle (to distinguish this from the original RSU $v$, we denote it as RSU $v'$), and then RSU $v'$ completes the delivery of the result to the original vehicle.

Additionally, after offloading a task, the requesting vehicle must wait for the return of the computation result. If the result arrives within the corresponding *Tmax*, the corresponding original task is considered successfully completed within *Tmax*. Otherwise, the corresponding original task is deemed to have failed due to exceeding *Tmax*.

### 3.2 Case with a Starting Task and the Occurrence of an Unexpected Task

Considering the scenario where vehicle $j$ is providing computation resources to vehicle $i$ (i.e., vehicle $j$ is processing an offloaded task), vehicle $j$ encounters an unexpected task during this process.

When an unexpected emergency task occurs on vehicle $j$, vehicle $j$ must determine whether the remaining computation resources are sufficient to sequentially process the starting task and the unexpected emergency task. If the available resources are sufficient, it implies that the "starting task and unexpected emergency task" and "offloaded task" can be completed in parallel within their respective *Tmax* constraints. In this case, once vehicle $j$ completes the offloaded task, it immediately returns the result to vehicle $i$. If the resources are insufficient, a resource adjustment mechanism is introduced to ensure timely handling of the unexpected emergency task. Specifically, we introduce an adjustment ratio parameter $\beta_j$, which represents the proportion of computation resources that vehicle $j$ continues to allocate to vehicle $i$ during the occurrence of an unexpected emergency task. Accordingly, $(1 - \beta_j)$ represents the proportion of resources reclaimed by vehicle $j$ from vehicle $i$. Under this setting, vehicle $j$ adopts parallel processing to handle the "starting task", "offloaded task", and "unexpected emergency task" concurrently, striving to complete all tasks within their respective *Tmax* constraints. The resource reallocation mechanism is designed such that the reclaimed resources are just sufficient to complete the unexpected emergency task within its *Tmax*, without withdrawing more resources than necessary. Based on this rule, the corresponding $\beta_j$ can be determined.

If vehicle $j$ needs to reclaim a substantial portion of its computation resources due to the unexpected emergency task, resulting in insufficient resources to meet the demand of vehicle $i$, it first transfers a portion of the offloaded task to the nearest RSU $v$, following the procedure for handling offloaded tasks at RSU $v$ as described in Sect. 3.1. In this case, the "starting task", "remaining portion of the offloaded task", and "unexpected emergency task" are processed in parallel, each within their respective *Tmax* constraints. Once vehicle $j$ completes the remaining portion of the offloaded task, it immediately returns the result to vehicle $i$. If the computation resources reallocated to vehicle $i$ are still sufficient, then the "starting task", "offloaded task", and "unexpected emergency task" can all be processed concurrently within their respective *Tmax* constraints. Upon completion of the offloaded task, vehicle $j$ immediately returns the result to vehicle $i$. In the case where vehicle $j$ reclaims all computation resources (i.e., $\beta_j = 0$) and can complete the unexpected emergency task within its *Tmax*, it transfers both the remaining unprocessed portion of the offloaded task and the result of the partially processed task to the nearest RSU $v$, following the procedure for handling offloaded tasks at RSU $v$ as described in Sect. 3.1. The "starting task" and "unexpected emergency task" are then processed in parallel and completed within their respective *Tmax* constraints.

However, if the unexpected emergency task of vehicle $j$ is too large to be completed within its *Tmax* even after reclaiming all previously allocated computation resources (i.e., $\beta_j = 0$), vehicle $j$ will first forward the remaining portion of the offloaded task along with the result of the already-processed portion to the nearest RSU $v$, following the procedure for handling offloaded tasks at RSU $v$ as described in Sect. 3.1. Subsequently, vehicle $j$ transitions into the role of a requesting vehicle, denoted as vehicle $i'$ (to distinguish it from the previously mentioned vehicle $i$), and initiates a search for nearby supporting vehicles. If a suitable supporting vehicle is found (here denoted as vehicle $j''$), vehicle $i'$ offloads part of the unexpected emergency task to vehicle $j''$. If no such vehicle $j''$ is found, vehicle $i'$ will offload the same part of the unexpected emergency task to the nearest RSU $v$, again following the procedure for handling offloaded tasks at RSU $v$ as described in Sect. 3.1. After offloading, the subsequent response process follows the same procedure described in Sect. 3.1 for a requesting vehicle waiting for the return of task results. Meanwhile, the requesting vehicle will complete the calculation of the starting task.

When an unexpected non-emergency task occurs on vehicle *j*, priority is given to completing the offloaded task from vehicle *i*. The unexpected non-emergency task is offloaded to the nearest RSU *v*, following the procedure for handling offloaded tasks at RSU *v* as described in Sect. 3.1. Subsequently, the "starting task" and "offloaded task" are processed in parallel, with the aim of completing both within their respective *Tmax* constraints. Once vehicle *j* completes the offloaded task, it immediately returns the result to vehicle *i*. Afterward, vehicle *j* waits for the result of the unexpected non-emergency task to be returned. If vehicle *j* receives the result within the unexpected non-emergency task's *Tmax*, the task is considered to be successfully completed within *Tmax*. However, if the result does not arrive within *Tmax*, the task is deemed to have failed to complete successfully within *Tmax*.

### 3.3 Case Without a Starting Task and with a Possibility of an Unexpected Task Occurrence

In this case, vehicle *j* allocates all of its computation resources (i.e., $\alpha_j = 1$) to process the offloaded task from vehicle *i*. If no unexpected task occurs during the calculation of the offloaded task, the procedure follows the one described in Sect. 3.1, excluding the parts related to the starting task. If an unexpected task occurs, the procedure follows the one described in Sect. 3.2, likewise excluding the parts related to the starting task.

### 3.4 Selection of a Suitable Service Vehicle

To effectively manage communication and cooperation among vehicles, a one-to-one pairing mechanism is employed. The computation demands of all requesting vehicles are first sorted in descending order, and pairing is then performed sequentially based on this order. For requesting vehicles whose computation demand exceeds or equals the threshold *D*, the nearest suitable supporting vehicle is selected based on the shortest-distance criterion. In this process, vehicles that are currently providing support are temporarily removed from the candidate list to prevent being selected multiple times. For requesting vehicles with computation demand less than *D*, all currently available supporting vehicles are sorted in ascending order according to their available computation capacities. Each requesting vehicle then selects a suitable supporting vehicle based on this order. This ensures that vehicles with smaller computation demands do not occupy supporting vehicles with disproportionately high available resources.

By applying the above dynamic resource allocation strategy, the vehicle can more effectively handle the occurrence of the unexpected emergency task, enabling it to be processed appropriately within its *Tmax* constraint.

## 4 Communication and Computation Models

### 4.1 Communication Model

This study adopts the Shannon capacity formula to estimate the wireless transmission delay, which is given by:

$$T_{I,J} = \frac{Z_I}{C_J} \quad (1)$$

where $Z_I$ denotes the amount of data to be transmitted from source $I$ and $C_J$ represents the data transmission rate at the receiving node $J$, calculated as

$$C_J = B\log(1 + SINR_J) \tag{2}$$

where $B$ is the transmission (system) bandwidth and $SINR_J$ denotes the signal-to-interference-plus-noise ratio perceived at the receiver $J$.

After task offloading is completed, to accurately evaluate the overall communication delay, it is necessary to further consider the delay caused by result transmission. To this end, we introduce a result-to-input data size ratio $\gamma$ to represent the ratio of the result data size relative to the offloaded task data size. In other words, the result data size is given by $\gamma$ times the offloaded task data size.

### 4.2 Computation Model

In terms of the task processing delay, it can be categorized into delay in processing tasks by vehicles and delay in processing tasks by RSUs.

The task processing delay at a vehicle is given by

$$T_{X_{task}, Y^{vehicle}_{resources}} = \frac{X_{task}}{Y^{vehicle}_{resources}} \tag{3}$$

where $X_{task}$ denotes the size of a task processed by the vehicle, which may correspond to various types of tasks as discussed in Sect. 3; $Y^{vehicle}_{resources}$ represents the amount of computation resources allocated by the vehicle at that moment. Similarly, the task processing delay at an RSU is defined as

$$T_{F_{task}, Y^{RSU}_{resources}} = \frac{F_{task}}{Y^{RSU}_{resources}} \tag{4}$$

where $F_{task}$ denotes the size of a task handled by the RSU, which may include different types of tasks as described in Sect. 3; $Y^{RSU}_{resources}$ represents the RSU's available computation resources allocated for task calculation.

## 5 Simulation Results

We compare the performance of the proposed dynamic resource allocation strategy with a baseline approach in terms of the total task completion rate and the unexpected emergency task completion rate. To highlight the advantages of the proposed strategy, the baseline differs from our method in the following ways. In the baseline approach, requesting vehicles search for supporting vehicles randomly rather than using a prioritized selection mechanism. Moreover, when a supporting vehicle encounters an unexpected emergency task during the collaboration process, it does not reclaim the computation resources previously allocated to the requesting vehicle. Instead, the unexpected emergency task is directly offloaded to the nearest RSU for processing.

The simulation environment in this study is designed as a unidirectional four-lane road (indexed by $k$, $k = 1, 2, 3, 4$), where each lane is spaced 4 m apart. To simplify the simulation, we assume that each vehicle travels at a fixed speed determined by the lane in which it is generated (i.e., a vehicle's speed is based on the lane index). Specifically, the fixed speed for vehicles in lane $k$ is randomly selected from [60, 90] km/h. Vehicles are randomly generated across all four lanes within a 2700 m road segment. Each vehicle is equipped with a CPU that operates at a processing speed of 200 KB/s. The wired transmission rate between any two RSUs is set to 20 Mbps. To ensure complete coverage of the roadway, RSUs are deployed along one side of the road with an inter-RSU distance of 390 m. Each RSU has a circular service radius of 200 m and is configured with a computation capacity of 800 KB/s. The system bandwidth $B$ is set to 20 MHz. The threshold $D$ is set to 100 KB/s. The result-to-input data size ratio $\gamma$ is randomly selected from [0.05, 0.15]. The channel model used in our simulation considers the path loss, Rayleigh fading, and Doppler effect.

In addition, the *Tmax* for emergency tasks is set to 0.5 s, while the *Tmax* for non-emergency tasks is set to 3 s. We assume that the time interval between a vehicle transitioning to the idle state and the generation of its next starting task follows an exponential distribution, with the mean randomly selected from [0.15, 1] s. The probability that a starting task is emergency or non-emergency is set to 0.5. The data size of each starting emergency task is randomly drawn from [25, 150] KB, while that of each starting non-emergency task is randomly drawn from [100, 900] KB. The probability that vehicle $j$ generates an unexpected task is set to 0.7. If such a task occurs, it is categorized as emergency or non-emergency with equal probability (0.5). The data size of each unexpected emergency task is randomly drawn from [25, 75] KB, and that of each unexpected non-emergency task is randomly drawn from [90, 680] KB. Based on the respective *Tmax* settings for emergency and non-emergency tasks, the time interval between the moment vehicle $j$ receives an offloaded task and the occurrence of an unexpected emergency task is randomly sampled from [0.01, 0.3] s, while if an unexpected non-emergency task is to occur, the time interval is randomly sampled from [0.1, 2] s. Each simulation run spans a duration of 30 s, and the simulation is repeated 10,000 times. The average results across these 10,000 runs are used for evaluation. In the performance plots presented in the following two figures, the red curve represents the proposed strategy, while the blue curve corresponds to the baseline.

Figure 1 illustrates the comparison between the proposed strategy and the baseline in terms of the total task completion rate and the unexpected emergency task completion rate under varying numbers of vehicles. When the number of vehicles is relatively low (fewer than 50), the environment is less congested and computation resources are generally sufficient. As a result, both the proposed strategy and the baseline achieve over 99% in both total task completion rate and unexpected emergency task completion rate. However, when the number of vehicles increases to 90, the total task completion rate and unexpected emergency task completion rate of the baseline drop by 2.7% and 8.3%, respectively. This degradation is mainly due to the exhaustion of computation resources at certain RSUs, which prevents some tasks from being completed within their respective *Tmax* constraints, thereby reducing the task completion rate. In contrast, the proposed strategy maintains the total task completion rate above 99%, and the unexpected emergency task completion rate shows only a slight decline of 1.2%. Under high-load conditions with 216 vehicles, the total task completion rate and unexpected emergency task completion rate achieved by the proposed strategy are 92.3% and 86%, respectively, while those of the baseline are 84.5% and 48%. These results demonstrate that the proposed strategy significantly outperforms the baseline under high-load conditions, particularly in handling unexpected emergency tasks. This is attributed to the proposed strategy's ability to dynamically allocate and reclaim computation resources among vehicles in real time, and to offload tasks appropriately to RSUs, thereby achieving superior adaptability and efficiency.

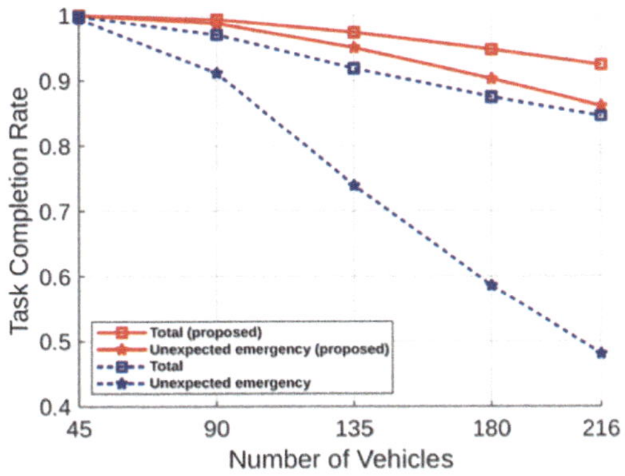


**Fig. 1.** Comparison of the proposed strategy and the baseline in terms of the total task completion rate and the unexpected emergency task completion rate under varying numbers of vehicles.

Figure 2 presents a comparison between the proposed strategy and the baseline in terms of the total task completion rate and the unexpected emergency task completion rate under different RSU computation capacities, with the number of vehicles fixed at 180. As the RSU computation capacity increases from 0 to 900 KB/s, the total task completion rate and unexpected emergency task completion rate of the proposed strategy improve to 96% and 92.3%, respectively, while those of the baseline reach 89.3% and 64.9%. It is evident that, under the same resource conditions, the proposed strategy achieves significantly higher total task completion rate and unexpected emergency task completion rate than the baseline, demonstrating its superior resource utilization efficiency. As the RSU computation capacity continues to increase up to 1600 KB/s, both performance metrics under the proposed strategy continue to improve, but the rate of improvement gradually tapers off. This is because the overall system resources are approaching saturation, allowing the total task completion rate and unexpected emergency task completion rate to exceed 99%, indicating that the system has reached near-optimal operational performance under the proposed strategy.

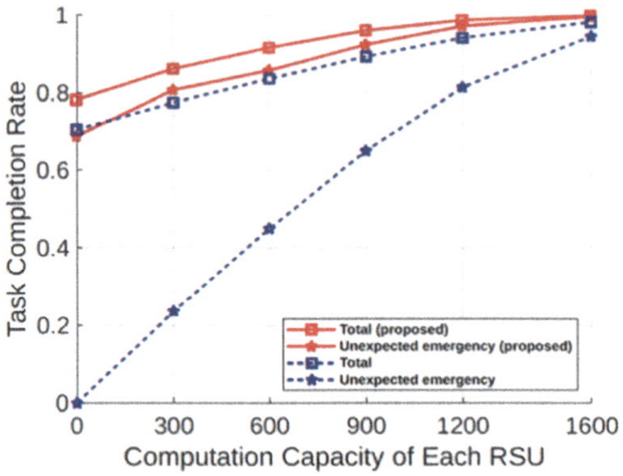


**Fig. 2.** Comparison of the proposed strategy and the baseline in terms of the total task completion rate and the unexpected emergency task completion rate under different RSU computation capacities (with 180 vehicles).

## 6 Conclusion

This study successfully proposes an efficient dynamic resource allocation strategy tailored for handling unexpected emergency tasks in VEC network environments. The strategy is designed to improve the vehicle's ability to respond to unexpected emergency situations and enhance its computational flexibility. This enables the vehicle to adjust the allocation and use of its own computation resources flexibly and immediately when faced with the offloaded task and the unexpected emergency task simultaneously. The strategy incorporates several key mechanisms, including the selection of service vehicles and the real-time allocation and reclamation of computation resources. Simulation results confirm the performance advantages of the proposed strategy. In particular, under high-load conditions, the proposed strategy outperforms the baseline by about 7.8% in total task completion rate and 38% in unexpected emergency task completion rate. Overall, the proposed dynamic resource allocation strategy demonstrates superior task handling performance and efficient resource utilization, effectively strengthening the real-time responsiveness of vehicles operating in dynamic VEC environments.

## References

1. Florin, R., Ghazizadeh, A., Ghazizadeh, P., Olariu, S., Marinescu, D.C.: Enhancing reliability and availability through redundancy in vehicular clouds. IEEE Trans. Cloud Comput. **9**(3), 1061–1074 (2021)
2. Tang, S., et al.: VECFrame: a vehicular edge computing framework for connected autonomous vehicles. In: Proc. 2021 IEEE Int. Conf. Edge Computing, Chicago, IL, USA, pp. 68–77 (2021)
3. Men, R., Fan, X., Shan, A., Yuan, G.: Fuzzy logic based binary computation offloading scheme in V2X communication networks. IEEE Access **12**, 45507–45518 (2024)
4. Yang, C., Liu, Y., Chen, X., Zhong, W., Xie, S.: Efficient mobility-aware task offloading for vehicular edge computing networks. IEEE Access **7**, 26652–26664 (2019)
5. Bavana, B.N., Kirthiga, S.: Performance analysis of task offloading in mobile edge computing using multi-objective optimization algorithm. In: Proc. 2023 IEEE 4th Annual Flagship India Council Int. Subsections Conf., Mysore, India, pp. 1–6 (2023)

6. Zheng, Y., Zhou, H., Chen, R., Jiang, K., Cao, Y.: SAC-based computation offloading and resource allocation in vehicular edge computing. In: Proc. 2022 IEEE Conf. Computer Communications Workshops, New York, NY, USA, pp. 1–6 (2022)
7. Liu, J., Liu, N., Liu, L., Li, S., Zhu, H., Zhang, P.: A proactive stable scheme for vehicular collaborative edge computing. IEEE Trans. Veh. Technol. **72**(8), 10724–10736 (2023)

# Fuzzy System for Norovirus Transmission Risk Assessment on Cruise Ships Based on Sensor Data

Georgia Sovatzidi, Georgios Triantafyllou, and Dimitris K. Iakovidis(✉)

Department of Computer Science and Biomedical Informatics, University of Thessaly, 35131 Lamia, Greece
{gsovatzidi,gtriantafylloy,diakovidis}@uth.gr

**Abstract.** Cruise ships are a popular means of vacation for millions of people every year. However, they present significant challenges, especially in terms of addressing the risk of outbreaks on board. Among various pathogens, norovirus is the most prevalent cause of gastrointestinal illness on cruise ships, posing serious health risks to both passengers and crew. Onboard recreational facilities such as swimming pools and pool parties can create conditions where standard hygiene protocols are not consistently enforced, increasing the risk of norovirus transmission. Motivated by these challenges, this paper presents a novel short-term risk assessment system for norovirus transmission in swimming pools of cruise ships using data from multiple sensors, and a fuzzy inference model. Traditional data-driven models are typically limited by the unavailability of training data in such environments. Instead, the proposed approach adopts a knowledge-based design, utilizing domain expertise and literature data, while incorporating sensor-driven inputs, in alignment with a recently introduced smart ship design. The system was validated using a real documented outbreak and tested under various simulated scenarios of a swimming pool derived from cruise ship schematics, demonstrating its practical relevance and adaptability across configurations.

**Keywords:** Norovirus · Waterborne disease · Risk assessment · Fuzzy logic · Knowledge-based system · Sensor-driven system

## 1 Introduction

Norovirus is a leading cause of gastroenteritis outbreaks on cruise ships, where confined spaces and shared amenities facilitate rapid transmission [1]. Despite the fact that transmission of infectious agents through water is generally rare under normal circumstances, norovirus outbreaks on cruise ships are a significant public health concern [1, 2]. Moreover, norovirus is highly contagious, as even asymptomatic individuals can shed large quantities of the virus, with studies reporting that a single gram of feces may contain up to 100 billion viral particles [3], while only a few particles are sufficient to cause an infection [4]. Therefore, asymptomatic cases can be detrimental to the spread of a highly infectious virus such as norovirus, since they might not get detected by ship personnel [5].


P. Pichappan et al. (Eds.): RTIS 2025, LNNS 1936, pp. 422–433, 2027.
https://doi.org/10.1007/978-3-032-23396-7_38

Cruise ships include a variety of recreational activities, such as pool parties, that can create situations where standard safety protocols may not always be strictly followed [6]. Recreational water treated with chlorine levels above 1 mg/L can eliminate more than 99% of norovirus particles within seconds [7]. Nevertheless, outbreaks of norovirus have been attributed to chlorine levels below the safety guidelines that are designated by the Center for Diseases Control (CDC), such as in [8]. In this incident, reduced chlorine levels can lead to a norovirus outbreak caused by an asymptomatic case. Specifically, the chlorine concentration dropped to a critical level ($<1$ mg/L), placing it in a "gray zone" where disinfection proved ineffective, allowing the virus to thrive and spread [8]. Considering the above, it can be inferred that a smart system capable of monitoring the presence of waterborne infectious diseases in swimming pools could provide valuable assistance in reducing the risk of disease transmission onboard. Smart technologies, including smart sensors, have been proven to be useful tools for monitoring the transmission of infectious diseases [9]. These technologies can be utilized to identify early symptoms of a disease and automate the monitoring process, while providing valuable information to ship crew and manufacturers. This results in more informed and adaptive policy-making when considering safety guidelines related to waterborne diseases.

Data-driven approaches, such as machine learning models, are commonly used in health-related risk prediction tasks, yet they require extensive training datasets. As such, their application in assessing short-term disease transmission scenarios on cruise ships is constrained by the limited availability of outbreak data. In contrast, knowledge-based approaches offer a viable alternative in data-scarce environments. These methods leverage domain expertise, *e.g.*, from domain experts and/or literature-derived insights to support decision-making and can be further enhanced through the integration of sensor-driven data, providing a more context-aware basis for risk assessment. Fuzzy logic is a powerful tool for handling uncertainty, incomplete data, and imprecise medical information [10]. Only a limited number of studies have been presented aiming to deal with waterborne infectious diseases using fuzzy approaches. Recent paradigms include, an adaptive neuro-fuzzy inference system was proposed for predicting norovirus in drinking water supply, based on water quality parameters such as water pH, turbidity, conductivity, temperature and rain [11]. A multi-criteria decision-making model that integrates Fuzzy Analytic Hierarchy Process (FAHP) with geospatial analysis was presented aiming to identify zones vulnerable to waterborne diseases [12]. However, the application of fuzzy logic to short-term disease transmission on cruise ships remains an open challenge. To this end, considering the need for early prevention of norovirus spread in crowded environments with water facilities, such as in cruise ships, a novel system based on sensor data and knowledge-based fuzzy reasoning is proposed.

The system aims to assess the short-term transmission risk of norovirus in swimming pools, which can be very crowded, especially during the summertime. The proposed system utilizes data derived from different smart sensors, *e.g.*, RNA sensors and chlorine sensors, to assess the risk of waterborne diseases. These sensors are integrated into a novel smart ship design proposed in the context of the project HS4U, which investigates methods for smart ship design, to provide enhanced passengers' health safety. The proposed fuzzy rule-based system considers several risk factors based on data provided by

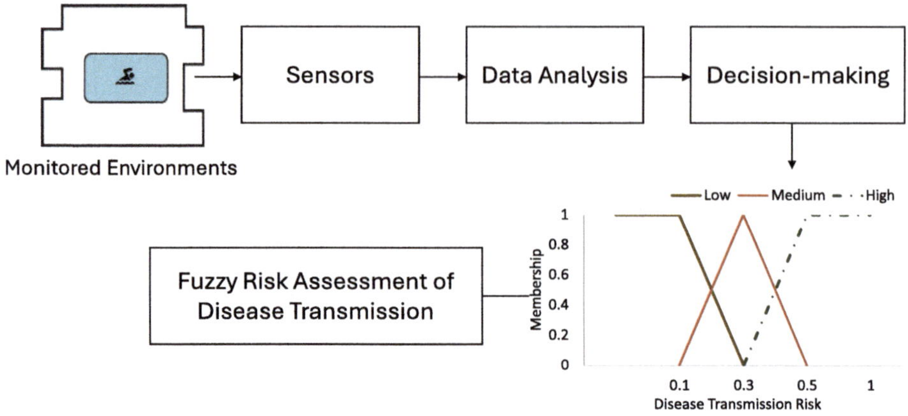


**Fig. 1.** Overview of the proposed system.

available smart sensors and information derived from the literature [9]. Compared to standard epidemiological studies that focus on long-term disease transmission and generic population dynamics [13, 14], this study analyzes the short-term effects of identified risk factors using a sensor-driven, fuzzy inference system, tailored to context-specific and operationally relevant conditions. To the best of our knowledge, this is the first time that a fuzzy rule-based system combined with smart sensors has been used to assess the risk of waterborne disease transmission in swimming pools. The main contributions of this study include the following:

- A novel knowledge-based system assessing the short-term risk of waterborne diseases in swimming pools
- Identification of risk factors that are early indicators of waterborne infectious diseases and analysis of their impact on short-term transmission.
- Experiments for validating the proposed system on a real-case study and for different scenarios with various swimming pool configurations.

The remainder of this paper is organized into three sections. The proposed system is described in Sect. 2. Section 3 includes the experimental setup, and the experiments conducted in this research are presented in Sect. 4. Finally, conclusions as well as future research directions are discussed in Sect. 5.

## 2 Methods

An overview of the proposed system is presented in Fig. 1. As illustrated, sensors, such as RNA sensors and chlorine sensors, are utilized to monitor areas of interest within the ship (monitored environments). The system is also linked with the ship's information system that can provide further details of the passengers occupying each monitored environment. The collected data are analyzed, and the decision-making process is then performed, aiming to infer the risk of waterborne disease transmission.

The resulting waterborne disease transmission risk is then linguistically characterized using defined fuzzy sets, as presented in this section.

**Table 1.** Risk factor analysis and their impact on waterborne disease transmission.

| Risk Factors | Impact on Transmission | |
|---|---|---|
| 1. Chlorine Levels (mg/L) | Lower ↑ | Higher ↓ |
| 2. Median Age (years) | Lower ↑ | Higher ↓ |
| 3. Occupancy (number of passengers) | Lower ↓ | Higher ↑ |
| 4. Time of Exposure (min) | Lower ↓ | Higher ↑ |

↓ Reduces risk, ↑Increases risk

## 2.1 Risk Factor Analysis

Based on the available sensors included in the smart ship design proposed in [9], and information retrieved from the literature [15–18], the following risk factors were identified: i) chlorine levels (CL), ii) median age (MA), iii) time of exposure (ET) of passengers to the virus, and iv) occupancy (OCC), *i.e.*, number of passengers in a specific area. These factors constitute important indicators of norovirus transmission in water [18] and their impact is summarized in Table 1.

Chlorine is the most important countermeasure against waterborne disease transmission in swimming pools. According to the literature, 99.2% of norovirus particles are eliminated after 4.2 s of contact with treated water that contains chlorine levels >1 mg/L. However, concentrations below 1 mg/L have not been thoroughly investigated and, according to [8], are considered ineffective in preventing the transmission of highly infectious diseases like norovirus in swimming pool environments. In the investigated outbreak [8], chlorine levels were reported to vary between 0.5 and 1mg/L, which were designated as a gray area. In addition, the CDC has also designated as safe, chlorine levels between 1 and 3 mg/L [17]. Chlorine concentrations above 3 mg/L are considered more effective at eliminating waterborne pathogens; however, levels exceeding 5 mg/L are deemed hazardous to human health [17]. Furthermore, the age of the passengers is correlated with the risk of waterborne disease transmission. Based on the outbreak analyzed in [8], the risk of infection was notably higher among individuals under the age of 16. Moreover, several studies have also highlighted that children are more susceptible to water-mediated infections, since they consume more water while swimming, compared to adults [16, 18]. Furthermore, the exposure time of individuals in waters that are contaminated with pathogens is associated with higher risk of infection [15]. Therefore, in the cases of norovirus transmission, the longer the exposure of the individuals the higher the transmission risk of the virus. In addition, the occupancy, *i.e.*, the total number of passengers, is also a risk factor when investigating water-mediated transmission of infectious disease. In detail, the World Health Organization (WHO) has designated specific guidelines to ensure safety and suggests a maximum capacity of bathers for swimming pools. The WHO dictates the maximum capacity of a pool can be calculated based on the surface area and depth of the pool as follows [7]:

$$B = \frac{L * W}{b_l} \tag{1}$$

| No | CL | MA | OCC | ET | $R_f$ |
|---|---|---|---|---|---|
| 1 | LOW | LOW | LOW | LOW | MEDIUM |
| 2 | LOW | LOW | LOW | LOW | MEDIUM |
| 3 | LOW | LOW | LOW | LOW | MEDIUM |
| 4 | LOW | LOW | LOW | MEDIUM | HIGH |
| 5 | LOW | LOW | LOW | MEDIUM | HIGH |
| 6 | LOW | LOW | MEDIUM | LOW | MEDIUM |
| 7 | LOW | LOW | MEDIUM | LOW | MEDIUM |
| 8 | LOW | LOW | MEDIUM | LOW | MEDIUM |
| 9 | LOW | MEDIUM | LOW | LOW | MEDIUM |
| 10 | LOW | MEDIUM | LOW | LOW | MEDIUM |
| 11 | LOW | MEDIUM | LOW | LOW | MEDIUM |
| 12 | LOW | MEDIUM | MEDIUM | MEDIUM | HIGH |
| 13 | LOW | MEDIUM | MEDIUM | MEDIUM | HIGH |
| 14 | LOW | MEDIUM | MEDIUM | MEDIUM | HIGH |
| 15 | LOW | MEDIUM | MEDIUM | HIGH | HIGH |
| 16 | LOW | MEDIUM | MEDIUM | HIGH | HIGH |
| 17 | LOW | MEDIUM | MEDIUM | HIGH | HIGH |
| 18 | LOW | MEDIUM | HIGH | MEDIUM | HIGH |
| 19 | LOW | MEDIUM | HIGH | MEDIUM | HIGH |
| 20 | LOW | MEDIUM | HIGH | MEDIUM | HIGH |
| 21 | LOW | MEDIUM | HIGH | MEDIUM | HIGH |
| 22 | LOW | MEDIUM | HIGH | HIGH | HIGH |
| 23 | LOW | MEDIUM | HIGH | HIGH | HIGH |
| 24 | LOW | MEDIUM | HIGH | HIGH | HIGH |
| 25 | LOW | MEDIUM | HIGH | HIGH | HIGH |
| 26 | LOW | MEDIUM | HIGH | HIGH | HIGH |
| 27 | LOW | HIGH | MEDIUM | MEDIUM | MEDIUM |
| 28 | LOW | HIGH | MEDIUM | MEDIUM | MEDIUM |
| 29 | LOW | HIGH | MEDIUM | MEDIUM | MEDIUM |
| 30 | MEDIUM | LOW | LOW | LOW | LOW |
| 31 | MEDIUM | LOW | LOW | LOW | LOW |
| 32 | MEDIUM | LOW | LOW | LOW | LOW |
| 33 | MEDIUM | LOW | LOW | LOW | LOW |
| 34 | MEDIUM | LOW | LOW | MEDIUM | LOW |
| 35 | MEDIUM | LOW | LOW | MEDIUM | LOW |
| 36 | MEDIUM | LOW | MEDIUM | LOW | LOW |
| 37 | MEDIUM | LOW | MEDIUM | LOW | LOW |
| 38 | MEDIUM | LOW | MEDIUM | LOW | LOW |
| 39 | MEDIUM | MEDIUM | LOW | LOW | LOW |
| 40 | MEDIUM | MEDIUM | LOW | LOW | LOW |
| 41 | MEDIUM | MEDIUM | LOW | LOW | LOW |
| 42 | MEDIUM | MEDIUM | LOW | LOW | LOW |
| 43 | MEDIUM | MEDIUM | LOW | LOW | LOW |
| 44 | MEDIUM | MEDIUM | LOW | LOW | LOW |
| 45 | MEDIUM | MEDIUM | LOW | LOW | LOW |
| 46 | MEDIUM | MEDIUM | LOW | LOW | LOW |
| 47 | MEDIUM | MEDIUM | MEDIUM | MEDIUM | LOW |
| 48 | MEDIUM | MEDIUM | MEDIUM | MEDIUM | LOW |
| 49 | MEDIUM | MEDIUM | MEDIUM | MEDIUM | LOW |
| 50 | MEDIUM | MEDIUM | MEDIUM | MEDIUM | LOW |
| 51 | MEDIUM | MEDIUM | MEDIUM | MEDIUM | LOW |
| 52 | MEDIUM | MEDIUM | MEDIUM | MEDIUM | LOW |
| 53 | MEDIUM | MEDIUM | MEDIUM | HIGH | LOW |
| 54 | MEDIUM | MEDIUM | MEDIUM | HIGH | LOW |
| 55 | MEDIUM | MEDIUM | MEDIUM | HIGH | LOW |
| 56 | MEDIUM | MEDIUM | HIGH | MEDIUM | LOW |
| 57 | MEDIUM | MEDIUM | HIGH | MEDIUM | LOW |
| 58 | MEDIUM | MEDIUM | HIGH | MEDIUM | LOW |
| 59 | MEDIUM | MEDIUM | HIGH | HIGH | MEDIUM |
| 60 | MEDIUM | MEDIUM | HIGH | HIGH | MEDIUM |
| 61 | MEDIUM | MEDIUM | HIGH | HIGH | MEDIUM |
| 62 | MEDIUM | HIGH | LOW | LOW | LOW |
| 63 | MEDIUM | HIGH | LOW | LOW | LOW |
| 64 | MEDIUM | HIGH | MEDIUM | MEDIUM | LOW |
| 65 | MEDIUM | HIGH | MEDIUM | MEDIUM | LOW |
| 66 | MEDIUM | HIGH | MEDIUM | MEDIUM | LOW |

RISK LEVELS: LOW, MEDIUM, HIGH

**Fig. 2.** Fuzzy rules for estimation of waterborne disease transmission on cruise ships.

where $B$ is the maximum capacity of the pool, $L$ is the length of the pool, $W$ is the width of the pool. The bather load $b_l$ can be defined as follows:

$$b_l = \begin{cases} 2.2\,m^2\,per\,person, & for\,depth\; < 1.0\,m \\ 2.7\,m^2\,per\,person, & for\;1.0\; < depth\; < 1.5\,m \\ 4.0\,m^2\,per\,person, & for\,depth\; > 1.5\;m \end{cases} \tag{2}$$

The risk regarding the waterborne transmission for the validation scenarios is calculated based on (3).

$$R_{GT} = \frac{P}{T} \tag{3}$$

where $P$ are the number of infected passengers, and $T$ is the total number of passengers in a pool.

## 2.2 Decision-Making Process

The processed data are assessed using a decision-making module, which evaluates the waterborne disease transmission risk in the monitored environment based on the identified risk factors. Each monitored environment is then assigned with a risk level, such as "Low" (L), "Medium" (M), and "High" (H). In this paper, a decision-making system based on fuzzy logic is proposed, aiming to perform the risk assessment of norovirus in swimming pools of cruise ships. The proposed system is developed based on Mamdani fuzzy inference, which is used as a baseline yet effective methodology [19, 20]. Unlike other methods, the Mamdani's method does not require extensive data and provides explanatory results that can be easily understood.

To define the transmission risk based on Mamdani's method, the following process is performed; the input variables are initially fuzzified using membership functions designed considering domain knowledge, as described in the following section. This fuzzification process transforms crisp input values into degrees of membership, enabling the system to handle data uncertainty and imprecision. The relationships between the identified risk factors and the output risk are inferred, using 'IF-THEN' rules. In this study, the fuzzy rule base was constructed by systematically combining the selected input factors. Thus, a comprehensive set of fuzzy rules was generated through a combination of the considered factors as well as expert knowledge and information derived from literature. Finally, a total number of 66 rules have been elicited and empirically selected by domain experts (Fig. 2). The activation of these fuzzy rules depends on the specific input values and the logical operators. Each rule contributes to the generation of output membership functions, denoted as ($M$ ), which are aggregated and defuzzified, resulting in a crisp final output, using the following equation:

$$M_j = \vee_{i=1}^{n}(A(x) \wedge B(z) \wedge C(d) \wedge D(k_i)) \tag{4}$$

where "$\bigvee$" and "$\wedge$" correspond to the fuzzy union and intersection operations. Furthermore, $j = 1, \ldots, M$, where $M$ is the total number of activated fuzzy rules; the variables $x$, $z$, $d$, $k$ are the input risk factors, *i.e.*, chlorine levels, median age, time of exposure of passengers to the virus, and occupancy, respectively. Each variable is defined by a corresponding fuzzy set $A_i$, $B_i$, $C_i$, $D_i$, respectively, with $i = 1, \ldots, n$. An indicative example of a fuzzy rule is the following: "**IF** Chlorine Level is *Low* **AND** Median Age is *Low* **AND** Occupancy is *Low* **AND** Time of Exposure is *Low* **THEN** Risk is *Medium*."

## 3 Experimental Setup

To assess the risk of waterborne disease transmission, scenarios with sensors placed in pool areas (Subsect. 3.1) were considered. The effectiveness of the defined fuzzy rules was validated using a norovirus outbreak scenario reported in the literature (Subsect. 3.2). Furthermore, fuzzy sets were defined for the linguistic characterization of the inferred transmission risks (Subsect. 3.3).

### 3.1 Sensors

The deployment and impact of the sensors was simulated to evaluate the effectiveness of the proposed system, in different experimental scenarios. The simulation of sensor deployment was guided by cruise ship schematics to ensure a representative spatial configuration of pool areas. This approach reflects typical onboard conditions relevant to health risk assessment, while maintaining flexibility in how sensor integration could be realized in practice, and aligns with the broader smart ship design principles explored in the HS4U project [21]. Furthermore, the following three sensors were considered: a) an RNA sensor was deployed in the examined pool to detect the presence of pathogens, providing information regarding potential contamination events; b) a chlorine sensor was used to monitor the concentration levels of chlorine in the water; c) a Radio-Frequency Identification (RFID) sensor was implemented to collect information of passengers, including the number of people in a monitored environment, their time duration of exposure, and their median age. The combined use of these sensors allowed for a more comprehensive and effective analysis of the identified risk factors influencing short-term waterborne disease transmission.

### 3.2 Swimming Pool Scenarios

The generated fuzzy rule set was validated based on the norovirus outbreak reported in a swimming pool area described in [8]. The dimensions of the pool of this validation scenario were 25 m length, 12.5 m width and 1.2 m depth. Subsequently, the surface area of the pool was 312.5 $m^2$ and $b_l$ for a pool with 1.2 m depth is 2.7 $m^2$, thus using these parameters in (1), (2) the maximum capacity of the pool, in terms of number of passengers, is equal to 115.

The effectiveness of the proposed system was also evaluated in various test scenarios, using the same risk factors and a set of different pool characteristics, compared to the validation scenarios. Specifically, the size of the current pool was 16.5 m length, 8.5 m width and 1.6 m–1.8 m depth. Given that the surface area of the pool is 140.25 $m^2$ and the $bl$ for a pool with an average depth of 1.7 m is 4.0 $m^2$, the maximum capacity, based on (1) and (2), is calculated to be 35 passengers. The resulting fuzzy sets for the pools used in the experiments are presented in the following subsection.

### 3.3 Fuzzy Set Construction

Based on the analysis conducted in Subsect. 2.1, four risk factors were identified according to the available sensors and information provided by the considered smart ship design. Each risk factor is represented by fuzzy sets that correspond to linguistic values, such as "Low", "Medium", and "High". A higher degree of membership towards a specific fuzzy set signifies an increased or decreased risk of the transmission of waterborne disease, based on the risk factor analysis (II-A). To linguistically characterize the transmission risk, the fuzzy sets presented in Tables 2 and 3, were defined.

**Table 2.** Fuzzy sets of the risk factors for the pool of the validation scenarios.

| Fuzzy Set | CL | MA | OCC | ET | RF |
|---|---|---|---|---|---|
| Low | [0, 1] | [1, 33] | [2, 59] | [15, 45] | [0, 0.3] |
| Med | [0.5, 1.5] | [17, 49] | [30, 87] | [30, 60] | [0.1, 0.5] |
| High | [1, 3] | [33, 65] | [59, 115] | [45, 120] | [0.3, 1] |

**Table 3.** Fuzzy sets of the risk factors for the pool of the testing scenarios.

| Fuzzy Set | CL | MA | OCC | ET | $R_F$ |
|---|---|---|---|---|---|
| Low | [0, 1] | [1, 33] | [2, 12] | [15, 45] | [0, 0.3] |
| Med | [0.5, 1.5] | [17, 49] | [6, 23] | [30, 60] | [0.1, 0.5] |
| High | [1, 3] | [33, 65] | [12, 35] | [45, 120] | [0.3, 1] |

**Table 4.** Validation results based on the case study [8].

| No | CL | MA | OCC | ET | $R_F$ | $R_{GT}$ |
|---|---|---|---|---|---|---|
| 1 | 0.29 | 25 | 58 | 90 | 0.78 | 0.79 |
| | (L) | (L) | (M) | (H) | (H) | (H) |

# 4 Experiments and Results

## 4.1 Validation Scenarios

To evaluate the effectiveness of the defined fuzzy rules, different scenarios comprising various values of the identified risk factors were examined for the two types of swimming pools. Experiments were conducted for all combinations of the four risk factors, each characterized by three possible linguistic values, resulting in a total of 81 different scenarios. A representative subset of these scenarios is presented in Tables 5 and 6. The parentheses contain the respective linguistic values, indicating the membership of a risk factor to a respective fuzzy set (L)ow, (M)edium, or (H)igh (Tables 2 and 3). Using the outbreak data reported in [8] as a reference, the defined fuzzy rules were applied to validation scenarios; only the participants who attended the Halloween party were considered, as the scope of the present study is to assess the short-term risk of waterborne disease transmission. Subsequently, it was reported that 58 participants attended the Halloween party, which lasted 120 minutes. Out of the 58 participants, 46 were reported to be infected, resulting in a disease transmission risk of 0.79, using (3), linguistically characterized as "High" (Table 2). The median age of participants was equal to 25, and during that time the chlorine levels were reported to be 0.29 mg/L. Furthermore, participants spent 60 to 120 minutes inside the pool, thus a median time of 90 minutes of exposure time was considered. Based on Table 4, it can be observed that for a chlorine level of 0.29 mg/L, a median age of 25, occupancy of 58 participants, and 90 minutes of exposure, the system inferred a risk equal to $R_F = 0.78$ ("High"), which was in close agreement with the ground truth, *i.e.*, $R_{GT} = 0.79$ ("High"). Additional experiments were conducted to observe the response of the system for other scenarios of detection of norovirus in the same swimming pool (Table 5). Regarding the chlorine level as a risk factor, its correct regulation significantly reduces the risk of disease transmission [8], as also confirmed by the experiments conducted in this study.

**Table 5.** Risk Assessment of the norovirus disease transmission on cruise ships (validation scenarios).

| No | CL | MA | OCC | ET | $R_F$ |
|---|---|---|---|---|---|
| 1 | 0.50 (L) | 20 (L) | 100 (H) | 80 (H) | 0.76 (H) |
| 2 | 3 (M) | 5 (L) | 55 (M) | 20 (L) | 0.15 (L) |
| 3 | 3 (M) | 55 (H) | 55 (M) | 20 (L) | 0.15 (L) |
| 4 | 0.70 (L) | 40 (M) | 102 (H) | 110 (H) | 0.71 (H) |
| 5 | 0.80 (M) | 30 (M) | 34 (L) | 110 (H) | 0.53 (H) |
| 6 | 0.70 (L) | 40 (M) | 102 (H) | 40 (M) | 0.62 (H) |
| 7 | 0.80 (M) | 30 (M) | 34 (L) | 45 (M) | 0.19 (L) |
| 8 | 0.70 (L) | 40 (M) | 102 (H) | 20 (L) | 0.59 (H) |
| 9 | 0.80 (M) | 30 (M) | 34 (L) | 20 (L) | 0.14(L) |
| 10 | 0.50 (L) | 20 (L) | 100 (H) | 20 (L) | 0.59 (H) |

**Table 6.** Risk Assessment of the norovirus disease transmission in a swimming pool of a cruise ship (testing scenarios).

| No | CL | MA | OCC | ET | $R_F$ |
|---|---|---|---|---|---|
| 1 | 0.50 (L) | 20 (L) | 21 (M) | 80 (H) | 0.76 (H) |
| 2 | 3 (M) | 5 (L) | 21 (M) | 20 (L) | 0.15 (L) |
| 3 | 3 (M) | 65 (H) | 10 (L) | 20 (L) | 0.15 (L) |
| 4 | 0.70 (L) | 40 (M) | 9 (L) | 110 (H) | 0.57 (H) |
| 5 | 0.80 (L) | 30 (M) | 31 (H) | 110 (H) | 0.59 (H) |
| 6 | 0.90 (M) | 10 (L) | 15(M) | 45 (M) | 0.26 (M) |
| 7 | 0.92 (M) | 30 (M) | 15 (M) | 45 (M) | 0.15 (L) |
| 8 | 0.50 (L) | 20 (L) | 21 (M) | 20 (L) | 0.53 (H) |
| 9 | 1 (M) | 5 (L) | 21(M) | 40 (M) | 0.21 (M) |
| 10 | 1( M) | 65 (H) | 21 (M) | 40(M) | 0.19(L) |

Specifically, adequate chlorine levels of 3 mg/L result to a "Low" risk of transmission, even in scenarios of a "Medium" pool occupancy with passengers of a "Low" median age (scenario no. 2), factors that have been shown to be associated with a "High" risk of transmission (Table 1). In addition, a "Low" median age increases the transmission risk especially Furthermore, "High" exposure time result to "High" transmission risk (scenario no. 5), especially when combined with "High" occupancy (scenario no. 1 and 4).

### 4.2 Experiments on Swimming Pools of Cruise Ships

Multiple experiments were also conducted on a ship pool of a cruise ship, described in III-B, using various risk factor combinations (Table 6). The experiments revealed that "Low" chlorine levels consistently resulted in "Medium" and "High" risk values of disease transmission, even when the median age was "Low" or "Medium" (no 1, 4). For "Medium" chlorine levels, the system inferred "Low" risk value of waterborne disease transmission (no. 2, 3). For scenarios with improperly regulated chlorine levels, such as "Low" and "High" chlorine levels (no. 1, 4), norovirus transmission is "High". Furthermore, "High" exposure time result to "High" waterborne transmission risk (no. 4, 5). By comparing the results between the validation pool and the ship pool, we can see that the proposed fuzzy rule-based approach adapts well to different physical settings. The experiments in the ship's swimming pool further emphasize that the system is robust across varying pool sizes and configurations, providing a practical tool for real-time risk management. The results of this study are consistent with known epidemiological patterns of waterborne disease transmission reported in the literature [6, 13], highlighting the effectiveness of the proposed system. Specifically, as presented in Tables 5 and 6, the risk increases with the increase in the duration of passengers spent in the contaminated water (time exposure) and the occupancy of the area, whereas the transmission risk can be decreased as the chlorine levels and the median age are increased.

## 5 Discussion and Conclusions

This paper presented a fuzzy risk assessment system based on sensor data for short-term norovirus transmission on cruise ships. The proposed rule-based system considered several risk factors based on data provided by available smart sensors, such as RNA and RFID sensors, in alignment with a recently introduced smart ship design [21]. To address the challenges of imprecise and incomplete data, the system incorporated information from domain experts and the literature, to predict the short-term transmission risk of norovirus onboard, using fuzzy rules. The experiments were conducted in two different pools with various characteristics, and the results were validated using a case study of an epidemic outbreak reported in the literature. Furthermore, the results confirmed the importance of the identified risk factors for the transmission of waterborne diseases on board.

Future work can focus on incorporating additional risk factors beyond waterborne exposure, such as fomite transmission, aerosol spread, and direct person-to-person contact. Moreover, although simulation-based validation offers controlled insights, real-world studies involving sensor deployments and pilot testing on operational ships will be essential to account for environmental noise, calibration errors, and potential sensor failures. Thus, future work may also aim to the integration of the proposed sensor-driven system into a real operational cruise ship setting to support real-time monitoring and automated health safety interventions. Such integration could also facilitate future extensions of the proposed fuzzy knowledge-based system that may also incorporate data-driven components, enabling hybrid modeling of complex transmission scenarios based on large-scale datasets. Privacy and logistical issues arising in large-scale implementations should also be considered.

**Acknowledgment.** This paper is supported by the European Union's Horizon Europe Research and Innovation Actions programme under grant agreement No 101069937, project name: HS4U (Healthy Ship 4U). Views and opinions expressed are those of the author(s) only and do not necessarily reflect those of the European Union or the European Climate, Infrastructure, and Environment Executive Agency. Neither the European Union nor the granting authority can be held responsible for them (https://hs4u.eu/).

## References

1. Fisher, J.J., Almanza, B.A., Behnke, C., Nelson, D.C., Neal, J.: Norovirus on cruise ships: motivation for handwashing? Int. J. Hospitality Manage. **75**, 10–17 (2018)
2. Mouchtouri, V.A., et al.: Systematic literature review and meta-analysis on preventing and controlling norovirus outbreaks on cruise ships, 1990 to 2020: calling for behaviour change strategies of travellers. Eurosurveillance **29**(10), 2300345 (2024)
3. Atmar, R.L., et al.: Norwalk virus shedding after experimental human infection. Emerg. Infect. Dis. **14**(10), 1553 (2008)
4. Hall, A.J.: Noroviruses: the perfect human pathogens? J. Infect. Dis. **205**(11), 1622–1624 (2012)
5. Wang, J., Gao, Z., Yang, Z., Liu, K., Zhang, H.: Global prevalence of asymptomatic norovirus infection in outbreaks: a systematic review and meta-analysis. BMC Infect. Dis. **23**(1), 595 (2023)
6. Bonadonna, L., La Rosa, G.: A review and update on waterborne viral diseases associated with swimming pools. Int. J. Environ. Res. Publ. Health **16**(2), 166 (2019)
7. W.H. Organization and others, Guidelines for safe recreational water environments. Volume 2: Swimming pools and similar environments. World Health Organization (2006)
8. Paranthaman, K., Pringle, E., Burgess, A., Macdonald, N., Sedgwick, J.: An unusual outbreak of norovirus associated with a Halloween-themed swimming pool party in England, 2016. Eurosurveillance **23**(44) (2018)
9. Triantafyllou, G., Kalozoumis, P.G., Cholopoulou, E., Iakovidis, D.K.: Disease spread control in cruise ships: monitoring, simulation, and decision making. In: The Blue Book: Smart Sustainable Coastal Cities and Blue Growth Strategies for Marine and Maritime Environments, pp. 93–141. Springer (2024)
10. Jiang, Q., et al.: Intelligent monitoring for infectious diseases with fuzzy systems and edge computing: a survey. Appl. Soft Comput. **123**, 108835 (2022)
11. Mohammed, H., Hameed, I.A., Seidu, R.: Comparison of adaptive neuro-fuzzy inference system (ANFIS) and Gaussian process for machine learning (GPML) algorithms for the prediction of norovirus concentration in drinking water supply. In: Transactions on Large-Scale Data-and Knowledge-Centered Systems XXXV, pp. 74–95 (2017)
12. Ajmal, U., Jamal, S., Ahmad, W.S., Ali, M.A., Ali, M.B.: Waterborne diseases vulnerability analysis using fuzzy analytic hierarchy process: a case study of Azamgarh city, India. Model. Earth Syst. Environ. **8**(2), 2687–2713 (2022)
13. Rooney, R.M., et al.: A review of outbreaks of waterborne disease associated with ships: evidence for risk management. Publ. Health Rep. **119**(4), 435–442 (2004)
14. Mari, L., Casagrandi, R., Bertuzzo, E., Rinaldo, A., Gatto, M.: Conditions for transient epidemics of waterborne disease in spatially explicit systems. Roy. Soc. Open Sci. **6**, 181517 (2019)
15. Pintar, K., Fazil, A., Pollari, F., Charron, D., Waltner-Toews, D., McEwen, S.: A risk assessment model to evaluate the role of fecal contamination in recreational water on the incidence of cryptosporidiosis at the community level in Ontario. Risk Anal.: Int. J. **30**(1), 49–64 (2010)

16. Schets, F.M., Schijven, J.F., de Roda Husman, A.M.: Exposure assessment for swimmers in bathing waters and swimming pools. Water Res. **45**(7), 2392–2400 (2011)
17. Sheet, F.: Operating Public Swimming Pools (2024)
18. Arias, C., et al.: Epidemiological and clinical features of norovirus gastroenteritis in outbreaks: a population-based study. Clin. Microb. Infect. **16**(1), 39–44 (2010)
19. Ahmad Shukri, F.A., Isa, Z.: Experts' judgment-based Mamdani-type decision sys-tem for risk assessment. Math. Prob. Eng. **2021**, 1–13 (2021)
20. Mamdani, E.H., Assilian, S.: An experiment in linguistic synthesis with a fuzzy logic controller. Int. J. Man Mach. Stud. **7**(1), 1–13 (1975)
21. Rassia, S.T.: The Blue Book: Smart Sustainable Coastal Cities and Blue Growth Strategies for Marine and Maritime Environments. Springer (2024)

# AI-Driven Literary Research Analytics: Building a Digital Repository of Indonesian Scholarly Trends (2000–2024)

Mohammad Rokib(✉), Parmin, Moh. Arif Susanto, Syamsul Sodiq, Diding W. Rohaedi, Moh. Mudzakkir, and Indar Sabri

Universitas Negeri Surabaya, Surabaya 60213, Indonesia
{mohammadrokib,parmin,mohsusanto,syamsulsodiq,didingrohaedi, mudzakkir,indarsabri}@unesa.ac.id

**Abstract.** The exponential growth of academic publications in the humanities necessitates intelligent systems for tracking, analyzing, and interpreting research trends. This study introduces a cutting-edge, AI-driven framework for processing bibliographic and thematic metadata from over two decades (2000–2024) of scholarly work in Indonesian literary studies. The dataset comprises a curated corpus of theses, dissertations, peer-reviewed articles, and conference proceedings sourced from institutional repositories and national indexing services. The system employs state-of-the-art machine learning algorithms and natural language processing (NLP) to automate the classification of research topics, extract methodological patterns, and map epistemological paradigms. Key techniques include topic modelling (LDA), hierarchical clustering, keyword co-occurrence networks, and temporal trend detection. All outputs are integrated into a dynamic digital repository, featuring interactive dashboards, visual analytics, and exportable datasets for further scholarly use. Results reveal dominant themes—such as postcolonialism, gender, and eco-criticism—shifts in methodological preferences (e.g., from structuralism to discourse analysis), and evolving paradigmatic alignments across institutions and periods. The system not only uncovers hidden structures in literary scholarship but also serves as a meta-research tool for policy, curriculum development, and academic planning. By bridging AI-based data analytics with humanities research, this project advances the field of digital humanities in Southeast Asia and contributes to broader global efforts in computational cultural analytics. The framework is modular, scalable, and transferable to other linguistic or regional literatures, offering a replicable model for data-driven meta-research in the Global South.

**Keywords:** AI-driven Research Analytics · Digital Literary Studies · Digital Humanities · Scholarly Trend Mapping · *Sastra Indonesia* · Data Visualization

## 1 Introduction

The application of AI and machine learning in digital humanities has gained significant momentum, especially for bibliometric and meta-research analysis [1]. However, in Southeast Asia, notably in Indonesia, scholarly meta-analysis remains underrepresented.


P. Pichappan et al. (Eds.): RTIS 2025, LNNS 1936, pp. 434–440, 2027.
https://doi.org/10.1007/978-3-032-23396-7_39

While such studies have tracked trends in digital humanities using bibliometric tools like VOSviewer, focusing on large-scale, interdisciplinary domains, systematic analysis of Indonesian literary scholarship metadata from 2000 to 2024 is still lacking [2].

This study addresses that gap by introducing an AI-driven system designed to process heterogeneous metadata drawn from Indonesian literary research. Indonesian academic publications often suffer from disparity and lack of standardization—spanning theses, dissertations, journal articles, and conference papers—with inconsistent metadata quality, missing abstracts, and disparate formats across platforms. Moreover, while tools such as MITAO enable non-technical scholars to apply topic modeling in humanities, they have not been tailored to the Indonesian scholarly landscape.

The primary objective of this study is to develop a scalable, AI-powered digital repository that systematically ingests metadata from scholarly publications in Indonesian literary studies spanning 2000 to 2024. The proposed system integrates structured natural language processing (NLP) and machine learning pipelines to extract quantitative insights on thematic concentrations, methodological approaches, and epistemological paradigms in literary research. Specifically, the framework automates metadata processing using state-of-the-art techniques such as Latent Dirichlet Allocation (LDA) for topic modeling, hierarchical clustering for methodology classification, and keyword co-occurrence network analysis—adapted for both Bahasa Indonesia and English metadata sources.

Through this framework, the platform is designed to analyze three critical dimensions: dominant research topics, prevailing methodological preferences, and paradigm shifts over time. In addition, the system supports interactive dissemination through a web-based interface featuring dashboards, visual analytics, and downloadable data summaries. These functions collectively address the absence of standardized infrastructures for tracking scholarly trends in the humanities, particularly within the Indonesian context.

This research contributes in three significant ways. First, it offers a robust AI-driven framework that resolves metadata heterogeneity using scalable and reproducible data processing methods. Second, it delivers the first comprehensive and explainable mapping of Indonesian literary meta-research over a 24-year period. Third, it advances digital humanities infrastructure by offering visualization tools that facilitate data-informed insights for academic researchers, curriculum planners, and institutional policy makers.

## 2 Related Work

Meta-research in the humanities and literary studies has gained increasing scholarly attention with the emergence of bibliometric and data-driven approaches. Judijanto *et al.* [3] conducted a bibliometric analysis of digital humanities research between 2000 and 2024, identifying dominant global themes such as digital libraries, big data, and artificial intelligence, while also noting the technological and infrastructural disparities in developing regions. In the Indonesian context, Mardiani *et al.* [4] utilized bibliometric techniques to examine the influence of digital technologies on arts and culture scholarship, revealing fragmented coverage and limited regional representation in research outputs.

Natural Language Processing (NLP) and Machine Learning (ML) techniques have been increasingly applied to bibliometric data, providing novel tools for classification, entity recognition, and semantic clustering. Suissa *et al.* [5] demonstrated the efficacy of deep neural networks for genre detection and text classification in digital humanities corpora, while Liu *et al.* [6] emphasized the transformative impact of generative AI models—such as ChatGPT and DALL·E 3—on document preservation and analysis in the humanities. These studies underscore the growing intersection between computational technologies and scholarly metadata processing.

Despite the global momentum, digital humanities research in Southeast Asia remains underdeveloped. Suryani *et al.* [7] conducted a bibliometric exploration of digital humanities in education and found minimal contributions from Southeast Asian scholars, despite rapid growth in the field since 2018. Other studies identified critical gaps in institutional readiness for digital humanities data curation in Indonesia, including the absence of integrated repository infrastructure and the lack of technical expertise.

However, no prior studies to date have proposed or implemented a fully data-intensive, AI-powered framework specifically tailored to analyze Indonesian literary research metadata. There is a notable absence of comprehensive systems that apply topic modeling, hierarchical clustering, or co-occurrence network analysis to map research trends, methodologies, and paradigms in Indonesian literary studies. This gap highlights the urgency and significance of the present study, which aims to fill this void by developing a scalable, intelligent repository for the analysis and visualization of Indonesian literary research from 2000 to 2024.

## 3 Dataset and Data Collection

The primary data sources for this study include Indonesian institutional repositories—such as SINTA (Science and Technology Index), Garuda (Garba Rujukan Digital), and university libraries—as well as scholarly outputs comprising journal articles, undergraduate theses, master's dissertations, doctoral theses, and conference proceedings. SINTA and Garuda serve as national aggregators of academic metadata and are widely used within bibliometric research in Indonesia [2, 8, 9].

Data selection criteria span four essential dimensions: (1) publication date between 2000 and 2024; (2) relevance to Indonesian literary studies; (3) availability of essential metadata (i.e., title, abstract, and author); and (4) inclusion of methodological and paradigmatic descriptors when available. This ensures that the dataset is both temporally comprehensive and thematically focused on Indonesian literary scholarship.

The metadata schema extracted includes the following fields: Title, Abstract, Keywords, Methodology, Paradigm (if explicitly stated), Institution, and Year of Publication. Metadata extraction combines automated harvesting—using APIs or web scraping tools—and manual vetting, particularly for methodological and paradigmatic entries where explicit labeling is inconsistent. Similar metadata-based workflows have previously been applied in other Indonesian bibliometric studies[4].

This multi-tiered dataset construction strategy enables robust analysis and supports the deployment of advanced NLP and machine learning techniques in subsequent stages of the research pipeline.

# 4 Methodology

## 4.1 Preprocessing

The first phase involves metadata cleaning and standardization. All records are parsed into unified BibTeX/JSON formats, removing duplicates and normalizing field values (e.g., standardized institution na10mes). Similar practices have been implemented successfully in metadata-rich bibliometric workflows []. Tokenization and stemming of Bahasa Indonesia metadata are performed using established NLP libraries such as Sastrawi and NLTK to prepare textual elements for downstream analysis.

## 4.2 NLP and Machine Learning Techniques

Our system applies multiple state-of-the-art computational methods:

1. Topic Modeling (LDA): We utilize Latent Dirichlet Allocation to identify latent research themes across the corpus, leveraging its capacity to infer topic distributions [10, 11].
2. Hierarchical Clustering & K-Means: These clustering methods group documents based on methodological and paradigmatic similarity, enabling quantitative identification of patterns in research approaches.
3. Co-occurrence Network Analysis: Co-occurrence networks are built from keywords and methods to reveal semantic and conceptual linkages. Such networks have proven effective in mapping literary and social contexts.

Temporal Trend Analysis: We apply time-series models to track shifts in thematic and methodological prevalence over the 2000–2024 timeframe, supporting trend detection and academic forecasting.

## 4.3 Repository and Visualization System

Processed metadata and analytical outputs are stored in a NoSQL backend (MongoDB) complemented by PostgreSQL for structured querying. The front-end visualization suite is built using interactive tools like Plotly and D3.js, integrated into dashboards powered by Grafana or Streamlit. The system supports dynamic filtering of topics, method-based searches, and distribution tracking across institutions and years—characteristics that align with best practices for digital humanities metadata visualizations [12].

# 5 Result and Discussion

## 5.1 Trend Analysis of Literary Research in Indonesia

Our automated topic modeling and trend analysis over the 2000–2024 metadata corpus revealed that the most prominent research themes are postcolonial studies, gender studies, environmental/ecocriticism, and digital literature. A notable example is the rise of postcolonial ecocritical themes, as recent scholarship highlights the intersection of environmental and colonial discourse in Indonesian literature. The upward trend in gender studies is corroborated by bibliometric analyses that map growing attention to gender and spirituality in literary research. Meanwhile, studies on indigenous digital writing and the digitization of vernacular narratives have seen a sharp uptick since 2020.

Institutionally, major universities in Java—particularly Universitas Indonesia and Universitas Gadjah Mada—account for the highest volume of publications in Indonesian literary research, aligning with national patterns of scholarly production. Regional contributions, while increasing, remain significantly lower, indicating a centralized distribution of research activity.

### 5.2 Methodological and Paradigmatic Pattern

Our findings indicate a notable shift in methodological approaches. In the early 2000s, structuralism dominated literary analyses in Indonesia, with researchers focusing on close reading and textual structures. Since the mid-2010s, however, cultural studies and hermeneutics have gained prominence, reflecting scholars' increasing interest in socio-cultural and interpretive dimensions. More recently, from around 2020 onward, the use of qualitative digital humanities methodologies—including textual mining and digital archiving—has begun to emerge as a new paradigm.

Quantitatively, hierarchical clustering shows a strong alignment between institutions and methodological preferences. State universities with stronger Social Sciences and Humanities faculties tend to adopt cultural studies and hermeneutic frameworks, whereas digital humanities approaches are more concentrated in universities affiliated with centers for digital culture and Information Technology. Co-occurrence network graphs further highlight inter-method linkages, especially between gender, postcolonial, and ecocritical frameworks, suggesting interdisciplinarity in contemporary literary research.

### 5.3 Visualization and Repository Functionality

To ensure accessibility, interpretability, and interactivity of the processed data, the system integrates an interactive web-based dashboard that allows users to explore Indonesian literary research trends through visual analytics. The dashboard is built using Plotly Dash and Streamlit frameworks, with backend support from MongoDB and PostgreSQL, enabling real-time querying, filtering, and visualization of the full metadata corpus as in Fig. 1.

One of the core components is the Topic-Paradigm Connection, which visualizes the association among major research themes (e.g., postcolonialism, structuralism, environment, digital literature). This connection is represented through a chord diagram and bipartite network graph, where node thickness indicates the frequency of co-occurrence many major keywords. For example, feminism show strong links with all keywords except semiotics.

A co-occurrence network graph illustrates the most frequently paired keywords and methodological labels, generated using *NetworkX* and rendered through D3.js. The resulting network reveals conceptual clusters, such as the four-part relationship among all keywords, suggesting emerging interdisciplinary zones in Indonesian literary discourse. Clusters are color-coded and arranged using a force-directed layout for clarity and interaction.

Temporal analysis is delivered through interactive time-series plots, showing the distribution of specific research topics and paradigms over the 2000–2024 timeline.

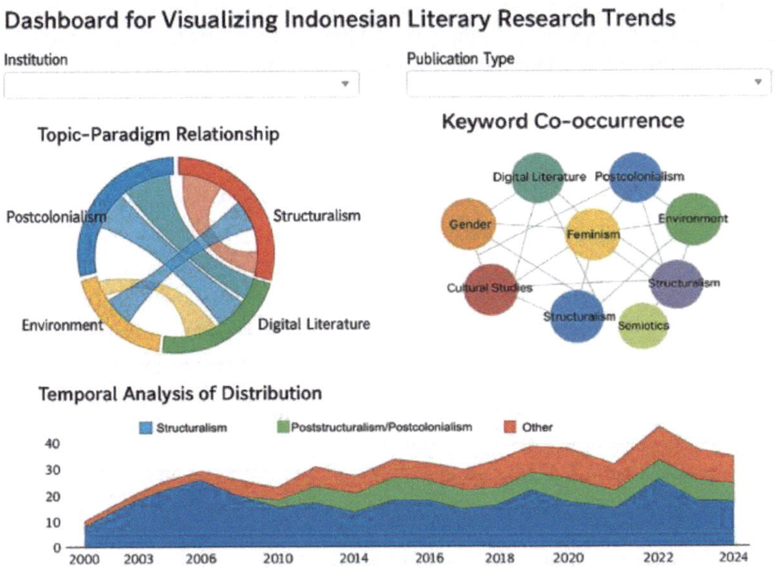


**Fig. 1.** Visualization of research paradigm and trend between 2000 and 2024

These visualizations use area charts and stacked bar graphs, allowing users to observe the relative rise or decline of approaches such as structuralism (declining after 2010) and other such as digital literature (increasing post-2020). Users can apply filters for institution, year range, or publication type (e.g., journal article vs. thesis) to refine the visualization outputs.

The repository also includes a searchable interface where users can explore publications based on thematic clusters, institutional affiliation, or methodological tags. Exportable reports (CSV, PDF) and embedded links to original documents (via Garuda or SINTA) further enhance usability for researchers and librarians. Together, these functionalities demonstrate the repository's potential as both a research tool and a meta-analytical infrastructure for tracking scholarly evolution in the Indonesian humanities.

## 6 Conclusion

This study presents an AI-driven framework for processing and analyzing Indonesian literary research metadata from 2000 to 2024, offering a scalable model for mapping scholarly trends in the humanities. By integrating machine learning and NLP techniques, the system enables the identification of thematic, methodological, and paradigmatic patterns and delivers them through an interactive digital repository designed for academic exploration and institutional insight. While the platform offers robust capabilities for visualizing and querying data, a key limitation lies in the inconsistent or implicit documentation of research methods and paradigms across many publications, which may affect analytical completeness.

**Acknowledgment.** This research was supported by the Faculty of Languages and Arts and the Center of Excellence for Arts and Culture (PUI Seni Budaya), Universitas Negeri Surabaya. The authors gratefully acknowledge their institutional support and resources provided for the development of this project.

## References

1. Spinaci, G., Colavizza, G., Peroni, S.: A map of Digital Humanities research across bibliographic data sources. Digital Sch. Humanit. **37**(4), 1254–1268 (2022)
2. Laksmi, L., Suhendra, M.F., Shuhidan, S.M., Umanto, U.: The readiness to implement digital humanities data curation of four institutional repositories in Indonesia. Digit. Libr. Perspect. **40**(1), 80–95 (2024)
3. Judijanto, L., Rohmah, S., Salfin, S.: Analisis bibliometrik perkembangan riset digital humanities. Sanskara Ilmu Sosial dan Humaniora **2**(01), 43–54 (2024)
4. Mardiani, E., Rukmana, A.Y., Maqfirah, P.A.-V., Nuswantoro, P., Uhai, S.: Bibliometric study on the influence of digital technology in the field of arts and culture. Eastasouth J. Soc. Sci. Humanit. **1**(02) (2024)
5. Suissa, O., Elmalech, A., Zhitomirsky-Geffet, M.: Text Analysis Using Deep Neural Networks in Digital Humanities and Information Science (2023)
6. Liu, J., Wang, Z., Xie, J., Pei, L.: From ChatGPT, DALL-E 3 to Sora: How has Generative AI Changed Digital Humanities Research and Services? (2024)
7. Suryani, I., Mustafa, W.A., Khamis, N., Warsidi, W., Zubir, F.: Two decade of scholarly landscape: a bibliometric exploration of digital humanities in education research. Int. J. Res. Innovation Soc. Sci. **IX**(IIIS), 290–302 (2025)
8. Tambunan, S.T.B., Widyaningdyah, A.U.: The journey of managerial accounting research in SINTA 4-Indexed journals. Jurnal Akuntansi Kontemporer **16**(1), 13–32 (2024)
9. Blei, D.M., Ng, A.Y., Jordan, M.I.: Latent Dirichlet Allocation. J. Mach. Learn. Res. **3**(Jan), 993–1022 (2003)
10. Jelodar, H., et al.: Latent Dirichlet allocation (LDA) and topic modeling: models, applications, a survey. Multimedia Tools Appl. **78**(11), 15169–15211 (2019)
11. Bornmann, L., Haunschild, R., Hug, S.E.: Visualizing the context of citations referencing papers published by Eugene Garfield: a new type of keyword co-occurrence analysis. Scientometrics **114**(2), 427–437 (2018)
12. Rokib, M., Supratno, H., Sodiq, S., Yuanita, A., Saing, M.E., Shofiyuddin, H.: Parmin: AI-Based Annotation for Learning Indonesian Internet Literature History. In: Proceedings of the 2024 IEEE Technology & Engineering Management Conference - Asia Pacific (TEMSCON-ASPAC), pp. 1–5 (2024)

# Designer Concept Based Classification for Explainable Artificial Intelligence

Andrew Swinn(✉) and Marek Grzes

School of Computing, University of Kent, Canterbury, UK
{a.swinn,m.grzes}@kent.ac.uk

**Abstract.** We propose a novel approach to explainable artificial intelligence (XAI) that uses concept channels to build neuro-symbolic classifiers. The concept channels use neural techniques to extract interpretable representations from raw data. Symbolic techniques are then used to reason over them. By doing this, we leverage the strengths of the two worlds, namely the ability of neural learning models to discover hidden patterns and to tolerate uncertainty in data, and the innate explainability and interpretability of symbolic programs.

This work focuses on the classification task and introduces a new concept-based neural architecture. Using background knowledge, neural channels are each trained to detect the presence (or absence) of a single concept. Together, they yield a concept probability vector which, when transformed by a class concept map, gives a class probability vector from which predictions can be made. Re-application of the trained neural channels to the training data offers additional background knowledge in the form of class concept mappings, as well as a set of concept probabilities over which Naïve Bayes reasoning is performed.

The transformation between concepts and classes bridges the gap between black box neural representations and the reasoning required to produce a classification. Furthermore, because each concept channel only detects one concept, it can produce its own relevance map showing where its concept was found.

## 1 Introduction

Artificial intelligence (AI) is transforming many aspects of our lives. AI is increasingly able to diagnose medical conditions as reliably as doctors [1], Large Language Models achieve remarkable performance on most natural language processing tasks, and a personalized e-learning experience can be delivered with minimal human input [2]. The success of AI is largely driven by neural networks, which are notoriously difficult to interpret or explain. However, explanations are essential in high-stakes or safety-critical domains, such as healthcare and autonomous vehicles [3].

Explainable AI has become an important branch of both fundamental and applied research, with many approaches and techniques used [4–6]. Interpretability enables the plausibility with which a classification or prediction can be


P. Pichappan et al. (Eds.): RTIS 2025, LNNS 1936, pp. 441–455, 2027.
https://doi.org/10.1007/978-3-032-23396-7_40

assessed. Without interpretability, the danger of misclassification is always present. Uninterpretable classifiers carry risk: making commercial or safety decisions learned from poor data or spurious examples raises the question of who is responsible if they get it wrong [5]? Without interpretability, it is difficult to trust a model, and if users do not trust a model or a prediction, they will not use it [7].

An approach to XAI is to design and build models that are explainable [8]. Such white or transparent box models including decision trees, linear models, and logic programs are considered explainable as long as their size is sufficiently small [4]. DeepProbLog [9] uses neural networks to learn the predicates for probabilistic Prolog. Another white box method with a symbolic component [10] embraced a rule-based interpretable classifier using Bayesian probabilistic analysis. These models provide transparency because it is possible to follow their decision-making processes.

In AI, human knowledge is often assumed or required in the form of theories [11] or heuristics [12]. Heuristics, designed using human understanding of a problem, can speed up convergence to a good or potentially optimal solution (e.g. admissible heuristics in A*), or well-vetted theories allow for answering causal questions from observational data [11]. In this paper, we show that a certain level of human input allows for accurate interpretability of machine learning models.

Deep neural networks are not symbolic, and there is no easy way to interpret their decision-making processes. The network itself is made up of a set of large multidimensional tensors that denote the weights associated with the edges that connect the nodes (artificial neurons). Accounting for individual weights, or explaining the function of any given node (or group of nodes), is the subject of ongoing research, e.g., using mechanistic interpretability [13].

Neural networks operate in high-dimensional vector spaces that cannot be immediately mapped to interpretable representations. It is difficult to know which features (pixels) of an image cause which activations in a network, and it is difficult to know which activations contribute to an overall classification. The problem is exacerbated by the existence of poly-semantic neurons which carry more than one type of knowledge, and because individual (groups of) features may be common across many classes. To illustrate, Fig. 1 gives 10 examples of digits in the MNIST digit test set that all have a common feature: the value of pixel $(14, 14) = 252$.

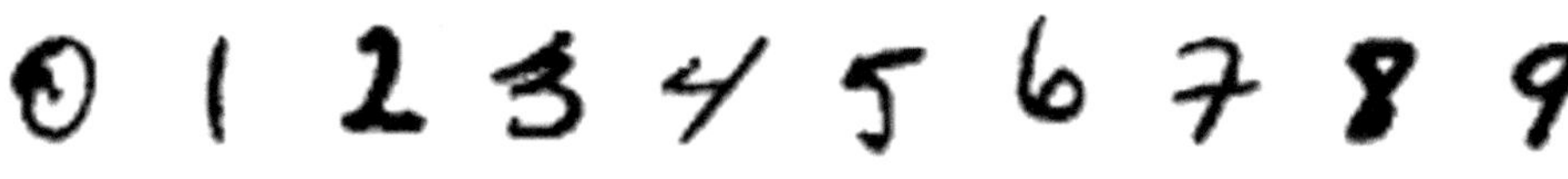

**Fig. 1.** Digits with a common pixel

It is the relationship between pixel values as much as the individual pixel values themselves that carry meaning; the information carried by an individual pixel depends on whether the pixel is part of a horizontal line or part of a vertical line.

This research offers an approach to classification that is designed to be, at least partly, explainable; it couples neural deep learning components with symbolic components to improve the quality of classification explanations. The symbolic components use human-understandable rules to reason over the outputs of the neural components. The neural components offer some degree of explainability by producing relevance maps and detecting concepts.

Our idea is to train a small neural model to map a highly dimensional image feature vector to a much lower dimensional concept vector and perform symbolic reasoning over this concept vector. Our approach differs from other concept-based approaches to explainability because we use human background knowledge to design the concepts, and then guide the neural components to detect them. This then leads to human-understandable reasoning. We also explore a sampling-based approach to see how human background knowledge performs against random background knowledge.

## 2 Related Work

The explainability problem arises because of the sub-symbolic representation of knowledge within a neural network. Artificial neural networks are trained by adjusting the weights between large numbers of neurons (nodes) in a network. Even small networks often contain large numbers of neurons; the comparatively small LeNet5 network [14] classified the MNIST digit dataset with a network composed of $6 \times 10^5$ nodes. Large language models have 10's or 100's of billions of nodes. The internal representation of an input vector is made up of the node activations across the network. With so many nodes, the mapping between the input vector, the internal representation, and the output classification is unintelligible to humans.

Several post hoc approaches have been proposed to explain how neural networks and other black box models make a certain decision. LIME [7] produces local linear explanations by computing a simple model that approximates the given model near a data point. SHAP [15] gives feature rankings based on their marginal contributions to a classification. There are also techniques for deep multi-layered neural networks that produce saliency maps highlighting which part of an image contributes most to a classification [16,17]. These techniques can be applied to existing pre-trained neural models but do not guide the models themselves to be inherently explainable.

A second set of approaches to explainability has been to integrate neural techniques with symbolic techniques to create an architecture which exploits the best characteristics of both types of classifiers. Symbolic techniques use explicit representations that are understandable by a human [18]. One symbolic technique is naïve Bayes reasoning which uses evidence, prior probabilities and likelihoods. A second symbolic approach is linear regression using data to find linear relationships. There are different neuro-symbolic architectures, including [Symbolic

→ Neural → Symbolic] in which neural processing is embedded within a symbolic system, and [Neural[Symbolic]] in which symbolic reasoning is embedded within the neural system. Two approaches to neuro-symbolic integration emerge [19]: learning for reasoning in which a neural component preprocesses input data to make it more suitable for symbolic reasoning (e.g., neural processing could reduce dimensionality before a symbolic tree search), and reasoning for learning in which symbolic preprocessing is performed to aid the neural learning. Typically, neuro-symbolic integration has four components [20], each of which is present in our approach:

1. translation of symbolic (background) knowledge into the network,
2. learning of additional knowledge from examples (and generalisation) by the network,
3. executing the network (i.e. reasoning),
4. symbolic knowledge extraction from the network.

Complimenting these approaches to explainability are a set of concept-based methods. Concepts are emerging as human understandable characteristics of highly dimensional feature vectors in which the individual features have no intrinsic meaning (illustrated by Fig. 1). The semantic meaning of such a data example is carried by human-understandable concepts, not by the machine-understandable features. A growing number of concept-based approaches are being explored [21]. They can be grouped together using three different types of concept-based explanations:

1. **class-concept** relations in which concepts (e.g. 'loop') are mapped to a subset of classes (e.g. a loop commonly found in the 0, 6, 8 and 9 classes of the MNIST Digits dataset) [22–25];
2. **node-concept** associations where individual network nodes are chosen to represent concepts (training them using concept loss and backpropagation) [26–29];
3. **concept visualizations** - using techniques like saliency maps applied to individual concepts[30–32].

Concept-based explanations can be defined as functions $g$ mapping machine explanations (vectors in a machine explanation space $E_m$) to human understandable explanations (vectors in the human explanation space $E_h$) [22]:

$$g : E_m \rightarrow E_h. \tag{1}$$

This definition and the fact that machine-based explanations and human-based explanations can be quite different reveals the difficulty in producing successful XAI. While machines can be very successful in imitating and even outperforming humans, they can do so based on evidence and reasoning that is completely different from that used by humans. By using such reasoning, they are vulnerable to confounding factors [11], as well as adversarial attacks (undetectable input perturbations designed to change a classification). Moreover, explanations based on such artificial reasoning can also be unnecessarily complex, making them less persuasive to a critical human audience.

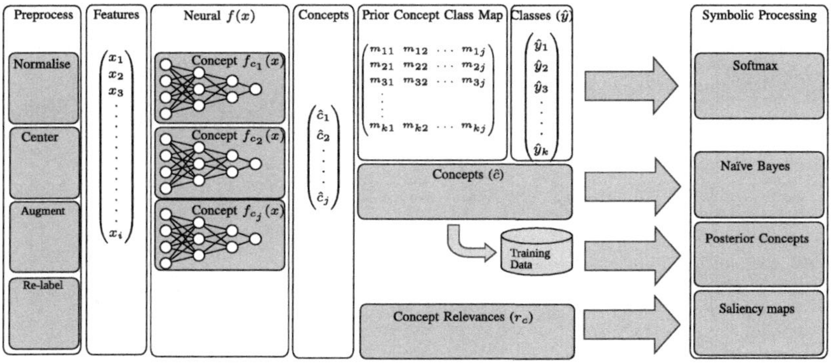


**Fig. 2.** Designer Concepts Architecture

Another complexity hidden by Eq. 1 is that, for a given model, there can be many different explanations suited to different humans with different ways of understanding the world or to different scientific theories [11].

Concepts are similar to but different from latent features. In order to enhance explainability, concepts must be visible (and ideally obvious) to humans. Constraining representations to express a single well-defined concept in a single network node is difficult; for example, concept methods are known to experience concept leakage [33] or a concept whitening module may be needed [34] to attempt to disentangle concept representations. Our use of concept channels is intended to avoid the need for whitening and prevent concept leakage altogether.

Finally, the original motivation for this work was the intuition that neural networks, because of their dependence on random initialisations and huge amounts of training data, are very difficult to test. The same network architecture trained on the same data will produce a different set of connecting weights. Although work to develop mechanistic interpretability of neural networks is ongoing, the role of individual nodes in a neural network is subject to luck [35]. Given these problems, it seemed that any approach that reduces the number of nodes in a neural part of the system ought to be easier to mechanistically interpret and so easier to fully test.

## 3 Method

The architecture used to implement the designer concepts approach is shown in Fig. 2. Input data is preprocessed (normalised, centred, augmented etc.) to create a feature vector ($x \in \mathbb{R}^n$) suitable for neural processing. The feature vector is then propagated through the parallel feedforward neural concept channels $f(x) = \big(f_{c_1}(x), f_{c_2}(x) \dots f_{c_m}(x)\big)$ to give a concept probability vector ($\hat{c} \in \mathbb{R}^m$). The concept probability vector is then symbolically translated to an unnormalised class prevalence vector $\hat{y} \in \mathbb{R}^k$, and then to a class probability vector ($\hat{p}(y) \in \mathbb{R}^k$) using the softmax. The model provides three outputs:

**Table 1.** Prior background knowledge

| Digit | Cross | T Junction | Loop | LineUp | LineDown | TopLeft | BottomLeft | TopRight | BottomRight |
|---|---|---|---|---|---|---|---|---|---|
| 0 | | | x | | | | | | |
| 1 | | | | x | x | | | | |
| 2 | | | | | | x | | | x |
| 3 | | | | | | x | x | | |
| 4 | | x | | x | x | | | | |
| 5 | | | | | | | x | x | |
| 6 | | x | x | x | | | | | |
| 7 | | | | | x | x | | | |
| 8 | x | | x | | | | | | |
| 9 | | x | x | | x | | | | |

1. Class prediction ($\hat{p}(y)$): the class probability vector lets us predict the most likely class.
2. Concept prediction ($\hat{c}$): the concept probability vector is provided directly from the neural channels. Exposing the concept vector allows us to see which concepts have been detected (or missed).
3. Concept relevances: concept relevances produce saliency maps similar to layer-wise or concept relevance propagation [17,23]. Our relevances differ from those techniques in that they are produced directly from the knowledge representations within the neural channels.

### 3.1 Neural Training

To train the neural channels, a human-generated prior class concept map ($\mathcal{M} \in \mathbb{R}^{k \times m}$), based on background knowledge (Table 1) is used to map labelled training data to concept vectors, $y \to c$ (e.g. all 0's, 6's, 8's and 9's have a loop). After training, it is used to translate predicted concept vectors back to class probabilities ($\hat{c} \to \hat{y} \to \hat{p}(y)$).

Each of the neural concept channels implements a function $f_{c_n}(x)$ mapping the normalised input vector to a single concept probability. The channels have been constrained to be entirely independent of each other to prevent concept leakage [33] and were kept as simple as possible. For more complex classification tasks, the channels could be built upon a shared pre-trained feature detector or have customised neural architectures based on the nature of their target concept.

Each channel implements a convolutional layer $[g_c * x]$ followed by a pooling layer ($\mathcal{P}_{max}$) and fully connected layer $w_c$ followed by a sigmoid activation. The overall concept activation is defined by Eq. 2[1].

$$f_{c_n}(x) = \sigma\Big(\sum_{i,j} \big(w_{c_n(i,j)} \cdot \mathcal{P}_{max}([g_{c_n} * x])_{(i,j)}\big)\Big) \tag{2}$$

Concept relevances $r_{c_n}(x)$ are produced by taking the product of the pooling layer $\mathcal{P}_{Max}\big([g_c * x]\big)$ and the weights matrix $(w_c)$, and retaining the positional indices $(i, j)$. The positional indices $(i, j)$ corresponding to the nodes in the pooled layer are scaled up to match the pixel coordinates of the input image. In this way, the concept relevances, $r_{c_n}(x)_{(i,j)}$, show exactly which regions of an image contribute to a concept identification:

$$r_{c_n}(x)_{(i,j)} = w_{c_n(i,j)} \cdot \mathcal{P}_{max}([g_{c_n} * x])_{(i,j)}. \tag{3}$$

The concept activation Eq. 2 can be restated in terms of concept relevance:

$$f_{c_n}(x) = \sigma\big(\sum_{i,j} r_{c_n}(x)_{(i,j)}\big). \tag{4}$$

Each of the three neural outputs implements a differentiable function, so each of the three outputs can have a loss function associated with it.

1. Class predictions: a categorical loss function is used to direct the network to produce correct class predictions over $k$ classes:

$$\mathcal{L}_{Class}(x) = -\sum_{c=1}^{k} y_{o,c} \log(p_{o,c}).$$

2. Concept predictions: a regression loss[2] can be used to direct each of the $m$ concept channels predict the presence of concepts directly (using the background knowledge contained in the concept feature map):

$$\mathcal{L}_{Concept}(x) = \sum_{i=1}^{m} (\hat{c}_i - c_i)^2.$$

3. Concept relevances: a focus loss can be used to encourage the channel classifiers to find hotspots:

$$\mathcal{L}_{Focus}(x) = \frac{\mathcal{P}_{avg_{total}}(r_c^+(x))}{\mathcal{P}_{max_{total}}\big(\mathcal{P}_{avg_{n\times m}} r_c^+(x)\big)}.$$

The focus loss function $\mathcal{L}_{Focus}(x)$ is evaluated on the 2 dimensional relevance $r_c(x)$. It operates on the positive contributions in the relevance (allowing negative contributions to remain diffuse), and by guiding the network to minimise the

[1] i, j represent the position of each node in the pooled output.
[2] A binary cross entropy loss could also have been used.

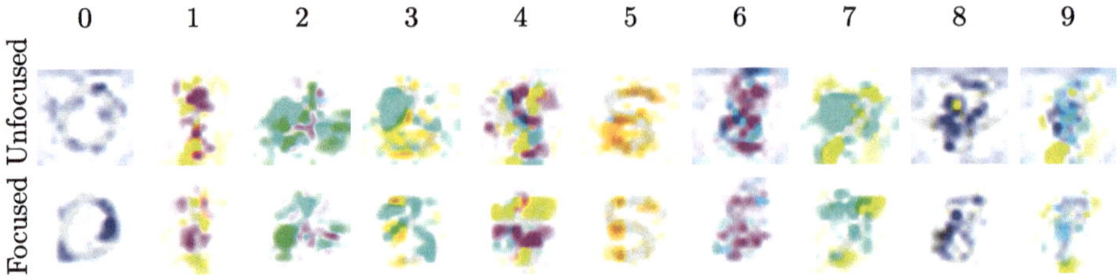


**Fig. 3.** Unfocused and focused relevances

ratio of average temperature to maximum temperature, it promotes relevance with a clear focus and discourages relevance with a diffuse focus. The shape of the average pooling kernel ($n \times m$) is used to determine the sharpness of the focus. Figure 3 gives examples of unfocused and focused relevance maps. The focused maps show far less 'noise', particularly on the 0, 6, 8 and 9 examples.

The modular nature of the architecture, the embedding of human background knowledge, and the ability to choose where to backpropagate losses gives a good deal of control over the neural representations learned by the concept channels. It lets us make those representations explainable.

### 3.2 Symbolic Processing

The output of our neural concept predictors is processed by three distinct symbolic approaches, mapping concepts to the final classes of the target classification task.

The first approach uses the prior class concept map $\mathcal{M}$ (Table 1) to translate from the concept probability vector to a class prevalence vector $\hat{y}$ ($\hat{c} \rightarrow \hat{y}$) followed by the softmax function (Eq. 5) to convert the logit $\hat{y}$ vector into a class probability vector $\hat{p}(y)$ used to select the most probable class.

$$\hat{p}(y_i) = \frac{e^{\hat{y}_i}}{\sum_i e^{\hat{y}_i}} \tag{5}$$

The second, naïve Bayes, approach uses Eq. 6 for Bernoulli-distributed predictors to find the most probable class. The prior class probabilities, $p(y_i)$, and the conditional concept likelihoods, $p(c_j|y_i)$, are found by re-applying the trained concept channels onto the training data. The concept evidence, $p(c_j)$, is found using the trained concept channels on the test data, where each concept channel gives the probability that each concept is present in each test data item. The probability of a concept being absent is represented as $p(\bar{c}_j)$.

$$p(y_i|c) \propto p(y_i) \prod_j \left[ p(c_j|y_i)^{p(c_j)} + (1 - p(c_j|y_i))^{p(\bar{c}_j)} \right] \tag{6}$$

A human understandable basis for the classification can be found by inspecting the concept vector ($\hat{c}$).

**Table 2.** Testing Accuracies

| Filters | Parameters | Accuracies (%) by Loss Type Explain | Prediction | Concept | Both |
|---|---|---|---|---|---|
| 1 | 2205 | None | 91.5 | **93.7** | 92.4 |
| 2 | 4410 | | 94.8 | **96.3** | 95.7 |
| 4 | 8820 | | 96.0 | 97.0 | **97.2** |
| 8 | 17640 | | 97.0 | **97.7** | 97.5 |
| 16 | 35280 | | 97.0 | **97.4** | 97.3 |
| 1 | 2205 | Focus | 91.0 | **93.8** | 93.1 |
| 2 | 4410 | | 94.9 | **95.9** | 95.6 |
| 4 | 8820 | | 96.2 | **97.2** | 96.6 |
| 8 | 17640 | | 96.4 | **97.3** | 97.2 |
| 16 | 35280 | | 96.5 | **97.6** | 96.6 |

The final 'Posterior Concepts' approach develops and extends the background knowledge, analysing the relationship between the learned concepts and the class attribute of the training data. The trained concept channels are re-applied to the training data to get a set of new concept vectors ($\hat{c}$). Each of these new concept vectors will have been detected in a subset of the training images and is mapped to the most common label of the images in that subset. These mappings to the most common class constitute our third classification method. This technique allows for classifications to be explained by showing the similar items in the training data. It also provides a degree of confidence in the classification based on the strength of the majority. It could also allow for unrepresentative training data to be removed from the training data set.

# 4 Implementation and Experiments

## 4.1 Data and Background Knowledge

The MNIST Digits dataset [36] was chosen for the experiments incorporating background knowledge into a network. It was partly chosen because it has been well studied; comparing results against other studies is straightforward [14,37]. It was also chosen because of the simplicity of deriving and expressing the prior background knowledge (1's contain a line pointing up towards the top of the digit, and a line pointing towards the bottom of the digit. 6's contain a loop and a line pointing up while 9's contain a loop and a line pointing down). The full set of background knowledge is given in Table 1. The prior class concept map $\mathcal{M}$ translates the background knowledge into a matrix. This matrix implements the symbolic translation of probabalistic concept vectors $\hat{c}$ into probabalistic class vectors $\hat{y}$:

$$\hat{y} = \mathcal{M} \cdot \hat{c}. \tag{7}$$

### 4.2 Training the Neural Component

The neural components were created with different numbers of $7 \times 7$ convolutional filters followed by a $2 \times 2$ pooling layer and trained with losses applied in combinations to the predictions, concepts and relevance outputs over 10 epochs. Classification accuracies were measured by comparing labels ($y_i$) and the result of the softmax function (Eq. 5) applied to the class vector ($\hat{y}$) output of the architecture. Table 2 shows the accuracies on test data.

The table shows that very small models (1 or 2 convolutional filters) perform relatively poorly, but with 4 or more filters the models classify consistently. It shows that models trained just using a concept loss perform marginally better (approx. 1%) than those models trained with just a class prediction loss. This can be understood because concept losses are applied directly to the concept channels rather than indirectly (where the losses must be first backpropagated through the symbolic prior class concept map).

Using both prediction and concept losses seems to result in an accuracy somewhere between the two. There is very little variance between the accuracies of the focused and unfocused relevance maps, and across the different prediction types suggesting little trade-off between explainability and accuracy.

### 4.3 Searching for Background Knowledge

A second experiment sought to remove the dependence on prior background knowledge by finding an alternative prior class concept map to that specified in Table 1. Two approaches to the search were considered: a tree search in which concept maps were expanded by one concept at a time, and a random approach, where class concept maps were randomly guessed. The idea of this search was to establish the extent to which the overall performance of the architecture depends on good background knowledge. The accuracy of the classifiers with randomly guessed prior concept maps was lower than the classifier with the handcrafted prior concept map but not by much. Averaged over 2624 separate random experiments, the classification accuracy with search methods was $94.9 \pm 0.8\%$. The very low variance of these classifiers militates against using a tree search because it is lower than the variance caused by the random initialisation of neural networks. While the tree search would follow the best path, the result would be statistically no different from random sampling.

To maintain the quality of the random prior concept maps, they were sampled with different numbers of concepts and different levels of sparsity (the probability of a concept mapping to a class). There was also an admissibility constraint applied to the prior concept map - no two classes could be mapped to the same set of concepts.

There was a weak correlation between accuracy and the number of concepts (4 concepts gave a mean accuracy of 93.5%, while 8 concepts gave an accuracy of 95.2%). The sparsity parameter made no difference. This experiment indicates that many concept maps are feasible on this data, and domain knowledge is necessary to determine the right (most useful) one, which agrees with the challenges of scientific enquiry, in general [11].

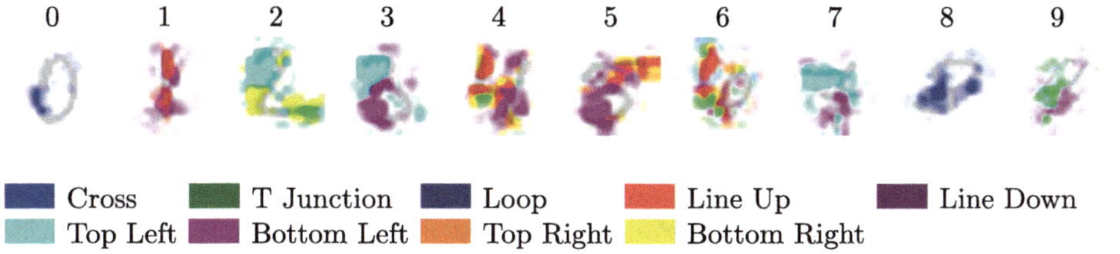


**Fig. 4.** Saliency maps ($r_c$)

### 4.4 Comparative Accuracies

Overall, the accuracies compare well to many techniques but lag behind the top performers. LeNet5 [14] achieved an accuracy of 99% on the MNIST dataset. In their review of MNIST classifiers [37] compare 5 neural architectures that achieve testing accuracies ranging from 98.4% to 99.6% - between 1% and 2% higher than our approach. They report 13 non-neural approaches which all failed to reach the accuracy of our approach. Comparing the accuracies with and without the custom focus loss suggests that applying a focus loss has no overall effect on predictive accuracies. Some classifiers perform marginally better, while others perform marginally worse than their unfocused counterparts.

## 5 Explanation Results

### 5.1 Explanations Using Background Knowledge

The architecture allows classifications to be made using the three symbolic approaches: using the softmax function applied to the class vector $\hat{y}$, using naïve Bayes prior and posterior probabilities, and using the posterior class concept map. The techniques give very similar accuracies ($97.5 \pm 0.3\%$). For comparison, a classifier using the LeNet5 convolutional architecture [14] was trained over 10 epochs using the same data. As well as providing a baseline for statistical comparisons of overall accuracy, the LeNet results allow us to analyse misclassifications of individual data items.

Analysis showed a substantial overlap between the misclassifications made by the different neuro-symbolic techniques (166 data examples are misclassified regardless of how the classification was made). It also lets us calculate the number of data items that were misclasified by at least 2 of the symbolic techniques (228). What was unexpected was that nearly half of the examples misclassified by the LeNet5 architecture (37 out of 96) were correctly classified by at least 1 of our classifiers.

Inspecting the input images, there was no obvious human detectable characteristic which led to a LeNet5 or neuro-symbolic misclassification.

#### 5.1.1 Softmax Explanations

The softmax classifier uses the class probability vector $\hat{p}(y)$ (obtained from $\hat{y}$

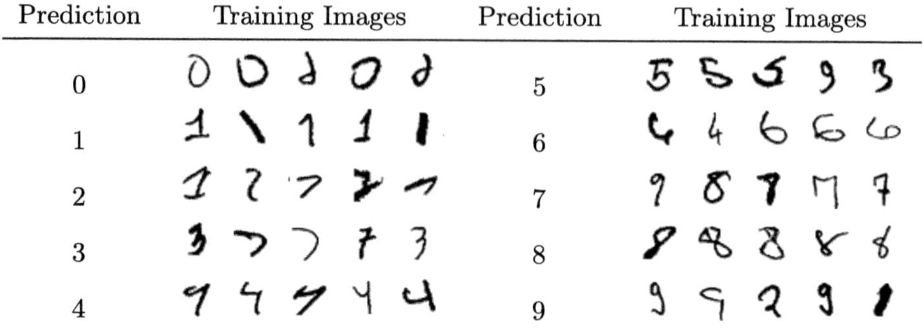


**Fig. 5.** Posterior Class Concept Map

using softmax) to choose the most likely class. This is the most successful technique but is the most difficult to explain because there is no direct link back to the training data. Instead, explaining these predictions relies on the relevance maps ($r_c$) which reveal which part of an image each of the concept channels is detecting (Fig. 4). From the images, it is clear which part of an image is contributing to the identification of a concept. The '2' has a yellow area corresponding to the bottom-right concept and a cyan area corresponding to the top-left. This concept is working as intended. The red Line Up concept is clearly present in the '1', the '4' and the '6'. The '4', '6' and '9' all have areas of green corresponding to a T junction in roughly the right areas.

However, it is difficult to see where the loop was identified in the '6' and '9', and there is no sign of the cross in the '8'. It might be because a faint dark blue looks like a clear navy blue.

#### 5.1.2 Naïve Bayes Classification Explanations

The Naïve Bayes classification uses Bayes rule (Eq. 6) to find the most probable class. Inspecting the probabilities showed values of $p(c_j|y_i) \leq 0.02$ (where the background knowledge indicates the lack of a concept) and $p(c_j|y_i) \geq 0.95$ (when a concept should be present); the neural classifiers are very good at detecting what they are told to detect (Table 1). The prior probabilities of the digits $p(y_i) \approx 0.10$ because the training data is balanced across classes.

#### 5.1.3 Posterior Concept Map Explanations

The Posterior Concept Map is derived by re-applying the trained neural architecture onto the training data to give a concept probability vector $\hat{c}$ to each training example. This allows training data items which do not fall into the definitions described by the prior background knowledge to be used, adding to the background knowledge. In the training data, there were 105 different concept probability vectors (95 new concept probability vectors not found in the prior concept map). In the Posterior Concept Map, for each concept probability vector, the predicted label is the most common label of those training examples sharing the concept probability vector $\hat{c}$. Figure 5 gives examples of images with concept configurations not found in the prior concept map. Each row contains digits sharing a concept configuration. The '0' row shows images with the loop and an additional top-left concept. The '7' row shows images with the line down and top left concept, and an additional loop concept.

The most striking thing about the images in table 5 is the number of those images that are clearly ambiguous. In the '0' row, there are two mirrored 6's. In the 4's row half of the images could easily be a 4 or a 9.

### 5.2 Sensitivity of Concept Activation Vectors

Finally, the concepts used in the initial prior class concept map (Table 1) were used to see if they could be found in the LeNet5 network. Using the fully connected hidden layer with 128 nodes and with the presence or absence of concepts denoted by the prior class concept map, a simple neural network was trained to search for Concept Activation Vectors (CAVs) [22]. Interestingly, our prior concepts were indeed found in the LeNet5 architecture. However, when the experiment was repeated with all possible concept maps (there are $2^{10}$ possible concept maps - each corresponding to a different subset of the 10 digits) most of them could be found as well (demonstrating the high recall and low precision of CAVs).

## 6 Discussion

This novel approach to building explainable classifiers presents the opportunity to build neural classifiers that have explainable reasoning, and which have a comparatively low number of neurons organised into distinct channels. A channel comprising a few thousand, or even a few hundred neurons ought to be much more tractable for other explainability techniques (e.g. automated circuit discovery [13]). It offers a direct view back onto the training data through the posterior class concept explanations - which would provide concrete grounds for a classification and would allow unhelpful, or erroneous training data to be identified and removed. While the prior class concept map is implemented as a fixed layer in a neural layer allowing errors to be back-propagated through to the concept channels, it could be updated (i.e. learned) through back-propagation, or updated with posterior concept mappings during the training process.

More generally, the use of background knowledge to structure classification subsystems could be applied to any set of machine learning techniques. However, in domains where background knowledge is unavailable or in a form difficult to represent, this technique may replace one unexplainable classifier with a different unexplainable classifier.

Finally, the real ambiguity of many of the data items revealed by this approach suggests that the very high claims of accuracy made by some classifiers could be misleading, especially if they exceed expert human classifiers on such a simple task.

## References

1. Baker, A., et al.: A comparison of artificial intelligence and human doctors for the purpose of triage and diagnosis. Front. Artif. Intell. **3** (2020)
2. Jin, J., Kim, M.: GPT-empowered personalized e-learning system for programming languages. Appl. Sci. **13**(23), 12773 (2023)
3. Kouvaros, P., Kyono, T., Leofante, F., Lomuscio, A., Margineantu, D., Osipychev, D., Zheng, Y.: Formal analysis of neural network-based systems in the aircraft domain. In: Huisman, M., Păsăreanu, C., Zhan, N. (eds.) FM 2021. LNCS, vol. 13047, pp. 730–740. Springer, Cham (2021). https://doi.org/10.1007/978-3-030-90870-6_41
4. Freitas, A.A.: Comprehensible classification models: a position paper. SIGKDD Explor. Newsl. **15**, 1–10 (2014)
5. Guidotti, R., Monreale, A., Ruggieri, S., Turini, F., Fosca, G., Pedreschi, D.: A survey of methods for explaining black box models. ACM Comput. Surv. **51** (2018)
6. Zhang, Y., Tino, P., Leonardis, A., Tang, K.: A survey on neural network interpretability. IEEE Trans. Emerg. Top. Comput. Intell. **5**, 726–742 (2021)
7. Ribeiro, M.T., Singh, S., Guestrin, C.: "Why Should I Trust You?": explaining the predictions of any classifier. In: Proceedings of the 22nd ACM SIGKDD International Conference on Knowledge Discovery and Data Mining, KDD 2016, New York, NY, USA, pp. 1135–1144. Association for Computing Machinery (2016)
8. Rudin, C.: Stop explaining black box machine learning models for high stakes decisions and use interpretable models instead. Nat. Mach. Intell. **1**, 206–215 (2019)
9. Manhaeve, R., Dumancic, S., Kimmig, A., Demeester, T., De Raedt, L.: DeepProbLog: neural probabilistic logic programming. In: Bengio, S., Wallach, H., Larochelle, H., Grauman, K., Cesa-Bianchi, N., Garnett, R. (eds.) Advances in Neural Information Processing Systems, vol. 31, Curran Associates, Inc. (2018)
10. Letham, B., Rudin, C., McCormick, T.H., Madigan, D.: Interpretable classifiers using rules and Bayesian analysis: building a better stroke prediction model. Ann. Appl. Stat. **9**, 9 (2015)
11. Pearl, J.: Causality, 2 edn. Cambridge University Press (2009)
12. Russell, S.J., Norvig, P.: Artificial Intelligence: A Modern Approach, 3rd edn. Pearson (2014)
13. Conmy, A., Mavor-Parker, A., Aengus, L., Heimersheim, S., Garriga-Alonso, A.: Towards automated circuit discovery for mechanistic interpretability (2023)
14. LeCun, Y., Bottou, L., Bengio, Y., Haffner, P.: Gradient-based learning applied to document recognition. Proc. IEEE **86**(11), 2278–2324 (1998)
15. Lundberg, S.M., Lee, S.-I.: A unified approach to interpreting model predictions. In: Proceedings of the 31st International Conference Neural Information Processing Systems, NIPS 2017, Red Hook, NY, USA, pp. 4768–4777. Curran Associates Inc. (2017)
16. Simonyan, K., Vedaldi, A., Zisserman, A.: Deep inside convolutional networks: visualising image classification models and saliency maps. arXiv preprint arXiv:1312.6034 (2013)
17. Sebastian, B., Binder, A., Grégoire, M., Frederick, K., Klaus-Robert, M., Wojciech, S.: On pixel-wise explanations for non linear classifier decisions by layer-wise relevance propagation. PLoS ONE **10**, 1–46 (2015)
18. Sarker, M.K., Zhou, L., Hitzler, A., Pascal, E.: Neuro-symbolic artificial intelligence: current trends (2021)

19. Yu, D., Yang, B., Liu, D., Wang, H., Pan, S.: A survey on neural-symbolic learning systems (2023)
20. Besold, T.R., et al.: Neural-symbolic learning and reasoning: a survey and interpretation (2017)
21. Poeta, E., Ciravegna, G., Pastor, E., Cerquitelli, T., Baralis, E.: Concept-based explainable artificial intelligence: a survey (2023)
22. Kim, B., et al.: Interpretability beyond feature attribution: quantitative testing with concept activation vectors (TCAV). Technical report, Proceedings of the 35th International Conference on Machine Learning (2018)
23. Achtibat, R., et al.: From attribution maps to human-understandable explanations through concept relevance propagation. Nat. Mach. Intell. **5**, 1006–1019 (2023)
24. Crabbé, J., van der Schaar, M.: Concept activation regions: a generalized framework for concept-based explanations. In: Proceedings of the 36th International Conference on Neural Information Processing Systems (2022)
25. Goyal, Y., Feder, A., Shalit, U., Kim, B.: Explaining classifiers with causal concept effect (CaCE) (2019)
26. Bau, D., Zhou, B., Khosla, A., Oliva, A., Torralba, A.: Network dissection: quantifying interpretability of deep visual representations. In: Conference on Computer Vision and Pattern Recognition, pp. 6541–6549 (2017)
27. Fong, R., Vedaldi, A.: Net2Vec: quantifying and explaining how concepts are encoded by filters in deep neural networks (2018)
28. Zhou, B., Bau, D., Oliva, A., Torralba, A.: Interpreting deep visual representations via network dissection. IEEE Trans. Pattern Anal. Mach. Intell. **41**, 2131–2145 (2019)
29. Koh, P.W., et al.: Concept bottleneck models (2020)
30. Ghorbani, A., Wexler, J., Zou, J., Kim, B.: Towards automatic concept-based explanations. Technical report (2019)
31. Yeh, C.-K., Kim, B., Arık, S., Li, C.-L., Pfister, T., Ravikumar, P.: On completeness-aware concept-based explanations in deep neural networks. Technical report (2020)
32. Zhou, B., Sun, Y., Bau, D., Torralba, A.: Interpretable basis decomposition for visual explanation. Technical report (2018)
33. Mahinpei, A., Clark, J., Lage, I., Doshi-Velez, F., Pan, W.: Promises and pitfalls of black-box concept learning models. arXiv preprint arXiv:2106.13314 (2021)
34. Chen, Z., Bei, Y., Rudin, C.: Concept whitening for interpretable image recognition. Nat. Mach. Intell. **2**, 772–782 (2020)
35. Frankle, J., Carbin, M.: The lottery ticket hypothesis: finding sparse, trainable neural networks (2019)
36. LeCun, Y., Cortes, C., Burges, C.: The MNIST database of handwritten digits (1998)
37. Kadam, S.S., Adamuthe, A.C., Patil, A.: CNN Model for image classification on MNIST and fashion-MNIST dataset. J. Sci. Res. (2020)

# Green Aware Evaluation of VADER-IT

Maria Chiara Martinis[1,2](✉), Chiara Zucco[1,2], and Mario Cannataro[1,2]

[1] Data Analytics Research Center, University "Magna Græcia", 88100 Catanzaro, Italy
{martinis,chiara.zucco,cannataro}@unicz.it
[2] Department of Medical and Surgical Sciences, University "Magna Græcia", 88100 Catanzaro, Italy

**Abstract.** Sentiment analysis, a key component of natural language processing (NLP), is crucial for determining the emotional tone expressed in textual data, thereby facilitating the understanding of public opinion across various domains. It has evolved into an important tool in fields where understanding human responses is essential, including marketing, healthcare, and social media monitoring. At the same time, the increasing environmental impact of artificial intelligence (AI) has raised significant concerns about the carbon emissions and energy requirements of computer models, especially those used in natural language processing (NLP) applications. This study compares two methods for Italian sentiment analysis: VADER-IT, a sentiment analysis model specifically designed for the Italian language, and the original VADER tool developed for English. In this last instance, Italian sentences are translated into English using the Google Translate API before being analyzed. Using the domain-specific dataset "Malattie della pelle" from the QSalute site, we evaluate both approaches in terms of execution time and carbon footprint (measured with the CodeCarbon framework). The findings show a clear distinction: the VADER pipeline takes more than 287 s and generates 0.002774 kg of emissions, while VADER-IT completes the analysis in just 0.02 s with negligible $CO_2$ emissions.

**Keywords:** First keyword · Second keyword · Another keyword

## 1 Introduction

Following concerns about an increase in energy consumption and carbon emissions associated with contemporary computational models, interest in green-aware artificial intelligence (AI) has increased recently [12]. The influence of AI technologies on the environment evolves in tandem with their increasing prevalence and complexity. Green AI, a new research path that encourages the adoption of models based on sustainability, efficiency, and ecological responsibility in addition to accuracy and performance, was born out of this knowledge [1]. In this context, Green NLP (Green Natural Language Processing) focuses on reducing the environmental impact of language technologies, which are often implemented on a large scale and used in real time in various fields. A key


P. Pichappan et al. (Eds.): RTIS 2025, LNNS 1936, pp. 456–464, 2027.
https://doi.org/10.1007/978-3-032-23396-7_41

task within NLP is sentiment analysis, which aims to detect and quantify the emotional tone in written texts. Sentiment analysis plays a crucial role in applications such as healthcare, marketing, customer feedback, and social media monitoring [4]. However, an issue arises when utilising tools like VADER, which was created for English, to perform sentiment analysis on non-English texts, such as those in Italian [8]. Translating the input text into English and then using VADER for sentiment classification is a typical method in these situations. This process can yield results, but it also incurs additional computing costs, lengthens execution times, consumes more energy, and may result in less accurate translations. Due to this, the method is not the best from both a linguistic and environmental perspective. An alternative and potentially more environmentally friendly solution is to utilise a sentiment analysis tool specifically designed for the target language. In this study, we focus on VADER-IT, an Italian adaptation of the original VADER model, designed to work directly with Italian input [9]. By eliminating the translation phase, VADER-IT promises faster execution, lower computational requirements, and reduced energy consumption, without relying on external APIs or additional processing. The goal of this research is to conduct an ecological comparative evaluation between these two sentiment pipelines: VADER: using Google Translate to process Italian text; VADER-IT: direct analysis of Italian text with a localized model. Using a dataset of patient reviews from the QSalute portal entitled "Malattie della pelle, extracted from the QSalute portal", we measure and compare the execution time, carbon emissions (via CodeCarbon), and sentiment results for both methods. In addition, we analyse how these metrics evolve as the dataset size increases (50, 100, 200,..., 3200+ entries) to understand the computational impact on growing workloads better. This would be useful to highlight the advantages of these models while also reinforcing the design choice for energy savings.

## 2 Green AI and Sentiment Analysis

An essential task in natural language processing (NLP) is sentiment analysis, sometimes referred to as opinion mining, which is the automatic extraction and categorization of emotions expressed in textual data. Determining whether a text conveys a positive, negative, or neutral sentiment is its main objective. Around 2004, when user-generated content, such as online reviews and forums, became more widely available, systematic computational analysis became feasible, marking the beginning of sentiment analysis research. A key contribution came from Hu and Liu [7], whose work on extracting opinion features from customer reviews laid the foundation for many subsequent developments. Early efforts focused on binary or ternary classification. Additional interesting aspects emerged as the method developed, including emotion labelling, which distinguishes between distinct states of emotions (e.g., joy, sorrow, rage, or fear), and intensity analysis, which calculates the sentiment's strength (e.g., highly negative, mildly positive). In the healthcare sector, where numerous health-related ideas are disseminated through personal blogs, social media, and medical review sites, this type of analysis has also proven highly successful in recent years. These texts offer important insights into patient experiences and public perceptions of medical care, having been gathered outside of official clinical channels. Accurate analysis of this data can enhance healthcare's responsiveness and quality [6]. The high awareness of the environmental impact of artificial intelligence models has led to the emergence of Green-Aware AI, a research area that focuses on energy efficiency and ecological sustainability in the development and implementation of models. With the widespread adoption of natural language processing systems and the increasing computational intensity, particularly with the use of deep learning techniques, their energy consumption and carbon emissions have become key evaluation criteria, alongside accuracy and robustness. Sentiment analysis is a particularly relevant example in this regard because it is extensively used in various domains and frequently applied to enormous amounts of textual data, which can have a substantial environmental impact. Lexicon-based methods are preferred due to their scalability and reduced environmental cost, as they require less computing power and are therefore more suitable for environmental sustainability goals and green-conscious concepts [5]. From a methodological standpoint, sentiment analysis approaches can be categorised into two main categories: machine learning and lexicon-based methods. Machine learning models, particularly those utilising supervised or deep learning algorithms, are trained on large, labelled datasets and can offer high performance. However, they are computationally expensive and often less interpretable. In contrast, lexicon-based methods rely on sentiment lexicons, which are predefined dictionaries where each word is associated with a sentiment polarity (positive, negative, or neutral) or a sentiment score, typically in the range $[-1, 1]$, indicating the degree of affective intensity. These models are lightweight and require a minimal amount of training data, making them particularly suitable for real-time analysis and eco-friendly applications where energy efficiency is a concern. Among lexicon-based tools, VADER (Valence Aware Dictionary and sEntiment Reasoner) [8] is one of the most widely used for English texts, especially in informal contexts such as social media. Its adaptability, speed, and linguistic heuristics make it effective in many fields, although it is limited when applied outside the English language. This limitation motivates the development and use of language-specific tools, such as VADER-IT, an adaptation of the original VADER for the Italian language, which can offer not only greater accuracy but also a lower environmental cost by eliminating the need for translation steps. The advantage of employing linguistically localized models over pipelines is that they require fewer computational phases. Machine translation, in particular, may result in a considerable resource overhead. By comparing VADER-IT, designed for the Italian language, with the original VADER applied through translation, this study provides a comprehensive evaluation of sentiment analysis in Italian.

Structure of the article. The remainder of this paper is organized as follows: Sect. 2 provides background on Green AI and Sentiment Analysis. Section 3 details the Methodology and Experimental Setup. Finally, Sect. 4 presents and discusses the Results.

## 3 Methodology

We experimentally compared two distinct processing pipelines, the original VADER model for English and the VADER-IT model, which has been modified for Italian, to assess the computational and environmental effects of sentiment analysis on Italian language data. Because both approaches were applied to the same dataset under the same circumstances, their performance and carbon emissions could be evaluated equally and, most importantly, reproducibly. The dataset used in this study comprises 3,289 Italian-language patient reviews from the "Malattie della pelle" category of the QSalute site https://www.qsalute.it/. These are real textual data related to patient satisfaction with the hospitals where they spent their hospitalisation period. The original model, called VADER, is based on extracting sentiment from texts written in the English language. The Google convert library, which interfaces with Google Translate, was used to automatically convert each Italian text into English because it is not made to handle non-English input. The VADER analyzer, which provides an emotion score between $[-1, 1]$, was then given the translated phrases. The second model, known as VADER-IT, utilises a modified version of VADER specifically tailored for Italian. This model is built on two specific language resources: an Italian sentiment lexicon and an emoji sentiment lexicon. It is implemented in accordance with the VADER scheme, adjusting the original five criteria for the Italian language. VADER-IT avoids translation cost by processing Italian inputs directly, in contrast to the VADER paradigm. To assess environmental sustainability, both experiments were instrumented using the CodeCarbon tracker, a Python-based library that estimates $CO_2$ emissions by monitoring system energy consumption and processor activity. The tracker was configured to collect data at one-second intervals and send the results to a local directory. For both models, we extracted:

- The total execution time (in seconds) from start to completion.
- The total $CO_2$ emissions (in kilograms) produced during sentiment analysis.

An ecological comparison of the two methods is made possible by this setup, which emphasises the efficiency and ecological effect tradeoffs between localised and translation-based models. To further validate the efficiency of VADER-IT, an additional analysis was conducted to measure its performance as a function of input text length. The dataset was divided into seven groups based on the number of words per review, ranging from very short texts (0–50 words) to very long ones (501–1000 words). For each group, the total processing time and the cumulative $CO_2$ emissions were calculated.

All experiments were conducted on a workstation equipped with an Intel Core i7-12700 CPU (12 cores), 32 GB RAM, running in Python 3.10. The CodeCarbon library was configured with local tracking, ensuring consistent monitoring of energy consumption across all runs.

### 3.1 VADER vs VADER-IT

One of the most important tools in this field is VADER (Valence Aware Dictionary and sEntiment Reasoner) [8], a rule-based sentiment analysis system that combines a highly specialized lexicon with a set of syntactic and grammatical heuristics to interpret sentiment in short text inputs. One of the main advantages of VADER is that it does not require training. This feature makes it particularly suitable for analysing clinical studies where a large labelled dataset may not be readily available. From a technical standpoint, VADER relies on a human-validated sentiment lexicon derived from established resources such as LIWC [10], ANEW [3], and General Inquirer. In addition to the lexicon, VADER incorporates five key rule-based heuristics that regulate sentiment intensity based on textual cues. These include:

- Punctuation;
- Capitalization;
- Degree modifiers;
- Contrastive conjunctions;
- Negations.

These heuristics allow VADER to capture subtle changes in sentiment that would normally be lost in standard bag-of-words models. We used the methods suggested in [13], which is based on the original English model with language-specific resources, to adapt sentiment analysis in Italian using the VADER model. The Italian sentiment lexicon Sentix [2] was included at the lexical level. By mapping the SentiWordNet annotations to the MultiWordNet synsets [11], this resource ensures semantic consistency across languages and extends them to Italian. The VADER model includes the interpretation of emoji emotion in addition to lexical resources. We used the list of emojis selected with Italian explanations found https://emojiterra.com/it/lista-it/ for the Italian version, although the original implementation included a list of emojis in English. This enabled VADER-IT to accurately associate the emotional content in social media and informal texts with the corresponding sentiment values in the target language. Since the semantic effect of capital letters and punctuation (e.g., exclamation points) as sentiment intensifiers is largely language-independent, only three of VADER's five core heuristics required adaptation for Italian: negation, contrastive particles, and intensifiers/modifiers. The set of negation terms was translated from English and enriched by exploiting related synsets in MultiWordNet, ensuring comprehensive coverage of negative constructions. In addition to assigning a recurring polarity score as in the original, VADER-IT additionally generates a discrete emotion class label: neutral, negative, or positive. Because of this, it is especially well suited for classification tasks in limited settings, like medical reviews or diagnostic feedback. It is interesting that when used with non-English materials, VADER-IT eschews the computationally taxing translation step that the English VADER requires. Consequently, it exhibits notably reduced energy consumption and execution time, rendering it a more environmentally friendly and sustainable substitute for real-time sentiment analysis on Italian content. This aligns with the objectives of green-aware natural language processing, where linguistic localisation, model simplicity, and low computational overhead are essential for minimizing environmental impact.

## 4 Result

The analysis was structured in two distinct phases to assess the extent to which VADER-IT could be identified as a green model. First, a direct comparison was carried out between VADER-IT and the original version of VADER (applied after automatic translation of the dataset from Italian to English). Second, an investigation into the environmental scalability of VADER-IT was conducted as a function of text length. In both cases, the evaluation metrics used were execution time and $CO_2$ emissions, measured using the CodeCarbon library. In the first experiment, the behavior of the two models was analyzed under realistic conditions. Since the dataset, composed of healthcare reviews written in Italian, is not directly compatible with the original VADER model, which is limited to English, it was necessary to introduce a preliminary automatic translation step using Google Translate. This step introduced a significant computational burden, affecting both performance and energy efficiency. In contrast, VADER-IT, being specifically adapted for the Italian language, enabled sentiment analysis to be performed directly, without intermediate processing. The results highlight a clear discrepancy between the two approaches. VADER-IT completed the entire analysis in approximately 0.02 s, producing a negligible amount of emissions (0.000000 kg of $CO_2$). The VADER pipeline, on the other hand, required over 287 s to process the same data, generating around 0.002774 kg of $CO_2$. This gap highlights the computational and environmental costs associated with using nonlocalized language tools and confirms the benefits of adopting models tailored to the target language. Figures1 and 2 visually illustrate this difference, showing the marked increase in both execution time and emissions due to the translation step in the VADER pipeline.

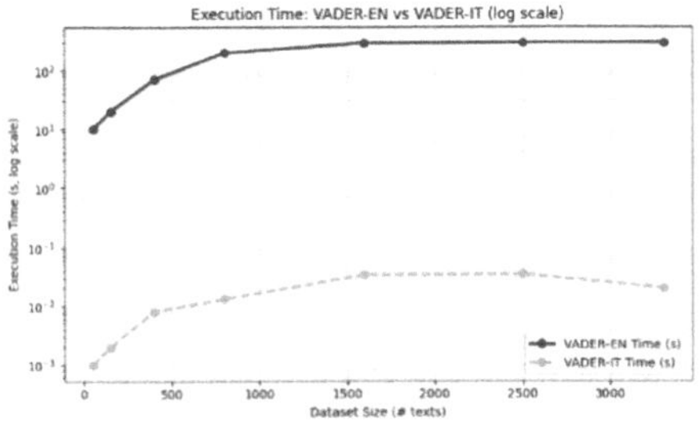


**Fig. 1.** Analysis in terms of time in VADER and VADER-IT.

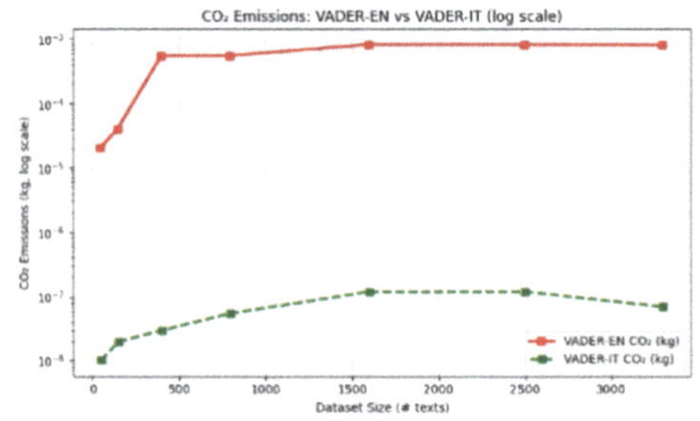


**Fig. 2.** Analysis in terms of $CO_2$ in VADER and VADER-IT.

To further validate the efficiency of VADER-IT, an additional analysis was conducted to assess its performance as a function of input text length. The dataset was divided into seven groups based on the number of words per review, from very short texts (0–50 words) to very long ones (501–1000 words). For each group, total processing time and cumulative $CO_2$ emissions were computed. The results, presented in Fig. 3, show that the groups with the highest execution times correspond to lengths between 51 and 150 words, which were also the most numerous. In contrast, the groups containing the longest texts, while potentially more demanding from a computational perspective, maintained low execution times (under 0.1 s) and minimal emissions (below 0.0000003 kg of $CO_2$).

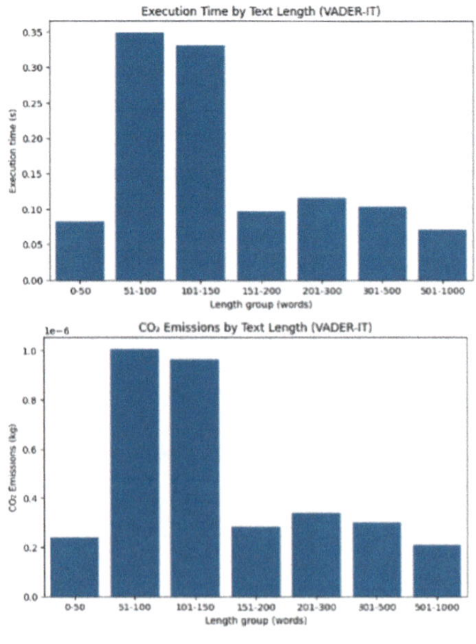


**Fig. 3.** Scalability of VADER-IT in terms of execution time and $CO_2$ emissions grouped by text length.

These observations suggest that VADER-IT's computational load is more affected by the number of instances processed than by the length of individual texts. This confirms the model's stability and efficiency even when handling extended and complex inputs. Moreover, the consistently low environmental impact supports its deployment in real-time or large-scale analysis contexts. The peak detected in processing time and $CO_2$ emissions for reviews of 51–150 words is primarily linked to the dataset's distribution. Since most reviews fell within this range, the system had to handle a larger workload. On the other hand, the scarcity of very long reviews meant their impact on emissions remained modest. Overall, the results underscore the importance of adopting models natively designed for the target language. VADER-IT proves to be an efficient and low-impact solution, capable of adapting to various levels of data complexity without compromising sustainability. The extremely low $CO_2$ consumption—on the order of micrograms demonstrates how the VADER-IT model fully aligns with the principles of Green NLP.

**Acknowledgments.** This work was funded by the Next Generation EU- Italian NRRP, Mission 4, Component 2, Investment 1.5, call for the creation and strengthening of 'Innovation Ecosystems', building 'Territorial RD Leaders' (Directorial Decree n. 2021/3277)- project Tech4You- Technologies for climate change adaptation and quality of life improvement, n. ECS0000009. This work reflects only the authors' views and opinions, neither the Ministry for University and Research nor the European Commission can be considered responsible for them.

We acknowledge the support of the PNRR project FAIR - Future AI Research (PE00000013), Spoke 9 - Green-aware AI, under the NRRP MUR program funded by the NextGen- erationEU. Progetto PRIN 2022NZNZH8 "Role of coffee circulating metabolites in the regulation of the epicardial adipose tissue (EAT) pro-atherogenic properties: a fat-depot specific approach to reduce coronary atherosclerosis", CUP: F53D23007160006. PIANO NAZIONALE DI RIPRESA E RESILIENZA (PNRR), Missione 4 "Istruzione e Ricerca" - Componente C2 Investimento 1.1, "Fondo per il Programma Nazionale di Ricerca e Progetti di Rilevante Interesse Nazionale (PRIN)".

## References

1. Anderson, T., Sarkar, S., Kelley, R.: Analyzing public sentiment on sustainability: a comprehensive review and application of sentiment analysis techniques. Nat. Lang. Process. J. **8**, 100097 (2024)
2. Basile, V., Nissim, M.: Sentiment analysis on Italian tweets. In: Proceedings of the 4th Workshop on Computational Approaches to Subjectivity, Sentiment and Social Media Analysis, pp. 100–107. Association for Computational Linguistics, Atlanta (2013). https://www.aclweb.org/anthology/W13-1614
3. Bradley, M.M., Lang, P.J.: Affective norms for English words (anew): instruction manual and affective ratings. Tech. rep., Technical report C-1, the center for research in psychophysiology (1999)
4. Bucur, C., Tudorica, B., Andrei, J.V., Dusmanescu, D., Paraschiv, D., Teodor, C.: Sentiment analysis of global news on environmental issues: insights into public perception and its impact on low-carbon economy transition. Front. Environ. Sci. **12**, 1360304 (2024)

5. Caratù, M., Brescia, V., Pigliautile, I., Biancone, P.: Assessing energy communities' awareness on social media with a content and sentiment analysis. Sustainability **15**(8), 6976 (2023)
6. Denecke, K., Deng, Y.: Sentiment analysis in medical settings: New opportunities and challenges. Artif. Intell. Med. **64**(1), 17–27 (2015)
7. Hu, M., Liu, B.: Mining and summarizing customer reviews. In: Proceedings of the tenth ACM SIGKDD International Conference on Knowledge Discovery and Data Mining, pp. 168–177 (2004)
8. Hutto, C., Gilbert, E.: VADER: a parsimonious rule-based model for sentiment analysis of social media text. In: Proceedings of the International AAAI Conference on Web and Social Media, vol. 8, pp. 216–225 (2014)
9. Martinis, M.C., Zucco, C., Cannataro, M.: An Italian lexicon-based sentiment analysis approach for medical applications. In: Proceedings of the 13th ACM International Conference on Bioinformatics, Computational Biology and Health Informatics, pp. 1–4 (2022)
10. Pennebaker, J.W., Boyd, R.L., Jordan, K., Blackburn, K.: The development and psychometric properties of LIWC2015. Tech. rep. (2015)
11. Pianta, E., Bentivogli, L., Girardi, C.: MultiWordNet: developing an aligned multilingual database. In: First International Conference on global WordNet, pp. 293–302 (2002)
12. Thakur, D., Guzzo, A., Fortino, G., Piccialli, F.: Green federated learning: a new era of green aware ai. ACM Comput. Surv. **57**(8), 1–36 (2025)
13. Zucco, C., Paglia, C., Graziano, S., Bella, S., Cannataro, M.: Sentiment analysis and text mining of questionnaires to support telemonitoring programs. Information **11**(12), 550 (2020)

# Optimized UWB Positioning: A Server-Based System with Obstacle Detection and Material-Dependent Runtime Correction

Jan-Gerrit Jaeger(✉) and Dietmar Tutsch

School of Electrical, Information and Media Engineering, University of Wuppertal, Wuppertal, Germany
{jgaeger,tutsch}@uni-wuppertal.de

**Abstract.** Ultra-Wideband (UWB) enables precise localization of objects. In environments with few obstacles, UWB delivers reliable results, but obstacles in the line of sight (NLOS) lead to significant deviations. This paper presents a server-based UWB system that significantly improves position accuracy by detecting obstacles and modeling material-specific signal propagation times. A multi-stage process specifically corrects the error-prone distance measurements. The results show a significant reduction in position errors in complex environments.

**Keywords:** Indoor Positioning · Ultra-Wideband (UWB) · Real-Time Locating System (RTLS) · NLOS Identification · NLOS Mitigation

## 1 Introduction

While GPS dominates outdoor positioning, it fails indoors due to signal attenuation. Similarly, indoor methods like odometry or RFID are limited by environmental factors such as floor conditions. Therefore, this work focuses on UWB for robust positioning in complex environments. The goal is to improve accuracy in the presence of obstacles using a novel architecture that separates data acquisition from processing. A central server acts as a passive logger for raw anchor data, while a client handles the calculation logic, including a multi-stage correction algorithm for signal attenuation. This separation offers high flexibility: attenuation models can be adjusted and applied to stored datasets post-hoc, allowing for refinement without requiring new measurements (Fig. 1).

## 2 Fundamentals and State of the Art

Ultra-Wideband (UWB) is a radio technology used for high-precision real-time localization operating in the 3.1 to 10.6 GHz range. By utilizing short pulses


P. Pichappan et al. (Eds.): RTIS 2025, LNNS 1936, pp. 465–476, 2027.
https://doi.org/10.1007/978-3-032-23396-7_42

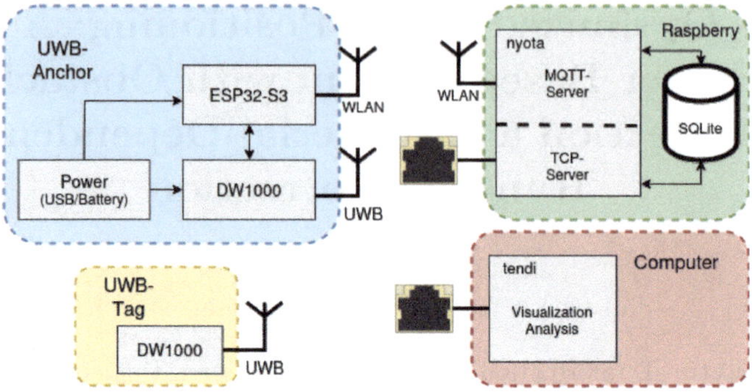


**Fig. 1.** The architecture of the overall system.

and Time-of-Flight (ToF) measurements, UWB theoretically offers high accuracy even in complex environments. While reliable in open spaces, the presence of obstacles leads to signal attenuation and propagation delays, resulting in distance measurement errors. The extent of this error depends on the obstacle's thickness and material composition; dense or conductive materials (e.g., concrete) cause significantly higher attenuation than materials like wood. The propagation speed v of the wave in a non-magnetic medium is determined by the relative permittivity $\epsilon_r$. Compared to the speed of light in a vacuum $c$, the speed is reduced as follows:

$$v = \frac{c}{\sqrt{\epsilon_r}}$$

For Time-of-Flight (ToF) based UWB localization, this means that a signal requires more time to penetrate an obstacle with a permittivity $\epsilon_r > 1$ than for the same distance in air. If this slowdown is not compensated, it leads to a positively biased time-of-flight value and thus to an overestimation of the measured distance [4]. Permittivity is not a fixed constant but depends on the material itself as well as the frequency of the wave. The measurement values used for modeling in this work are taken from the study by Zhekov et al. (2012). The publication provides dielectric properties of common building materials in the gigahertz range relevant to UWB. For example, according to Zhekov, dry concrete has an $\epsilon_r$ value of approx. 5.0 at 6 GHz, while the value for plasterboard is only approx. 2.5 [9]. The correct modeling of these material-specific values is therefore the basis for compensating for NLOS-related time-of-flight errors. Various approaches exist to compensate for NLOS errors in UWB localization. These range from the identification of erroneous measurements by analyzing channel parameters, to probabilistic filters like the Kalman filter, to machine learning models that estimate the error directly from the signal properties. A key difference in these methods often lies in the system architecture. While many more complex models are executed on powerful edge or server systems, recent research such as the InSight framework [3] focuses on running optimized models directly on the resource-constrained UWB tags. In contrast, the solution presented in this paper deliberately chooses a server- and client-based architecture to keep the mobile tags as simple and energy-efficient as possible. The computationally intensive correction is offloaded to the client, where a novel, deterministic, and geometry-based approach is used. By using a 3D environment map and a hybrid correction model that switches between a physics-based model and an empirical "black box" factor depending on the obstacle type, a comprehensible and flexible error compensation is achieved, which also enables robust localization in poorly covered scenarios.

# 3 System Architecture and Hardware Design

The presented indoor positioning system consists of several sensor nodes. The sensor nodes are divided into two categories, Anchor and Tag. Both types of nodes are equipped with UWB modules, which are intended to enable the most precise possible time-of-flight measurement of the signals. The modules are DWM1000 modules from the company DecaWave [7]. The modules have a UWB ceramic antenna that enables communication via UWB radio waves, an integrated circuit for controlling the antenna, and a crystal oscillator for time measurement. Communication with the module is done via the Serial Peripheral Interface. According to the manufacturer, these modules should achieve an accuracy in distance measurement between two sensors of up to 10 cm under optimal conditions. The anchors are positioned at known locations in three-dimensional space. These serve as reference points for position determination. The DWM1000 module is installed on a custom-developed board that also contains a microcontroller and a power supply. The microcontroller used is an ESP32-S3, which handles communication with the module. It also manages the data of the objects to be localized, so that new modules can be registered and modules no longer in range can be removed. The ESP32-S3 also sends the measurement data to the server via WLAN. MQTT is used as the transmission protocol to transfer the data. The board has a dual power supply to operate the modules via both USB and a battery; for this purpose, there is a voltage regulation circuit on the board. Several LEDs on the board serve as status indicators and for simplified diagnosis of the module. The mobile units of the localization system are the tags. These are compact sensor nodes that are attached to the objects to be localized and move freely in space. Each tag is provided with a fixed, unique ID, which it periodically sends to the surrounding anchors at an interval of 100 milliseconds. These signals enable the anchors to perform the time-of-flight measurement. Since the sole task of the tag is to send the ID, the requirements for computing power are minimal. For this reason, a low-power microcontroller of the type Arduino Pro Mini is used as the control unit, which exclusively manages the communication with the UWB module. This minimalistic hardware design allows for energy-efficient operation and a compact form factor for the tags. The distance between an anchor and a tag is determined by measuring the signal's Time-of-Flight (ToF). This system uses the Single-Sided Two-Way Ranging (SS-TWR) method. This approach was chosen because, unlike alternative methods such as Time Difference of Arrival (TDoA) [6], it does not require highly precise and complex time synchronization between the individual anchors and tags. The measurement process takes place in three steps. An anchor initiates the measurement by sending a poll signal and storing a timestamp ($t_{poll}$). The tag receives this signal and, after a defined delay ($t_{reply}$), sends back a response signal. The original anchor receives the tag's response and again stores a timestamp ($t_{response}$). The total round-trip time of the signal ($t_{flight}$) can now be calculated by the anchor from these three timestamps. The geometric distance $d$ then corresponds to half the signal propagation time, multiplied by the speed of light $c$:

$$d = \frac{t_{response} - t_{poll} - t_{reply}}{2} \cdot c \quad (1)$$

Since the internal processing in the electronic components also causes a slight delay, this calculated distance must also be corrected by a system-specific calibration offset to accurately represent the actual distance. This offset is determined once through reference measurements at known distances [2]. After successful calculation, the anchor sends the determined distance value back to the tag. This information can be used by the tag, for example, to directly perform a plausibility check or to trigger its own simple actions based on the distance to specific anchors (Fig. 2).

**Fig. 2.** The hardware board of the developed UWB anchor (LoCaT Ver. 1.1).

A Raspberry Pi serves as the central data acquisition unit. On it, the server-side logic is executed by a C++ application, hereinafter referred to as **nyota**. The primary task of nyota is to receive and persist the measurement data sent by the anchors via WLAN and the MQTT protocol. For each incoming distance measurement, the application adds a precise server timestamp. This approach is fundamental to the system as it eliminates the need for a hardware-based Real-Time Clock (RTC) on the low-cost anchor and tag modules and ensures the synchronization of all measurement data on a uniform time basis. The data, consisting of tag ID, anchor ID, distance, and timestamp, are then stored in an SQLite database. This database solution was chosen as particularly suitable for use on a resource-constrained system like the Raspberry Pi due to its properties. As a serverless, file-based database, it requires no separate installation or configuration and is extremely lightweight. Nevertheless, it offers the full functionality of SQL for structured and efficient data queries. For security reasons, the network configuration of the system is segmented. The anchors communicate exclusively with the server application via WLAN. Client access to the stored data, however, is via a dedicated TCP interface on the physical LAN port of the Raspberry Pi. The recorded data can be retrieved in JSON format via this interface, thus ensuring a secure separation from the local sensor network. The client software, hereinafter called **tendi**, is an application developed in C++, which retrieves, processes, and visualizes the raw data provided by the server. The ImGui [1] library is used for the graphical user interface and data visualization. The application offers both a live view of the current positions of the tags and anchors, as well as a history of the positions. Furthermore, obstacles can be visualized and enabled or disabled. These are taken into account in real-time during position determination. The anchors can also be enabled or disabled to reduce the number of required distance measurements or to exclude inaccurate measurements from faulty anchors in advance. The fundamental method for calculating the position of a tag is trilateration. This geometric method uses the measured distances of the tag to several anchors with known positions. While in a two-dimensional space three anchors are theoretically sufficient, this system deliberately achieves overdetermination by using additional anchors to increase accuracy and robustness. This method is a common principle that is also used in satellite navigation (GPS) to improve position accuracy [5,8]. Each distance measurement $d_i$ to an anchor $i$ defines a circle around its position. In the ideal, error-free case, these circles would intersect at a single point. In practice, however, the distance measurements are always subject to errors, for example, due to signal attenuation or multipath propagation. This causes the circles not to intersect at a single point, but to overlap in an intersection area. At this point, the advantage of the overdetermined system becomes clear. The task of the positioning algorithm is now to estimate the most likely position of the tag from this ambiguous situation. The method of least squares is used for the solution. It determines the point $(x, y)$ that has the lowest total error by minimizing the sum of the squared differences between the measured distances and the distances to the estimated point. The error $E$ to be minimized is described by the following expression:

$$E(x, y) = \sum_{i=1}^{n} (\sqrt{(x - x_i)^2 + (y - y_i)^2} - d_i)^2 \rightarrow \min$$

Here, $(x_i, y_i)$ is the position of the i-th anchor and $d_i$ is the measured distance to it. By using a larger number of anchors, measurement errors of individual modules are reduced, as the method finds an optimal compromise between all measurements, thus enabling more stable and precise position determination. The size of the resulting intersection area, or the residual error from the least-squares optimization, simultaneously serves as an important quality metric for the position determination. A small, compact intersection area indicates consistent and accurate distance measurements, while a large area suggests low reliability of the calculated position. The determined permittivities were assigned to the objects and stored in the JSON file. The JSON file contains the following information for each obstacle:

- **Name of the obstacle:** Unique identifier
- **Faces:** An array of faces, which together form the complete 3D body of the obstacle. Each individual face in this array is a separate polygon, defined by its own array of vertices. Each vertex is stored as an [x, y, z] coordinate array. The order of the points (clockwise) is crucial for the geometric analysis of each individual face, as it defines its surface.
- **Material:** Material type (e.g., “Concrete”, “Wood”, “Glass”)
- **Permittivity:** Relative permittivity for the material

This detailed JSON structure enables the client software tendi to perform a precise geometric analysis for each tag-anchor connection. Using ray-casting, a vector is cast between the estimated tag position and the anchor position. This ray is then tested for a collision with every single face of all defined 3D obstacles. The distance between this entry and exit point corresponds exactly to the path that the signal travels within the obstacle. This value, multiplied by the material's attenuation coefficient, yields the distance error to be corrected, thus enabling a highly accurate adjustment of the NLOS measurements.

## 4 Multi-stage Procedure for Position Determination

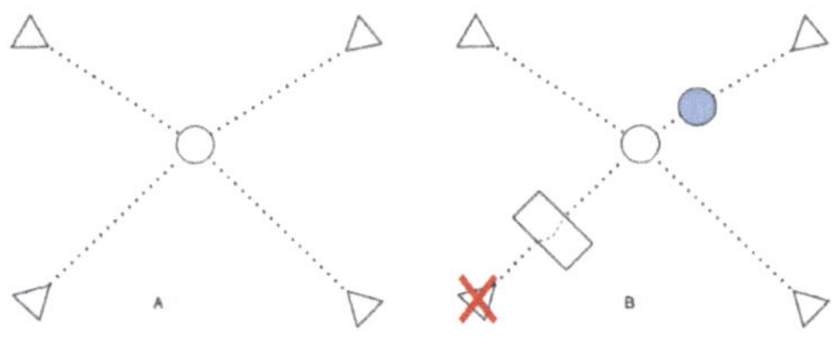


**Fig. 3.** Algorithm for Case A, where an NLOS connection is excluded. The position is recalculated in the second step with the remaining LOS connections.

To significantly improve the accuracy of the position determination, the presented system applies a multi-stage correction procedure. This uses the previously captured room geometry to identify and specifically correct erroneous measurements. The process is divided into the following steps:

### 4.1 Initial Position Determination

First, an initial base position of the tag is calculated using trilateration and the least-squares method. In this first pass, the distance measurements of all available anchors are considered to obtain an initial approximation of the position.

### 4.2 Obstacle Detection and NLOS Classification

Based on the initial position, a Line-of-Sight (LOS) check is performed for each connection to an anchor. The direct path between the tag and the anchor in three-dimensional space is analyzed to check if it is blocked by one or more mapped obstacles. The result is a classification of each connection as either:

- **LOS (Line-of-Sight):** Clear line of sight, the measurement is considered reliable.
- **NLOS (Non-Line-of-Sight):** Blocked line of sight, the measurement is potentially erroneous.

### 4.3 Adaptive Recalculation of the Position

Based on the classification, an adaptive recalculation is now performed according to a fixed logic:

*Case A: Exclusion of NLOS Connections (Preferred Method).* If the number of remaining LOS connections is sufficient (at least four for 3D positioning), all measurements identified as NLOS are discarded. The position is then recalculated exclusively with the reliable LOS data. This procedure eliminates the largest sources of error and leads to the most accurate position determination (Fig. 3).

*Case B: Correction of NLOS Connections.* If the number of LOS connections alone is not sufficient to determine a position, the NLOS measurements must be corrected. For this, a damping model is applied to estimate the error caused by an obstacle and subtract it from the measured distance ($d_{\text{corrected}} = d_{\text{measured}} - d_{\text{error}}$). When modeling, it must be noted that the signal traverses the obstacle twice due to the Two-Way-Ranging method, so the damping effect must be considered twice.

### 4.4 System Requirements and Damping Model

The effectiveness of the procedure depends on the quality of the previously collected data. This data is stored in a JSON file and used by the client software tendi for evaluation:

- **Room Geometry:** The exact positions and dimensions of all relevant obstacles (walls, furniture, etc.) must be known. Inter-nested obstacles must also be taken into account.
- **Damping Model:** Creating a damping model is a complex step, as signal attenuation depends on factors such as material, thickness, and frequency. A practical approach is to use an internal material database. The JSON map then only specifies the material type (e.g., "Concrete," "Drywall") and the thickness of the obstacle. The software can then look up the specific attenuation coefficient and calculate the distance error.

### 4.5 (Optional) Iterative Optimization

With a sufficient number of anchors, an iterative procedure is also possible. After the first corrected position calculation, the new, more accurate position can serve as a starting point for a new NLOS check. This process of recalculation and correction can be repeated until the change in position falls below a certain threshold or a maximum number of iterations is reached.

## 5 Evaluation and Results

While signal attenuation is physically attributable to a material's relative permittivity, real-world obstacles (e.g., a PC or a cabinet) rarely consist of a homogeneous substance, making exact physical modeling impractical.

For this reason, a pragmatic, empirical approach was chosen to determine an effective correction factor (F) for typical, complex obstacles by treating them as a "black box". This factor is derived by comparing a measurement with an obstacle present ($d_{\text{measured}}$) to a reference measurement on the same path without the obstacle ($d_{\text{reference}}$):

$$F = \frac{d_{\text{reference}}}{d_{\text{measured}}}$$

The corrected distance is then simply calculated by multiplication: $d_{\text{corrected}} = d_{\text{measured}} \cdot F$.

To empirically determine these factors, a calibration experiment was conducted. Two UWB modules were placed in a laboratory environment at a fixed distance of 3.0 m to establish a stable reference path. The procedure involved two stages: first, a baseline measurement was taken without any obstruction to determine $d_{\text{reference}}$ after an initial offset calibration. Second, each test object was placed centrally in the direct signal path. For each configuration, 500 measurements were recorded, and a stable mean value for $d_{\text{measured}}$ was calculated after removing statistical outliers.

The resulting, system-specific correction factors are presented in Table 1. As expected, a factor less than 1.0 indicates that the obstacle increases the signal's time of flight, requiring the measured distance to be shortened for correction.

**Table 1.** Empirically determined correction factors (F) for various obstacles.

| Obstacle | Correction Factor (F) |
|---|---|
| Files (approx. 62 cm of paper) | 0.57 |
| Desktop PC | 0.80 |
| Human Body | 0.81 |
| Wooden Cabinet | 0.92 |

The high correction requirements for paper- and wood-based obstacles observed in our experiments are consistent with the findings of Zhekov [9], who also demonstrated significant dielectric properties for these material classes, which lead to a strong influence on UWB signals. The key advantage of this approach is that it bundles all complex effects—from the material mixture and internal geometry to the dual signal path of the TWR method—into a single factor, $F$. All complex effects, from the material mixture and internal geometry to the dual signal path of the TWR method, are bundled into a single, practical value. Unlike a purely physical model, which is based on a single relative permittivity and would be ineffective with complex composite objects, this 'black box' factor offers greater practical generalisability for specific object classes (e.g. 'desktop PC' or 'wooden cabinet'). To achieve maximum flexibility and accuracy, a hybrid correction model was implemented. The JSON file describing the obstacles allows the definition of two different correction methods. For geometrically simple obstacles made of homogeneous materials, a physics-based model is used, which considers the material's attenuation coefficient and the exact signal path length within the object. For complex composite objects like a PC, whose exact physical modeling would be impractical, a lump-sum, empirically determined correction factor $F$ can be stored instead.

The software then decides on an object-specific basis which correction method to apply. To test the improved positioning algorithm, the laboratory environment described in the previous section was used. All internal obstacles relevant to the test were precisely measured and their geometry was stored in the JSON file for the client application tendi. The surrounding walls of the room were considered as system boundaries for this scenario, but were not included as active obstacles in the correction model. Four anchors were placed in a rectangular formation in the room. Their exact 3D coordinates were determined beforehand and permanently stored in the firmware of the respective ESP32 modules. This number provides the necessary redundancy (overdetermination) for the least-squares method and ensures that a position can still be calculated from the remaining three anchors even if one NLOS connection fails or is deliberately discarded (Case A). A single tag was used for the measurement, positioned at a precisely measured reference point. Its exact target position was measured with a laser distance meter at the point (x=3.0 m, y=2.0 m) and at a height of 1.5 m to ensure a precise reference value for the error analysis. The evaluation of the measurement data at the defined reference point shows a significant improvement in position accuracy through the multi-stage correction procedure. For the analysis, the measurement data were evaluated once without (Baseline) and once with the correction algorithm (Optimized). Table 2 summarizes the key metrics for a challenging test scenario.

**Table 2.** Comparison of positioning accuracy for the baseline and the two optimized cases.

| Metric | Baseline | Optimized (Case B) | Optimized (Case A) |
|---|---|---|---|
| RMSE | 0.31 m | 0.22 m | **0.16 m** |
| Max. Error | 0.55 m | 0.45 m | **0.22 m** |

*Test Scenario 1: Case A (Exclusion).* At the first reference point, the signal path to only one of the four anchors was deliberately blocked by positioning a wooden cabinet ($F = 0.92$). The algorithm was able to discard this single NLOS connection (anchor 3D3E) and calculate a highly precise position from the remaining three pure LOS connections. As can be seen in Table 2, the highest accuracy was achieved here with an RMSE of only 16 cm.

*Test Scenario 2: Case B (Correction).* To evaluate the correction mechanism, a more constrained scenario was created using only three anchors. At the second reference point, the signal path to one of these three anchors was blocked by a wooden cabinet. Since discarding this single NLOS measurement would leave only two anchors, the system was forced to apply the correction using the empirical factor (Case B) (Fig. 4).

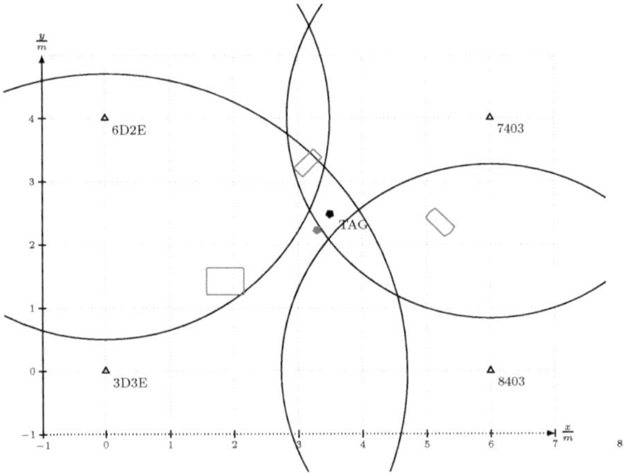


**Fig. 4.** Visualization of the correction effect in the application tendi for a “Case A” scenario.

*Analysis of Results.* The analysis of the results confirms the hierarchical strategy of the algorithm. The highest accuracy was achieved in scenarios where enough LOS connections were available to discard all measurements classified as NLOS

(Case A). This is expected, as a direct, undisturbed measurement is, due to its physical nature, always more precise than a mathematically corrected estimate.

However, the key practical benefit of the system is demonstrated in the more challenging test scenarios where a pure LOS calculation was no longer possible. At the chosen reference point, the signal paths to two of the four anchors were blocked by wooden cabinets, forcing the algorithm to apply the correction using the empirical factor (Case B). Even in this demanding scenario, the root mean square error (RMSE) was reduced by 29% from 0.31 m to 0.22 m (Table 2). This proves the system's ability to provide a robust and significantly more accurate position even in poorly covered environments, where standard methods without error correction would yield unusable results.

It must be noted, however, that the system's effectiveness is directly dependent on the completeness and accuracy of the obstacle map. An unmodeled object causing an NLOS condition would not be detected by the algorithm. The corresponding measurement would be incorrectly treated as LOS and incorporated into the position calculation without correction, leading to reduced accuracy. The investigation of methods to detect such dynamic or unknown obstacles is therefore an essential point for future work.

### 5.1 System Performance and Computational Load

To evaluate the computational performance of the client application tendi independent of the physical UWB hardware limitations, a simulation was conducted. A test scenario was created that simulated the data stream from 50 tags, each providing updates at the system's target rate of 10 Hz (100 ms interval).

The application, written in C++, proved to be highly efficient. Even under this simulated high-load condition, a standard desktop PC was able to process the incoming data, including the core ray-casting algorithm for NLOS detection, without performance degradation. This confirms that the client-side processing architecture is computationally well-suited for high-update-rate tracking in scenarios with a moderate number of tags.

## 6 Summary and Outlook

In this work, a server-based UWB positioning system for optimizing localization accuracy in complex indoor environments was presented. The core problem of erroneous distance measurements with blocked line of sight (NLOS) was addressed by a multi-stage correction algorithm. The system uses a hybrid correction model that applies physical parameters for known, homogeneous materials and empirically determined correction factors for complex composite objects. The evaluation has proven the effectiveness of this approach; in challenging NLOS scenarios, the root mean square error (RMSE) of positioning could be reduced from 0.31 m to 0.22 m. The results also show that the hierarchical strategy of the algorithm, which prefers pure LOS-based calculations, achieves the highest accuracy of up to 16 cm RMSE. The experiments conducted confirm that overdetermination by using more than the minimally necessary three anchors is crucial for the robustness and accuracy of the system. A higher anchor density increases the probability of obtaining a sufficient number of pure LOS connections for each position, which enables the most precise position determination (Case A). For a practical implementation, however, a classic trade-off between accuracy and cost arises. While more anchors per square meter improve performance, they also increase installation effort and system costs. Further research could therefore analyze the optimal ratio of anchor density to achievable accuracy for different use cases (e.g., an open warehouse vs. a convoluted office) to determine an economically viable configuration.

**Acknowledgments.** The project on which this publication is based was funded by the European Union under grant number EFRE-20800274. Responsibility for the contents of this publication rests with the author.

## References

1. Cornut, O.: Dear ImGui: bloat-free graphical user interface for C++ with minimal dependencies (2024). https://github.com/ocornut/imgui. Accessed 17 June 2025
2. DecaWave: APS014 Application Note: Antenna delay calibration of DW1000-Based products and systems. Tech. Rep. APS014, DecaWave (2018). version 1.2
3. Gallacher, M., Stocker, M., Baddeley, M., Römer, K., Boano, C.A.: InSight: enabling NLOS classification, error correction, and anchor selection on resource-constrained UWB devices. In: 20th International Conference on Embedded Wireless Systems and Networks: EWSN 2023 (2023)
4. Guvenc, I., Sahinoglu, Z., Molisch, A.F., Orlik, P.: Analysis of undetected direct path in time of arrival based UWB indoor geolocation. In: IEEE Workshop on Wireless Ad Hoc and Sensor Networks, pp. 536–541 (2005)
5. Hofmann-Wellenhof, B., Lichtenegger, H., Wasle, E.: GNSS - Global Navigation Satellite Systems: GPS, GLONASS, Galileo, and more. Springer Vienna, Vienna (2008)
6. Le-Tien, A., Van-Tam, P., Le, T.M.H., Nguyen, V.D., Dhaou, M.R.T.B.: FlexTDOA: robust and scalable time-difference of arrival localization using ultra-wideband devices. In: 19th ACM/IEEE International Conference on Information Processing in Sensor Networks (IPSN), pp. 321–322 (2020)
7. Qorvo: DWM1000 Datasheet (2024). https://www.qorvo.com/products/p/DWM1000#documents. Accessed 14 July 2025
8. Teunissen, P.J.G.: Least-squares estimation of the integer GPS ambiguities. In: General Meeting of the International Association of Geodesy (IAG), Beijing, China (1993)
9. Zhekov, S.S., Franek, O., Pedersen, G.F.: Dielectric properties of common building materials for ultrawideband propagation studies. In: 2012 6th European Conference on Antennas and Propagation (EUCAP), pp. 1–5 (2012). https://doi.org/10.1109/EuCAP.2012.6206089

# Improving BI-RADS Risk Prediction via Ordinal Machine Learning on Tabular CESM Data

Federica Saldano[1(✉)], Valeria Popello[1,3], Chiara Zucco[1,3], Marianna Milano[2,3], and Mario Cannataro[1,3]

[1] Department of Medical and Surgical Sciences, University "Magna Græcia", 88100 Catanzaro, Italy
federica.saldano@studenti.unicz.it, {valeria.popello,chiara.zucco,cannataro}@unicz.it

[2] Department of Experimental and Clinical Medicine, University "Magna Græcia", 88100 Catanzaro, Italy
m.milano@unicz.it

[3] Data Analytics Research Center, 88100 Catanzaro, Italy

**Abstract.** Breast cancer is one of the leading causes of female mortality worldwide, underscoring the need for early and accurate diagnosis. In this context, Contrast-Enhanced Mammography (CEM) has emerged as a promising technique for detecting lesions. A well-established system for assessing breast lesion risk is BI-RADS, with categories from 0 to 6, where higher scores indicate increasing malignancy risk. While ordinal regression methods have been shown to outperform conventional classification when predicting ordinal target variables, no previous study, to the best of our knowledge, has systematically compared these approaches for predicting BI-RADS categories in CEM. The present study aims to compare classification and ordinal regression models for predicting BI-RADS categories 1 to 5 by using a public tabular dataset derived from CEM imaging. We evaluate Support Vector Machine (SVM) and Random Forest (RF) models, implemented both as standard classifiers and as ordinal regressors. Results show that ordinal regression consistently outperformed classification for both SVM and RF models, with Support Vector Regression achieving an accuracy of 72.2%, a mean absolute error of 0.362, and a Quadratic Weighted Kappa (QWK) of 0.891, and Random Forest Regression achieving an accuracy of 70.8%, a mean absolute error of 0.356, and a QWK of 0.898. These findings serve as a proof of concept, suggesting the potential value of explicitly modeling the ordinal nature of BI-RADS in breast cancer risk assessment.

**Keywords:** Contrast Enhanced Mammography · Breast Cancer · Ordinal Regression · Machine Learning

F. Saldano and V. Popello—These authors contributed equally to this work.


P. Pichappan et al. (Eds.): RTIS 2025, LNNS 1936, pp. 477–490, 2027.
https://doi.org/10.1007/978-3-032-23396-7_43

## 1 Introduction

Breast cancer is one of the leading causes of women's mortality worldwide, with an estimated 2.3 million new cases diagnosed each year [1]. Therefore, early diagnosis is a key factor in improving the prognosis and significantly increasing survival rates.

Among the various imaging techniques available, contrast-enhanced mammography (CEM) is a technological advancement of traditional digital mammography, offering improved detection in patients with dense breasts, where the sensitivity of conventional mammography is reduced [2–4].

CEM involves the intravenous administration of an iodinated contrast medium (1.5 mL/kg, with a maximum dose of 150 mL) and the use of a dual-energy acquisition protocol [5]. The latter combines low-energy (LE) images, which are essential for morphological analysis of breast tissue, with high-energy (HE) images, typically acquired 2–3 minutes after the contrast medium is injected, for functional assessment.

The acquired images are then processed to generate recombined images that highlight areas of contrast accumulation as high-intensity areas related to vascular uptake. Combining morphological and functional information in this way enables more accurate differentiation between benign and malignant lesions, thanks to CEM's ability to detect tumour neoangiogenesis, a hallmark of malignancy [6].

Although magnetic resonance imaging (MRI) is considered the gold standard for tissue characterization [7], CEM offers numerous advantages, including lower cost, faster performance, compatibility with implanted medical devices, and the absence of contraindications related to claustrophobia or the use of gadolinium [8].

Risks associated with the use of iodinated contrast medium, such as allergic reactions or nephrotoxicity, are rare and can be easily managed in clinical practice [9].

The standardized clinical protocol for performing CEM involves acquiring the two standard mammography projections, cranio-caudal (CC) and mid-lateral oblique (MLO), for both breasts within ten minutes of administering the contrast medium, while maintaining constant compression of the tissue [10].

Figure 1 illustrates the schema for acquiring the CEM images.

In general, early and accurate diagnosis of breast cancer is crucial for improving patients' treatment and survival prospects. One of the most widely used tools for assessing breast lesions is the Breast Imaging-Reporting and Data System (BIRADS) classification, which is based on an ordinal scale that estimates the likelihood of malignancy.

This study examines the use of machine learning techniques, specifically Support Vector Machine (SVM) [11] and Random Forest (RF) [12], to predict the BIRADS index when it is treated as an ordinal variable. The aim

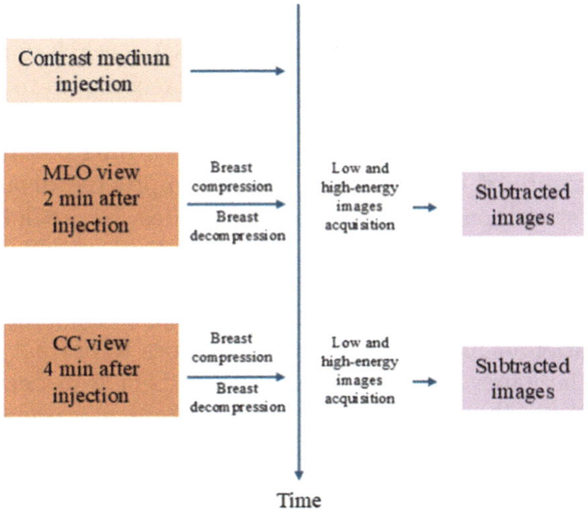


**Fig. 1.** The flow chart outlines a process for imaging after contrast medium injection. It starts with the injection, followed by two imaging views taken at different times: the MLO view at 2 minutes and the CC view at 4 minutes after injection. For each view, images are taken under two conditionsâĂŤbreast compression and breast decompression. These images undergo low and high-energy acquisition, which then leads to the creation of subtracted images. The chart shows a timeline from injection through imaging steps to final image processing, highlighting how different views and compression states contribute to obtaining detailed subtracted images for analysis.

is to demonstrate that ordinal regression models can be a viable alternative to traditional classification models by exploiting the ordinal nature of the target variable to make more accurate and interpretable predictions. The analysis was conducted using a public tabular dataset called CDD-CESM, derived from contrast-enhanced spectral mammography (CESM) images. The effectiveness of ordinal regression models in the medical field has been demonstrated in previous studies. For example, a multivariate ordinal regression model was successfully applied to predict the progression of Alzheimer's disease, resulting in a continuous index that was found to be significantly correlated with standard cognitive measures. Furthermore, it has been demonstrated that these models enable the complete exploitation of the information contained in ordinal data, thereby overcoming the limitations of dichotomisation in classical logistic regression [13]. In the field of epidemiology, it has also been demonstrated that ordinal models can exploit the information contained in ordinal data, overcoming the limitations of dichotomisation in classical logistic regression while guaranteeing interpretable and statistically robust models [14]. In oncology, studies in clinical settings have demonstrated the effectiveness of ordinal logistic regression in BI-RADS classification by identifying relevant predictive factors, such as age, tumour location, level of education, and familiarity with cancer [15,16].

More complex approaches have extended the use of ordinal regression to high-dimensional data by integrating genome-wide DNA methylation information to predict tumour stage, as well as applying penalty techniques to handle the large number of variables [17]. In light of this evidence, the present study contributes to the growing focus on statistical and machine learning methods that respect the ordered structure of clinical data, thereby improving diagnostic accuracy. Specifically, this study compares SVM and RF models for predicting BI-RADS in CESM data, evaluating their performance using ordinal variable-specific metrics. The aim is to provide more accurate and interpretable tools for radiological breast cancer diagnosis, supporting clinical decision-making and promoting the informed application of machine learning techniques to ordinal data.

The rest of the paper is organised as follows: Sect. 2 presents the materials and methods, where the dataset used in this study, the preprocessing phase, the machine learning model we used and the evaluation metrics are described in detail, Sect. 3 outlines the results obtained from the proposed model, and finally, Sect. 4 discusses the conclusions.

## 2 Materials and Methods

This section describes the dataset, preprocessing steps, machine learning models implemented, and the evaluation metrics used to assess model performance.

### 2.1 Dataset

This study utilized a publicly available CESM dataset from The Cancer Imaging Archive (TCIA) [18].

This dataset includes high-resolution CESM images. These were collected by the Department of Radiology at the National Cancer Institute, Cairo University, Egypt, between January 2019 and February 2021. The dataset comprises a total of 2,006 images, acquired from 326 female patients aged between 18 and 90 years. Eight images are available for each patient: the low-energy image and the corresponding subtracted image for each side of the breast in the two central projections (CC and MLO). An example of a 45-year-old patient with a malignant lesion, classified as BIRADS 5 on the left side and shown in the cranio-caudal view, is presented in Fig. 2. However, in some cases, only four images were included due to mastectomy or image quality issues. All images were anonymised and identified by unique patient codes. Radiological annotations were manually performed by experienced radiologists, who followed the guidelines of the American College of Radiology (ACR) for their definition. These annotations include information such as breast composition, shape, density and margin of the mass. Additionally, each image is associated with a BIRADS (Breast Imaging Reporting and Data System) radiological index, which indicates the level of malignancy of the lesion [19].

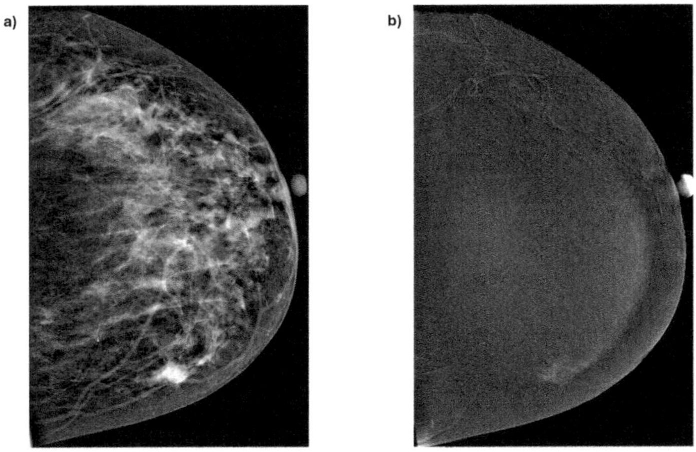


**Fig. 2.** Two side-by-side breast images compare different imaging techniques. The left image shows a dense breast with many white lines and spots, indicating fibrous tissue and possible abnormalities. The right image shows the same breast with a clearer, less dense appearance, highlighting a distinct, lighter area near the lower edge that may represent a lesion or tumor. This comparison illustrates how the second technique can improve visibility of abnormalities in dense breast tissue, aiding in more accurate detection.

## 2.2 Preprocessing

The first phase of the analysis involved a preliminary exploration of the dataset to visualize the distribution of the BIRADS target class. This class refers to the standardized classification system developed by the American College of

**Table 1.** BIRADS classification

| BIRADS categories | Description | Clinical Implications |
|---|---|---|
| 0 | Incomplete | Need for additional imaging evaluation |
| 1 | Negative | No abnormalities detected |
| 2 | Benign | Benign findings (e.g., cysts) |
| 3 | Probably benign | Short-term follow-up recommended |
| 4 | Suspicious abnormality | Biopsy should be considered |
| 5 | Highly suggestive of malignancy | Appropriate action should be taken promptly |
| 6 | Known biopsy-proven malignancy | Breast cancer diagnosis already established |

Radiology (ACR) for assessing the malignancy of suspected breast lesions. The system is divided into seven categories, ranging from 0 to 6, which reflect different degrees of risk and guide subsequent clinical decisions, from further investigation to confirming a breast cancer diagnosis [20,21].

In Table 1, provide a detailed description of all BIRADS categories.

The visualization of the BIRADS classes distribution within the dataset was key to understanding and assessing the balance of classes in this work. Specifically, the preliminary analysis revealed six classes in the dataset, ranging from 1 to 6. However, class 6 was excluded from the study due to its low sample size. Figure 3 shows the BIRADS classes distribution in detail.

### 2.3 Machine Learning Models

In this paper, two fundamental machine learning approaches were implemented for the classification and regression of BIRADS classes: Support Vector Machine (SVM) and Random Forest (RF). The selection of both models was based on their proven effectiveness in managing high-dimensionality data, nonlinear relationships, and class imbalance. The theoretical foundations and implementation choices adopted for each model are described in detail below.

**Support Vector Machines (SVM).** Support Vector Machines (SVMs), introduced by Vapnik [22], are a class of supervised learning algorithms based on statistical learning theory and structural risk minimization. These algorithms find wide application in both classification (SVC) and regression (SVR) tasks due to their robustness and strong generalization capabilities. A key strength of SVMs lies in their capacity to model complex, nonlinear decision boundaries through the use of kernel functions. These functions implicitly project the original input space into a higher-dimensional feature space where a linear separation may exist.

In the classification approach, the objective of the Support Vector Classifier (SVC) is to identify the optimal hyperplane that maximizes the margin between different classes. When data are not linearly separable in the original space, a methodology referred to as the "kernel trick" is employed, wherein the dot product in the feature space is replaced with a suitable kernel function, thus

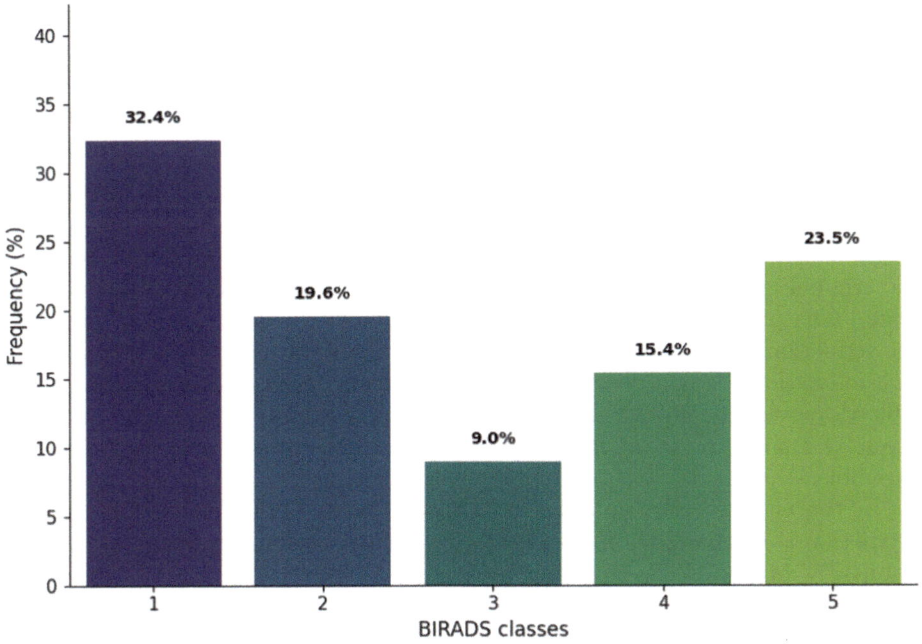


**Fig. 3.** Frequency percentages of five BIRADS classes are compared using vertical bars in different shades of blue and green. Class 1 has the highest frequency at 32.4%, followed by class 5 at 23.5%. Classes 2, 4, and 3 have frequencies of 19.6%, 15.4%, and 9.0% respectively. This chart highlights how often each BIRADS class occurs, showing that class 1 is the most common and class 3 the least.

allowing non-linear classification. For the BIRADS classification task, an RBF (Radial Basis Function) kernel was selected due to its effectiveness in modelling complex decision boundaries [11]. Given the class imbalance in the dataset, the SMOTENC algorithm (Synthetic Minority Over-sampling Technique for Nominal and Continuous) was applied [23]. This variant of SMOTE is tailored to datasets with both categorical and continuous features, and it generates synthetic samples while preserving the integrity of categorical variables.

Preprocessing steps included standardisation of the numerical features, along with ordinal and one-hot encoding of categorical variables. The model was validated using stratified 5 K-fold cross-validation to preserve the original class distribution in each fold [24]. The main hyperparameters, such as the regularisation parameter C (set to 1.0) and the gamma coefficient of the RBF kernel, were optimised to balance the trade-off between model complexity and generalisation. Performance evaluation was conducted using standard classification metrics such as accuracy, F1-score, Cohen's kappa index, and the multiclass ROC analysis.

In the regression approach, the Support Vector Regressor (SVR) was employed to predict the BIRADS classes using an $\varepsilon$-insensitive cost function, where a margin $\varepsilon = 0.1$ are not penalized. This formulation enhances model robustness to minor fluctuations in the target variable [25]. The predicted values were subsequently rounded and constrained to the interval [1, 5] to facilitate direct comparison with the original ordinal labels. As in classification, the RBF kernel demonstrated its efficacy in capturing the non-linear structure of the data.

**Random Forest (RF).** Random Forests, introduced by Breiman in 2001 [12], are ensemble learning algorithms that combine a large number of independent decision trees. The construction of each tree begins with a bootstrap sample of the original dataset and a randomly selected subset of predictor features at each node. This procedure, known as "bagging" (bootstrap aggregating), facilitates the reduction of the model's variance, thereby mitigating the overfitting that is characteristic of single decision trees, and enhancing the generalization ability [26]. Moreover, the stochastic nature of feature selection at each split promotes diversity among the trees, improving the overall robustness and stability of the ensemble.

In the classification approach (Random Forest Classifier, RFC), the final prediction is obtained through a majority voting mechanism, wherein the most frequently predicted class among all trees is selected [12]. For this study, an ensemble of 100 decision trees was employed, with the `class_weight` parameter set to `''balanced''` to address class imbalance and improve sensitivity to minority classes. As with SVM models, preprocessing included standardisation of numerical features and appropriate encoding of categorical variables. Model validation was carried out using stratified 5-fold cross-validation [24], which ensures that the distribution of classes is preserved across folds, providing a reliable estimation of performance. Evaluation metrics included accuracy, precision, recall, and F1-score. In addition, multiclass ROC curve analysis was conducted to assess the model's discriminative power further.

The Random Forest Regressor (RFR) is a prediction model of the target variable by aggregating the average predictions of all individual decision trees. The construction of these trees is analogous to the classificatory case, employing bootstrap sampling and random feature selection. The regression approach allows the modelling of complex continuous relationships between the input features and the target variable, which rigid classification models may overlook [27]. The final continuous predictions were subsequently discretised to align with the original ordinal class labels (range 1–5). The performance of the model was evaluated using regression metrics such as mean absolute error (MAE) and mean squared error (MSE), to quantify the accuracy of the continuous predictions. These were complemented by the traditional classification metrics that had been applied.

### 2.4 Evaluation Metrics

Evaluation metrics are fundamental tools for analysing and understanding the effectiveness of predictive models [28]. The choice of appropriate metrics depends on the nature of the problem (classification or regression) and the characteristics of the dataset [29].

**Classification Metrics:** For SVC and RFC, the following metrics were used:

- Accuracy
$$Accuracy = \frac{TP + TN}{TP + TN + FP + FN} \quad (1)$$
Measures the proportion of correct predictions to the total samples. It is simple and intuitive, but can be misleading with unbalanced classes [30]. Where $TP$, $TN$, $FP$, and $FN$ represent true positives, true negatives, false positives, and false negatives, respectively.
- Precision
$$Precision = \frac{TP}{TP + FP} \quad (2)$$
Measures the quality of positive predictions, i.e. the fraction of instances correctly classified as positive compared to all cases that the model has labelled as such. This metric is useful when a false positive has critical consequences.
- Recall (Sensitivity)
$$Recall = \frac{TP}{TP + FN} \quad (3)$$
Indicates the model's ability to correctly identify all positive instances, which is crucial when it is essential to minimise false negatives.
- F1-Score
$$F1 = 2 \times \frac{Precision \times Recall}{Precision + Recall} \quad (4)$$
It is the harmonic mean of precision and recall, which provides a balanced measure when both types of error are important.
- Weighted Kappa
$$\kappa_w = 1 - \frac{\sum_{i,j} w_{ij} O_{ij}}{\sum_{i,j} w_{ij} E_{ij}} \quad (5)$$
Measures the correlation between the predicted and actual labels, considering the degree of disagreement. It is useful for multiclass ordinal problems, as it penalises misclassifications according to their severity. The matrix $O_{ij}$ represents the observed agreement, $E_{ij}$ represent the expected agreement by chance and $w_{ij}$ is the weights matrix, tipically using linear or quadratic weights depending on the ordinal distance between classes [31].

- Confusion Matrix
The confusion matrix is a tabular representation showing the counts of $TP$, $TN$, $FP$, and $FN$, allowing detailed error analysis.
- ROC Curve and AUC
The Receiver Operating Characteristic (ROC) curve plots the relationship between the true positive rate (Recall) and the false positive rate at various classification thresholds [32]. The Area Under the Curve (AUC) quantifies the model's ability to discriminate between classes. In clinical and diagnostic contexts, this tool is widely used to assess the reliability of clinical tests, as discussed in [33].

**Regression Metrics:** For SVR and RFR, predictions were evaluated using the following metrics before rounding:

- Mean Absolute Error (MAE)

$$MAE = \frac{1}{n}\sum_{i=1}^{n}|y_i - \hat{y}_i| \tag{6}$$

  Represents the average absolute differences between predicted and actual values, expressed in the target variable's units.
- Mean Squared Error (MSE)

$$MSE = \frac{1}{n}\sum_{i=1}^{n}(y_i - \hat{y}_i)^2 \tag{7}$$

  Calculates the average of squared errors, which penalizes larger errors more heavily.
- Root Mean Squared Error (RMSE)

$$RMSE = \sqrt{\text{MSE}} \tag{8}$$

  Provides error metrics in the same units as the target variable, making it interpretable for clinical applications [34].

Although BI-RADS scores are typically treated as categorical labels, they are inherently ordinal, indicating increasing suspicion of malignancy. By modelling the target as a continuous variable (SVR, RFR), we capture this ordinal structure, potentially improving predictive performance and consistency. Predictions are then discretized to match the standard BIRADS labels, enabling comparison with traditional classification models.

## 3 Results

The performance of models based on classification (SVC, RFC) and ordinal regression (SVR, RFR) was evaluated using accuracy, mean absolute error (MAE), F1-score and Weighted Kappa, with 5-fold cross-validation.

As shown in Table 2, the SVR ordinal regression model outperformed the SVC classifier in most metrics. In particular, it achieved significantly higher accuracy (72.2% vs 68.1%) and a lower MAE (0.362 vs 0.398), indicating a better ability to capture the ordinal structure of the data. Although the F1-score values were almost equivalent (0.661), the SVR model showed better agreement between predictions and actual values, as evidenced by a higher weighted kappa value (0.891 vs 0.883).

A similar behaviour is observed in the Random Forest models, as shown in Table 3. Again, the RFR ordinal regression model outperformed the RFC classifier in all metrics. The RFR recorded a 7.9% relative improvement in accuracy

**Table 2.** Performance Comparison of SVM-Based Classification vs. Ordinal Regression

| Model | Accuracy ($\mu \pm \sigma$) | MAE ($\mu \pm \sigma$) | F1-Score ($\mu \pm \sigma$) | Weighted $\kappa$ ($\mu \pm \sigma$) |
|---|---|---|---|---|
| SVC | 0.681 ± 0.019 | 0.398 ± 0.032 | 0.661 ± 0.012 | 0.883 ± 0.016 |
| SVR | **0.722 ± 0.007** | **0.362 ± 0.024** | 0.661 ± 0.007 | 0.891 ± 0.015 |

**Abbreviations:** SVC = Support Vector Classifier; SVR = Support Vector Regressor; MAE = Mean Absolute Error.
Performance metrics reported as mean ± standard deviation across 5-fold cross-validation.

**Table 3.** Performance Comparison of Random Forest-Based Classification vs. Ordinal Regression

| Model | Accuracy ($\mu \pm \sigma$) | MAE ($\mu \pm \sigma$) | F1-Score ($\mu \pm \sigma$) | Weighted $\kappa$ ($\mu \pm \sigma$) |
|---|---|---|---|---|
| RFC | 0.656 ± 0.023 | 0.457 ± 0.036 | 0.659 ± 0.019 | 0.854 ± 0.018 |
| RFR | **0.708 ± 0.022** | **0.356 ± 0.026** | 0.714 ± 0.017 | 0.898 ± 0.012 |

**Abbreviations:** RFC = Random Forest Classifier; RFR = Random Forest Regressor; MAE = Mean Absolute Error.
Performance metrics reported as mean ± standard deviation across 5-fold cross-validation.

(70.8% vs. 65.6%) and a 22.1% reduction in MAE (0.356 vs. 0.457). It also achieved a higher F1-score (0.714 vs 0.659) and greater robustness in predictive concordance, as evidenced by the Weighted Kappa value (0.898 vs 0.854), indicating greater reliability in ordinal tasks.

It was shown that both SVR and RFR consistently outperformed their respective classification models, particularly in reducing MAE, a crucial metric in ordinal problems where the magnitude of error is significant. Furthermore, the lower standard deviations in the SVR metrics (e.g. accuracy: ±0.007) indicate greater stability than the SVC (±0.019), a trend also observed in the Random Forest models. Although the F1-scores are similar in the case of SVMs, the marked improvement of RFR over RFC reinforces the hypothesis that ordinal regression is particularly advantageous in contexts with unbalanced distributions between classes.

## 4 Conclusions

The present study investigates the application of machine learning models, in particular Support Vector Machine and Random Forest, for the prediction of BIRADS classes from tabular data obtained from contrast-enhanced mammography (CEM) images. The aim of this study is to perform a comparative assessment between standard classification approaches and ordinal regression models, given the inherently ordinal nature of the BI-RADS score. Experimental results indicate that ordinal regression models (Support Vector Regression and Random Forest Regression) consistently yield better performance than their classification counterparts (Support Vector Classifier and Random Forest Classifier) across several evaluation metrics, i.e. accuracy, mean absolute error, and quadratic weighted kappa. Explicitly modeling the ordinal structure of the BI-RADS score with regression-based approaches appears to reduce large prediction errors and improve alignment with clinical expectations. Additionally, ordinal regression methods exhibit reduced variance in cross-validation, suggesting enhanced robustness across different data partitions.

While these findings are preliminary, they offer a proof of concept suggesting that ordinal modeling may provide advantages in the prediction of BI-RADS categories. Further studies will be required to assess the robustness and generalizability of these observations, particularly through validation on larger and more diverse datasets. Future work will explore whether the observed benefits persist when applying ordinal regression directly to imaging data. This may contribute to a more comprehensive understanding of the potential role of ordinal modelling in real-world diagnostic settings.

**Acknowledgment.** We would like to acknowledge the Breast Cancer Unit at the Renato Dulbecco University Hospital, Magna Graecia University of Catanzaro, Italy, for their support and collaboration throughout the study.

This work was funded by the Next Generation EU- Italian NRRP, Mission 4, Component 2, Investment 1.5, call for the creation and strengthening of 'Innovation Ecosystems', building 'Territorial R&D Leaders' (Directorial Decree n. 2021/3277)- project Tech4You- Technologies for climate change adaptation and quality of life improvement, n. ECS0000009. This work reflects only the authors' views and opinions, neither the Ministry for University and Research nor the European Commission can be considered responsible for them.

We acknowledge the support of the PNRR project FAIR - Future AI Research (PE00000013), Spoke 9 - Green-aware AI, under the NRRP MUR program funded by the NextGenerationEU.

## References

1. Siegel, R.L., Giaquinto, A.N., Jemal, A.: Cancer statistics, 2024. CA: Cancer J. Clin. **74**(1), 12–49 (2024)
2. Jochelson, M.S., Lobbes, M.B.: Contrast-enhanced mammography: state of the art. Radiology **299**(1), 36–48 (2021)
3. Mann, R.M., et al.: Breast cancer screening in women with extremely dense breasts recommendations of the European society of breast imaging (eusobi). Eur. Radiol. **32**(6), 4036–4045 (2022)
4. Popello, V., Zucco, C., Conti, G., Alessio, C., Cannataro, M.: A software pipeline for multimodal and network-based analysis of breast cancer data. In: 2024 Fourth International Conference on Digital Data Processing (DDP), pp. 63–68. IEEE (2024)

5. Zanardo, M., et al.: Technique, protocols and adverse reactions for contrast-enhanced spectral mammography (CESM): a systematic review. Insights Imaging **10**, 1–15 (2019)
6. Dromain, C., et al.: Dual-energy contrast-enhanced digital mammography: initial clinical results. Eur. Radiol. **21**, 565–574 (2011)
7. Mann, R.M., et al.: Breast MRI: eusobi recommendations for women's information. Eur. Radiol. **25**, 3669–3678 (2015)
8. Lobbes, M., Heuts, E., Moossdorff, M., Van Nijnatten, T.: Contrast enhanced mammography (CEM) versus magnetic resonance imaging (MRI) for staging of breast cancer: the pro CEM perspective. Eur. J. Radiol. **142**, 109883 (2021)
9. Najjar, R.: Clinical applications, safety profiles, and future developments of contrast agents in modern radiology: a comprehensive review. iRADIOLOGY **2**(5), 430–468 (2024)
10. Chung, W.-S., Tang, Y.-C., Cheung, Y.-C.: Contrast-enhanced mammography: a literature review of clinical uses for cancer diagnosis and surgical oncology. Cancers **16**(24), 4143 (2024)
11. Cortes, C., Vapnik, V.: Support-vector networks. Mach. Learn. **20**(3), 273–297 (1995)
12. Breiman, L.: Random forests. Mach. Learn. **45**(1), 5–32 (2001)
13. Doyle, O.M., et al.: Predicting progression of Alzheimer's disease using ordinal regression. PLoS ONE **9**(8), e105542 (2014)
14. Armstrong, B.G., Sloan, M.: Ordinal regression models for epidemiologic data. Am. J. Epidemiol. **129**(1), 191–204 (1989)
15. Bustan, M.N., Tiro, M.A., Annas, S., Adiatma, A.: Analysis of ordinal logistic regression model on breast cancer diagnosis by Birads mammography. Indian J. Public Health Res. Dev. **10**(1), 1199–1203 (2019)
16. Muhammad, T., Susilaningrum, D., Tanggo, V.V.C.: Study of ordinal regression method for supporting factors investigation in stages condition of breast cancer patients in Dr. Soetomo Surabaya hospital, Indonesia. EKSAKTA: Berkala Ilmiah Bidang MIPA **22**(3), 220–229 (2021)
17. Gentry, A.E., Jackson-Cook, C.K., Lyon, D.E., Archer, K.J.: Penalized ordinal regression methods for predicting stage of cancer in high-dimensional covariate spaces. Cancer Inform. **14**, CIN–S17 277 (2015)
18. Khaled, R., et al.: Categorized digital database for low energy and subtracted contrast enhanced spectral mammography images [dataset]. https://doi.org/10.7937/29kw-ae92, 2021, the Cancer Imaging Archive. Available at: https://www.cancerimagingarchive.net/collection/cdd-cesm/. Accessed 29 May 2025
19. Khaled, R., Helal, M., Alfarghaly, O., et al.: Categorized contrast enhanced mammography dataset for diagnostic and artificial intelligence research. Sci. Data **9**, 122 (2022)
20. Rao, A.A., Feneis, J., Lalonde, C., Ojeda-Fournier, H.: A pictorial review of changes in the bi-rads fifth edition. Radiographics **36**(3), 623–639 (2016)
21. Spak, D.A., Plaxco, J., Santiago, L., Dryden, M., Dogan, B.: Bi-rads® fifth edition: a summary of changes. Diagn. Interv. Imaging **98**(3), 179–190 (2017)
22. Vapnik, V.: The Nature of Statistical Learning Theory. Springer, Cham (2013)
23. Chawla, N.V., Bowyer, K.W., Hall, L.O., Kegelmeyer, W.P.: Smote: synthetic minority over-sampling technique. J. Artif. Intell. Res. **16**, 321–357 (2002)
24. Kohavi, R., et al.: A study of cross-validation and bootstrap for accuracy estimation and model selection. In: IJCAI, Montreal, Canada, vol. 14, no. 2, pp. 1137–1145 (1995)

25. Smola, A.J., Schölkopf, B.: A tutorial on support vector regression. Stat. Comput. **14**(3), 199–222 (2004)
26. Chen, X., Ishwaran, H.: Random forests for genomic data analysis. Genomics **99**(6), 323–329 (2012)
27. Frank, E., Hall, M.: A simple approach to ordinal classification. In: De Raedt, L., Flach, P. (eds.) ECML 2001. LNCS (LNAI), vol. 2167, pp. 145–156. Springer, Heidelberg (2001). https://doi.org/10.1007/3-540-44795-4_13
28. Sokolova, M., Lapalme, G.: A systematic analysis of performance measures for classification tasks. Inf. Process. Manage. **45**(4), 427–437 (2009)
29. Gunawardana, A., Shani, G.: A survey of accuracy evaluation metrics of recommendation tasks. J. Mach. Learn. Res. **10**(12) (2009)
30. Hernández-Orallo, J., Flach, P., Ferri, C.: A unified view of performance metrics: Translating threshold choice into expected classification loss. J. Mach. Learn. Res. **13**(1), 2813–2869 (2012)
31. Yilmaz, A.E., Demirhan, H.: Weighted kappa measures for ordinal multi-class classification performance. Appl. Soft Comput. **134**, 110020 (2023)
32. Fawcett, T.: An introduction to ROC analysis. Pattern Recogn. Lett. **27**(8), 861–874 (2006)
33. Pepe, M.S.: The Statistical Evaluation of Medical Tests for Classification and Prediction. Oxford University Press (2003)
34. Chai, T., Draxler, R.R.: Root mean square error (RMSE) or mean absolute error (MAE)?-arguments against avoiding RMSE in the literature. Geosci. Model Dev. **7**(3), 1247–1250 (2014)

# GAT for Brain Connectivity Classification

Luca Barillaro[1,4(✉)], Marianna Milano[3,4], Giuseppe Agapito[2,4], and Mario Cannataro[1,4]

[1] Department of Medical and Surgical Sciences, University "Magna Græcia" of Catanzaro, 88100 Catanzaro, Italy
[2] Department of Law, Economics and Social Sciences, University "Magna Græcia" of Catanzaro, 88100 Catanzaro, Italy
[3] Department of Experimental and Clinical Medicine, University "Magna Græcia" of Catanzaro, 88100 Catanzaro, Italy
[4] Data Analytics Research Center, University "Magna Græcia" of Catanzaro, 88100 Catanzaro, Italy
{luca.barillaro,m.milano,agapito,cannataro}@unicz.it

**Abstract.** The classification of functional magnetic resonance imaging (fMRI) data plays a crucial role in the research of neurological disorders and clinical diagnostics. In this study, we present a lightweight classification pipeline, utilizing an open dataset, based on Graph Attention Networks (GAT), which incorporates connectivity features derived from fMRI networks using graph theory. Furthermore, we investigate interpretability possibilities by including analyses based on visualizing attention weights and gradient-based feature attribution. Our research highlights the importance of graph-based learning in fMRI analysis, with additional clinical literature validation.

**Keywords:** Functional Magnetic Resonance Imaging · Machine learning · Deep learning · Graph Neural Networks

## 1 Introduction

Functional Magnetic Resonance Imaging (fMRI) [11] has become a powerful tool for studying brain connectivity, enabling researchers to explore neural interactions in both healthy individuals and those with mood disorders.

By capturing time-dependent fluctuations in blood-oxygen-level-dependent (BOLD) [12] signals, fMRI allows for the reconstruction of functional brain networks. Regions of interest (ROIs) are represented as nodes, with their interdependencies forming edges within a connectivity graph. Analyzing these networks yields valuable insights into altered connectivity patterns associated with mood diseases, including depression.

In recent years, Graph Neural Networks (GNNs) have garnered significant attention for their ability to model complex, non-Euclidean structures such as


P. Pichappan et al. (Eds.): RTIS 2025, LNNS 1936, pp. 491–501, 2027.
https://doi.org/10.1007/978-3-032-23396-7_44

brain connectivity graphs. Unlike traditional machine learning techniques, GNNs take advantage of spatial dependencies within the data, making them particularly adapt at analyzing the intricate organization of brain networks.

Specifically, Graph Attention Networks (GATs) incorporate an attention mechanism that allows the model to dynamically assess the importance of different node connections. This capability makes GATs particularly well-suited for identifying critical brain regions and understanding how functional interactions contribute to neurological disorders.

This study introduces a deep learning pipeline based on GATs to classify depression patients versus healthy controls using fMRI-derived brain networks. The proposed approach integrates functional connectivity data with graph-based learning, facilitating accurate classification while also enhancing interpretability of the most influential brain regions.

Our study is focused on:

- Implementing a full pipeline for fMRI graph classification using GATs, including ROI-level and global graph features.
- Integrating soft interpretability mechanisms: (i) attention weight visualization and (ii) gradient-based feature attribution, enhancing transparency.
- Using SMOTETomek to address class imbalance and evaluating performance using stratified cross-validation.
- Demonstrating hardware feasibility by deploying the pipeline on a Jetson AGX Orin edge device, a novel aspect in GAT-based neuroimaging studies.

The remainder of this paper is organized as follows: in Sect. 2 we present trends in fMRI analysis, including machine and deep learning approaches, as well as graph neural network approaches. Section 3 introduces materials and methods discussing the dataset, preprocessing pipeline, and model implementation. Results are presented in Sect. 4 where classification performance, interpretability findings, and the identification of key brain regions are discussed. Moreover, it explores the implications of our findings for depression research and discusses functional brain network analysis. Finally, Sect. 5 presents conclusions and future directions.

## 2 Related Work

The analysis and classification of fMRI data have evolved considerably, progressing from traditional machine learning techniques to more recent deep learning and graph-based approaches. In this section, we briefly survey the major trends in fMRI classification.

### 2.1 Traditional Machine Learning Approaches

Early approaches to fMRI classification relied on models such as Support Vector Machines (SVMs), Random Forests, and k-Nearest Neighbors. These models operate on handcrafted features such as region-wise connectivity measures, statistical activation maps, or atlas-based summaries. While effective to some extent, they often lack the capacity to capture the complex spatial and temporal dependencies inherent in fMRI data. Moreover, their limited interpretability and dependence on feature engineering restrict their applicability in clinical contexts.

### 2.2 Deep Learning on fMRI Data

Deep learning models, particularly Convolutional Neural Networks (CNNs) and Recurrent Neural Networks (RNNs), have shown strong performance in neuroimaging tasks [13,15] CNNs are useful for identifying spatial patterns in voxel-based data, while RNNs capture the temporal dynamics of BOLD time series. Despite their advantages, these models assume regular data structures and struggle to model the non-Euclidean nature of brain connectivity, which is more naturally represented as graphs.

### 2.3 Graph Neural Networks for Brain Connectivity

Graph-based learning has emerged as a powerful framework for modeling brain networks. In this representation, nodes correspond to regions of interest (ROIs), and edges capture functional or structural relationships. Graph Neural Networks (GNNs) [16] utilize message-passing mechanisms to aggregate local neighborhood information, preserving both local and global topological features.

Several studies have successfully applied GNNs to fMRI data. Li et al. [8] used GNNs to identify biomarkers from task-fMRI data, demonstrating the model's ability to learn from ROI-level graphs. Kim and Ye [7] evaluated Graph Isomorphism Networks (GINs) for resting-state fMRI, emphasizing permutation invariance and robustness. However, these approaches often rely on fixed weighting schemes for neighbor contributions, limiting flexibility in feature aggregation.

### 2.4 Graph Attention Networks and Interpretability

Graph Attention Networks (GATs) [14] enhance the GNN framework by introducing attention mechanisms that assign dynamic weights to neighbor nodes, improving model expressiveness and interpretability. This makes GATs especially appealing for fMRI studies, where understanding inter-regional influence is crucial.

Hu et al. [6] introduced GAT-LI, a model for fMRI classification using attention-based learning. Their method offered insights into brain network importance via learned attention scores.

## 3 Material and Methods

In this section, first, we describe the dataset used in our experiment, discussing its features, and then the software and libraries involved in our workflow. We then discuss the methodology by describing the network model, the analysis pipeline, and the strategies employed to achieve our goal.

### 3.1 Materials

We used a publicly available dataset on the popular platform OpenNeuro. In particular, we refer to "Resting state with closed eyes for patients with depression and healthy participants"[1] [2,3].

This dataset is composed of resting-state fMRI (rs-fMRI) data from 72 individuals, where 51 patients are diagnosed with mild depression and the remaining 21 are control subjects. In this dataset, the gender distribution is: 19 males (6 controls) and 53 females (15 controls).

Each scanning session in the dataset consists of 100 scans with a repetition time (TR) of 2.5 s.

These data have been preprocessed using fMRIPrep [5], a comprehensive tool for fMRI data preparation, with a voxel size set at $2 \times 2 \times 2$ mm.

From the computational point of view, our attention was focused on using low-consumption devices. For this reason, we trained and tested our model on an edge device from NVIDIA; the Jetson AGX ORIN Developer Kit was employed. This ensured low power consumption of the pipeline.

For replicability purposes, we highlight, in Table 1, programming languages, libraries, and packages versions used in our work. We highlight that the Jetson device requires its own version of PyTorch to ensure full functionality.

**Table 1.** Libraries and packages versions

| Library/Package | Version |
|---|---|
| Python | 3.10.12 |
| Pytorch | 2.4 nv |
| Pytorch Geometric | 2.4 |
| Scikit-learn | 1.5.2 |
| NumPy | 1.26.1 |
| Seaborn | 0.13.2 |
| NetworkX | 3.4.2 |
| Pandas | 2.2.3 |

### 3.2 Methods

As introduced earlier, our pipeline is based on different software and libraries, depending on the various phases of the workflow.

First of all, we are dealing with medical imaging, so the first required step is preprocessing. Then, to obtain suitable data for our deep learning model, we have to transform such data into network data by obtaining a meaningful representation in a graph form.

[1] https://openneuro.org/datasets/ds002748/versions/1.0.5.

Thus, our pipeline comprises software to finalize the fMRI preprocessing and then extract the graph representation. For this purpose, we used the CONN toolbox, which relies on MATLAB and SPM. Using this toolbox, we can import preprocessed data, perform some further preprocessing steps, and conclude with an analysis that extracts graph theory results.

The workflow of this pipeline is clarified in Fig. 1.

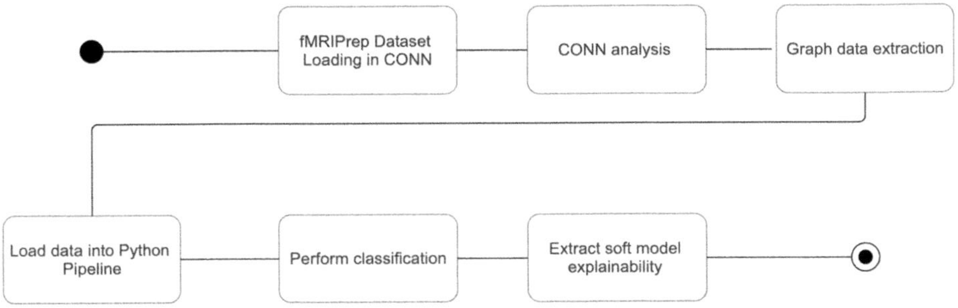


**Fig. 1.** Pipeline workflow.

Our pipeline begins with loading fmriprep preprocessed data and using CONN pipeline [9] to apply additional steps, including functional smoothing (8 mm FWHM), denoising (via anatomical CompCor [1,4,10], motion regressors, and band-pass filtering between 0.008–0.09 Hz), and scrubbing of outlier volumes. Functional connectivity matrices were then computed by averaging BOLD signals within predefined ROIs and calculating pairwise Pearson correlations.

At the end of the CONN pipeline, we were able to utilize fMRI-derived connectivity matrices. Each subject's brain network is represented as a graph, which is structured as follows:

- Nodes: Represent 164 brain regions (ROIs).
- Edges: Represent functional connectivity values (correlation between time series of ROIs).
- Node Features: Include local graph measures (e.g., efficiency, centrality).
- Global Features: Include overall network properties (e.g., path length, clustering coefficient).

This information is stored by CONN into a single MATLAB file (*.m*) to be processed to extract relevant data for our purposes. Thus, we used some simple MATLAB scripts to perform data extraction (handling also NaNs detection and correction) and create appropriate comma-separated values (*csv*) files.

From the CONN output, we retrieved and used the following data:

- Adjacency matrices: one per patient, representing functional connectivity between brain regions
- Region of interest (ROI) features: at node level, representing graph node attributes, containing connectivity-based measures

- Global features: these contain connectivity properties at graph level

In addition, since we are dealing with a binary classification, we include two label vectors to identify our classes. We used 0 for the 21 control cases, and 1 for the 51 disease patients.

Choosing the GAT model was motivated by previous attempts with other simpler models, such as decision trees and simple MLPs, the results of which are presented in the following section.

These models were evaluated using the same resampling strategy and cross-validation. Although some shallow models showed reasonable performance, they consistently underperformed in handling both classes symmetrically, particularly due to the presence of imbalance.

The remaining part of the pipeline was performed using Python, under the convenient structure of a Jupyter notebook. Data loading was performed using the popular Pandas and Numpy libraries. By doing so, we loaded our CSV files into dataframe objects.

After that, we were able to convert this data into a PyTorch geometric data object with the goal of having a graph structure. This step converted adjacency matrices into edge lists, node features into feature tensors, while global features were kept away to be integrated later in the model.

A critical phase was represented by the imbalanced dataset, after many attempts, in our solution we employed, on training data, a hybrid resampling strategy that combines Synthetic Minority Over-sampling Technique (SMOTE) with Tomek Links Undersampling (SMOTETomek).

SMOTETomek first generates synthetic samples for the minority class by interpolating feature values between existing samples and their nearest neighbors. Then Tomek Links were used to remove overlapping majority class samples that were too close to minority class samples, thereby enhancing class separability.

The proposed Graph Attention Network v2 (GAT) model leverages attention mechanisms to assign different importance levels to different brain region connections. From the architectural point of view, it includes:

- Two GAT layers: they use a multi-head attention to assign different weights to each node's neighbors and are useful to capture non-uniform importance across connectivity patterns
- Graph pooling: these units aggregate node embeddings into a graph-level representation
- Fully connected layers: useful to integrate graph embeddings and global features, the final layer is a softmax activation function to predict one of the two classes of our binary classification

Moreover, optimization is implemented by using the Adam optimizer with L2 weight (set to $5e-04$) regularization, while, to ensure a good generalization, dropout (set to 0.5) layers are used.

The last important detail was the creation and usage of a class-wise focal loss with different gamma values for control cases and depression patients. This helped in reducing the penalty for control cases (the minority) by having a low gamma value, while on disease subjects, the higher gamma value prevents dominance of this class (the majority). The model is trained using a 5-fold stratified cross-validation with the goal to ensure a proportional representation of both classes in training and test sets. Other essential features of this phase are the use of early stopping to prevent overfitting, class weight adjustments to deal with the imbalanced dataset, and the use of common evaluation metrics such as accuracy, precision, recall, and F1-score. All of these metrics will be presented in the results section.

After model training, we aimed to propose some soft interpretability.

We used graph visualization to plot functional brain networks, where edge thickness is proportional to attention weights. The ultimate goal of the visualization is to provide an intuitive representation of how the model interprets alterations in brain connectivity in patients affected by depression. This will be presented in the results section.

Moreover, we extracted feature importance by a gradient-based analysis. After computing gradients of model predictions, we can reveal which ROI-level features are most influential in distinguishing between the two classes.

## 4 Results and Discussion

In this section, we analyze the key findings and potential implications from the results of the pipeline, focusing on classification and some interpretability.

The primary objective of the model was to classify fMRI brain networks into two categories: control (healthy individuals) and disease (depression patients). The model was trained and evaluated using 5-fold Stratified Cross-Validation to ensure robustness. This resulted in obtaining some common metrics, such as mean accuracy, and further ones from the popular Scikit-learn classification report. Accuracy reached a mean value of almost 67%, while detailed results of the classification report are shown in Table 2

**Table 2.** Classification report

| Class | Precision | Recall | F1-score |
|---|---|---|---|
| Control | 41% | 33% | 37% |
| Disease | 75% | 80% | 77% |

The classification results demonstrate a good mean accuracy (67%) after addressing class imbalance using SMOTETomek. While results for the majority class are better than control, this represents an improvement over standard ML approaches.

However, the relatively small dataset size and the impact of synthetic samples should be considered when interpreting the robustness of these results.

**Table 3.** Comparison with ML classifiers

| Model | Accuracy | F1 Control | F1 Disease |
|---|---|---|---|
| SVM (RBF Kernel) | 55% | 0% | 71% |
| Random forest | 58% | 20% | 62% |
| MPL (feedforward) | 59% | 25% | 65% |
| GAT (ours) | 67% | 37% | 77% |

In Table 3 we present a comparison of our model with some common ML and simple DL models to prove the relevance of our GAT model.

It is clear that the GAT model reaches the best mean accuracy and also the highest F1-score, but more importantly, with respect to other tested models, it is the most capable in dealing with the minority class.

The other aspect of our analysis was the attempt to show some soft model interpretability capabilities along with graph visualization. This was carried out by using graph attention weights and feature gradient analysis.

The graph attention weights analysis highlighted that some edges of the graph structure received higher attention, potentially indicating a biomarker value for disease classification. We computed a mean graph averaging attention weights across all cross-validation folds. The resulting graph is shown in Fig. 2. We show the top 5 nodes, representing key brain regions from our analysis.

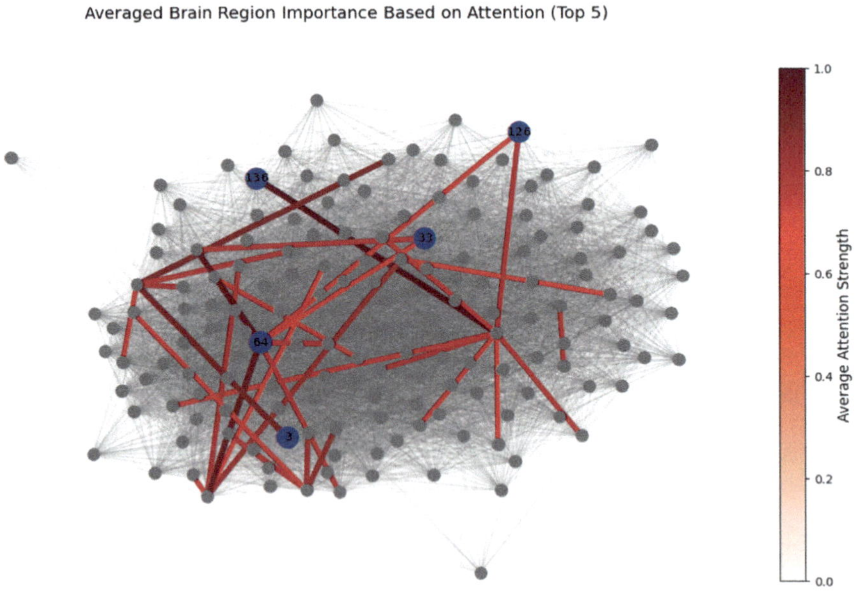


**Fig. 2.** Average brain region importance based on attention

From this representation, by labeling the nodes with the atlas representation from CONN, we identified the following regions along with their functional roles:

- PCC (Default Mode Network) (Node 3): Involved in self-referential processing and rumination. It is often implicated in default mode network hyperconnectivity and persistent rumination.
- Nucleus Accumbens (L) (Node 136): Plays a role in reward and motivation. Dysfunction in this area are linked to blunted reward response and anhedonia.
- Caudate (L) (Node 126): Associated with Cognitive/motivational control in the reward circuit. Linked to striatal dysfunction, altered corticostriatal connectivity.
- Frontal Pole (L) (Node 33): Involved in High-order cognitive control, valuation. Alterations in this region are linked to refrontal hypoactivity and impaired regulation.
- Postcentral Gyrus (R) (Node 64): Linked to somatosensory integration, bodily self-awareness. Thus producing abnormal sensory processing, links to somatic symptoms

The gradient-based analysis of node features, identified from all the 10 features, Degree and Betweenness Centrality were the most influential for classification. Other features, such as Global and Local Efficiency and Cost, showed lower contributions. These results are presented in Fig. 3.

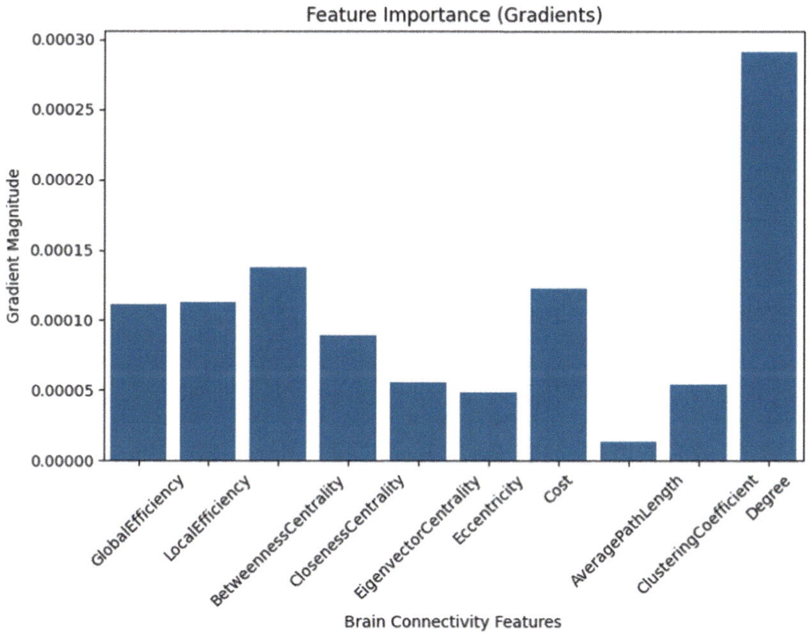


**Fig. 3.** Feature importance by gradients.

In particular, a higher or lower degree in DMN hubs (e.g., PCC) reflects disrupted integration of self-referential processing. Altered degree in striatum (caudate, accumbens) supports reward network dysfunction. While Betweenness Centrality deals with communications inside a node, in our case, this reflects changes in prefrontal and limbic hubs, thus indicating inefficient communication across cognitive and emotional networks.

## 5 Conclusions and Future Directions

This study classified depression vs. control fMRI brain networks using Graph Attention Networks (GATs) and incorporated soft interpretability through attention visualization. The extracted results align with previous research on depression-related brain connectivity changes.

Although results were achieved by using synthetic samples to balance the dataset, there is still room for improvements. The major problem was dealing with minority class, and further attempts will be made to improve overall classification performances, heavily influenced by scores on minority class.

Another key point to be addressed will be a further validation with larger and independent datasets. This would be beneficial to confirm the model's generalization and clinical relevance.

**Acknowledgments.** We acknowledge the support of the PNRR project FAIR - Future AI Research (PE00000013), Spoke 9 - Green-aware AI, under the NRRP MUR program funded by the NextGenerationEU.

## References

1. Behzadi, Y., Restom, K., Liau, J., Liu, T.T.: A component based noise correction method (CompCor) for BOLD and perfusion based fMRI. NeuroImage **37**(1), 90–101 (2007). https://doi.org/10.1016/j.neuroimage.2007.04.042. https://linkinghub.elsevier.com/retrieve/pii/S1053811907003837
2. Bezmaternykh, D.D., et al.: Brain networks connectivity in mild to moderate depression: resting state fMRI study with implications to nonpharmacological treatment. Neural Plasticity **2021**, 1–15 (2021). https://doi.org/10.1155/2021/8846097. https://www.hindawi.com/journals/np/2021/8846097/
3. Bezmaternykh, D.D., Melnikov, M.E., Savelov, A.A., Petrovskii, E.D.: Resting state with closed eyes for patients with depression and healthy participants (2021). https://doi.org/10.18112/OPENNEURO.DS002748.V1.0.5. https://openneuro.org/datasets/ds002748/versions/1.0.5
4. Chai, X.J., Castañón, A.N., Öngür, D., Whitfield-Gabrieli, S.: Anticorrelations in resting state networks without global signal regression. NeuroImage **59**(2), 1420–1428 (2012). https://doi.org/10.1016/j.neuroimage.2011.08.048. https://linkinghub.elsevier.com/retrieve/pii/S1053811911009657
5. Esteban, O., et al.: fMRIPrep: a robust preprocessing pipeline for functional MRI. Nat. Methods **16**(1), 111–116 (2019). https://doi.org/10.1038/s41592-018-0235-4. https://www.nature.com/articles/s41592-018-0235-4

6. Hu, J., Cao, L., Li, T., Dong, S., Li, P.: GAT-LI: a graph attention network based learning and interpreting method for functional brain network classification. BMC Bioinform. **22**(1), 379 (2021). https://doi.org/10.1186/s12859-021-04295-1
7. Kim, B.H., Ye, J.C.: Understanding graph isomorphism network for rs-fMRI functional connectivity analysis. Front. Neurosci. **14**, 630 (2020). https://doi.org/10.3389/fnins.2020.00630. https://www.frontiersin.org/article/10.3389/fnins.2020.00630/full
8. Li, X., Dvornek, N.C., Zhou, Y., Zhuang, J., Ventola, P., Duncan, J.S.: Graph neural network for interpreting task-fMRI biomarkers. In: Shen, D., et al. (eds.) MICCAI 2019. LNCS, vol. 11768, pp. 485–493. Springer, Cham (2019). https://doi.org/10.1007/978-3-030-32254-0_54
9. Nieto-Castanon, A.: Handbook of functional connectivity Magnetic Resonance Imaging methods in CONN. Hilbert Press (2020). https://doi.org/10.56441/hilbertpress.2207.6598. https://www.hilbertpress.org/link-nieto-castanon2020
10. Nieto-Castanon, A.: Preparing fMRI data for statistical analysis (2022). https://arxiv.org/abs/2210.13564
11. Ogawa, S., Lee, T.M., Kay, A.R., Tank, D.W.: Brain magnetic resonance imaging with contrast dependent on blood oxygenation. Proc. Natl. Acad. Sci. U.S.A. **87**(24), 9868–9872 (1990). https://doi.org/10.1073/pnas.87.24.9868
12. Shibasaki, H.: Human brain mapping: hemodynamic response and electrophysiology. Clin. Neurophysiol. **119**(4), 731–743 (2008). https://doi.org/10.1016/j.clinph.2007.10.026. https://www.sciencedirect.com/science/article/pii/S1388245707006578
13. Suk, H.I., Lee, S.W., Shen, D.: Deep ensemble learning of sparse regression models for brain disease diagnosis. Med. Image Anal. **37**, 101–113 (2017). https://doi.org/10.1016/j.media.2017.01.008. https://linkinghub.elsevier.com/retrieve/pii/S1361841517300166
14. Veličković, P., Cucurull, G., Casanova, A., Romero, A., Liò, P., Bengio, Y.: Graph attention networks. In: International Conference on Learning Representations (2018). https://openreview.net/forum?id=rJXMpikCZ
15. Vieira, S., Pinaya, W.H., Mechelli, A.: Using deep learning to investigate the neuroimaging correlates of psychiatric and neurological disorders: methods and applications. Neurosci. Biobehav. Rev. **74**, 58–75 (2017). https://doi.org/10.1016/j.neubiorev.2017.01.002. https://linkinghub.elsevier.com/retrieve/pii/S0149763416305176
16. Wu, Z., Pan, S., Chen, F., Long, G., Zhang, C., Yu, P.S.: A comprehensive survey on graph neural networks. IEEE Trans. Neural Netw. Learn. Syst. **32**(1), 4–24 (2021). https://doi.org/10.1109/TNNLS.2020.2978386. http://arxiv.org/abs/1901.00596

# Alzheimer's Disease Detection Methods: Current Advancements and Future Directions

M. Manda Sridhar[1](✉), V. S. Hariharan[2], Pavani Reddy Andru[3], Sree Kumar Vaggu[4], and Ravi Kiran Chaganti[5]

[1] Department of Computer Science and Engineering (Data Science), Balaji Institute of Technology and Science, Narsampet, Warangal, India
mandasridhar550@gmail.com
[2] Balaji Institute of Technology and Science, Narsampet, Warangal, India
[3] Ophthalmologist, Warangal, India
[4] Department of Ophthalmology, Regional Eye Hospital, Warangal, India
[5] Specialist in Radiology, Govt. Civil Hospital, Wardhannapet, Warangal, India

**Abstract.** Alzheimer's disease (AD) is one of the leading causes of dementia, and it is a progressive neurodegenerative disorder that leads to death. This disease causes cognitive decline, changes in behavior, and memory loss. It is predicted that by 2050, AD will affect 1 in 85 people globally. Medically or clinically, there is no accurate diagnosis available for this disease, and AD detection is still challenging for the medical field. To improve quality of life, early detection of AD is essential, thereby enabling management for a more extended period. Over the past decade, deep learning models have become widely used for image processing, and impressive results have been reported in AD classification. For this reason, deep learning has attracted considerable attention for detecting AD. Nevertheless, this technology for AD detection remains challenging for classification because it requires highly discriminative features to represent distinct yet similar image patterns. A structured literature review was conducted on the latest technologies contributing to AD detection. Although not following the formal PRISMA guidelines, the evaluation systematically categorises studies by imaging modality and AI technique. There are different methods to detect Alzheimer's Disease, for example, Imaging Techniques like MRI (Magnetic Resonance Imaging), PET (Positron Emission Tomography), Retinal Imaging (Fundus Photography) and other methods, such as Electroencephalogram (EEG), Blood-Based Biomarkers, Cerebrospinal Fluid (CSF) Analysis, Neuropsychological Tests and Behavioural and Lifestyle Assessments. While many diagnostic methods exist for Alzheimer's disease, the Retinal Fundus imaging is suggested due to its non-invasive nature, affordability, and its best suited for early detection.

**Keywords:** Alzheimer's Disease · Machine Learning · Neuroimaging · Retinal Imaging · Biomarkers · Deep Learning · CNN · SVM


P. Pichappan et al. (Eds.): RTIS 2025, LNNS 1936, pp. 502–510, 2027.
https://doi.org/10.1007/978-3-032-23396-7_45

# 1 Introduction

Alzheimer's Disease (AD) is severe and complex, and it is a progressive neurodegenerative disease, characterised by behavioural changes, memory impairment, and cognitive decline. Affecting millions worldwide, AD represents a significant public health challenge, with its prevalence expected to rise due to ageing populations, as noted by the World Health Organisation (WHO). Despite extensive research, no definitive cure exists, underscoring the critical importance of early detection and intervention for managing symptoms and improving the quality of life for affected individuals.

Common symptoms of AD include memory loss, mood changes, personality shifts, and difficulty in performing daily tasks. These symptoms progressively impair an individual's capability to lead an independent life, underscoring the importance of early diagnosis.

Recent advancements in neuroimaging techniques, such as "Magnetic Resonance Imaging (MRI) and Positron Emission Tomography (PET)", have provided valuable insights into the morphological and functional changes in the brains of AD patients. However, the complexity and volume of neuroimaging data necessitate advanced analytical tools. Deep learning, a subset of artificial intelligence, has emerged as a powerful method for automating the extraction of meaningful patterns from such data. Techniques such as Convolutional Neural Networks (CNNs) and other architectures enhance the accuracy and efficiency of AD detection by identifying biomarkers that are imperceptible to traditional analyses.

Complementing neuroimaging, retinal imaging techniques are gaining traction as promising noninvasive methods for early AD detection. Evidence suggests that retinal characteristics, such as peripapillary retinal nerve fiber layer (RNFL) and macular thickness measured through Optical Coherence Tomography (OCT), reflect neurodegenerative changes associated with AD and Mild Cognitive Impairment (MCI). By leveraging advanced technologies, such as CNNs, to analyse retinal fundus images captured with high-definition cameras, researchers aim to identify early markers of AD, thereby enabling timely interventions and personalised care.

Additionally, researchers have explored innovative biomarkers, including specific DNA methylation patterns, that could enable the diagnosis of AD from blood samples. Such approaches, as demonstrated by Blanca Acha Santamaría's work, offer a less invasive and cost-effective alternative to traditional methods, such as lumbar punctures and brain imaging. Early detection via blood-based biomarkers has the potential to transform AD management, providing accessible diagnostic options and enabling early therapeutic interventions.

This paper systematically reviews the state-of-the-art approaches for Alzheimer's disease (AD) detection, focusing on imaging techniques, deep learning models, and emerging biomarkers. By summarising the findings from various studies, we aim to highlight the methodologies employed, datasets utilised, techniques applied, results achieved, and limitations identified by different researchers. This survey does not propose new methodologies but instead consolidates and analyses existing work to provide a comprehensive understanding of the current advancements, by identifying research gaps and potential areas for future exploration. This study adopts a structured literature review framework. Unlike a formal systematic review conducted in accordance with

PRISMA, it does not employ predefined inclusion/exclusion criteria. Still, it focuses on synthesising significant peer-reviewed research published between 2017 and 2024 across key imaging and AI-based diagnostic modalities.

## 2 Methodology of Review

Literature was retrieved from IEEE Xplore, PubMed, ScienceDirect, SpringerLink, and Google Scholar. The search combined terms such as "Alzheimer's Disease detection," "retinal imaging," "MRI," "PET," "deep learning," and "machine learning."

Inclusion criteria: (i) peer-reviewed publications between 2017 and 2024, (ii) studies involving AD or MCI detection, and (iii) approaches using neuroimaging, retinal imaging, or biomarkers.

Exclusion criteria: (i) non-English papers, (ii) non-peer-reviewed preprints, and (iii) works focusing solely on drug therapies or clinical trials.

This ensures methodological transparency and minimises selection bias.

### 2.1 Related Work

*Advances in Machine Learning and Retinal Imaging for Alzheimer's Disease Detection*

Alzheimer's disease (AD) is a progressive neurodegenerative disorder marked by cognitive decline, neuronal loss, and complex pathological heterogeneity. Early and accurate diagnosis remains a major clinical challenge. Over the past two decades, interdisciplinary research has increasingly leveraged artificial intelligence (AI), neuroimaging, and noninvasive biomarkers—particularly retinal imaging— to improve detection, particularly in the pre-symptomatic stages.

### 2.2 Early Clinical and Epidemiological Foundations

Foundational work by Lam et al. [1] highlighted the clinical and pathological heterogeneity among patients with AD, underscoring the limitations of single biomarkers and advocating a multidimensional diagnostic approach that integrates clinical, imaging, and molecular data. Brookmeyer et al. [2] projected a threefold rise in global AD cases by 2050 due to ageing populations, emphasising the urgent need for scalable early diagnostic strategies. Complementing this, Winblad [3] called for global collaboration to prioritise prevention and early detection, shifting focus from pharmacological trials to computational and biomarker-based tools. Economically, Wong [4] estimated AD-related costs exceeded \$1 trillion globally in 2020, reinforcing the value of cost-effective, AI-driven screening systems.

Rathore [5] reviewed early machine learning (ML) applications in neuroimaging, demonstrating that automated pattern recognition in MRI and PET scans could significantly enhance diagnostic accuracy. Meanwhile, den Haan [6] proposed retinal thinning—particularly of the retinal nerve fibre layer (RNFL)—as a potential non-invasive biomarker. However, they noted that a lack of direct molecular correlations limited its diagnostic certainty.

### 2.3 Deep Learning in Neuroimaging

The integration of deep learning (DL) has revolutionised AD analysis using neuroimaging data. Ebrahimighahnavieh [7] found that convolutional neural networks (CNNs) achieved up to 92% accuracy in distinguishing AD from healthy controls using the ADNI dataset and ~83% accuracy in mild cognitive impairment (MCI) classification. However, issues like small sample sizes, limited data augmentation, and poor generalizability to real-world clinical settings persisted.

Al-Shoukry [8] evaluated various ML and DL models on the OASIS and NACC datasets, reporting cross-validation accuracies of over 90% with CNNs, Random Forests, and support vector machines (SVMs). Yet, they warned that reliance on isolated biomarkers might overlook broader physiological indicators. Tian [9] applied SVMs to retinal vasculature data from the UK Biobank, achieving 84% accuracy but requiring manual biomarker extraction, which hinders scalability.

These studies established CNNs as the leading architecture for detecting subtle brain morphological changes but exposed consistent limitations: insufficient dataset diversity, lack of external validation, and minimal longitudinal follow-up.

### 2.4 Retinal Imaging as a Window into Neurodegeneration

Retinal imaging has emerged as a promising, non-invasive tool for AD detection. Cheung [10] analysed optical coherence tomography (OCT), OCT angiography (OCTA), and fundus photography, identifying significant correlations between RNFL thinning, microvascular changes, and cognitive decline. However, inconsistent imaging protocols and the absence of standardised datasets limited reproducibility.

Wisely [11] trained a CNN on multimodal OCT and fundus autofluorescence (FAF) images from 159 participants, achieving an AUC of 0.841 for symptomatic AD detection. While demonstrating DL's sensitivity to subtle retinal changes, the study highlighted the challenges posed by data scarcity and demographic imbalance. Kavitha [12] used ensemble ML models (Random Forest, SVM, Gradient Boosting) on the OASIS dataset, achieving 83% validation accuracy with a Voting classifier—yet lacking external validation and a diverse feature set.

Standardization efforts also progressed: Delaby [13] harmonized cerebrospinal fluid (CSF) biomarker reporting for β-amyloid and tau across international labs, improving diagnostic consistency. Separately, Cheung [14] developed an EfficientNet-B2 model trained on nearly 13,000 retinal photos, achieving AUROC values of 0.73–0.93, supporting retinal screening for large-scale AD risk assessment—though the dataset size remained modest for deep learning demands. Krishna [15] enhanced PET-based classification using ensemble learning, though data imbalance remained a concern.

### 2.5 Multimodal and Non-Traditional Approaches

Beyond traditional imaging, novel diagnostic modalities have gained traction. Ahn [16] used a ResNet50 model to analyze voice recordings from Mini-Mental State Examination interviews, achieving 93.7% accuracy with perfect specificity—suggesting speech analysis as a low-cost, remote screening tool, though limited by participant diversity.

Kim [17] applied MobileNetV3 to 85,711 UK Biobank retinal fundus images, attaining an AUC of 0.927. Despite strong performance, the model proved sensitive to image quality and environmental variables. Feng [18] proposed a hybrid 3D-CNN with a Feature-Selective Bi-LSTM to fuse MRI and PET data, capturing spatial and temporal patterns and achieving 95% accuracy on the ADNI dataset. This highlighted the power of multimodal integration but stressed the need for longitudinal validation to track progression.

Interpretability has also become a priority. Yousefzadeh [19] introduced the LAVA framework, combining CNNs, guided backpropagation, and AutoMorph to visualize neuron-level retinal biomarkers across 48,388 fundus images. While producing explainable prediction maps, the authors emphasized the need for clinical expert validation. Dumitrascu [20] applied U-Net segmentation to fundus images, achieving 97.7% accuracy for right-eye images but lower performance for left eyes, revealing data asymmetry. In a follow-up, O. M. Dumitrascu [21]. Compared self-supervised networks ADRET and ADVAS on 178,803 unlabeled fundus images; ADRET reached 98.9% accuracy, underscoring the potential of self-supervised learning to enhance scalability with limited labeled data.

### 2.6 OCTA and Microvascular Biomarkers

High-resolution retinal angiography, particularly OCTA, is now recognized as a frontier for early AD detection. Turkan [22] reviewed DL applications in OCT/OCTA, reporting F1-scores of 0.86–0.89 and noting a shift from handcrafted features to end-to-end models that directly interpret volumetric data. Still, public OCT/OCTA datasets remain scarce, and longitudinal imaging is lacking.

Kwapong [23] compared Swept-Source OCTA (SS-OCTA) and Spectral-Domain OCTA (SD-OCTA) for the detection of microvascular changes in early-onset AD. Their models achieved AUCs of 0.913 and 0.991, respectively, affirming OCTA's diagnostic potential—though cross-sectional designs couldn't establish causal disease progression. Xie [24] introduced OCTA-Net, which was trained on the ROSE-1 dataset and achieved >94% AUC in distinguishing AD/MCI from controls. While the results were promising, the lack of risk-adjusted and longitudinal data limits clinical translation.

### 2.7 Machine Learning Models in AD Research

Several ML/DL architectures have been applied to AD detection:

- Autoencoders (AE): Useful for dimensionality reduction and feature extraction; the bottleneck layer captures compressed data representations.
- Restricted Boltzmann Machines (RBM): Generative models effective in feature learning and image processing, but computationally intensive.
- Deep Neural Networks (DNN): Excel with large, vector-based datasets and nonlinear pattern recognition, though prone to slow training and overfitting.
- Deep Polynomial Networks (DPN): Effective with small samples but limited by hierarchical feature concatenation.

- Recurrent Neural Networks (RNN): Ideal for longitudinal/sequential data but suffer from vanishing/exploding gradients.
- Convolutional Neural Networks (CNN):
  - o *2D CNNs*: Efficient for local 2D image features but miss 3D spatial context.
  - o *3D CNNs*: Better capture volumetric brain structures but demand significant computational resources.

AI-driven approaches—especially those integrating neuroimaging, retinal biomarkers, and multimodal data—hold transformative potential for early, non-invasive AD diagnosis. CNNs, ensemble methods, and self-supervised models have demonstrated high accuracy across diverse datasets. However, persistent challenges remain: limited dataset diversity, poor generalizability, lack of longitudinal validation, and insufficient standardisation. Future progress depends on larger, harmonised, multimodal datasets, interpretable models, and rigorous clinical validation to bridge the gap between research and real-world deployment.

## 3 Survey of Methods

### 3.1 Neuroimaging Approaches

"Magnetic Resonance Imaging (MRI) and Positron Emission Tomography (PET)" are widely recognised as gold-standard tools for diagnosing Alzheimer's Disease (AD). MRI provides detailed structural images that enable the identification of brain atrophy, particularly in the hippocampus. PET imaging is invaluable for detecting amyloid-beta plaques and tau protein accumulation, hallmark features of AD.

Deep learning models have significantly enhanced the analysis of neuroimaging data, automating the detection of disease-specific patterns. CNNs are particularly effective, with 2D CNNs analyzing individual slices of MRI and PET scans and 3D CNNs capturing volumetric relationships. Hybrid models integrating CNNs with "Recurrent Neural Networks (RNNs)" have been employed to analyse longitudinal data, providing insights into disease progression.

Despite high diagnostic accuracy (often exceeding 90%), challenges remain. These include the need for large, well-annotated datasets, computational demands, and the high costs of neuroimaging. Nevertheless, neuroimaging combined with deep learning continues to play a pivotal role in advancing early detection of AD.

Most CNN-based studies employed 3D architectures (e.g., ResNet-50, 3D-CNN, FSBi-LSTM) trained on ADNI datasets with 70/15/15 train-validation-test splits. Hyperparameters such as learning rate ($10^{-4}$–$10^{-5}$), batch size (8–32), and optimisers (Adam, SGD) were commonly used. Reported accuracies above 90% were often obtained on balanced datasets, but statistical validation (AUC, p-values, or confidence intervals) was rarely reported, limiting reproducibility.

### 3.2 Retinal Imaging Techniques

Retinal imaging has emerged as a promising non-invasive and cost-effective tool for Alzheimer's Disease (AD) diagnosis. Optical Coherence Tomography (OCT) and Optical Coherence Tomography Angiography (OCTA) allow for the measurement of retinal nerve fibre layer (RNFL) thickness and microvascular density, which have been linked to neurodegeneration in AD and Mild Cognitive Impairment (MCI).

Machine learning techniques, particularly CNNs, have enhanced the analysis of retinal images, enabling the identification of subtle structural and vascular changes. Models incorporating attention mechanisms focus on regions of interest, such as the macula and optic disc, improving feature extraction. Studies using these methods have reported high diagnostic accuracies, with some achieving AUC values exceeding 0.90.

Challenges in retinal imaging include variability in image quality, small and heterogeneous datasets, and the lack of standardised imaging protocols. Addressing these issues could further improve the clinical applicability of retinal imaging in AD detection.

Although many reviewed models achieve high AUC (>0.90), this metric does not necessarily translate into clinical effectiveness. In low-prevalence populations (~5–10%), even models with 90% sensitivity/specificity may yield positive predictive values (PPV) below 50%. Future work should therefore evaluate real-world diagnostic utility using PPV, NPV, and cost-effectiveness analyses, comparing modalities like MRI, retinal OCT, and blood biomarkers head-to-head.

In retinal imaging-based DL studies, architectures like EfficientNet-b2, MobileNetV3, and U-Net variants achieved AUC values up to 0.93. However, few works reported cross-validation or statistical significance testing, and none examined real-world performance in low-prevalence clinical settings.

## 4 Categorising AD Detection Approaches

Based on the reviewed studies, AD detection approaches can be categorised into:

(i) Neuroimaging-based models (MRI, PET, EEG),
(ii) Retinal imaging-based models (OCT/OCTA, fundus), and
(iii) Molecular biomarkers (blood, CSF, DNA methylation).

Integration of these modalities via multimodal deep learning or ensemble fusion can enhance diagnostic precision while balancing accuracy, cost, and scalability. A conceptual roadmap is proposed in which retinal imaging serves as a first-line, noninvasive screening tool, followed by confirmatory neuroimaging or biomarker assays.

## 5 Conclusion

This survey paper provided a comprehensive review of recent advancements in AD detection, focusing on neuroimaging, retinal imaging, and machine learning models. Retinal imaging, enabled by tools such as OCT and OCTA, has significant potential as a non-invasive, accessible diagnostic modality. Deep learning models, particularly CNNs, have further enhanced the ability to detect subtle neurodegenerative changes linked to AD.

Despite these advancements, significant challenges remain. Limitations such as small datasets, variability in imaging quality, lack of standardised protocols, and the interpretability of AI models highlight the need for continued research. Addressing these challenges requires the development of large, diverse datasets, particularly from underrepresented populations in low-and middle-income countries, which account for over 60% of dementia cases worldwide. Furthermore, integrating multimodal data—including neuroimaging, retinal imaging, and emerging biomarkers like DNA methylation—can improve diagnostic accuracy and enable early intervention.

The global burden of Alzheimer's disease is staggering. Affecting more than 55 million people globally, this number is projected to rise to 139 million by 2050. Early detection through non-invasive methods like retinal imaging and explainable AI could play a pivotal role in reducing the socioeconomic impact of the disease, which cost an estimated USD 1.3 trillion in 2019.

Future efforts should prioritise the development of explainable AI models to build clinical trust and promote adoption. By overcoming these challenges, we can move closer to transforming Alzheimer's diagnosis, improving patient outcomes, and reducing the burden on caregivers and healthcare systems worldwide.

Future research should focus on developing multimodal diagnostic pipelines combining retinal, neuroimaging, and molecular data using interpretable AI frameworks. Establishing open-access, demographically diverse datasets and reporting standardized performance metrics (AUC, PPV, NPV) will enable clinical translation. Comparative studies assessing cost, sensitivity, and scalability across modalities are also needed to guide practical adoption in healthcare systems.

## References


1. Lam, B., Masellis, M., Freedman, M., Stuss, D.T., Black, S.E.: Clinical, imaging, and pathological heterogeneity of the Alzheimer's disease syndrome. Alzheimer's Res. Therapy **5**(1), 1–14 (2013)
2. Brookmeyer, R., Johnson, E., Ziegler-Graham, K., Arrighi, H.M.: Forecasting the global burden of Alzheimer's disease. Alzheimer's Dementia **3**(3), 186–191 (2007)
3. Winblad, B., et al.: Defeating Alzheimer's disease and other dementias: a priority for European science and society. Lancet Neurol. **15**(5), 455–532 (2016)
4. Wong, W.: Economic burden of Alzheimer disease and managed care considerations. Am. J. Managed Care **26**(8 Suppl), S177–S183 (2020)
5. Rathore, S., Habes, M., Iftikhar, M.A., Shacklett, A., Davatzikos, C.: A review on neuroimaging-based classification studies and associated feature extraction methods for Alzheimer's disease and its prodromal stages. Neuroimage **155**, 530–548 (2017)
6. den Haan, J., Verbraak, F.D., Visser, P.J., Bouwman, F.H.: Retinal thickness in Alzheimer's disease: a systematic review and meta-analysis. Alzheimer's Dement.: Diagn. Assess. Dis. Monit. **6**, 162–170 (2017)
7. Ebrahimighahnavieh, M.A., Luo, S., Chiong, R.: Deep learning to detect Alzheimer's disease from neuroimaging: a systematic literature review. Comput. Methods Programs Biomed. **187**, 105242 (2020)
8. Al-Shoukry, S., Rassem, T.H., Makbol, N.M.: Alzheimer's disease detection by using deep learning algorithms: a mini-review. IEEE Access **8**, 77131–77141 (2020)
9. Tian, J., et al.: Modular machine learning for Alzheimer's disease classification from retinal vasculature. Sci. Rep. **11**(1), 238 (2021)
10. Cheung, C.Y., Mok, V., Foster, P.J., Trucco, E., Chen, C., Wong, T.Y.: Retinal imaging in Alzheimer's disease. J. Neurol. Neurosurg. Psychiatry **92**(9), 983–994 (2021)

11. Wisely, C.E., et al.: Convolutional neural network to identify symptomatic Alzheimer's disease using multimodal retinal imaging. Br. J. Ophthalmol. **106**(3), 388–395 (2022)
12. Kavitha, C., Mani, V., Srividhya, S., Khalaf, O.I., Tavera Romero, C.A.: Early-stage Alzheimer's disease prediction using machine learning models. Front. Public Health **10**, 853294 (2022)
13. Delaby, C., et al.: Clinical reporting following the quantification of cerebrospinal fluid biomarkers in Alzheimer's disease: an international overview. Alzheimer's Dementia **18**(10), 1868–1879 (2022)
14. Cheung, C.Y., et al.: A deep learning model for detection of Alzheimer's disease based on retinal photographs: a retrospective, multicentre case–control study. Lancet Digital Health **4**(11), e806–e815 (2022)
15. Krishna, K.S., Sarma, K.J., Bamane, K.D., Prasad, J., Tiwari, M., Karthikeyan, T.: Alzheimer disease detection using AI with deep learning based features with development and validation based on data science. J. Adv. Zool. (2023)
16. Ahn, K., et al.: Deep learning of speech data for early detection of Alzheimer's disease in the elderly. Bioengineering **10**(9), 1093 (2023)
17. Kim, H.K., Lim, Y.J., Sunwoo, M.H.: Efficient deep retinal fundus image-based network for Alzheimer's disease diagnosis using mobile device applications. IEEE Access (2024)
18. Feng, C., et al.: Deep learning framework for Alzheimer's disease diagnosis via 3D-CNN and FSBi-LSTM. IEEE Access **7**, 63605–63618 (2024)
19. Yousefzadeh, N., Tran, C., Ramirez-Zamora, A., Chen, J., Fang, R., Thai, M.T.: Neuron-level explainable AI for Alzheimer's disease assessment from fundus images. Sci. Rep. **14**(1), 7710 (2024)
20. Dumitrascu, O., et al.: Retinal color fundus photography and deep learning can discriminate Alzheimer's disease (P8-9.006). Neurology **102**(Suppl 1), 2716 (2024)
21. Dumitrascu, O.M., et al.: Color fundus photography and deep learning applications in Alzheimer disease. Mayo Clin. Proc.: Digital Health **2**(4), 548–558 (2024)
22. Turkan, Y., et al.: Automated diagnosis of Alzheimer's disease using OCT and OCTA: a systematic review. IEEE Access (2024)
23. Kwapong, W.R., et al.: Choriocapillaris reduction accurately discriminates against early-onset Alzheimer's disease. Alzheimer's Dementia (2024)
24. Xie, J., et al.: Deep segmentation of OCTA for evaluation and association of changes of retinal microvasculature with Alzheimer's disease and mild cognitive impairment. Br. J. Ophthalmol. **108**(3), 432–439 (2024)

# Author Index




P. Pichappan et al. (Eds.): RTIS 2025, LNNS 1936, pp. 511–513, 2027.
https://doi.org/10.1007/978-3-032-23396-7

**Y**



**Z**

GPSR Compliance

*The European Union's (EU) General Product Safety Regulation (GPSR) is a set of rules that requires consumer products to be safe and our obligations to ensure this.*

*If you have any concerns about our products, you can contact us on ProductSafety@springernature.com*

In case Publisher is established outside the EU, the EU authorized representative is:

Springer Nature Customer Service Center GmbH
Europaplatz 3
69115 Heidelberg, Germany

**Batch number: 10755034**

Printed by Printforce, the Netherlands